W9-BUS-774
WITHDRAWAL

WORLD DIRECTORY OF ENERGY INFORMATION
VOLUME 2: Middle East, Africa and Asia/Pacific

WORLD DIRECTORY OF ENERGY INFORMATION

VOLUME 2: Middle East, Africa and Asia/Pacific

Compiled by Cambridge Information and
Research Services Limited

Facts On File, Inc.

**460 Park Avenue South,
New York, N.Y. 10016**

WORLD DIRECTORY
OF ENERGY
INFORMATION
Volume 2: Middle East, Africa and Asia/Pacific

Copyright © 1982 by GOWER PUBLISHING COMPANY LIMITED

All rights reserved. No part of this book may be reproduced or utilized in any form or by any means, electronic or mechanical, including photocopying, recording or by any information storage and retrieval systems, without permission in writing from the Publisher.

Published in the United Kingdom in 1982 by Gower Publishing Company Limited,
Croft Road, Aldershot, Hants, England.

Published in the United States of America in 1982 by Facts On File, Inc.,
460 Park Avenue South, New York, N.Y. 10016

Compiled by Cambridge Information and Research Services Limited,
Sussex House, Hobson Street, Cambridge CB1 1NJ

Edited by Christopher Swain BA and Andrew Buckley BA

The compilers and publishers have used their best efforts in collecting and preparing material for inclusion in the *World Directory of Energy Information.* They do not assume, and hereby disclaim, any liability to any party for any loss or damage caused by errors or omissions resulting from negligence, accident or any other cause.

Library of Congress Cataloging in Publication Data
Main entry under title:
World Directory of Energy Information.
Includes indexes.
Contents: Vol. I Western Europe
Vol. II Middle East, Africa and Asia/Pacific

1. Power resources—Information Services—Directories.
I. Swain, Christopher. II: Buckley, Andrew Robert.
III. Cambridge Information and Research Services Ltd.
TJ163.17. W67 333.79 81-754
ISBN 0-87196-563-1 Vol. I AACR2
ISBN 0-87196-602-6 Vol. II

Printed in Great Britain

Ref.
TJ
163.17
W67
v.2

Contents

Illustrations

The country reviews in Part Two contain illustrations showing overall energy trends and levels of production and/or consumption of oil, coal, gas and electricity. Tables are also included showing the patterns of primary fuel supply and energy consumption, net import/export movements of the fuels and analyses of sources of imports and destinations of exports where appropriate.

Introduction

The first volume in the World Directory of Energy Information series, published last year, brought together basic information on energy in Western Europe, an industrialised region with high levels of energy consumption per head and dependent on outside sources for more than half of its energy supplies. This second volume in the series contains similar information for the Middle East, Africa and Asia/Pacific. These regions include the vast majority of the world's developing countries, where energy consumption per head is very low but where there is a wealth of untapped energy resources.

The Western World relies on these countries to make good its energy deficit. American and West European companies are active throughout the area to mobilise new energy resources to meet that demand. But energy is also a critical asset for the developing countries themselves. It is essential to the achievement of national economic and social development programmes. Government agencies and local companies are therefore equally active in developing resources and supply systems.

Each country covered in this volume faces its own energy problems, policy options and supply arrangements. In most the pattern of supply is already complex and undergoing rapid change. Oil and gas fields are being found and developed, coal mines opened up, hydro-electric schemes built and more nuclear power stations commissioned in those countries with large energy requirements such as Japan, Korea and South Africa.

The appearance of Volume 2 is therefore timely, in view of the energy resource development under way not only in the Middle East but also in Africa, Asia and Australasia.

The information is structured in four main parts. Parts One and Two contain summary information and analysis for the three regions separately—viz the Middle East, Africa and Asia/Pacific—followed by detailed reviews of the energy situation in 64 selected countries. Parts Three and Four are reference sections devoted to energy organisations and publications.

Part One is entitled International Framework. Aggregate and comparative information on energy production, trade and consumption is given in eight maps and tables, illustrating the wide range of situations to be found within the individual countries and the distribution of activity in energy production. A checklist of the major international energy resource projects highlights the inter-dependence of producer and consumer countries.

Part Two, Country Reviews, looks in detail at the 64 countries—15 in the Middle East, 30 in Africa and 19 in the Asia/Pacific area. A set of key energy indicators is followed by sections dealing with energy market trends, the supply industries and energy trade.

Part Three contains profiles of 489 organisations involved in the exploration, production, transmission and distribution of energy as well as relevant official bodies. Organisations are grouped within each country under the following heads:

—government departments and official agencies
—enterprises (public and private sector companies)
—professional institutions and trade associations.

In many instances recent financial information is included to supplement details on operational activities. An index of organisations can be found at the end of Part Three.

Part Four is a bibliography of nearly 800 publications on energy matters at either the national or trans-national level. The bibliography includes many statistical sources as well as journals, books and other publications dealing with general, political, economic and technical aspects. Indexes are included which cross-reference the publications by subject matter and country.

ACKNOWLEDGEMENTS

The editors wish to acknowledge the invaluable assistance of government departments, official agencies, energy supply companies and industry associations in countries covered in this book. Statistical information has been derived from these organisations and from the Organisation for Economic Co-operation and Development, the Organisation of Arab Petroleum Exporting Countries and United Nations agencies.

Part One: International Framework

International Framework

The Middle East, Africa, Asia and the Pacific together encompass a population of some 2,860 million, or approximately 70 per cent of the world total. As a whole the area is a net exporter of energy, thanks mainly to the abundance of oil in the Middle East and North Africa. The Asia/Pacific region on the other hand is deficient in energy to the extent of close to 400 million tonnes of oil equivalent, which is balanced by inflows of oil and gas from the Middle East, whilst exports of North African oil and gas go largely to Western Europe.

Net exports from the area, taken as a whole, currently amount to around 700 million tonnes of oil equivalent. This major outflow is made possible by low levels of energy consumption. In the Asia/Pacific region consumption per head is only one-third of the world average and in Africa the level is less than one-fifth. Even in the Middle East average energy consumption remains at little more than half of the world average, although most oil and gas exporting countries have been increasing consumption at a rapid rate.

The pressure that could be exerted on international energy supplies can be vividly illustrated by assessing the change in net availability of energy, if consumption per head of population in this area were to approach the world average level. Such a change would entail a three-fold increase in demand even without taking into account the continuing and inexorable growth in population under way in many countries. In this event the Middle East would still remain a substantial net exporter of energy, but Africa would become a net importer to the tune of 300 million tonnes of oil equivalent and the Asia/Pacific deficit would increase to 2,860 million tonnes of oil equivalent. Taken as a whole the area would have a deficit of over 2,400 million tonnes of oil equivalent instead of the current surplus.

The Distribution of Energy Resources

This latent demand for energy focuses attention on the ability of individual countries in the area to raise output. Table 1 identifies which forms of indigenous energy are important in each country. The analysis identifies some 13 countries lacking any significant energy production. These are not for the most part populous countries, but the group does include Ethiopia, a country of over 30 million people, which is clearly in a vulnerable position. Sudan, with a population of around 19 million, has at least the promise of developing oil or gas and Senegal has plans for hydro-electric development.

But in general only countries with at least two significant indigenous energy sources can be considered to have a sound basis of energy supply. A closer examination of Table 1, then, highlights Japan's continuing energy problems: one of the world's largest consumers of energy, but heavily dependent on imports. Korea and Taiwan are in a similar position and are following the Japanese path towards nuclear power as a major energy source. A different picture is presented by Bangladesh, a country with substantial volumes of natural gas for current purposes, but scarcely adequate for the development of a nation of 90 million people. It is, therefore, perhaps a matter of some reassurance that China, India and Indonesia have several strings to their bows, with substantial reserves of at least three energy forms.

Energy Resource Development

In many African and Asian countries traditional non-commercial fuels, such as firewood, charcoal, rice husks and dung, still play an important part for the majority of the population. This tends to obscure the potential demand for commercial energy and in all areas the development of modern infrastructures and industrial and service activities goes hand in hand with rising consumption of oil, gas, coal and electricity. Indeed, the use of firewood will need to be reduced in some countries because of the threat of deforestation and the greater value of wood in non-fuel markets.

It is evident that many countries are devoting great efforts to increasing indigenous supplies of energy, exploiting known resources which either have become economic owing to rising real energy prices or are considered strategically important for security or balance of payments reasons. The scope for non-conventional energy is also widely appreciated: the foremost examples being the geothermal and dendro-thermal programmes in the Philippines and the widespread interest in solar applications. All of this is additional to the more intensive use of hydro-electric potential and the growing interest in nuclear power.

The inter-dependence of energy consuming and producing countries is borne out by Table 2, which identifies a number of major energy resource development projects, in operation, under construction or at the stage of detailed planning. They cover coal, gas, hydro-electricity and uranium. These projects have been, or are being, developed essentially as co-operative ventures between producer and consumer interests, with supply contracts as the basis for investment in production facilities. In many instances the purchasing/ consuming interests take a share in the project and inject finance.

The development of coal for export markets is one of the most active fields, with projects in Australia and South Africa, both of which expect to rival the United States in the international coal trade. Additional South African export capacity seems likely to serve mainly Western Europe, but Australian exports will be absorbed largely within Asia.

The movement of gas presents particular problems. An international trunk pipeline has linked Iran to the Soviet Union for more than a decade, but it has only recently become feasible to lay pipeline at depths to enable exports across the sea to Western Europe. The LNG option typically involves much greater investment, but has nevertheless been taken by Japan, in an effort to draw on abundant unutilised gas resources in several countries of the Middle East and Asia.

Hydro-electric projects are of immense importance in some countries of Asia and Africa, serving both water management and power generation purposes. Two notable international projects have taken place in Africa at Cabora Bassa, on the Limpopo River in Mozambique, and at Kariba, on the Zambezi between Zambia and Zimbabwe. The Cabora Bassa dam supplies around 10 per cent of South African consumption as well as providing a base for Mozambique's economy.

Uranium prospecting and development has been a notable feature of the Australian minerals boom. Although the long established Mary Kathleen mine has an uncertain future because of its high cost of production, several major new deposits are being worked to supply fuel for the nuclear power programmes of Asia and Western Europe. Africa is also an important area of uranium resources, with export production in Gabon, Niger and South Africa and a new deposit due to come into production in Algeria.

Principal Energy Resource Locations

Figures 3, 4 and 5 show the widespread existence of significant actual and potential energy production in the Middle East, Africa and Asia/Pacific regions respectively. These regions include several of the most promising and active areas of offshore oil and gas development, including the South China Sea and Yellow Sea. Oil or gas has been shown to be more general in West Africa than thought previously and new fields have been found recently off the coasts of India, Malaysia and Tanzania. Indonesia, until now largely content to use up oil and gas reserves is turning attention to its extensive coal resources as the base for a national electrification programme. Coal is already a key energy input in two of the advanced economies, Australia and South Africa, as well as the principal energy source in China.

Energy Production, Consumption and Trade

Figure 6 illustrates the energy supply situation in the Middle East. Energy production is of the order of 1,000 million tonnes of oil equivalent per annum, (whereas consumption is only 120 million tonnes and of this Iran and Turkey account for more than half). Energy consumption per head of population is exceedingly high in some of the small oil producing states, notably Bahrain, Kuwait, Qatar and the United Arab Emirates. Israel also has above-average consumption, despite dependence on energy imports. In contrast, however, a number of countries still have very low average levels of energy consumption, including some which are currently net exporters of energy.

Energy production in Africa (Figure 7) is three times as great as consumption, but the production is very unevenly distributed. Only Algeria, Libya and Nigeria are major exporters, whilst Egypt, which currently exports some oil, is one of the more populous countries, with a rapidly rising demand for energy. Figure 7 identifies the many countries which are heavily dependent on energy imports. A number of others are in approximate balance only because internal consumption is at very low levels. The exception is South Africa, which has a high demand for energy, but is largely self-sufficient as a result of the intensive exploitation of its indigenous resources.

Figure 8 shows that only Indonesia is a major energy exporter in the Asia/Pacific area. Exports from Australia are important, but not of major significance in the context of the area's actual and potential demand for energy. The Asia/Pacific area is already 30 per cent dependent on energy imports, and has a very large capacity to absorb additional resources. Aside from the anomalous position of Brunei, levels of energy consumption per head are above the world average only in Australia, Japan and New Zealand. Of the other more populous countries only Korea and Taiwan approach this level. Enormous potential demand for energy lies in Bangladesh, China, India, Indonesia, Pakistan, and to a lesser extent Burma, the Philippines and Thailand. These eight countries have a total population in excess of 2,000 million, but in China consumption per head is only one-third of the world average, and in none of the others is it greater than one-sixth.

TABLE 1 THE DISTRIBUTION OF ENERGY PRODUCTION IN THE MIDDLE EAST, AFRICA AND ASIA/PACIFIC

The table identifies those countries which have significant production of individual energy forms in terms of ability to meet a high proportion of internal demand or to export to world markets. Countries with major hydro-electric or nuclear power plants are also noted.

	Oil	Coal	Gas	Hydro-Electricity	Uranium	Nuclear Power	Geothermal Energy
Algeria	*		*		*		
Angola	*			*			
Australia	*	*	*	*	*		
Bahrain	*		*				
Bangladesh			*	*			
Brunei	*		*				
Burma	*		*				
Cameroon	*			*			
China (People's Republic of)	*	*		(*)			
Congo	*						
Egypt	*		*	*			
Ethiopia							
Gabon	*		*	*	*		
Ghana				*			
Hong Kong							
India	*	*		*	*	*	
Indonesia	*	(*)	*				
Iran	*		*	*			
Iraq	*		*				
Israel							
Ivory Coast	*			*			
Japan						*	
Jordan							
Kenya				*			*
Korea (Republic of)		*				(*)	
Kuwait	*		*				
Lebanon							
Liberia				*			
Libya	*		*				
Malagasy Republic				*			
Malawi				*			
Malaysia	*		*	*			
Mali							
Mauritania							
Morocco		*					
Mozambique		*		*			
New Zealand	*	*	*	*			*
Niger		*			*		
Nigeria	*		*	*			
Oman	*		*				

	Oil	Coal	Gas	Hydro-Electricity	Uranium	Nuclear Power	Geothermal Energy
Pakistan			*	*			
Papua New Guinea/ Pacific Islands				*			*
Philippines		(*)		*			*
Qatar	*		*				
Saudi Arabia	*		*				
Senegal							
Sierra Leone							
Singapore							
South Africa		*			*	(*)	
South Yemen (PDR)							
Sri Lanka				*			
Sudan							
Syria	*			*			
Taiwan				*		(*)	
Tanzania				*			
Thailand			*	*			
Tunisia	*						
Turkey		*		*			
United Arab Emirates	*		*				
Upper Volta				*			
Yemen Arab Republic							
Zaire				*			
Zambia		*		*	*		
Zimbabwe		*		*			

Note: Asterisks in brackets represent known resources/energy projects currently under development or planned.

TABLE 2 INTERNATIONAL ENERGY RESOURCE PROJECTS

Listed below are 34 resource development projects in the Middle East, Africa, Asia and Australia, which highlight the interdependence of producer and consumer interests. In each case the involvement of trading or consuming businesses in the underwriting of production investment is an integral part of project development. This is particularly the case where very large capital investments are involved—in LNG plant and tankers, trunk pipelines, construction of dams and barrages or establishing production, processing and transportation facilities.

COAL

AUSTRALIA

Location: Blair Athol, Queensland
Start-up: 1984
Production: 3.0 million tonnes p.a. rising to 5.0 million
Participants: CRA (50.2%), Arco Coal (15.4%), Bunderberg Sugar (12.2%), ACI Resources (12.2%), Japan Electric Power Development Co (7%), Japan Coal Resources Development Co (3%)
Markets: Japanese electricity utilities and other major consumers.

Location: Drayton, New South Wales
Start-up: 1983
Production: 3.0 million tonnes p.a.
Participants: Shell (39%), CSR, AMP
Markets: Japanese and Korean electricity utilities and industrial companies.

Location: German Creek, Queensland
Start-up: 1982
Production: reaching 3.2 million tonnes by 1986
Participants: Austen & Butta Collieries (21.4%), Shell Australia (16.7%), Australian Consolidated Industries (13%), National Mutual Life Assurance (13%), Commonwealth Superannuation Fund (13%)
Markets: Coking coal users in Japan and Taiwan.

Location: Mount Thorley, New South Wales
Start-up: 1982
Production: 1.1 million tonnes p.a. rising to 4.5 million by 1986
Participants: R W Miller & Co (80%), Pohang Iron & Steel Co (20%)
Markets: Steam and coking coal users in Korea and other countries.

Location: Newlands, Queensland
Start-up: 1984
Production: 4.0 million tonnes p.a. of steam coal to be combined with increased output from Collinsville mines
Participants: Group led by MIM Holdings
Markets: Electricity utilities in Japan, South East Asia and Western Europe.

Location: Oaky Creek, Queensland
Start-up: 1983
Production: 2.3 million tonnes in 1983 rising to 3.0 million tonnes p.a. from 1984
Participants: Mount Isa Mines (79%), Estel Delstoffen (8.5%), Italsider (7.5%), Ensider (5%)
Markets: Coking coal consumers. The three West European participants have contracted to take 1.7 million tonnes p.a.

Location: Riverside, Queensland
Start-up: 1983
Production: 3.3 million tonnes p.a.
Participants: Operating company is Thiess Dampier Mitsui Coal Pty, in which shares are: BHP (58%), CSR (22%), Mitsui & Co (20%)
Markets: Japanese steel mills under 15 year contracts.

SOUTH AFRICA

Location: Ermelo, Transvaal
Start-up: Existing mine
Production: Eventually 3.75 million tonnes p.a.
Participants: Gencor, BP Southern Africa, CFP joint-venture (33.3% each)
Markets: Western Europe and other coal importing countries.

Location: Rietspruit, Transvaal
Start-up: 1978
Production: Eventually 5.0 million tonnes p.a.
Participants: Operating company is Rietspruit Open Cast Services (Pty), a joint-venture between Rand Mines and Shell South Africa
Markets: Overseas power stations.

GAS (PIPELINE AND LNG)

ALGERIA

Location: Arzew/Skikda
Start-up: 1965/1972
Production: 4.0 billion cubic metres p.a. under 25 year contracts
Participants: SONATRACH sale to Gaz de France
Market: Delivered into Gaz de France's system at Fos, St Nazaire and Le Havre.

Location: Arzew/Skikda
Start-up: 1976
Production: 4.5 billion cubic metres p.a. over 20 years
Participants: SONATRACH sale to ENAGAS
Markets: Spanish distributors in Barcelona region and other centres to the south and north-west.

Location: Arzew/Skikda
Start-up 1978
Production 10.0 billion cubic metres p.a. over 25 years
Participants: SONATRACH sale to El Paso Co
Markets: Natural gas distributors in southern USA. Sales suspended in 1981 following breakdown of price re-negotiations.

Location: Hassi R'Mel
Start-up 1982
Production: Up to 12 billion cubic metres p.a. via Trans Mediterranean Pipeline
Participants: SONATRACH, SNAM
Markets: Gas undertakings in Italy and other central and north-west European countries.

AUSTRALIA

Location: North-West Shelf, Western Australia
Start-up: 1984 for gas to Perth region, 1987 for LNG.
Production: 6.5 million tonnes p.a. Recoverable reserves estimated at 245 billion cubic metres
Participants: Woodside Petroleum (50%), BP (16.7%); Standard Oil Co of California (16.7%); BHP (8.3%), Shell (8.3%), Public (28.5%). BHP and Shell hold 43% of Woodside Petroleum
Markets: Contracts under negotiation with eight Japanese gas and electricity companies.

BRUNEI

Location: Lumut
Start-up 1972
Production: 7.5 billion cubic metres p.a. over 20 years
Participants: State of Brunei, Brunei Shell, Mitsubishi

INDONESIA

Location: Arun, Sumatra
Start-up: 1978
Production: 4.6 million tonnes p.a., to be increased to meet additional sales contracts
Participants: Pertamina, Mobil
Markets: Japanese gas and electricity utilities and Nippon Steel. Korea Electric Co will commence purchases in 1985.

Location: Badak, East Kalimantan
Start-up: 1977
Production: 3.3 million tonnes p.a. to be increased to meet additional sales contracts
Participants: Pertamina, Huffington, Inpex, Union Oil Co
Markets: Japanese utilities and industrial companies.

IRAN

Location: Bid Boland
Start-up: 1970
Production: 8-10 billion cubic metres p.a. for 15 years
Participants: National Iranian Gas Company and Soviet import agency
Markets: Main centres of USSR. Sales suspended 1980.

LIBYA

Location: Marsa El Brega
Start-up: 1972
Production: 3.9 billion cubic metres p.a. over 15-20 years
Participants: Libyan National Oil Co and Esso Libya. (Esso production assets taken over in 1982)
Markets: Spanish and Italian distributors.

MALAYSIA

Location: Bintulu
Start-up: 1983
Production: 6.0 million tonnes p.a. under a 20-year contract
Participants: Petronas (65%), Sarawak Shell (17.5%), Mitsubishi (17.5%)
Markets: Tokyo Gas Co and Tokyo Electric Power Co.

NIGERIA

Location: Bonny
Start-up: Not determined
Production: 8.0 billion cubic metres p.a. over 20 years
Participants: Nigerian National Petroleum Co (60%), Shell, AGIP, Elf-Aquitaine. Participation uncertain following the withdrawal of Phillips, the operator, and BP in 1981
Markets: Utilities in USA and Western Europe. Provisional contracts negotiated with BEB, Distrigaz, Enagas, Gasunie, Gaz de France, Ruhrgas, SNAM and Thyssengas.

UNITED ARAB EMIRATES

Location: Das Island, Abu Dhabi
Start-up: 1976
Production: 550 mcfd of gas, producing 2.2 million tonnes p.a. of LNG and 1.2 million tonnes p.a. of liquids from offshore fields
Participants: ADNOC (51%), BP, CFP, Bridgestone, Mitsui.

Location: Ruwais, Abu Dhabi
Start-up: 1982
Production: 4-5 million tonnes p.a. LNG and gas liquids from onshore fields
Participants: Operating company is Abu Dhabi Gas Industries Co (GASCO), in which shares are: ADNOC (68%), CFP (15%), Shell (15%), Partex (2%)
Markets: Japanese utilities and industrial companies.

HYDRO-ELECTRICITY

MOZAMBIQUE

Location: Cabora Bassa
Start-up: 1975
Production: 11,000 + GWh p.a. from 2,000 MW plant
Participants: Portuguese interests and Mozambique government
Markets: 9-10 TWh p.a. for South Africa as well as home market requirements.

ZIMBABWE

Location: Kariba
Production: 2,000 + GWh p.a. from 660 MW plant, located on Zimbabwe side of Kariba
Participants: Zambia and Zimbabwe governments in Central African Power Corporation
Markets: Zambia and Zimbabwe, to meet power demands of the mining industries.

URANIUM

AUSTRALIA

Location: Jabiluka, Northern Territory
Start-up: Not determined
Production: Initial production rate of 4,500 tonnes p.a. of uranium oxide. Reserves estimated at 200,000 tonnes
Participants: Pancontinental Mining (65%), Getty Oil (35%)
Markets: Contracts with overseas buyers yet to be negotiated. Buyers may also participate in eventual production plant.

Location: Jabiru, Northern Territory
Start-up: 1981
Production: Initial plant capacity is 3,000 tonnes p.a. of uranium oxide. Capacity may be doubled
Participants: The Ranger project is managed by Energy Resources of Australia Ltd, in which the shares are: EZ Industries (30.9%), Peko-Wallsend (30.5%), Commonwealth of Australia, Australian Atomic Energy Commission, Japanese and West European electricity utilities
Markets: Nuclear power plants in Belgium, Japan, Republic of Korea, Sweden, West Germany and also United States.

Location: Mary Kathleen, Queensland
Start-up: 1958
Production: 1958-63 and 1976-82. Capacity is 850 tonnes p.a.
Participants: CRA (51%), Australian Atomic Energy Commission (41.6%), Kathleen Investments (2.6%), and Australian public
Markets: 1958-63 sales of 4,090 tonnes to UK Atomic Energy Authority. 1976-84 overseas contracts totalling 4,800 tonnes.

Location: Nabarlek, Queensland
Start-up: 1980
Participants: Project is managed by Queensland Mines, in which Kathleen Investments has a 50 per cent interest. Project finance is provided by Japanese utilities
Markets: Kyushu Electric Power Co, Shikoku Electric Power Co.

Location: Yeelirrie, Western Australia
Start-up: 1986
Production: 2,500 tonnes p.a. of uranium oxide. Proven reserves are estimated at 46,000 tonnes
Participants: Western Mining Corporation Holdings (75%), Esso Australia (15%), Urangesellschaft Australia (15%)
Markets: Overseas electricity utilities, including West Germany.

GABON

Location: Mounana
Production: 1,100 tonnes p.a. of uranium oxide. Reserves in Franceville area estimated at 30,000 tonnes
Participants: Operating company is COMUF, in which shares are: Gabon State (25%), Cie de Mokta (32%), COGEMA (19%), Minatome (13%)
Markets: Nuclear fuel processors and consumers in France.

NIGER

Location: Akouta
Start-up: 1978
Production: Eventually 2,000 tonnes p.a. of uranium oxide

Participants: Operating company is COMINAK, in which shares are: Uraniger (31%), Commissariat à l'Energie Atomique (34%), ENUSA (10%), Japan Overseas Uranium Resources Development Co (25%)
Markets: Nuclear power companies in France, Japan and Spain.

Location: Arlit
Production: 1,800 tonnes p.a. of uranium oxide
Participants: Operating company is SOMAIR, in which shares are: ONAREM (33%), COGEMA (27%), AGIP, CFP, Urangesellschaft
Markets: Nuclear power companies in France, Italy and West Germany.

SOUTH AFRICA (NAMIBIA)

Location: Rössing
Start-up: 1975
Production: 5,000 tonnes p.a. of uranium oxide
Participants: RTZ (46.5%), CFP, Gencor, Industrial Development Corporation
Markets: UK Atomic Energy Commission and other West European nuclear power organisations.

FIGURE 3 ENERGY RESOURCES IN THE MIDDLE EAST

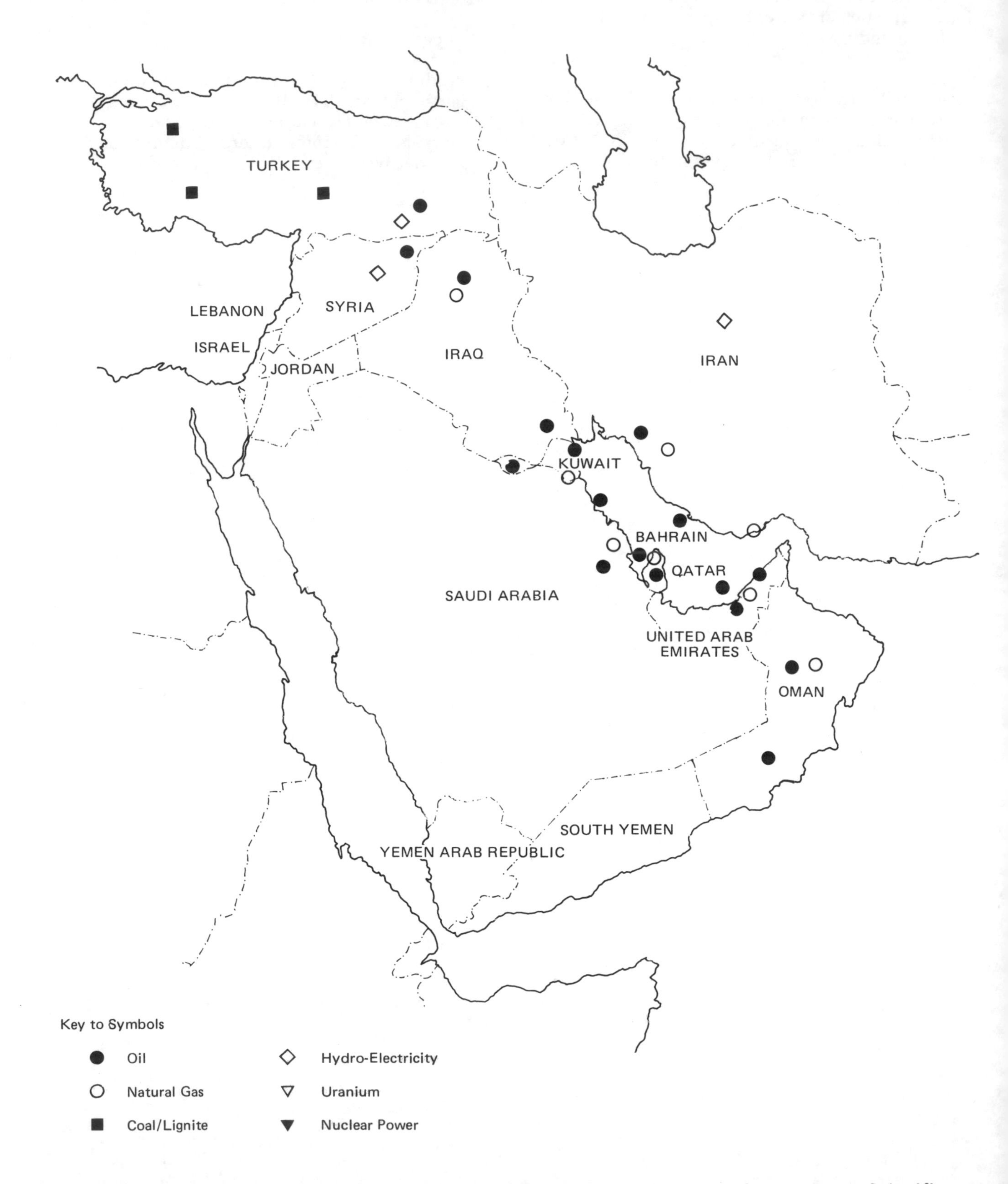

Note: Symbols indicate where indigenous production is important in the national context or of significance for international trade.

FIGURE 4 ENERGY RESOURCES IN AFRICA

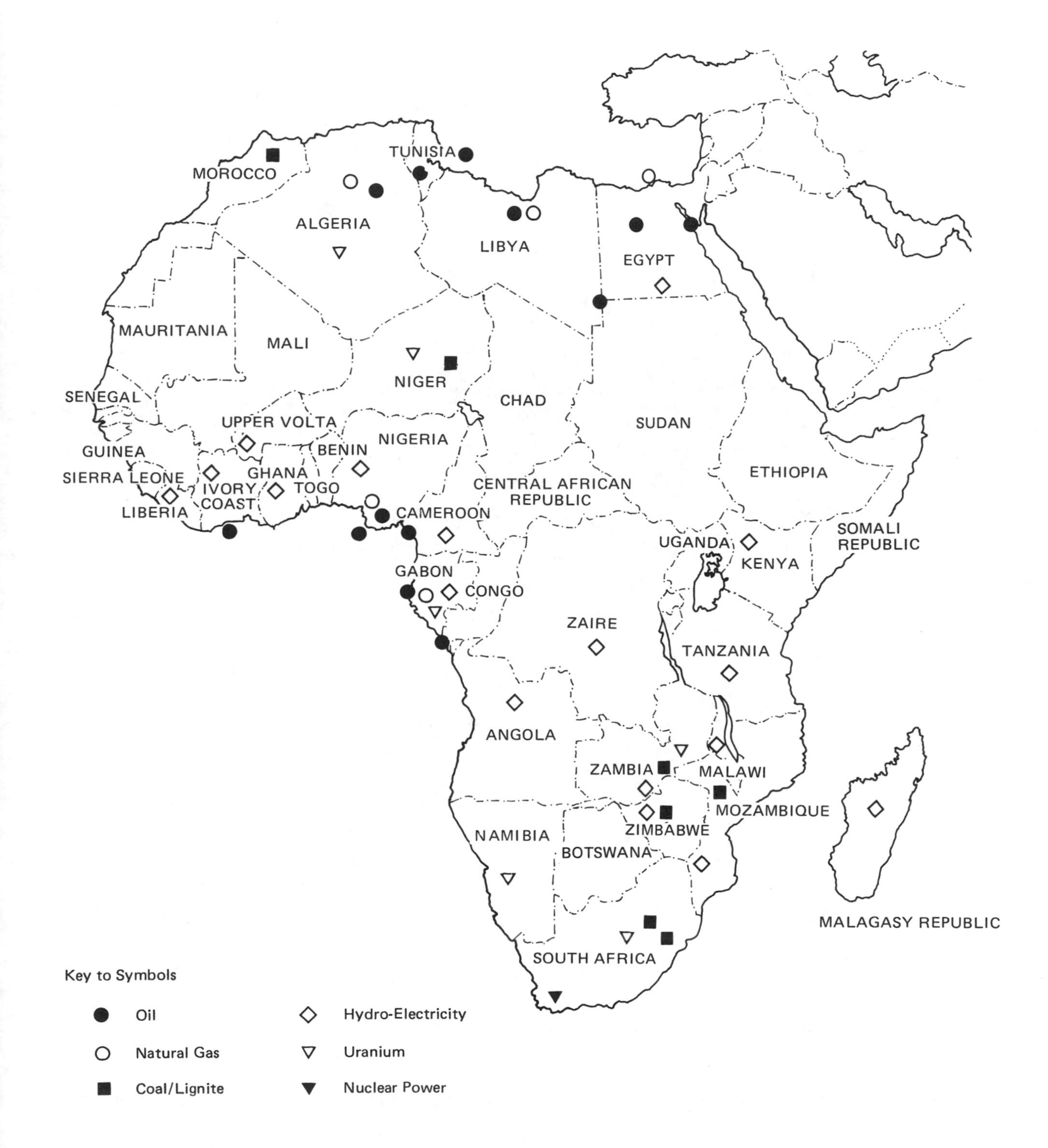

Note: Symbols indicate where indigenous production is important in the national context or of significance for international trade.

FIGURE 5 ENERGY RESOURCES IN ASIA/PACIFIC

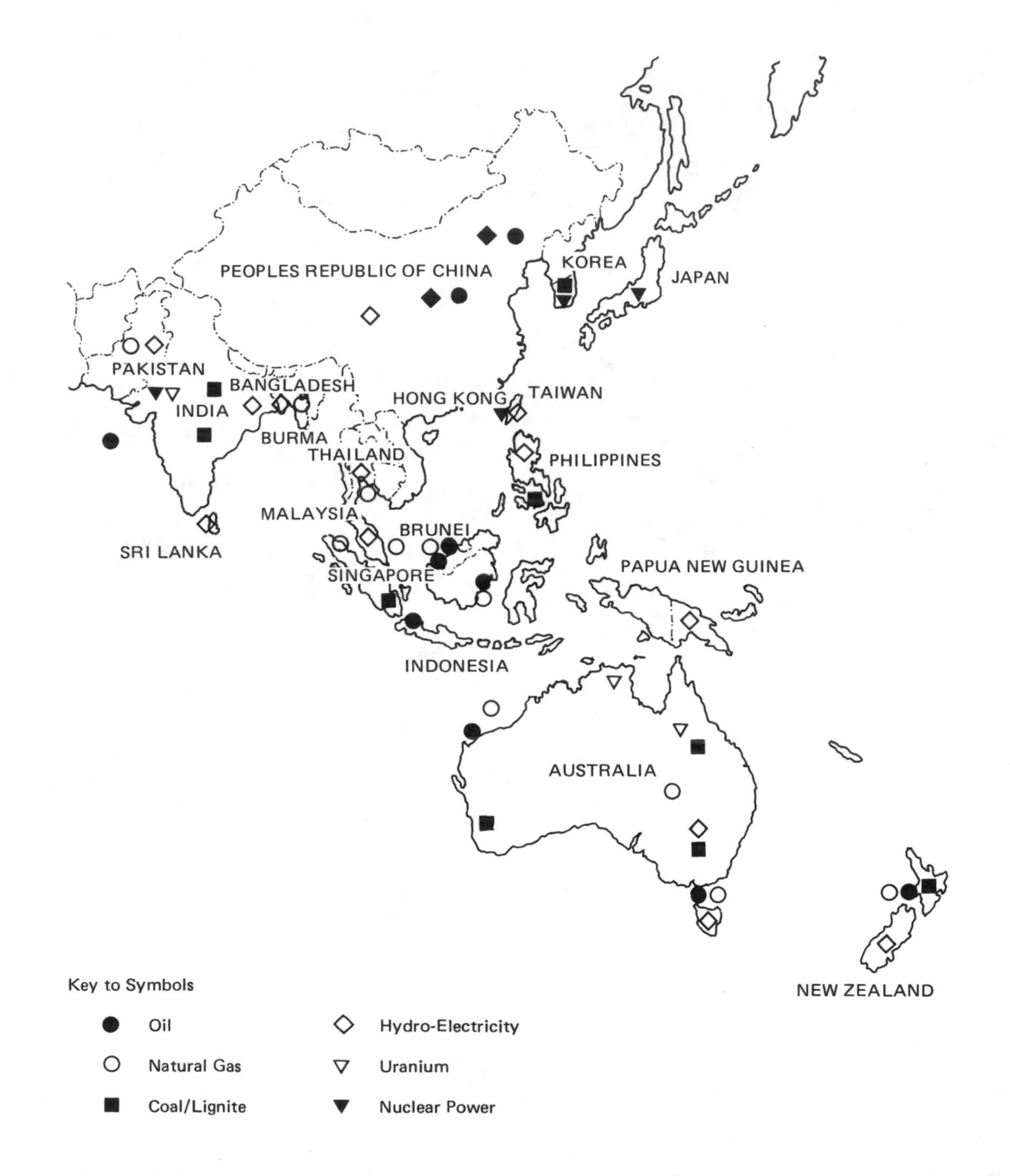

Note: Symbols indicate where indigenous production is important in the national context or of significance for international trade.

FIGURE 6 ENERGY IMPORTERS AND EXPORTERS IN THE MIDDLE EAST

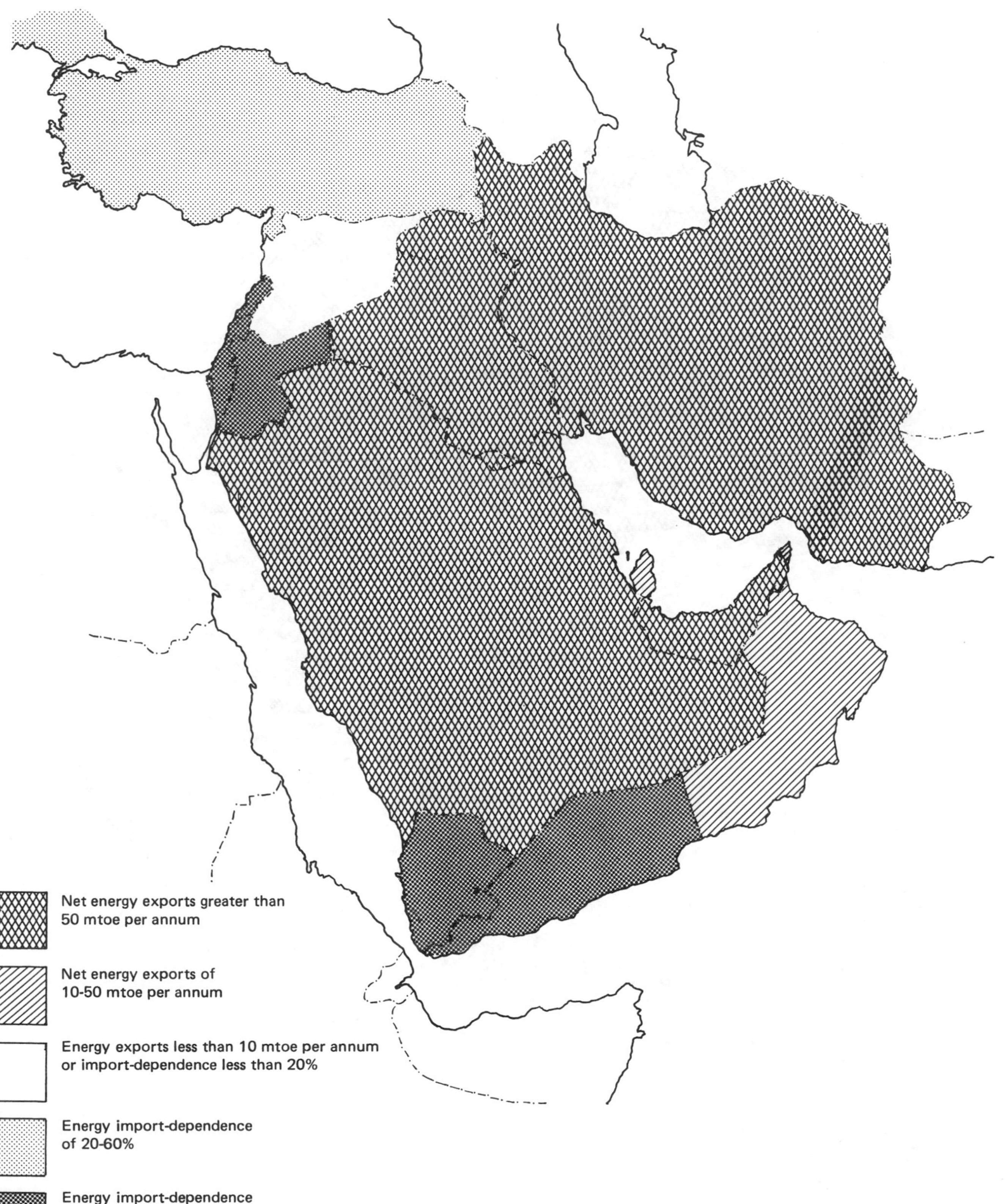

FIGURE 7 ENERGY IMPORTERS AND EXPORTERS IN AFRICA

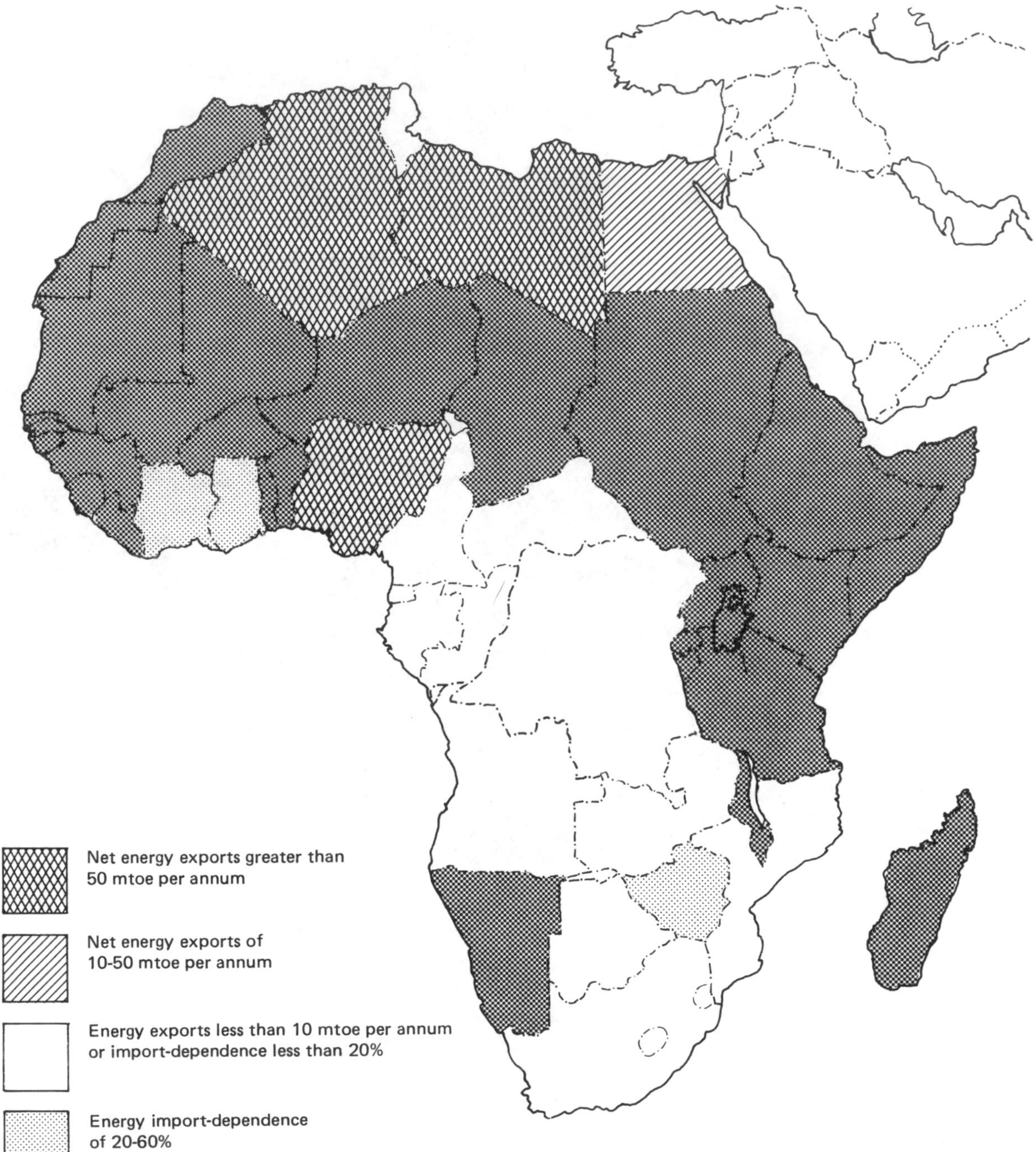

FIGURE 8 ENERGY IMPORTERS AND EXPORTERS IN ASIA/PACIFIC

Net energy exports greater than 50 mtoe per annum

Net energy exports of 10-50 mtoe per annum

Energy exports less than 10 mtoe per annum or import-dependence less than 20%

Energy import-dependence of 20-60%

Energy import-dependence greater than 60%

Part Two:
Country Reviews

Algeria

KEY ENERGY INDICATORS

Energy Consumption	
—million tonnes oil equivalent	8.1
Consumption Per Head	
—tonnes oil equivalent	0.44
—percentage of world average	27%
Net Energy Imports	nil
Oil Import-Dependence	nil

Algeria has very large energy resources, in the form of crude oil and, particularly, natural gas, which are the basis for ambitious general economic and social development programmes. The country remains primarily an energy exporter, with only limited value-adding activities and energy-based manufacturing. Consumption of energy per head of the population is still only one-quarter of the world average, but there exists a well developed government structure and commercial sector able to capitalise on these resources.

ENERGY MARKET TRENDS

Algeria was one of the principal beneficiaries of the sharp increase in oil prices during the 1970s. Additional revenue so generated was invaluable in underpinning investment in the economy, although a substantial proportion of this investment has been in energy production and transportation. In contrast, progress in developing manufacturing industry has been limited. Nevertheless, high rates of economic growth have been achieved, with an increase of more than 25 per cent in Gross Domestic Product in the period 1978-80, when many other economies were stagnating.

Pattern of Energy Supply

Primary Fuel Supply (percentage)	
Oil	57
Solid Fuel	2
Natural Gas	41
Primary Electricity	..
TOTAL	100

With such large indigenous resources of oil and gas available, these have assumed the predominant role in energy supply. Small quantitites of coal/coke are imported to meet special market requirements. Algeria is deficient in hydro-electric potential and a negligible contribution is obtained from this source.

ENERGY SUPPLY INDUSTRIES

State organisations have full control of the energy supply industries. The SONATRACH organisation is responsible for all aspects of oil and gas production and transportation. Distribution of gas and electricity to the internal market is also undertaken by a public monopoly agency. Foreign oil companies, particularly the French based ones, continue to play a role in oil and gas exploration and production in association with SONATRACH.

Oil

Oil Production

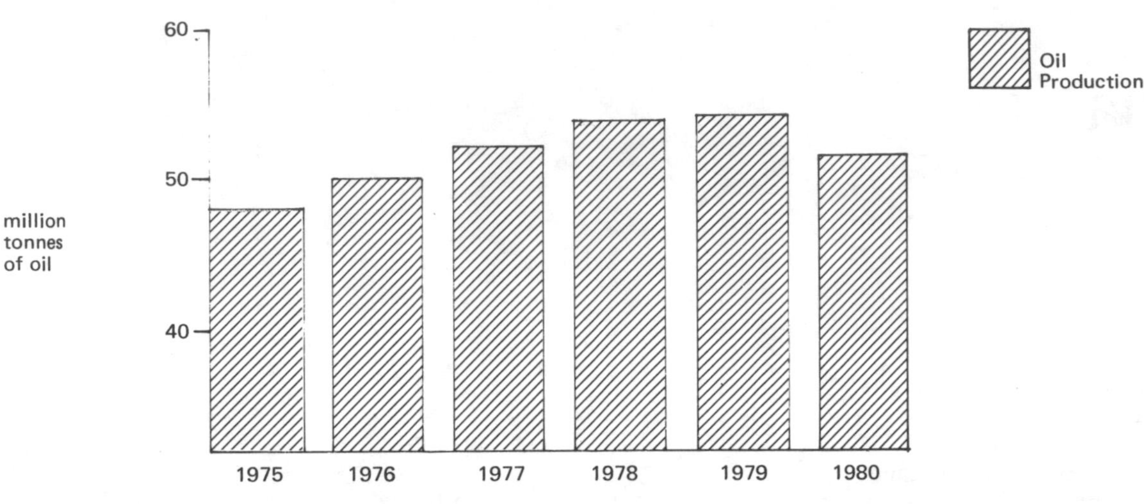

Production of crude oil has remained relatively stable in recent years, suffering a significant cut back only in 1981 in the face of generally weak market conditions. Between 1975 and 1978 output recovered to reach 54 million tonnes. Production remained at over 50 million tonnes until 1981, when it fell to around 40 million tonnes. Less than 10 per cent of output is consumed in Algeria.

The state organisation SONATRACH controls all aspects of the oil industry, with full responsibility for crude oil transportation and refining. SONATRACH handles 80 per cent of all crude oil disposals, except in the case of the French companies CFP and Elf-Aquitaine, which continue to enjoy a special relationship and acquire almost half of their production. Producing areas are in the north-eastern part of the country near the Libyan border and around Hassi Messaoud. Oil is piped to Arzew and Bejaia for export and also to the Algiers refinery. SONATRACH operates refineries at Algiers and Arzew, each of throughput capacity around 2.5 million tonnes per annum. A small plant is also in operation at Hassi Messaoud. A major new refinery is being constructed at Skikda which will not only meet additional Algerian demand for several years, but is also intended to increase the proportion of finished products in total oil exports.

Exploration programmes are being undertaken mainly by SONATRACH and the French groups. The Italian state oil company AGIP has exploration rights and Getty Oil and Hispanoil are involved in production with SONATRACH. Algerian reserves of crude oil are officially estimated at around 1,100 million tonnes.

Gas

Exploitation of natural gas is increasingly important as an earner of foreign exchange in addition to, and partly in substitution for, crude oil, production of which may have passed its peak. Production of natural gas has risen towards 40,000 million cubic metres per annum of which more than half is re-injected. Internal consumption accounts for around 10 per cent of production.

Algeria is fortunate in having substantial gas fields in addition to associated gas in the existing oil fields. The Hassi R'Mel field is the basis of Algeria's LNG export sales, which exceed 12,000 million cubic metres per annum. This level will be at least doubled as a result of the operation of the Trans-Mediterranean pipeline to Italy. Exports are made from liquefaction plants at Arzew and Skikda. Gas reserves are estimated to be as high as 3,700 billion cubic metres.

The state organisation SONATRACH is responsible for gas production, transport and sale of LNG. Within Algeria, a separate state undertaking, SONELGAZ, draws gas from SONATRACH's trunk pipelines, for sale to residential, commercial and domestic consumers. SONELGAZ and SONATRACH co-operate in the supply of LPG.

Electricity

Electricity production and distribution is carried out by the state organisation SONELGAZ. SONELGAZ operates the major part of generating capacity, which is mainly thermal plant with some hydro-electric plant. Hydro-electricity, however, forms only five per cent of electricity supplies. Total generating capacity is 1,225 MW, which produces over 5,000 GWh per annum. Production was only 3,800 GWh in 1975.

Uranium

Resources of uranium have been discovered at several locations in Algeria. At Abanker, Daira and Tingaoline production of uranium ore is planned at a level of 1,800 tonnes per annum. Proven reserves are estimated to be sufficient for a 15-20 year production operation. Uranium exploration and development falls under the control of the state agency ONAREM.

ENERGY TRADE

Net Imports/(Exports) 1975-80

	1975	1976	1977	1978	1979	1980
Crude Oil (million tonnes)	(40.6)	(43.4)	(47.3)	(47.1)	(47.1)	(45.0)

Crude oil forms the major element in Algeria's energy trade. Since 1975 exports have been in the range of 40-50 million tonnes per annum. Exports dropped below 40 million tonnes in 1981, because of weak export markets. Trade in oil products is very limited, but expected to increase with the commissioning of the export-orientated refinery at Skikda. Exports of natural gas are becoming increasingly important. LNG exports now exceed 12,000 million cubic metres of gas annually, with sales to several gas utilities in Western Europe and the United States. A similar volume of gas will be transported via the Trans-Mediterranean pipeline, which is being constructed in Algeria, Tunisia and Italy and will link Algeria's gas fields directly to the West European trunk pipeline system.

Angola

KEY ENERGY INDICATORS

Energy Consumption	
—million tonnes oil equivalent	0.9
Consumption Per Head	
—tonnes oil equivalent	0.13
—percentage of world average	8%
Net Energy Imports	nil
Oil Import-Dependence	nil

Angola has substantial proven reserves of oil and the prospect of additional discoveries. A large proportion of the oil is exported, with only limited use within the country. Angola also has hydro-electric resources which meet most of the demand for electricity. However, economic development is still very limited and consumption of commercial energy is only eight per cent of the world average.

ENERGY MARKET TRENDS

Consumption of commercial energy has reflected both the lack of development of the economy and the disrupted situation arising from internal and cross-border disturbances. Use of energy within the extractive industries has fluctuated considerably from year to year as output has varied. The impact of these trends has outweighed any upward progression in consumption from developments in the manufacturing or commercial sectors.

Pattern of Energy Supply

Primary Fuel Supply (percentage)	
Oil	65
Solid Fuel	–
Natural Gas	8
Primary Electricity	27
TOTAL	100

Oil accounts for two-thirds of energy supply. This reflects the relatively high demand for transport fuels and international bunkers. Some oil is used in power generation. Natural gas is utilised in a number of plants as fuel. Hydro-electricity is another important indigenous resource, but its high percentage share of total primary energy input serves only to underline the lack of industrial development.

ENERGY SUPPLY INDUSTRIES

Private sector companies, particularly foreign companies, have retained a considerable role in the oil industry in Angola. A state company has been set up, but it operates in partnership with foreign companies in exploration, production, refining and marketing. Outside expertise and investment is particularly important in exploration and development. Apart from some auto-production of electricity the state assumes responsibility for electricity supply.

Oil

Oil Production

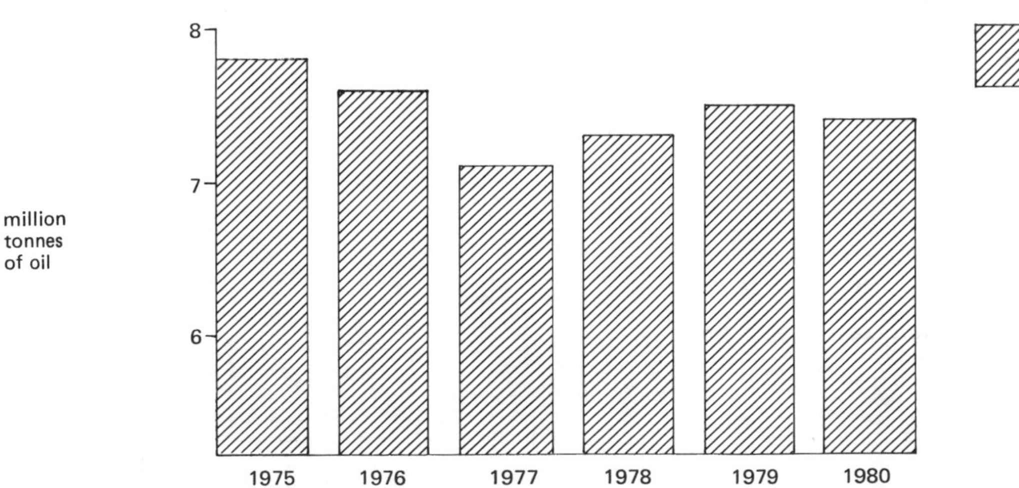

Production of crude oil has reached a plateau of 7-8 million tonnes, most of which is exported. The principal source of production is in the Cabinda enclave, where Cabinda Gulf, a subsidiary of Gulf Oil, discovered an offshore field in the mid 1960s. Other production takes place south of the Zaire River, where Texaco is operator.

A state oil company SONANGOL was established in 1976 to control exploration and production activity. SONANGOL participates in exploration and development groups with a number of foreign companies in addition to Gulf and Texaco. Actively involved are AGIP, CFP, Elf-Aquitaine, Mobil and others.

The Belgian company Petrofina has a particularly well-established position in Angola and participates with SONANGOL in PETRANGOL, which operates the country's only refinery and markets refined products throughout the country. The refinery is located at Luanda. Capacity is around 1.5 million tonnes per annum and meets most of the market's requirements, which total well under one million tonnes.

Electricity

Total electricity production is of the order of 1,400 GWh per annum, of which more than 1,000 GWh is derived from hydro-electric plant. The public supply system, operated under the aegis of the Ministry of Industry and Energy, has a generating capacity of 450 MW, of which a large proportion is hydro-electric.

ENERGY TRADE

Net Imports/(Exports) 1975-80

	1975	1976	1977	1978	1979	1980
Crude Oil (million tonnes)	(7.0)	(6.9)	(6.1)	(6.3)	(6.5)	(6.4)

Angola exports most of its production, as internal consumption is well under one million tonnes per annum. Exports have been maintained at above six million tonnes per annum, although there has been a downward trend in evidence since 1975. Angolan crude oil is of high quality, and competes with other West African and North African crude oils in North American and West European markets.

Australia

KEY ENERGY INDICATORS

Energy Consumption	
—million tonnes oil equivalent	77.0
Consumption Per Head	
—tonnes oil equivalent	5.27
—percentage of world average	325%
Net Energy Imports	nil
Oil Import-Dependence	31%

With a population of less than 15 million Australia has one of the highest levels of energy consumption amongst the developed economies. The development of the country's still uncharted wealth of natural resources has meant that Australia has had available the energy to match the demands of the extractive and beneficiating industries. The country is substantially self-sufficient in oil, although output has now begun to tail off, but major known deposits of natural gas are expected to be developed during the coming decade. A rapid growth in exports of coal has made Australia a net exporter of energy, and this position is likely to be further reinforced as more export-orientated projects come to fruition.

ENERGY MARKET TRENDS

Since 1975 economic growth has been irregular, but nevertheless consistently positive. Overall in the five-year period 1975-80 Gross Domestic Product rose by more than 13 per cent. But energy consumption rose at a substantially higher rate in every year except 1979, the total increase for the same period being nearly 22 per cent. This reflects the energy needs of certain sectors of the economy which are expanding strongly in Australia. However, the impact of conservation efforts may also be seen in a general downward trend in the relationship of growth of energy consumption to growth of GDP.

Trend of Energy Consumption

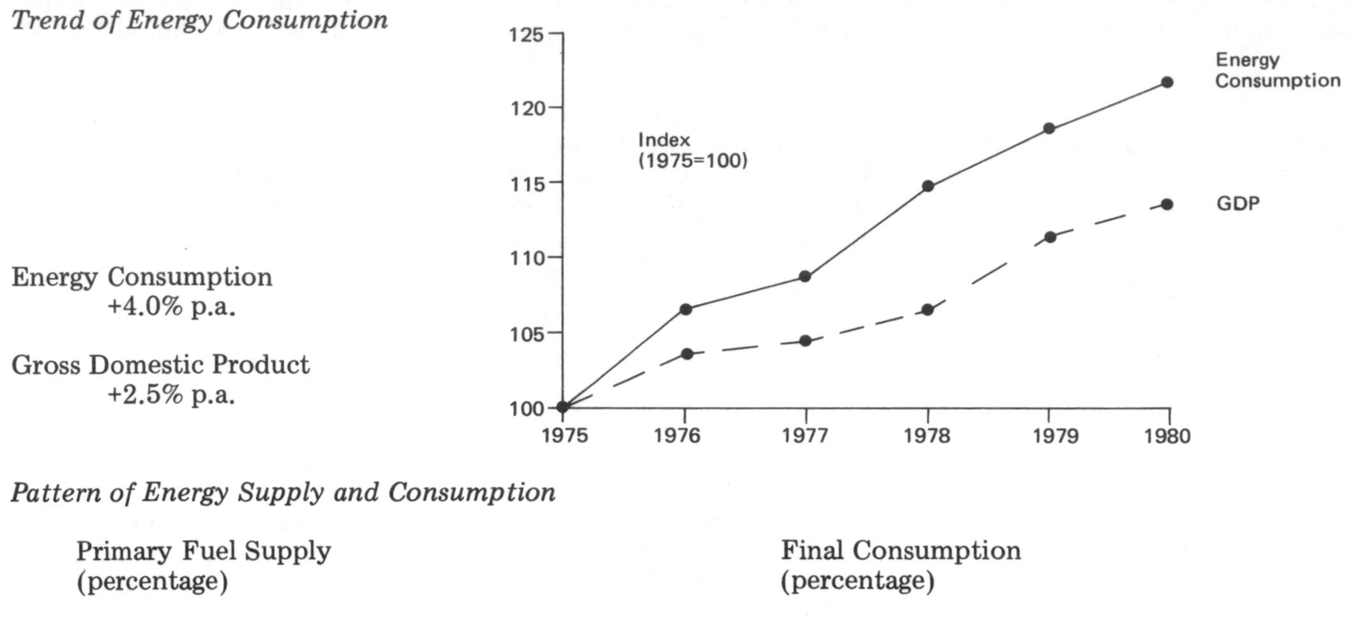

Pattern of Energy Supply and Consumption

Primary Fuel Supply (percentage)		Final Consumption (percentage)	
Oil	43	Industry	44
Solid Fuel	41	Residential	12
Natural Gas	10	Transport	31
Primary Electricity	6	Other	13
TOTAL	100	TOTAL	100

Oil is the largest individual source of energy for the Australian economy but is rivalled by the extensive use of black coal and brown coal, which are the predominant sources of fuel in power stations throughout the more populous states. The use of oil for such purposes is very small. However, the amounts of natural gas used for bulk heat and power may increase as new gas fields are developed to supply Western Australia and South Australia. Primary electricity generation consists of the hydro-electric resources exploited by the Snowy Mountains scheme and on a much smaller scale in Tasmania.

Industry absorbs a relatively large proportion of total final energy consumed, owing to the importance of highly energy intensive industries such as mineral extraction and processing, steelmaking and aluminium production. The transportation sector is also demanding in its use of oil products because of the long distances involved in inter-state and intra-state movements.

ENERGY SUPPLY INDUSTRIES

The Commonwealth Government plays a very limited role in energy supply and its activities are largely confined to indirect influence through powers over pricing, inter-state developments and foreign trade considerations. Individual states are closely involved with coal and electricity supply. Electricity generation is largely their responsibility and they are interlocked with the coal mining industry, either by virtue of integrated mining/power generation activities or as principal customers for new coal developments. However, coal exporting lies in the hands of the private sector, except insofar as energy and trade policies are affected, and the dynamic oil, gas and minerals exploration and development industries are entirely the preserve of private companies involved in energy supply, manufacturing, trading and investment finance.

Oil

During the last decade or more Australia has benefited greatly from the cushion of indigenous production of crude oil. Following the development of small fields at Moonie, Queensland, and at Barrow Island, off the north-west coast, the partnership of Esso and BHP (Broken Hill Proprietary) made a string of discoveries in the newly found oil province of the Gippsland Basin off the Victorian coast. Exploitation of this resource rapidly led to Australia achieving a substantial level of self-sufficiency in oil, and arrangements were devised whereby all refining and marketing companies handled indigenous crude oil as a priority, with resort to imported crude oil or petroleum products to make up the balance. Nevertheless, output has now stabilised at around 21 million tonnes per annum, while consumption has continued to rise to over 29 million tonnes.

Oil Production and Consumption

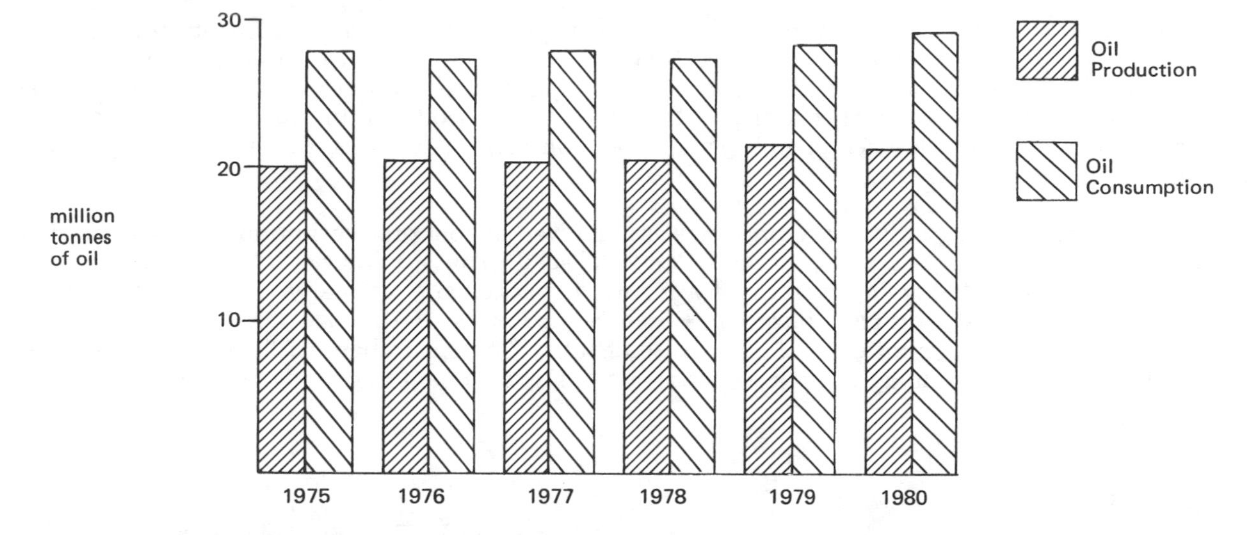

Major international oil companies are prominent in oil refining and marketing. Shell has refineries at Geelong, Victoria, and Clyde, New South Wales with a total capacity of more than eight million tonnes per annum. The Esso-Mobil partnership operates refineries at Altona, Victoria, and Port Stanvac, South Australia, under the name of Petroleum Refineries (Australia) Pty, also with a total capacity of eight million tonnes per annum. BP also has two refineries, including one at Westernport (Fremantle), the only one located away from the south-eastern part of the country. BP's total capacity is just under eight million tonnes per annum. Amoco, Total and Caltex, the joint-affiliate of Standard Oil Company of California and Texaco, each have a refinery, Amoco at Brisbane, the others in New South Wales. The only Australian based company involved in refining is Ampol, which has a three million tonnes per annum refinery at Lytton, Queensland. Ampol also has distribution and marketing operations.

Overall, Australian based companies retain a significant role in the oil industry and this is increasing as local effort is put into exploration and development work and even foreign controlled companies are pressed to increase the Australian content of their operations. The foremost Australian company continues to be BHP, through its interest in Gippsland production. Santos has been growing rapidly as a result of its involvement in development of the Cooper Basin fields.

Coal

Coal Production and Consumption

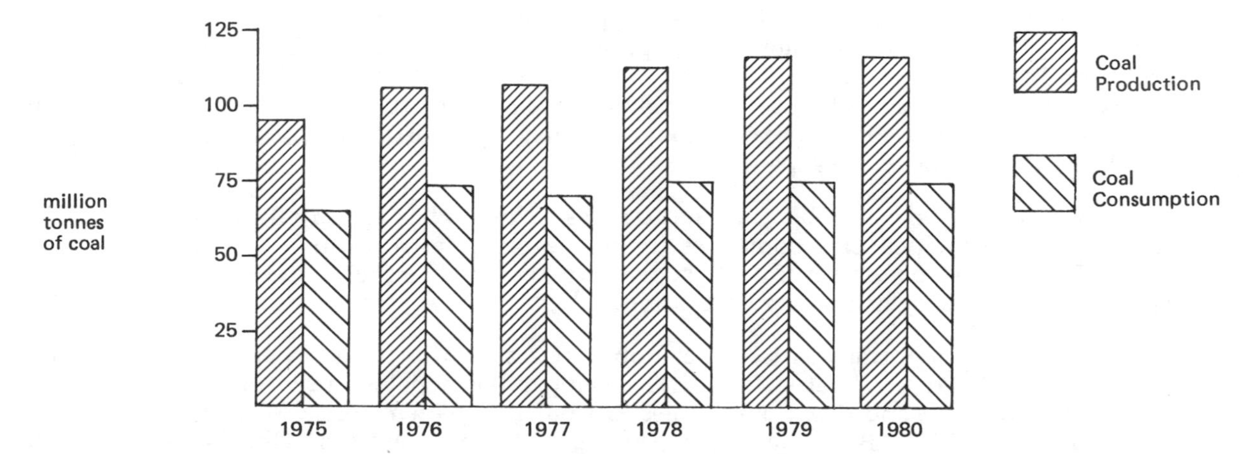

Australia has very large resources of black coal and brown coal. These are mainly located in the eastern part of the country, in New South Wales, Victoria and Queensland, although deposits are worked in other states too and large resources are being proved in the northern part of South Australia. The extensive deposits of black coal in New South Wales and Queensland are easily worked with around 50 per cent obtained by open-cast extraction, and are well located in relation to the main centres of population and to the coast for export markets. Victoria contains the overwhelming proportion of brown coal resources so far proved. These resources are particularly concentrated in the Latrobe Valley.

Total coal production is now around 120 million tonnes per annum, having risen from 95 million tonnes in 1975. Over 70 per cent of production is of hard coal. A large proportion of indigenous production is dedicated to power stations. In Victoria the State Electricity Commission is responsible for mining some 30 million tonnes per annum of coal, making it the largest coal producer in Australia. In most other instances coal is mined by private sector companies and consortia, supplying the electricity generating authorities under long term contracts.

A number of Australia's leading industrial companies are involved in coal production, including Broken Hill Proprietary (BHP), CSR, CRA, R W Miller and Peko-Wallsend. The subsidiaries of overseas and international companies are also becoming increasingly involved, particularly as many new developments are taking place with exports as a primary purpose. Foremost in terms of tonnage mined is currently the Australian affiliate of Utah International of the United States. The Australian affiliates of BP, Shell and several other international oil companies are also taking a close involvement in coal development.

Gas

Gas Production

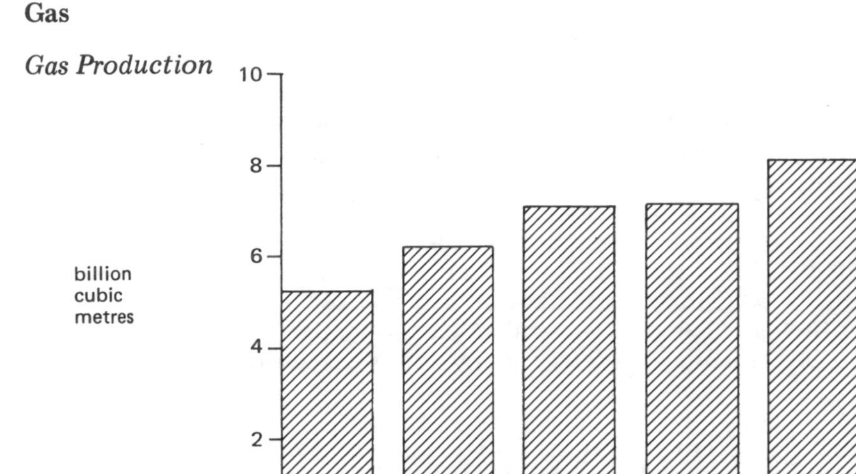

Discovery of natural gas has gone hand in hand with the discovery of oil. In fact Australian deposits of hydrocarbons have tended to be more promising as far as gas is concerned. Considerable quantities of gas became available with the development of the Bass Strait fields and gas is also of primary significance in the Cooper Basin of South Australia/Queensland and off the north-west coast of South Australia.

As a result the consumption of natural gas has been expanding, with trunk pipelines from the Cooper Basin to Adelaide and Sydney and a smaller scale supply becoming available in Brisbane from the Surat Basin fields. Total production rose from 5,300 million cubic metres in 1975 to 9,500 million cubic metres in 1980, and is expected to continue to rise. A pipeline is being constructed from the prolific Rankin field, off the north-west coast, to the Perth area, although the major proportion of these reserves is expected to be exported as liquefied natural gas to Japan.

Until the exploitation relatively recently of these natural gas fields, towns gas supply networks were very limited and outside the main cities LPG continues to be the principal form in which gas is distributed. Even now less than one-quarter of the natural gas is distributed to residential and commercial consumers, and the major part goes to industry and power generation.

Australian companies play an important part in the supply of gas. Santos has a 45 per cent interest in the Cooper Basin development, and is operator for it. The Cooper Basin fields are virtually the sole source of natural gas for South Australia and New South Wales. Output is expected to rise to around 5,300 million cubic metres per annum by 1987. BHP handled 2,300 million cubic metres of Gippsland gas in 1981. BHP is also a participant in the North-West Shelf gas development project. Local distribution of gas is carried out by state or local authority undertakings, or by private companies operating under franchise. Australian Gas Light Company, which operates in New South Wales, is the largest individual distributor.

Electricity

Electricity Production

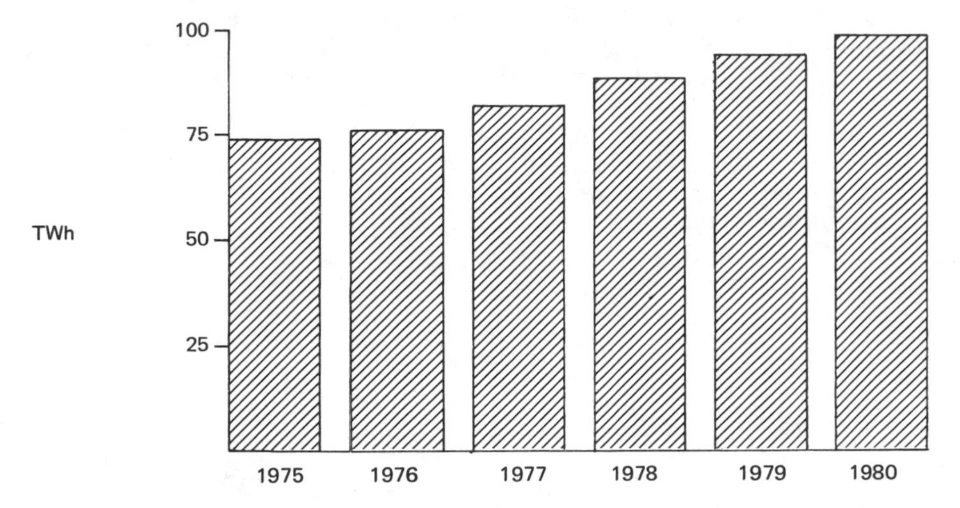

Demand for electricity has increased rapidly as a consequence of the further development of extracting and processing industries. From a level of 74 TWh in 1975 consumption by all sectors rose to 98 TWh in 1980. This overall increase of some 32 per cent is much higher than the increase in total primary energy consumption during that period, reflecting the coming on stream of individual projects in smelting etc, and a growth in demand inherent in a relatively healthy economy.

The generation and bulk supply of electricity for the public supply system lies almost entirely in the hands of state electricity commissions. The Electricity Commission of New South Wales sells bulk supplies to retail undertakings, which are largely under the control of local government bodies. These electricity supply authorities number around 40, with three city councils and over 30 groups of shires and/or municipal councils involved. Sales to industrial consumers and the railways are handled by the Commission.

In Victoria, where there is almost complete electrification, the State Electricity Commission is even more predominant. Not only does it control its coal fuel supplies, but it also handles all retail sales outside Melbourne. Other states have rather more complex supply situations. The Electricity Trust of South Australia has overall responsibility for unifying and co-ordinating the major part of electricity supply. The Trust operates power stations running on coal and natural gas.

The State Electricity Commission of Queensland has main responsibility for generation and transmission. Distribution is undertaken by seven boards with the southern-central network largely separate from the northern network. By contrast the State Energy Commission for Western Australia has broad responsibility for gas as well as electricity supply, although there remain numerous small local systems which cannot economically be connected to the state grid and which are run by the local public authority.

Australian electricity undertakings use coal for a high proportion of electricity generation. The overall proportion is around 75 per cent, but is 85 per cent in New South Wales and exceeds 90 per cent in Queensland. Oil and gas are also significant in South Australia and particularly Western Australia, although both are moving to increase the use of coal. Hydro-electric potential is limited and has been largely exploited already, except in Tasmania, where it is the main energy source. However, the largest hydro-electric scheme is in the Snowy Mountains, the capacity of which is 7.7 GW. The scheme is run by an authority jointly controlled by the states of Victoria and New South Wales, which take most of the power, and by the Commonwealth Territory, for which it is the primary source of electricity.

Uranium

It is estimated that Australia contains around 20 per cent of all low cost uranium reserves outside the centrally planned economies. Reasonably assured resources of uranium oxide, recoverable at less than $50 per pound, are put at some 300,000 tonnes. There are large deposits in Northern Territory and other significant deposits in Queensland, South Australia and Western Australia. On the other hand Australia accounts for only a small percentage of world production, so that major projects are under way to exploit the country's resources in order to meet the needs of nuclear power programmes in many parts of the world.

Uranium was extracted from the Mary Kathleen development in Queensland during the period 1958 to 1963 for export to the United Kingdom. Production operations recommenced in 1976. Two other deposits, both in Northern Territory are in the process of being developed. These are the Ranger and Nabarlek deposits along the East Alligator River. This area also includes a very large uranium deposit discovered at Jabiluka. Projects are being developed to exploit this and a deposit at Yeelirrie, Western Australia. Production of uranium oxide in 1980 was 1,840 tonnes, compared with 420 tonnes in 1976.

Overseas interests are prominent in the uranium developments, particularly those representing electricity generating companies in Japan and Western Europe. There is no nuclear power station in Australia currently, although one is being considered in Western Australia, which is remote from the main areas of coal and oil production in the south-east quarter of the country.

ENERGY TRADE

Net Imports/(Exports) 1975-80

	1975	1976	1977	1978	1979	1980
Coal (million tonnes)	(32.7)	(30.4)	(35.4)	(35.5)	(40.4)	(42.0)
Crude Oil (million tonnes)	8.3	8.4	9.0	8.9	8.6	6.6
Oil Products (million tonnes)	0.1	0.4	0.6	(0.8)	1.9	1.5
Uranium Oxide (tonnes)	..	36	1,545	1,114	1,317	1,131

Australia is a net exporter of energy, owing to the substantial contribution from indigenous resources towards meeting oil requirements and the great potential which the country has for exporting coal. The attractiveness of Australia's coal is enhanced by the great demands of the Japanese economy for imported energy. The net import requirement of oil has fallen back from the level of nine million tonnes in the mid 1970s, due largely to a pause in growth of demand and positive attempts to avoid dependence on imported oil. Imports are mostly in the form of crude oil since the country has adequate refining capacity.

The energy value of oil imports is far exceeded by that of coal exports, which have been rising steadily since 1976. However, export figures have stagnated to some extent as exports are mostly of coking coal and these therefore reflect the international recession in the steel industry. The underlying strength of the Australian coal industry lies in prospective demand for steam coal for many energy hungry countries in Asia and Europe. Exports of coal are expected to grow rapidly as finance is ploughed into production capacity and export terminals.

The country's net export position in energy will be further boosted when the Rankin field development moves ahead. This will involve the export of around 6,000 million cubic metres per annum of natural gas as LNG to Japan, equivalent to around 5.5 million tonnes of oil. Similarly, the country is expected to increase exports of uranium for overseas nuclear power stations during the 1980s and 1990s.

Sources of Imported Crude Oil

(percentage)	
Saudi Arabia	56
Indonesia	14
Iraq	9
Kuwait	8
United Arab Emirates	6
Other	7
TOTAL	100

Saudi Arabia accounts for well over half of Australia's needs for imported crude oil, and the Middle East as a whole supplies around 80 per cent of all imports. The balance comes mainly from Indonesia, an important oil-exporting country, and one relatively close to Australia. However, imports require to be generally heavier grades, which are produced by Saudi Arabia, Iraq and Kuwait.

Despite the extent of Australia's refining capacity and the balancing role of crude oil imports, there is nevertheless a significant trade in oil products. Of the total of three million tonnes imported in 1980 nearly half was brought in from the export refineries of Singapore. The balance was obtained principally from three other sources in the Middle East, namely Kuwait, Iran and Bahrain. Imports were to some extent offset by exports of 1.5 million tonnes, including movements to numerous locations in the islands of the Pacific area.

Destination of Coal Exports

(percentage)	
Japan	68
Europe	18
Republic of Korea	7
Taiwan	3
Other	4
TOTAL	100

Japan, with its voracious demand for all forms of transportable energy is by far the most important market currently for Australian exports of coal. Imports are mainly of coking coal for the Japanese steel industry and at present account for more than 80 per cent of total Australian shipments to that country. Other industrialising countries of the Far East, such as Taiwan and the Republic of Korea, also import Australian coal. But of increasing importance is the scope for shipping coal to Western Europe, where the market demand for coal in industrial and power generation sectors has been expanding rapidly. Apart from Japan, shipments to Europe, of both coking coal and steam coal, are now more significant than to all other countries combined.

ENERGY POLICIES

The formulation and implementation of energy policies is influenced by the federal nature of the Australian constitution, with limitations on the initiatives which can be taken at either state or Commonwealth level without broad agreement. This process of dialogue is encouraged through the existence of several bodies, including the National Energy Council. Since 1979 Australia has been a member country of the International Energy Agency.

Energy Supply

There is a general policy stance in favour of developing indigenous mineral and energy resources. This has led to investment in plant and facilities to exploit oil and gas for local purposes, and coal and uranium under export contracts. There are taxation and investment incentives for exploration and development of oil, gas and coal. Since 1978 the Commonwealth government has been moving crude oil prices towards parity with imported oil and oil discovered since 1976 is free of the levies which apply to the bulk of existing crude oil production. This also represents an important incentive for exploration and development. Government support is also being given to the project to develop the Rundle oil shale deposit.

Conservation

The principal planks of conservation policy are the movement towards pricing indigenous oil at international market levels and continuing information programmes on conservation practice. Up to the present time the government has set its face against mandatory controls on energy consumption. Thus it has established codes of practice for building construction and voluntary targets in the field of vehicle engine performance.

The Commonwealth government maintains a close watch on wholesale prices of petroleum products, the justification for which can be determined by the Prices Justification Tribunal, which was set up in 1973. State governments have the power to control prices, but only New South Wales sets prices for petroleum products. The Commonwealth government has established a three-tier price structure for liquefied petroleum gas, designed particularly to encourage its use as an automotive fuel.

Bahrain

KEY ENERGY INDICATORS

Energy Consumption	
—million tonnes oil equivalent	2.8
Consumption Per Head	
—tonnes oil equivalent	8.0
—percentage of world average	494%
Net Energy Imports	nil
Oil Import-Dependence	nil

The Sheikhdom of Bahrain has a population of only 350,000, but a substantial resource base in crude oil and associated gas. It is an international refining centre, with a major export refinery based on indigenous crude oil and increasing supplementary quantities from Saudi Arabia, including an offshore field which is effectively shared between Bahrain and Saudi Arabia. Natural gas, of which there is ample availability, forms the basic fuel for the country and is being used to attract additional industrial projects with export potential, such as fertiliser and methanol production.

ENERGY MARKET INDICATORS

The Bahrain economy is dominated by three energy intensive operations—the large export-orientated oil refinery of Bahrain Petroleum Company, the Alba aluminium smelting works and the power stations necessary to provide electricity for these activities as well as for other industrial, commercial and residential requirements. The path of energy consumption is therefore strongly linked to the export operations of BAPCO and the international aluminium market; these two activities over-riding any upward movement in local demand. They also account for the very high energy consumption per head of population.

Pattern of Energy Supply and Consumption

Primary Fuel Supply (percentage)	
Oil	29
Solid Fuel	–
Natural Gas	71
Primary Electricity	–
TOTAL	100

Consumption of Natural Gas (percentage)	
Electricity Generation	40
Oil Refining	36
Aluminium Smelting	22
Other	2
TOTAL	100

Bahrain's pattern of energy demand has developed on the basis of relatively large oil resources, and production is still three times as great as internal consumption. However, natural gas, which is also available in abundance from the oil fields, provides the main source of energy overall, enabling exports of oil products to be maximised and flaring of gas to be minimised. The availability of gas which would otherwise be flared has meant that it was an obvious fuel for use in electricity generation and in the oil refinery and that energy intensive aluminium smelting plant could be attracted to the island. Other uses of gas absorb a very small proportion of the available gas.

ENERGY SUPPLY INDUSTRIES

For many years the dominant industrial activity in Bahrain was the refinery of Bahrain Petroleum Company at Sitra, absorbing local production and linked by pipeline to Saudi Arabia, where the BAPCO parent companies, Standard Oil Company of California and Texaco, participate in oil exploration and production. The establishment of the state oil company, however, has removed from these companies their key role in oil production and their interest in local distribution, both of which activities have been taken over by BANOCO. BANOCO is also taking the leading role in promoting the use of associated natural gas.

Oil

Production of crude oil has been on a downward trend for a number of years and over the five years 1975-80 declined by 20 per cent to fall below 18 million barrels (less than 2.5 million tonnes). While this is nevertheless three times as great as local oil requirements, it represents a decline in capacity to earn foreign exchange. The BAPCO refinery now relies on imported Saudi Arabian oil for 80 per cent of its throughput. Output of finished products from the refinery in 1980 totalled 88 million barrels, including a high proportion of middle distillate products for markets in the Indian Ocean area and other Asian countries.

Since the establishment of the national oil company (BANOCO), BAPCO's interests have been confined to refining and export marketing of products. BANOCO now operates the oil fields and is responsible for supplying products to the local market. BAPCO is, however, involved as managing company for the gas separation plant of BANAGAS, which extracts LPG from associated gas for export to Japan.

Gas

Gas is becoming increasingly important as a fuel and feedstock for the Bahrain economy. Availability is increasing at the same time as the requirement for re-injection in the oil field is declining. Reserves of the Jebel Al-Dukhan field are estimated to be over 250 billion cubic metres, 70 times the 1980 output level. New major gas-using projects are planned, including ammonia and methanol plants utilising up to 70 million cubic feet per day. It will also be used as the principal fuel throughout the economy.

Electricity

Demand for electricity has been rising at a very rapid rate. From a level of 500 GWh in 1975 consumption almost tripled to 1,400 GWh in 1980. Generating capacity is around 530 MW. Power stations absorb some 40 per cent of the natural gas not otherwise flared or re-injected.

ENERGY TRADE

Bahrain imports around 10 million tonnes per annum of crude oil from Saudi Arabia, but this goes entirely into the BAPCO refinery, along with production from Bahrain's Jebel Al-Dukhan field. In 1980 the refinery was utilised at a high rate, with output totalling more than 11 million tonnes, of which almost 70 per cent was exported.

Bangladesh

KEY ENERGY INDICATORS

Energy Consumption	
—miillion tonnes oil equivalent	2.8
Consumption Per Head	
—tonnes oil equivalent	0.03
—percentage of world average	2%
Net Energy Imports	56%
Oil Import-Dependence	100%

Bangladesh is one of the more populous countries of Asia, but is at a very early stage of industrialisation. Consumption per head of commercial energy is only two per cent of the world average, and still of much less significance than non-commercial fuels. The country does, however, possess important gas resources, with large fields still to be developed and the promise of others still to be found. Large coal deposits are also known to exist and hydro-electric potential can be further exploited.

ENERGY MARKET TRENDS

Trend of Energy Consumption

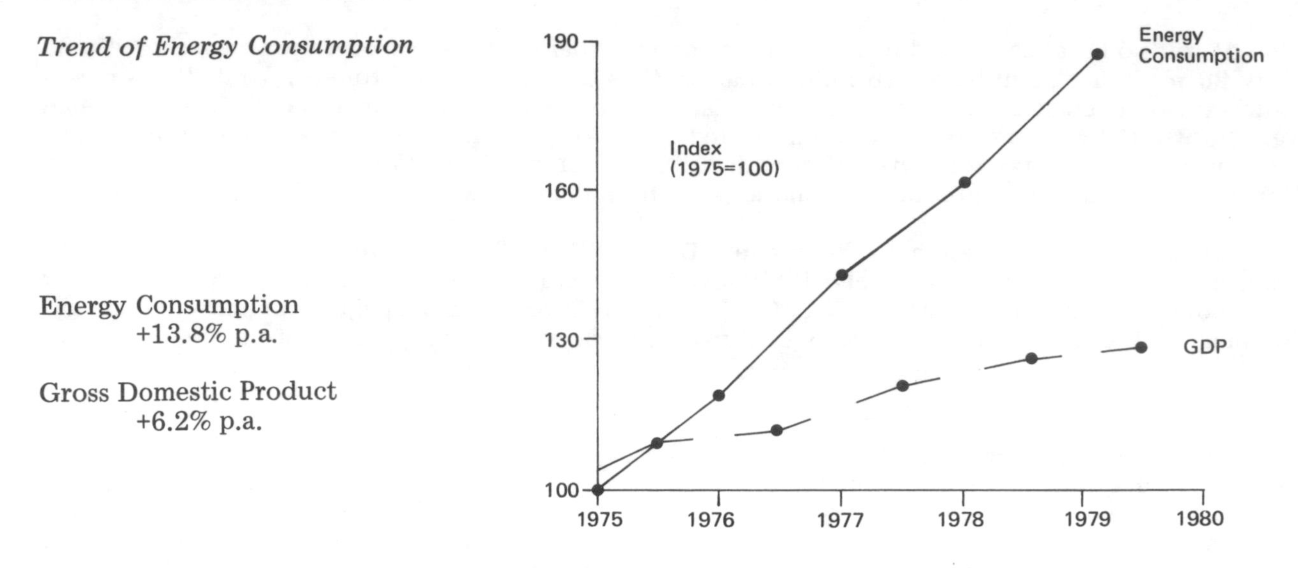

Energy Consumption
+13.8% p.a.

Gross Domestic Product
+6.2% p.a.

The availability of indigenous natural gas has enabled the Bangladesh economy to continue to make progress despite general difficulties in the world economy and a complete dependence on imports for oil. Economic growth in the five years to 1979-80 was around 30 per cent and positive in all years. Consumption of energy in total is estimated to have risen by a broadly similar figure, but consumption of commercial fuels rose by more than 90 per cent.

Pattern of Energy Supply and Consumption

Primary Fuel Supply (percentage)	
Oil	49
Solid Fuel	7
Natural Gas	39
Primary Electricity	5
TOTAL	100

Final Consumption (percentage)	
Industry	23
Residential	17
Transport	15
Electricity Generation	28
Other	17
TOTAL	100

Bangladesh's energy economy is being built on the twin pillars of locally produced natural gas and imported oil, with a rising share being met from natural gas. Almost 90 per cent of commercial fuel supplies are met by oil and gas. However, the total extent of use of commercial fuels is still much smaller than consumption of traditional fuels such as rice hulls, rice straw, cow dung etc. These sources are estimated to contribute more than six million tonnes of oil equivalent per annum, and usage has been rising continually since the early 1970s. The rate of increase is nevertheless much less than in the case of commercial fuels.

A high proportion of commercial fuel supply is absorbed in electricity generation, including a substantial amount of natural gas as well as fuel oil. Transport occupies a relatively important position, although low energy-intensive water transport remains important. Use of natural gas for fertiliser manufacture represents a significant element in total consumption of commercial fuels.

ENERGY SUPPLY INDUSTRIES

State corporations play an important part in the supply of oil, natural gas and electricity. Bangladesh Oil and Gas Corporation is concerned with exploration for oil and gas and development of the country's substantial gas reserves. Bangladesh Petroleum Corporation controls the importation and sale of oil, although private companies, including an affiliate of Burmah Oil Company, carry out distribution and marketing of products. Burmah also has a minority interest in the state-owned oil refinery.

Oil

Consumption of oil products now exceeds one and a half million tonnes per annum, increasing in successive years since 1977. In fact it was only in 1977 that consumption regained the level of 1970, having fallen back in the years after the oil supply crisis of 1973-74. The pattern of consumption is dominated by demand for middle distillates, especially kerosine and diesel oil, to meet the needs of the transport sector and of Bangladesh Power Development Board.

There is no production of crude oil in Bangladesh currently and exploration efforts in the Bay of Bengal have so far proved unpromising. Crude oil is therefore imported by the state oil company Bangladesh Petroleum Corporation, which is responsible for the supply of crude oil and finished products. The Corporation holds a majority interest in the country's oil refinery at Chittagong. This refinery is operated by Eastern Refinery Limited and has a capacity of around 1.5 million tonnes per annum.

Marketing of petroleum products is carried out by Burmah Eastern Limited, part of the UK based Burmah group, which has a 30 per cent shareholding in the refinery, and by two locally based companies Jamuna Oil Company and Meghna Petroleum Limited. These companies each supply approximately one-third of the Bangladesh market, with wholesale and retail sales of a full range of products.

Coal

Consumption of solid fuels in Bangladesh is largely confined to traditional fuels such as firewood, rice husks etc. Coal and/or coke is imported for industrial uses. Major reserves of coal have been proved in the Jamalganj area at some 700 million tonnes. But these are located at depth and studies are being carried out to see how these resources might be exploited.

Gas

Gas Production

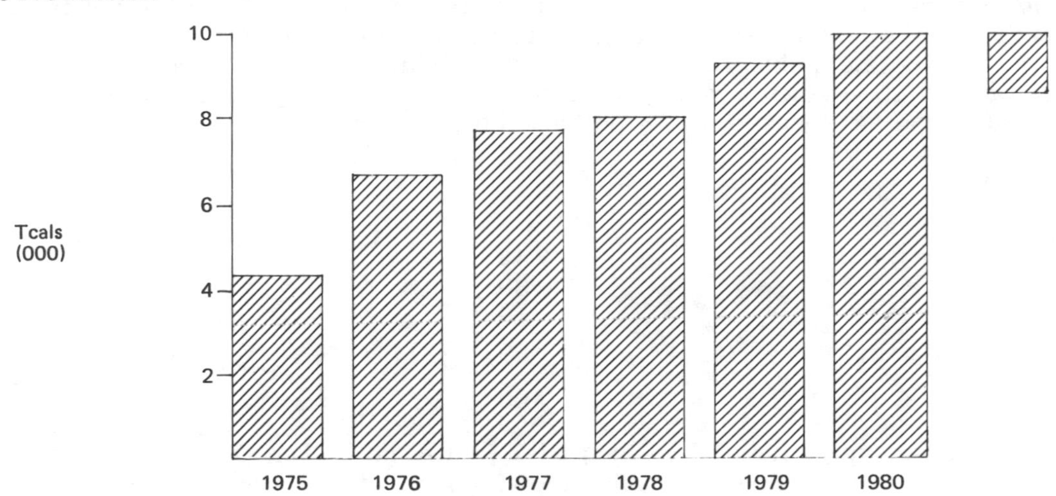

The period since 1974-5 has seen a rapid development of natural gas production. Output more than doubled in the five-year period to 10,000 teracalories in 1979-80, equivalent to 1,200 million cubic metres. This represented a recovery of the pre-1974 position and a further substantial advance. Around 70 per cent comes from the Titas gas field, which was developed during the 1970s, adding greatly to existing production in Sylhet. Exploration, production and transmission of natural gas is carried out by the state oil and gas company Bangladesh Oil and Gas Corporation (Petrobangla), through its subsidiaries Titas Gas Transmission and Jalalabad Gas Distribution System.

Natural gas is being utilised primarily as a fuel in electricity generation and as feedstock in the manufacture of fertiliser, each accounting for around 40 per cent of consumption. Most of the balance is used by a few industrial enterprises, notably the Chatak cement works. Use in commercial and residential situations is very limited.

Gas fields are all located in the eastern part of the country and of the three main industrial areas only Dacca is so far connected by pipeline. But the existing fields are capable of producing 240 million cubic feet per day and other fields await development, including the Bakhrabad field, which is the largest so far discovered. Natural gas is expected to be available in the Chittagong area by 1983, when there will be a further increase in demand. Official plans envisage natural gas consumption rising almost three-fold between 1979-80 and 1984-85.

Electricity

Consumption of electricity has been rising approximately in line with demand for commercial fuels in general, but very much faster than total energy consumption. Electricity generation between 1974-75 and 1978-79 rose by 60 per cent, twice the rate of growth of the economy. The supply of electricity is in the hands of two public enterprises, the Bangladesh Power Development Board (BPDB), which is responsible for generation, transmission and distribution and the Rural Electrification Board. The Bangladesh Water Development Board is also involved, by virtue of its concern with the overall development of water resources.

BPDB operates separate grids in the Eastern and Western Zones of the country as well as localised systems in both zones. Total generating capacity in 1980 was 750 MW, consisting of 494 MW of hydro-electric, steam and gas turbine capacity in the Eastern grid and 256 MW of steam, gas turbine and diesel capacity in the Western grid. The Board's 80 MW Karnafuli hydro-electric station accounts for 25-30 per cent of electricity generation and is being further extended. Largest power generating plant is currently at Khulna where the Board operates a combined 160 MW of steam and gas turbine plant, with a further 110 MW under construction.

ENERGY TRADE

Net Imports/(Exports) 1979

Coal (million tonnes)	0.3
Crude Oil (million tonnes)	1.1
Oil Products (million tonnes)	0.3

With only very limited industrial need for coal or coke and use of solid fuel in the domestic sector being met from traditional locally available fuel sources, Bangladesh's foreign trade is predominantly in crude oil and finished products. The entire requirement for crude oil to be processed at the Chittagong refinery is imported. This has risen to over one million tonnes per annum. In addition finished products have to be imported, particularly of high speed diesel, which the refinery is unable to produce in large enough quantities. Imports are partially offset by exports of naphtha and fuel oil.

Sources of Imported Crude Oil

(percentage)	
United Arab Emirates	43
Saudi Arabia	33
Iran	16
Iraq	8
TOTAL	100

A limited number of countries supply crude oil to Bangladesh, all from the Persian Gulf area. Principal source is the United Arab Emirates, with Saudi Arabia supplying most of the balance. Heavier crude oil grades from Iran and Iraq have also been imported, partly in order to provide feedstock for Eastern Refinery Limited's recently built bitumen plant.

Brunei

Brunei is essentially an energy producing country with very limited consumption of energy in other sectors of the economy. The population is little more than 200,000, but possesses substantial resources of both oil and gas. Energy production is of the order of 17 million tonnes of oil equivalent per annum, of which only 1½ million tonnes is consumed in Brunei. Brunei was one of the earliest countries to liquefy natural gas, which is exported under long-term arrangements with Japanese utilities. The state government participates in production projects, but relies on international companies for exploration and development work.

KEY ENERGY INDICATORS

Energy Consumption	
—million tonnes oil equivalent	1.4
Consumption Per Head	
—tonnes oil equivalent	6.7
—percentage of world average	414%
Net Energy Imports	nil
Oil Import-Dependence	nil

ENERGY MARKET TRENDS

The Brunei economy is at a low level of general development by international standards and consumption of energy has increased primarily as a result of operations of the oil and gas production industries and of the gas liquefaction plant. Energy consumption is of the order of one and a half million tonnes of oil equivalent per annum. Consumption of electricity has been rising relatively rapidly from a level of 230 GWh in 1975 to over 360 GWh by 1978.

ENERGY SUPPLY INDUSTRIES

The Brunei government has allowed private companies to take the initiative in exploring for oil and natural gas and in developing export projects. The government has, however, taken a share in the natural gas liquefaction project.

Oil

Oil Production

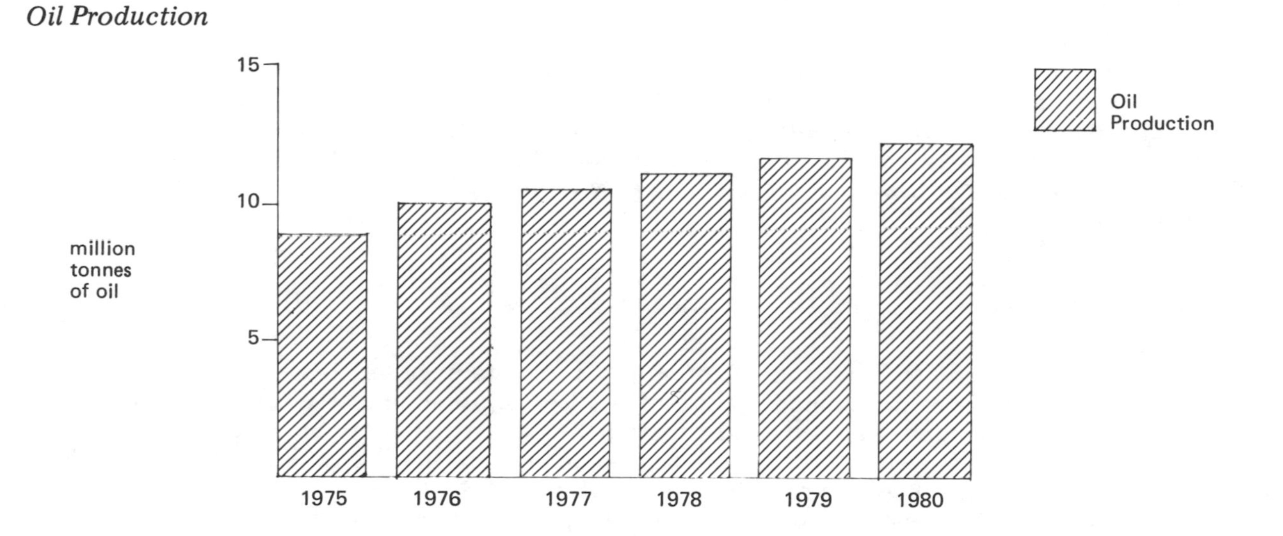

A key role in the oil industry in Brunei is played by the Royal Dutch/Shell group, which has been established in the area for many decades. Oil was discovered near the western border at Miri, in Sarawak, in 1910 by Shell, which has been involved continuously in the area since that period, with production from its Seria and other fields. Crude oil production in Brunei has maintained a steady rise since 1976 to a level of just over 12 million tonnes in 1980. The government has a 50 per cent participation in this production, most of which is exported to Japan. Oil is pumped to the export terminals at Lutong and Miri in Sarawak. Petroleum products are supplied from the Sarawak refinery of Shell Malaysia, as there is no refinery in Brunei.

Gas

Brunei was one of the first countries to become involved in the exportation of liquefied natural gas. The Lumut plant was brought on stream in 1972 to utilise the country's natural gas resources, which were far in excess of Brunei's own likely requirements. The Brunei government and Shell are partners in the production of the gas, which is sold to the liquefaction and shipment operation for transport in liquid form to Japan. Brunei is a major source of LNG for Japanese power companies, supplying one-third of that country's imports in 1980.

These consuming companies participate in the liquefaction plant as part of the long term underwriting of the investment involved in liquefaction and transportation. Total production of natural gas has now reached approximately five billion cubic metres per annum, at which level the Lumut plant is kept fully utilised. Local consumption of gas accounts for less than one per cent of the total.

Burma

KEY ENERGY INDICATORS

Energy Consumption	
—million tonnes oil equivalent	1.8
Consumption Per Head	
—tonnes oil equivalent	0.05
—percentage of world average	3%
Net Energy Imports	7%
Oil Import-Dependence	9%

Consumption of commercial fuels in Burma is less than two million tonnes of oil equivalent per annum for a population of more than 35 million. The country is at a low state of economic development and has a per capita consumption of only three per cent of the world average. So far Burma has maintained a high level of self-sufficiency in energy, having indigenous resources of oil, natural gas and hydro-electric potential. The state has a close control of all aspects of energy supply in line with its approach to industrial and economic development.

ENERGY MARKET TRENDS

Burmese governments have sought to limit imports of energy products in order to maintain a high degree of self-sufficiency. Growth has been sustained by use of traditional non-commercial fuels and by a slow expansion of production from indigenous oil fields. Output of primary energy increased by 47 per cent between 1975 and 1978 and there was a further increase of nine per cent in oil production in 1979. Added to this was a doubling of natural gas production between 1978 and 1980. Overall in the period 1975-76 to 1980-81 energy consumption is estimated to have risen by nearly 50 per cent.

Pattern of Energy Supply

Primary Fuel Supply (percentage)	
Oil	68
Solid Fuel	..
Natural Gas	23
Primary Electricity	9
TOTAL	100

Of the commercial fuel forms in use in Burma oil is of dominant importance, accounting for 68 per cent of total supply. Natural gas production has been increasing rapidly, particularly since 1978 and its share in the pattern of supply has increased from 15 per cent in 1975-76 to 23 per cent. Use of coal is negligible, but hydro-electric output has maintained a share of around nine per cent of primary energy supply.

ENERGY SUPPLY INDUSTRIES

Activities in the energy sector are closely controlled by the state, with public corporations responsible for oil supply, refining and exploration. The Myanma Oil Corporation is the state corporation responsible for production of oil and gas and the supply of crude oil to the refineries operated by Petrochemical Industries Corporation. The Electric Power Corporation is responsible for generation and transmission of electricity throughout the country.

Oil

Burma relies on output from local fields to provide supplies of crude oil to its refineries. Production has been on a rising trend and increased by 30 per cent between 1976-77 and 1980-81, despite a levelling off in production in 1980. Production is the responsibility of Myanma Oil Corporation, the state owned oil company, which is also undertaking an extensive programme of exploration for additional oil reserves. Burma has only one significant oil refinery, of throughput capacity around one million tonnes per annum. This is operated by another state corporation, Petrochemical Industries Corporation.

Gas

Production of natural gas has increased significantly in recent years, giving this fuel almost one-quarter of the total energy market. Output in 1976-77 was 240 million cubic metres rising to 280 million in 1978-79. Over the next two years output more than doubled to reach 570 million cubic metres in 1980-81.

Electricity

The use of electricity has been growing rapidly, reaching 1,200 GWh in 1980-81. This represents an increase of nearly 60 per cent compared with 1975-76. Capacity operated by the Electric Power Corporation and other state owned generating companies increased to a much lesser extent, as progress was made in cutting down transmission losses and existing capacity was more fully utilised. At the end of 1980-81 capacity of the Electric Power Corporation was 500 MW, with a further 220 MW operated by other undertakings. The Corporation has 170 MW of hydro-electric plant, with the balance consisting mainly of gas turbines and diesel sets.

Cameroon

KEY ENERGY INDICATORS

Energy Consumption	
—million tonnes oil equivalent	0.8
Consumption Per Head	
—tonnes oil equivalent	0.09
—percentage of world average	6%
Net Energy Imports	nil
Oil Import-Dependence	nil

The United Republic of Cameroon is still at a very early stage of development, with very limited use of commercial fuels. Consumption per head of population is only six per cent of the world average. However, the recent exploitation of oil has assured Cameroon of self-sufficiency for many years ahead, as well as providing a source of export earnings. The availability of oil, with the prospect of further discoveries, comes as an addition to the country's existing base of hydro-electric production and known reserves of natural gas, which have not yet been exploited.

ENERGY MARKET TRENDS

Total energy consumption in Cameroon is some 800,000 tonnes of oil equivalent per annum. Consumption has increased sharply since 1976, owing to the greater activity in the mining sector, as well as to additional development in the energy industries themselves.

Pattern of Energy Supply

Primary Fuel Supply (percentage)	
Oil	60
Solid Fuel	–
Natural Gas	–
Primary Electricity	40
TOTAL	100

Supply of commercial energy consists of oil and hydro-electricity. The latter has been developed in order to provide power for mining operations, principally the smelting of alumina. Oil meets the remaining energy requirements of this sector, as well as for transportation and the commercial sector.

ENERGY SUPPLY INDUSTRIES

The state is involved in the supply of both oil and electricity, but foreign companies play a large part in all aspects of oil exploration and development. Refining and marketing is undertaken by affiliates of international oil companies. Industrial energy-consuming companies dominate electricity generation, but the state-owned public supply company is increasing its importance as it develops the national grid.

Oil

Production of crude oil commenced in 1977 and by 1981 had built up to an annual rate of four million tonnes. Part of the output is being fed into the recently completed refinery of SONARA at Pointe Limboh, near Victoria, with the major part destined for export. Production is from fields operated by the French state group Elf-Aquitaine in partnership with Shell.

The SONARA refinery has a throughput capacity of two million tonnes per annum and produces all main products for the local market, which currently amounts to only 500,000 tonnes per annum. The state holds a majority share in the refinery, with participation by Elf-Aquitaine, CFP, Mobil and Shell, each of which is involved in product distribution and marketing.

Electricity

Consumption of electricity is dominated by the bauxite and iron ore mining industries. Total consumption is around 1,300 GWh per annum, but fluctuates with activity in these key industries. The mining companies largely produce their own electricity and the state owned public supply organisation SONEL operates less than a quarter of the country's 340 MW of generating capacity. Electricity supply to the mining complexes comes from hydro-electric plant. SONEL has depended on small thermal power plants, but is developing hydro-electric capacity at Song Loulou and Lagdo. Song Loulou capacity is to be eventually 290 MW, supplying power mainly to the alumina industry, but also reinforcing SONEL's grid.

ENERGY TRADE

Until the commissioning of the SONARA refinery all petroleum products had to be imported. Operation of the refinery, following on the development of the Kole oil field, means that Cameroon now exports quantities of crude oil and finished products, amounting to more than two million tonnes in 1980 and doubling by the mid 1980s.

China (People's Republic of)

KEY ENERGY INDICATORS

Energy Consumption	
—million tonnes oil equivalent	510.0
Consumption Per Head	
—tonnes oil equivalent	0.53
—percentage of world average	33%
Net Energy Imports	nil
Oil Import-Dependence	nil

The People's Republic of China is one of the world's major producers and consumers of energy, although its impact on the international scene has been negligible so far, as a result of political and economic policies which have concentrated on self-sufficiency and limited international trading links. It is only in the recent past that the scale of the United States coal industry has surpassed that of China and the country also has one of the largest integrated oil sectors. The scale of the country's population and intermediate state of development means that per capita consumption is still low by world standards. Economic development therefore presents massive problems of raising output of coal and oil, of exploiting hydro-electric potential, with possible nuclear power plant construction. Foreign companies have become involved in the problems of the coal industry and are poised to play an important part in offshore oil and gas development.

ENERGY MARKET TRENDS

Although the exact path of the Chinese economy over recent years is not clear, it seems that energy consumption has been rising steadily and that supply problems may indeed be acting as a brake on development. Between 1975 and 1980 energy consumption increased by one quarter and whilst this amounts to less than five per cent per annum, a low rate for a developing economy, it nevertheless meant demand for an additional 100 million tonnes of oil equivalent per annum between the two dates.

Pattern of Energy Supply and Consumption

Primary Fuel Supply (percentage)	
Oil	22
Solid Fuel	70
Natural Gas	4
Primary Electricity	4
TOTAL	100

Consumption of Coal (percentage)	
Industry	47
Metallurgy	11
Electricity Generation	20
Other	22
TOTAL	100

Coal is a cornerstone of the Chinese energy economy, accounting for around 70 per cent of total primary energy input. It is used widely as main fuel in all sectors of the economy, with the exception of road transport. Some 70-80 million tonnes per annum is consumed in production of electricity, for which it is the major fuel. Oil accounts for almost all of the balance of energy supply. So far other resources have not been used much, although the use of hydro-electric power is expected to increase considerably as massive new projects are implemented on the Yangtze and other rivers.

Almost one-half of coal supply is consumed by industry, with a further 11 per cent used in metallurgical operations. Electricity generation is also a large consumer of coal. Retail sales to domestic and commercial sectors account for 18 per cent and the railways approximately four per cent. Industry and power stations also consume around 30 per cent of oil supplies. A large proportion of petroleum demand consists of heating oil and diesel fuel, but gasoline is of relatively little importance.

ENERGY SUPPLY INDUSTRIES

All energy supply operations are controlled by the state, with numerous ministries and state agencies involved. Separate ministries are responsible for the coal industry, petroleum and electricity. There was also established in 1980 a State Energy Commission to provide overall co-ordination of planning and development. The Ministries themselves fulfil planning and administrative functions, as well as managing exploration, production, processing and supply operations.

Oil

Oil Production

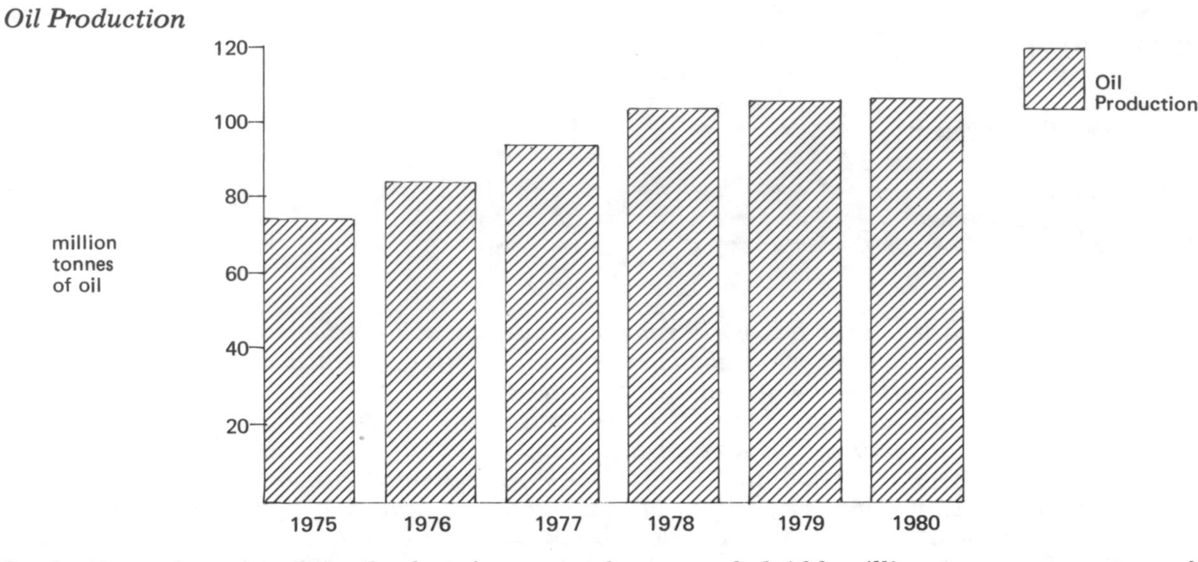

Production of crude oil in the last few years has exceeded 100 million tonnes per annum, but the level has stagnated after a relatively rapid expansion of the industry between 1975 and 1978. During that three-year period production rose by one-third. Plans for 1981 anticipated that there would be no further increase in that year either. Reported difficulties in maintaining output have added weight to moves to substitute coal in bulk heat-raising in power generation and in industry and to encourage exploration for further reserves.

China has numerous oil fields in production in several parts of the country, but the main ones are in the north-eastern provinces, where the Taching and Shengli fields are located. There are also several fields in the north-west of the country. Proven reserves have been estimated at around 7,000 million tonnes in onshore areas, with perhaps more than half located in the north and north-east. The extent of offshore reserves is more speculative, but has been put at the same order of magnitude. Interest in offshore exploration is noticeable, with many international oil companies participating with state agencies in initial seismic work. This covered an area of more than 400,000 square kilometres in the Yellow Sea and South China Sea and is expected to form the basis for bids in the next round of activity.

Activities in the oil sector are controlled by the Ministry of Petroleum Industry directly and through several state corporations. These include the National Oil and Gas Exploration and Development Corporation, the Petroleum Refining Corporation and the Petroleum Corporation of China, which was set up in 1977 to assist in the procurement of foreign technology for the oil industry. An additional body is being set up, designated the National Offshore Oil Corporation, to represent state interests in offshore exploration and production ventures.

Coal

China has one of the largest coal industries in the world, with output now well in excess of 600 million tonnes per annum. A peak level of output of 635 million tonnes was attained in 1979, but there was a fall in 1980 and 1981 production was expected to make only a partial recovery. The increase in output between 1975 and 1978 was almost 150 million tonnes, or approximately one-third. Production has been hampered by lack of transport and handling facilities as well as by lack of modern mining plant.

The Ministry of Coal Industry controls all mining operations and operates directly mines producing 55 per cent of total output. These are almost entirely deep-mining operations and include the country's main collieries. Other mines are operated by provincial and other local authorities and are usually very small. There are some 20 mines capable of producing five million tonnes or more per annum and the 10 leading mines in 1980 produced 143 million tonnes, or a quarter of national output.

The major mines are located in the north-eastern part of the country. This is also the area in which a high proportion of proven reserves are to be found. At present cumulative reserves are put at 640,000 million tonnes, spread over an area of half a million square kilometres in most of the provinces.

Coal Production

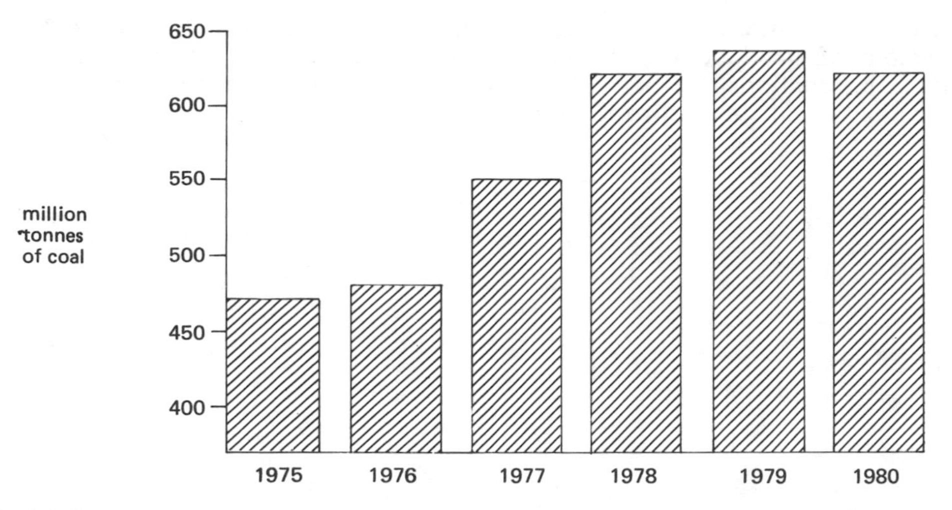

Electricity

Electricity production in 1980 is put at some 300 TWh, representing a rapid increase in recent years from 235 TWh in 1977. It is estimated that generating capacity is around 50 GW, consisting mainly of thermal power stations using coal and oil. Some oil-burning stations have already been converted to coal. Hydro-electric potential appears to have been little exploited so far, although a scheme is being completed for 2,700 MW at Gezhouba on the Yangtze River. A much larger scheme is under consideration which could yield over 100 TWh per annum.

The Ministry of Electric Power has departments concerned with all aspects of electricity generation and supply, including operations, hydro-electric plant, nuclear projects and power station construction. The operational divisions manage separate grids in the north, north-east, east, north-west and central regions of the country. Nuclear plant has not yet been built, but a decision may be made by 1983 to commence construction of the country's first plant, which will probably be located in Guangdong Province. This area is already deficient in electricity production capacity and is co-operating with China Light and Power, the producer of electricity in the New Territories of Hong Kong, to link grids and optimise power supply.

ENERGY TRADE

The energy trade of China is very limited, as the country has emphasised self-sufficiency in energy, but yet has not been able to produce significant quantities of oil or coal for export. Principal trading partner is Japan, with which China has a long-term trading agreement. Under this agreement up to 10 million tonnes per annum of oil is being exported to Japan. Although this is small within the context of Japan's total import requirement, it does provide further diversification of supply sources as well as the basis for bilateral trade in other products.

ENERGY POLICIES

Chinese energy policy aims to tackle the twin problems of increasing energy supply and at the same time conserving energy and introducing greater efficiency in use. Both aspects of policy involve very large programmes of investment and raise important political questions as to the extent of foreign involvement which will be allowed. This applies to the need to revitalise and expand the established oil fields and coal mines, to the possibility of exploiting uranium deposits and to constructing nuclear power stations. In each of these fields change appears to be very slow and subject to the risk of reversal in the face of economic or political obstacles.

Most headway has been made in the field of oil and gas exploration in offshore areas. In the last two years most of the leading international oil companies have participated in seismic programmes in the South China Sea and Yellow Sea. The machinery is being set up by the government for joint-venture exploration activity.

Attention is also being paid to the scope for unconventional fuels. The localised use of marsh gas is to be encouraged and the development of fuel forests promoted. Efforts are being put into utilisation of solar energy. These programmes will complement the main thrust of investment in renewable energy in the form of hydro-electric power.

The Chinese Authorities are also acutely aware of the need for improved efficiency and for energy conservation. Plans include replacement of small boilers and low and medium pressure generating plant by modern combined heat and power installations, further cuts in the ratio of energy consumption to output in the steel industry, and the introduction of standards for energy consuming goods.

Congo

KEY ENERGY INDICATORS

Energy Consumption	
—million tonnes oil equivalent	0.2
Consumption Per Head	
—tonnes oil equivalent	0.13
—percentage of world average	8%
Net Energy Imports	nil
Oil Import-Dependence	nil

Congo is a very large and underdeveloped country with a very low level of consumption of commercial energy forms per head of population. Energy consumption is mainly dependent on the activities of a limited number of industrial operations, including the level of activity in oil production. During the 1970s Congo began to produce crude oil in commercial quantities, but no major deposits have yet been discovered.

ENERGY SUPPLY INDUSTRIES

State organisations exist in the two key energy sectors of oil and electricity. The state oil company controls the national refinery and distribution operations, though in exploration and production it relies to a large extent upon foreign investment and expertise. The state electricity organisation operates hydro-electric plant and the distribution system.

Oil

Production of commercial quantities of crude oil commenced in 1972 and built up to a level of 2.5 million tonnes in 1974. Production difficulties and limited reserves caused a decline in succeeding years, but new fields have been developed since 1978 to raise production to 3.5 million tonnes per annum. The French oil company Elf-Aquitaine plays a leading part in production and exploration, in association with the Italian state oil company AGIP. Fields are located onshore and offshore.

The state oil company Hydro-Congo has been established to control the one million tonnes per annum refinery at Pointe Noire and internal distribution operations. Hydro-Congo is also responsible for exploration and development work by the foreign oil companies. Almost all of the country's crude oil production is exported, as internal market demand amounts to less than 200,000 tonnes per annum.

Electricity

Electricity is generated and distributed by the state-owned Société Nationale d'Energie, which operates hydro-electric plant as its main source of supply. Base of the company's supply is hydro-electric plant at Djoué and Moukoukoulou.

Egypt

KEY ENERGY INDICATORS

Energy Consumption	
—million tonnes oil equivalent	17.0
Consumption Per Head	
—tonnes oil equivalent	0.40
—percentage of world average	25%
Net Energy Imports	nil
Oil Import-Dependence	nil

Egypt has substantial energy resources at her disposal. Oil and gas have been discovered both onshore and offshore and there is the prospect of more to be found. This is additional to the established base of hydro-electric power being produced on the Nile at Aswan. However, Egypt's demand for energy, and for the foreign exchange which oil exports can earn, is very great, as it has a population of more than 40 million. Energy consumption per head of population is therefore still low by world standards.

ENERGY MARKET TRENDS

Trend of Energy Consumption

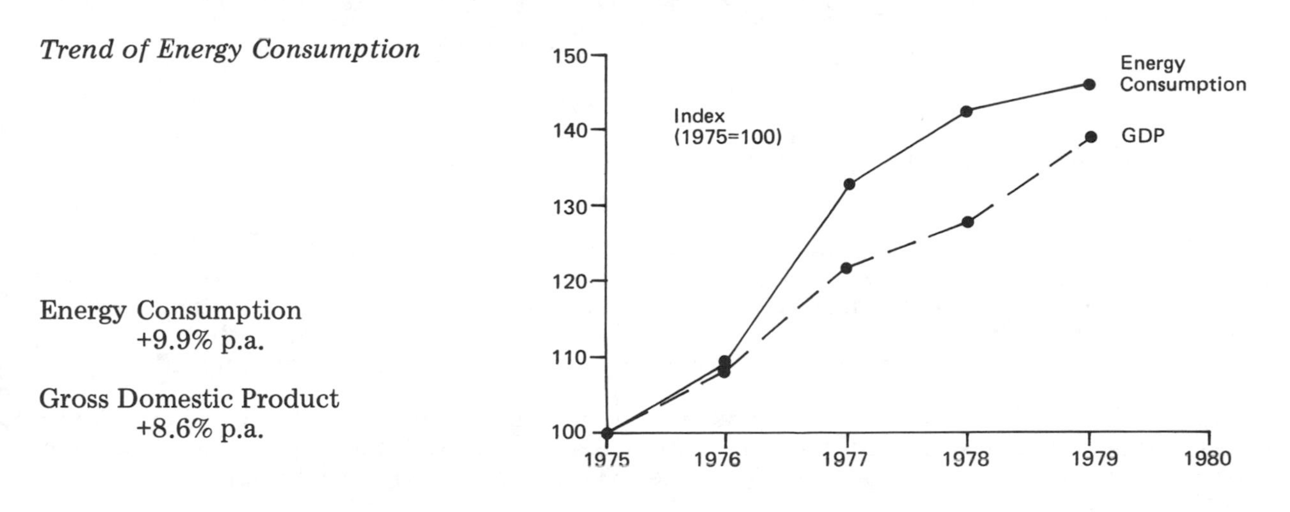

Energy Consumption
+9.9% p.a.

Gross Domestic Product
+8.6% p.a.

The availability of substantial oil and gas resources added to the existing base of hydro-electricity, has sustained economic development at a steady rate in recent years. Energy intensive industries have been established, including steelmaking and aluminium smelting, and efforts are being made to increase the size of the manufacturing sector. Gross Domestic Product has been increasing at over eight per cent per annum. Consumption of energy has risen slightly more quickly.

Pattern of Energy Supply

Primary Fuel Supply (percentage)	
Oil	74
Solid Fuel	3
Natural Gas	7
Primary Electricity	16
TOTAL	100

Oil meets around three-quarters of total energy supply, absorbing up to one half of national production. Hydro-electricity is the next most important form of energy, providing two-thirds of all electricity supplied.

Natural gas currently occupies a minor role, but exploitation of gas is increasing, and its share is expected to rise. Egypt has by contrast limited coal resources and its use is by and large confined to special requirements of steelmaking and other industries.

ENERGY SUPPLY INDUSTRIES

State corporations occupy central roles in the oil, gas and electricity supply industries. The Egyptian General Petroleum Corporation controls all aspects of oil and gas developments. Foreign companies are, however, important partners in the major producing fields and are encouraged to expand their exploration programmes. State agencies are responsible for operating the central power network and developing the Qattara Depression project.

Oil

Oil Production

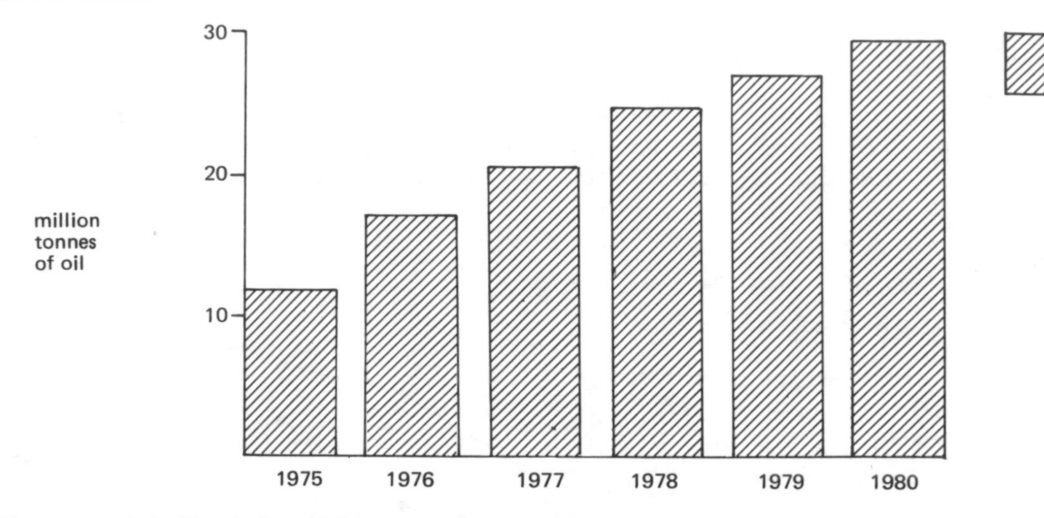

The second half of the 1970s saw the establishment of Egypt as an important exporter of crude oil as production climbed to 29 million tonnes in 1980 and nearly 33 million tonnes in 1981, almost three times the level of 1975. A major part of production comes from fields in the Gulf of Suez, where EGPC is partnered by Standard Oil of Indiana (Amoco). Production also takes place in the north-western part of the country, where Phillips Petroleum Company is operator in the Western Desert Petroleum Company.

Subsidiary companies of EGPC operate six oil refineries, with a total capacity of around 15 million tonnes per annum. Oil refineries are located at Alexandria, Mostord, Suez and Tanta, and supply products to state-owned distribution companies. Consumption of oil in the Egyptian market is some 13 million tonnes per annum, twice the level of 1975.

Gas

Gas Production

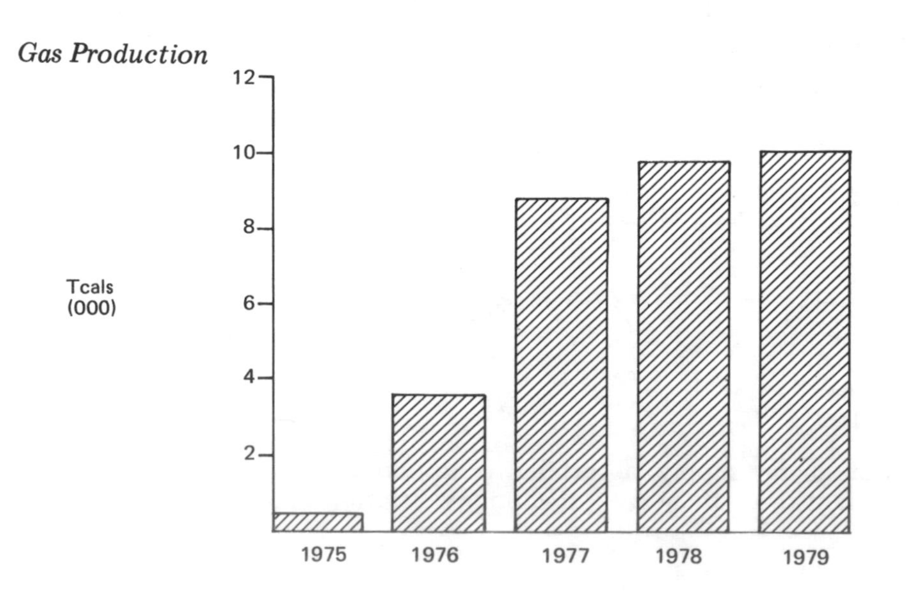

Production of natural gas has become significant only since 1975. Three gas fields are now being exploited mainly to feed fertiliser plants, but gas is also being distributed to residential consumers in parts of Cairo. Several manufacturing plants use gas and it is expected to become the main fuel in the Helwaun industrial area.

Gas is being produced in the Nile Delta (Abu Madi), the western desert (Abu Gharadig) and in Abu Qir Bay. Output has increased rapidly to a rate of over 1,000 million cubic metres per annum. An additional field offshore near Abu Madi is being developed by Elf-Aquitaine. Egypt also has potentially available to it associated gas which is currently flared.

Electricity

Electricity Production

TWh
15
10
5
1975 1976 1977 1978 1979
Electricity Production

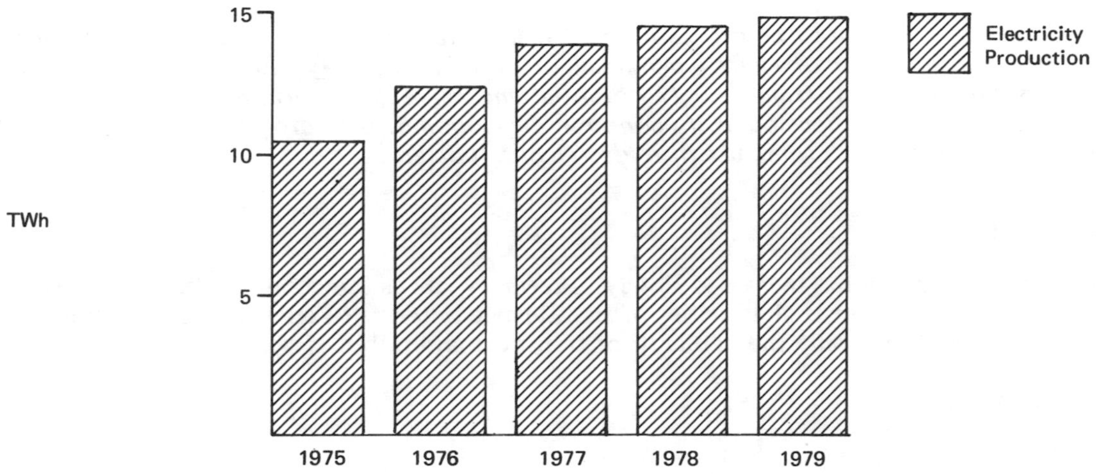

After a rapid rise in consumption of electricity between 1975 and 1978, the level of demand attained 14 TWh. A further surge in demand took this to around 18 TWh in 1980, approximately 80 per cent higher than in 1975. Hydro-electric power from the Aswan dam provides the major part of supplies, but additional demand is having to be met from thermal plant in the northern part of the country.

The Egyptian Electricity Authority is responsible for operating the Aswan plant and power stations forming the unified grid in the main populated area of the north. Some 15 stations are included in this system with a total generating capacity of 4,500 MW. A separate government agency, the Qattara Depression Authority, is examining the possibility of utilising the height difference between the Mediterranean and the Qattara Depression to generate hydro-electric power.

ENERGY TRADE

Net Imports/(Exports) 1975-1979

	1975	1976	1977	1978	1979
Coal (million tonnes)	1.3	0.9	1.1	0.9	1.0
Crude Oil (million tonnes)	(5.5)	(8.0)	(11.1)	(11.9)	(14.0)

The development of indigenous oil fields has made Egypt a substantial net exporter of energy. Exports have risen to some 14 million tonnes per annum. Although production is continuing to rise, internal consumption is also rising rapidly and net exports in future years may show a decline. Egypt lacks significant coal production and around one million tonnes of coal, mainly for coking purposes, is imported annually.

Ethiopia

KEY ENERGY INDICATORS

Energy Consumption	
—million tonnes oil equivalent	0.7
Consumption Per Head	
—tonnes oil equivalent	0.02
—percentage of world average	1%
Net Energy Imports	90%
Oil Import-Dependence	100%

Ethiopia is ranked as one of the poorest countries in the world. The state of its economic development is reflected in a very low consumption of energy per head of population, only one per cent of the world average. Ethiopia has significant geothermal energy potential and as yet undetermined reserves of oil and gas. These resources will need to be developed if economic development is to progress. At present most of the energy consumed is imported, with some use of hydro-electric power.

ENERGY MARKET TRENDS

Consumption of commercial energy has increased slowly in recent years as the economy has been disrupted by internal unrest. There are few industrial activities of significance in terms of energy consumption and the trend of consumption reflects principally the demands of transport and public services.

Pattern of Energy Supply

Primary Fuel Supply (percentage)	
Oil	90
Solid Fuel	-
Natural Gas	-
Primary Electricity	10
TOTAL	100

The Ethiopian economy relies heavily on oil, which accounts for 90 per cent of supplies of commercial energy. Oil is used in the industrial sector and to some extent in electricity generation, as well as for transportation and bunkers. The only indigenous supply of energy at present comes from hydro-electricity, the potential of which is considerable, as Ethiopia is the most mountainous country in Africa.

ENERGY SUPPLY INDUSTRIES

The post-imperial regimes in Ethiopia have followed strongly centralised policies. These have involved the nationalisation and state control of many industries and commercial activities. Supply of oil products and of electricity both fall under state control.

Oil

Oil is the principal energy commodity used in Ethiopia, although this amounts to only 600-700,000 tonnes per annum, for a population in excess of 30 million. Most products are obtained from the national refinery at Assab, on the Red Sea. The refinery has a crude oil processing capacity of around 750,000 tonnes per annum. There is some prospect of oil or natural gas being produced in due course, but there is no commercial production as yet.

Electricity

Production of electricity has been on a steadily upward path, although the rate of increase has on average been well under 10 per cent per annum. The level of production is in excess of 700 GWh per annum. More than half of production is from hydro-electric plant, of which there is around 200 MW capacity. The remaining 125 MW of thermal generating plant accounts for 45 per cent of production. The electricity supply industry is publicly controlled, with 85 per cent of generating capacity. Only 15 per cent is produced for own consumption by industrial organisations.

ENERGY TRADE

Net Imports/(Exports) 1975-79

	1975	1976	1977	1978	1979
Crude Oil (million tonnes)	0.5	0.6	0.6	0.7	0.6

Trade in energy consists mainly of imports of crude oil for the Assab refinery. These are in the range of 600-700,000 tonnes per annum. Imports increased to over 650,000 tonnes in 1978 from 520,000 tonnes in 1975, but subsequently fell back somewhat.

Gabon

KEY ENERGY INDICATORS

Energy Consumption	
—million tonnes oil equivalent	0.8
Consumption Per Head	
—tonnes oil equivalent	1.45
—percentage of world average	90%
Net Energy Imports	nil
Oil Import-Dependence	nil

Gabon has valuable energy resources in the form of oil, gas, uranium and hydro-electric potential. All are being exploited to some extent and consumption of energy per head of population has risen towards the world average. This is somewhat misleading as to the state of development of the Gabon economy, as the level of consumption is the result of a few energy-intensive industrial operations. The country nevertheless has a firm base for further economic development.

ENERGY MARKET TRENDS

Consumption of energy has moved erratically, but has generally followed a definite upward trend during the latter part of the 1970s. This pattern reflects the dominant effect of the fortunes of a very small number of industrial operations such as cellulose, paper and ammonia manufacture. The activities of COMUF, the uranium production company, have reached a plateau, but there has been an increase in the scale of oil refining, in order to export more oil in the form of finished products, which has increased internal consumption of energy.

Pattern of Energy Supply

Primary Fuel Supply (percentage)	
Oil	66
Solid Fuel	–
Natural Gas	23
Primary Electricity	11
TOTAL	100

Given the availability of oil it is not surprising that it accounts for as much as two-thirds of primary fuel supply. However, gas is also produced and it has made sense to utilise this as far as possible, both as fuel and as a feedstock for fertiliser manufacture. Hydro-electric power is a significant element in the fuel pattern, providing a large proportion of total electricity.

ENERGY SUPPLY INDUSTRIES

Private companies are active in oil, gas and uranium development. Companies include French-based companies and para-statal organisations, particularly in the uranium extraction and processing industry, and other West European and American Companies. In some cases companies have interests in both energy production and consumption. The state participates in all of the industries, but is content to encourage private companies as operators.

Oil

Oil Production

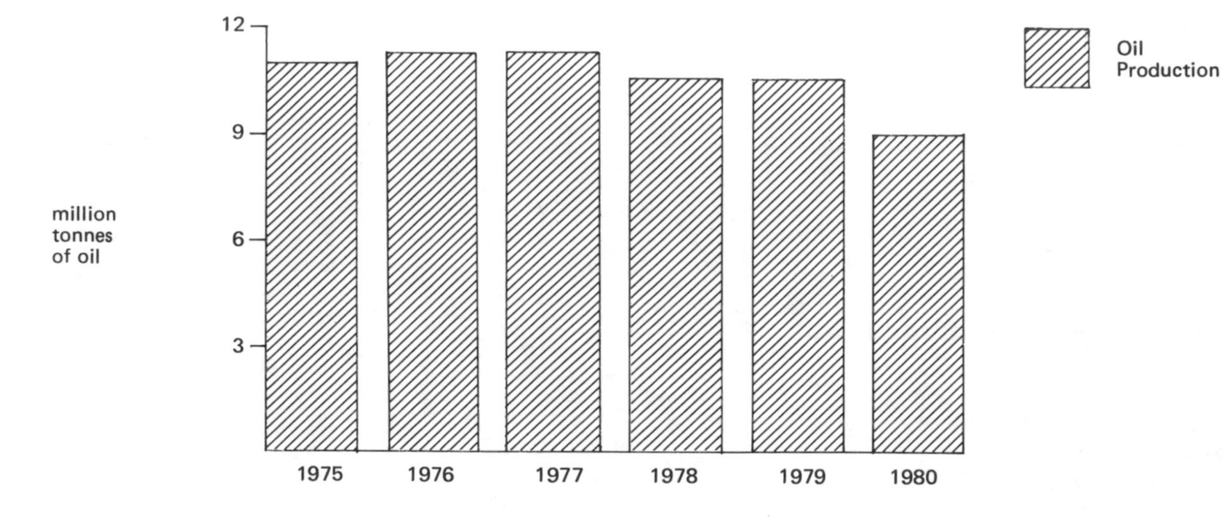

Production of crude oil exceeded 11 million tonnes per annum in the period 1975-77, since which time it has declined, mainly owing to slack conditions in the international oil market. Local market demand for oil products is only around 600,000 tons per annum, although crude oil is refined to supply bunkers and parts of neighbouring countries.

Elf-Gabon, an associate company of the French state oil group Elf-Aquitaine, operates the producing fields. The group is also closely involved as operator of the refining plant at Port-Gentil. The original SOGARA refinery, in which all of the marketing companies participate, has been supplemented with capacity financed by Elf-Aquitaine and the state, to increase total capacity to over two million tonnes per annum.

Electricity

Consumption of electricity increased rapidly up to 1977, but stagnated thereafter at around 450 GWh per annum. A major part of demand is attributable to a small number of mining and industrial activities, notably production of uranium oxide, cellulose and fertiliser, in addition to the refinery at Port-Gentil.

Power supply to all users is from the Société d'Energie et d'Eau du Gabon (SEEG) in which Elf-Gabon, SOGARA and COMUF participate with the state. Thermal power plants are located at Franceville and Port-Gentil, but SEEG produces 75 per cent of its power from hydro-electric plant.

Uranium

Gabon is an important source of uranium. Exploitation has been undertaken for many years by Compagnie des Mines d'Uranium de Franceville (COMUF) at Mounana. French uranium exploration, development and engineering companies have a 75 per cent share of COMUF. The Gabon state holds the remaining 25 per cent. Production of uranium oxide (yellow cake) is around 1,000 tonnes per annum and is supplied to French consumers. Reserves in the Franceville area are estimated to be around 30,000 tonnes, with the major part around Mounana and the rest at Mikouloungou. Reserves may be higher and exploration is continuing in the area.

ENERGY TRADE

Net Imports/(Exports) 1975-79

	1975	1976	1977	1978	1979
Crude Oil (million tonnes)	(10.4)	(10.6)	(10.2)	(8.8)	(7.0)

The major part of Gabon's crude oil production is exported, but as refining capacity has increased more is being processed locally enabling some finished products to be exported. Exports of crude oil have therefore declined from the levels of 10 million tonnes or more between 1975 and 1977 to seven million tonnes in 1979 and lower still in 1980. Gabon is also an exporter of uranium in the form of yellow cake. Exports, which are shipped to France, are around 1,000 tonnes per annum.

Ghana

KEY ENERGY INDICATORS

Energy Consumption	
—million tonnes oil equivalent	2.0
Consumption Per Head	
—tonnes oil equivalent	0.17
—percentage of world average	10%
Net Energy Imports	38%
Oil Import-Dependence	83%

Ghana has a low level of consumption of commercial energy per head of population, at 10 per cent of the world average. A significant proportion of this total is attributable to one major aluminium smelting plant. The country has valuable hydro-electric resources, which have been exploited for many years. Until recently it was dependent on foreign sources for oil, but exploration activity has shown that Ghana does have oil and natural gas resources in common with a number of other countries around West Africa, although the extent of these resources is still uncertain.

ENERGY MARKET TRENDS

Consumption of energy has been rising quite slowly. Since 1975 the average annual rate of increase has been only 9.2 per cent, with consumption of oil static most of the time and consumption of electricity rising slowly. The economy has experienced fluctuating fortunes depending on activities in key sectors, such as cocoa production, gold and diamond production and aluminium smelting, the last being particularly significant in its impact on electricity consumption. The economy has also suffered disruption from high inflation rates and foreign exchange crises, accompanied by political unrest.

Pattern of Energy Supply

Primary Fuel Supply (percentage)	
Oil	45
Solid Fuel	1
Natural Gas	-
Primary Electricity	54
TOTAL	100

In terms of energy value the use of hydro-electricity is of greatest significance among the commercial fuels. The scale of use of electricity is the outcome of the establishment of the Volta Aluminium Company's smelting plant, which was instrumental in bringing about the construction of the Akosombo dam on the Volta River. Oil accounts for virtually all of the remaining consumption of commercial energy. Some gas has been discovered recently along with oil, but is not yet being exploited.

ENERGY SUPPLY INDUSTRIES

The state occupies a central role in the supply of energy. It controls the production of hydro-electric power on the Volta River, and the distribution of electricity to final consumers. The Ghana Supply Commission controls purchases and sales of crude oil and oil products, and the state oil distributor is market leader. But private sector oil companies still play an important part in oil exploration and production and marketing of finished products.

Oil

Consumption of oil by the domestic market is around 850,000 tonnes per annum. The level has tended to stagnate in recent years, reflecting the general state of the economy and problems in financing imports of crude oil as well as other economic inputs. Oil was discovered offshore in 1978 and production of crude oil amounts to 150-200,000 tonnes per annum. Several companies are actively exploring to evaluate other discoveries, including Phillips Petroleum Company and the Italian state oil company AGIP, which have existing production interests in West Africa.

In 1976 the refining and marketing interests of AGIP were taken over by the state and formed the basis of the Ghana Oil Company. These include the country's only oil refinery at Tema, the capacity of which is around 1.3 million tonnes per annum. The refinery operates as Ghanaian Italian Petroleum Company. Crude oil for the refinery and any other imported oil products are supplied by the Ghana Supply Commission, which sells products to the local marketing companies. These include affiliates of most of the major international oil companies.

Electricity

Ghana has an important electricity industry based on the exploitation of the waters of the Volta River. The Volta River project was initiated in 1961 with the setting up of the Volta River Authority to construct a dam and manage hydro-electric operations. The Authority supplies almost all of the electricity produced in Ghana, also exporting some under long term agreements to Benin and Togo. The Authority supplies electricity in bulk to the Electricity Corporation of Ghana, which is responsible for distribution throughout the country.

Consumption of electricity has been on a rising trend, but the total level is very much affected by the activities of Volta Aluminium Company (VALCO), which absorbs more than 60 per cent of the Volta River production. Generating capacity is 900 MW, producing up to 4,600 GWh per annum. A second dam is being built at Kpong, with a capacity of 150 MW. The Ghanaian system is being further reinforced through integration with that of the Ivory Coast.

ENERGY TRADE

Apart from very small quantities of solid fuel, energy imports are almost entirely of crude oil. These amount to under one million tonnes per annum, for processing at the Tema Refinery, and come to a great extent from Nigeria. Some surplus products are exported. In addition there are small exports of electricity to Togo and Benin.

Hong Kong

KEY ENERGY INDICATORS

Energy Consumption	
—million tonnes oil equivalent	5.2
Consumption Per Head	
—tonnes oil equivalent	1.02
—percentage of world average	63%
Net Energy Imports	100%
Oil Import-Dependence	100%

Hong Kong is a highly developed economy, but, with the accent on commercial activity and light industry, energy demand is below the world average and the high density of population limits the use of transportation fuels. There are no indigenous energy resources either in Hong Kong or Kowloon and the country is entirely dependent on imports. Up to the present time oil has met almost all energy needs, but a major step in diversification will take place with the installation of coal-fired electricity generating capacity.

ENERGY MARKET TRENDS

Hong Kong has shown great resilience in the period since 1974, despite its complete dependence on foreign sources of energy. It has not developed energy intensive industrial activities, partly due to lack of energy resources, but also because of lack of space and other resources, and thus has not suffered the relative impact of high energy prices. Energy demand, partly in the form of petroleum products, and partly as electricity and manufactured gas, continues to rise at least as rapidly as Gross Domestic Product.

Almost all of Hong Kong's energy requirements are met directly or indirectly by oil. A very small quantity of solid fuel is also imported. Approximately half of total oil requirements is accounted for by the fuel oil needed to produce electricity. Imported naphtha is also the feedstock for production of town gas.

ENERGY SUPPLY INDUSTRIES

The supply of energy lies in the hands of private companies, although these are subject to government supervision in varying degrees. There is little involvement of the government in the importation and distribution of oil products, but electricity and gas companies operate under franchise from the government. In 1980 the government took the initiative of chartering a loaded VLCC tanker as emergency stockpile of fuel oil, pending construction of large storage capacity at the new power stations on Lamma Island.

Oil

The provision of oil is of great importance, but the government relies upon private companies and affiliates of the international oil companies to supply the market. Around 50 per cent of total oil imports is represented by fuel oil destined for the power stations of China Light and Power Company and Hong Kong Electric Company. China Light and Power, which produces the major part of all electricity in Hong Kong has long term arrangements with Esso Eastern, an affiliate of the Exxon group, for supply of fuel oil. Esso Eastern holds a 60 per cent interest in the power stations supplying China Light and Power.

Gas

Hong Kong has an active and expanding gas industry. Between 1975 and 1980 consumption more than doubled to reach 204 million cubic metres. The gas is produced by Hong Kong and China Gas Company and distributed to some 1.2 million customers both on Hong Kong Island and in the New Territories. The company has a modern gas manufacturing plant, based on naphtha reforming located at Ma Tan Kok in Kowloon. Hong Kong Island is supplied by submarine pipeline. The plant is capable of producing 2.3 million cubic metres per day.

Electricity

Consumption of electricity has been rising rapidly in recent years. Electricity sent out by the producing companies in 1980 totalled 11.7 TWh compared with only 7.7 TWh in 1976, an increase of more than 50 per cent. More than 70 per cent is consumed in Kowloon and the New Territories and 30 per cent on Hong Kong Island itself. The rate of increase in consumption has been similar for both areas.

Electricity is supplied by private companies operating under licence from the government. China Light and Power Company is sole supplier in Kowloon and the New Territories, while Hong Kong Electric Company is responsible for supply on Hong Kong Island, Ap Lei Chau and Lamma. In 1982 work is being completed on a submarine link between the two companies' supply systems. The China Light and Power system is also linked to to supply grid of Guangdong Province in the People's Republic of China.

China Light and Power obtains electricity from two associate companies, in which Esso Eastern holds a majority interest. These companies operate oil-fired power stations of 2,700 MW capacity and additional units of 1,400 MW, capable of firing on oil or coal, are being installed. It is planned that the next tranche of capacity, for commissioning in the late 1980s, should be coal-fired. Hong Kong Electric operates 1,060 MW of oil-fired plant at Ap Lei Chau and is developing a 1,700 MW dual fuel station on Lamma Island.

ENERGY TRADE

Hong Kong imports around seven million tonnes of oil products per annum to supply the domestic market and a substantial bunker trade, amounting to around one and a half million tonnes per annum. Fuel oil figures prominently in the pattern of imports. Sources of imports are primarily the export refineries of Singapore and also the People's Republic of China, which has a net export trade in finished products.

India

KEY ENERGY INDICATORS

Energy Consumption	
—million tonnes oil equivalent	126.5
Consumption Per Head	
—tonnes oil equivalent	0.19
—percentage of world average	12%
Net Energy Imports	26%
Oil Import-Dependence	64%

India's economic development has been largely built on the foundation of a large-scale coal industry and substantial hydro-electric potential. This has enabled the economy to develop with a large degree of energy self-sufficiency, although per capita energy consumption is still only 12 per cent of the world average. However, until recently, the continuance of economic growth has been threatened by inadequate indigenous resources of hydrocarbons. This situation has been relieved by the discovery of significant oil and natural gas deposits off the west coast, the fruits of a long programme of exploration by the state oil and gas organisation. This success, followed by further shows in other offshore areas, holds the prospect that India may be able to support economic growth at the same time as raising the level of self-sufficiency.

ENERGY MARKET TRENDS

The ability of the Indian economy to withstand the shocks arising from energy price increases during the early 1970s is evidenced by the continued economic growth which was achieved in the period 1975-79.

This may be attributed to the relatively high degree of energy self-sufficiency overall. Between 1975 and 1979 Gross Domestic Product rose by 28 per cent, although it fell back at the end of the decade. A noteworthy feature of the development of energy consumption was that it grew at a rate similar to GDP and without any significant further intensification of the use of energy.

Trend of Energy Consumption

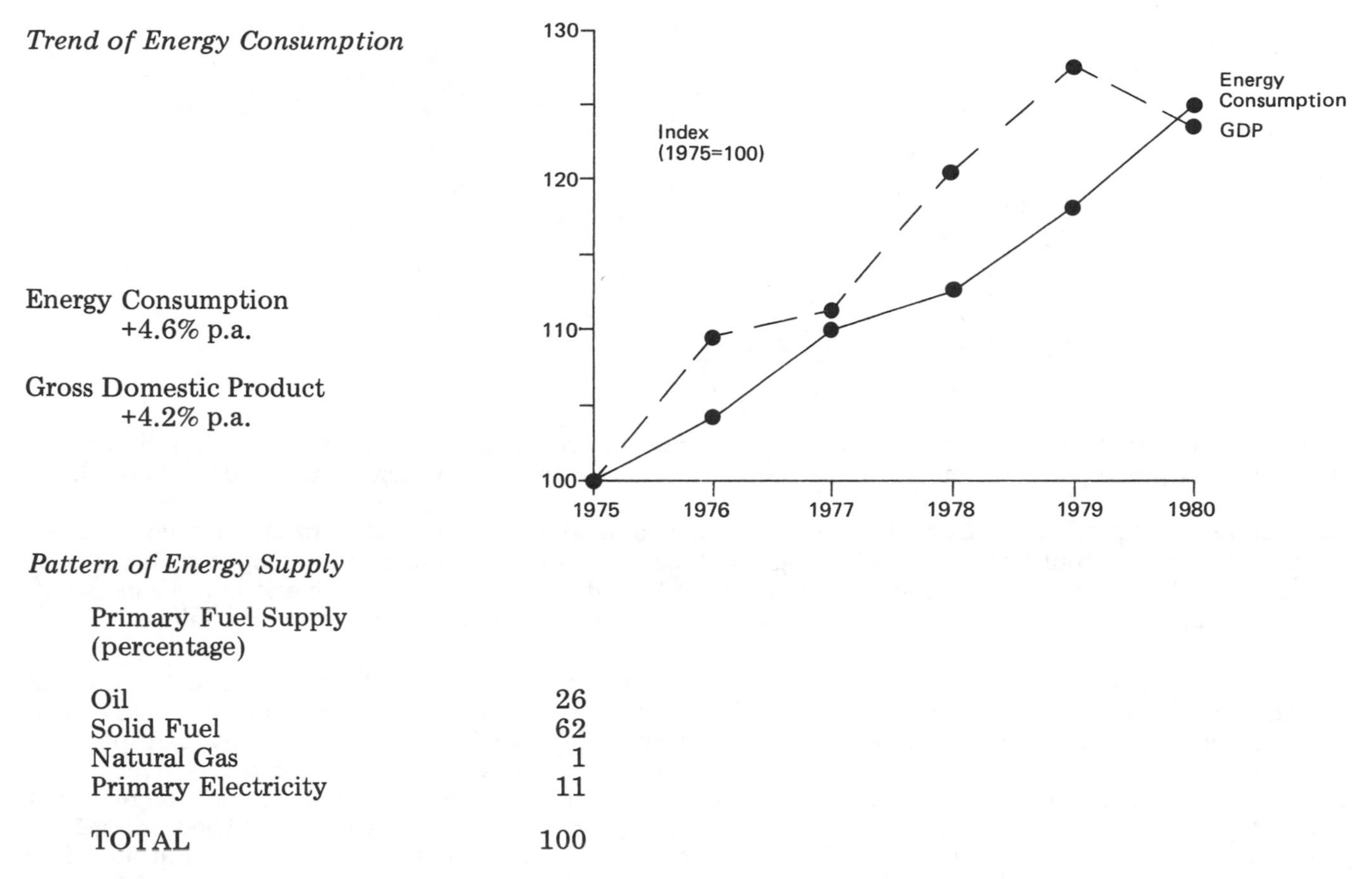

Energy Consumption
+4.6% p.a.

Gross Domestic Product
+4.2% p.a.

Pattern of Energy Supply

Primary Fuel Supply (percentage)	
Oil	26
Solid Fuel	62
Natural Gas	1
Primary Electricity	11
TOTAL	100

Coal still plays a key role in energy supply, accounting for more than 60 per cent of total supply of commercial fuels. Availability of both coking coal and steam coal has enabled the country to be virtually self-sufficient in solid fuel. Added to this is the considerable hydro-electric potential of the country, which means that electricity production has been almost entirely based on indigenous resources. Oil consumption has come to account for around one-quarter of energy supply, with a large demand for cooking and heating oils as well as essential transport fuel. Natural gas has until recently been of very limited significance, although its role will expand as offshore fields are exploited.

The largest sector using solid fuel is electricity generation, which burns around 35 million tonnes of coal and lignite. Railways use a further 12.5 million tonnes. Much of the balance is used by energy-intensive industries such as steelmaking and cement. Coal and lignite are also made into briquettes for domestic consumption. This sector also uses a substantial proportion of oil products and a large amount of non-commercial fuel, particularly firewood.

ENERGY SUPPLY INDUSTRIES

During the 1970s the role of the central government in the supply of all forms of energy became increasingly important as the demands of an industrialising economy became more complex and the scale of investments in energy much larger. During that decade the nationalisation, and subsequent rationalisation, of the coal industry was completed, leaving only a limited amount of production in private hands where output is basically dedicated to internal operations. The role of the international oil companies in supplying and refining crude oil was largely eliminated, and the burden of exploration for oil and gas shouldered almost entirely by the state Oil and Natural Gas Commission. In electricity supply state corporations are assuming a key role in providing co-ordinated supply and efficient generation at regional level.

Oil

Consumption of oil products has been rising steadily, reaching some 33 million tonnes in 1980, compared with only 25 million tonnes in 1976. In the past India has been largely dependent on external sources for

Oil Production and Consumption

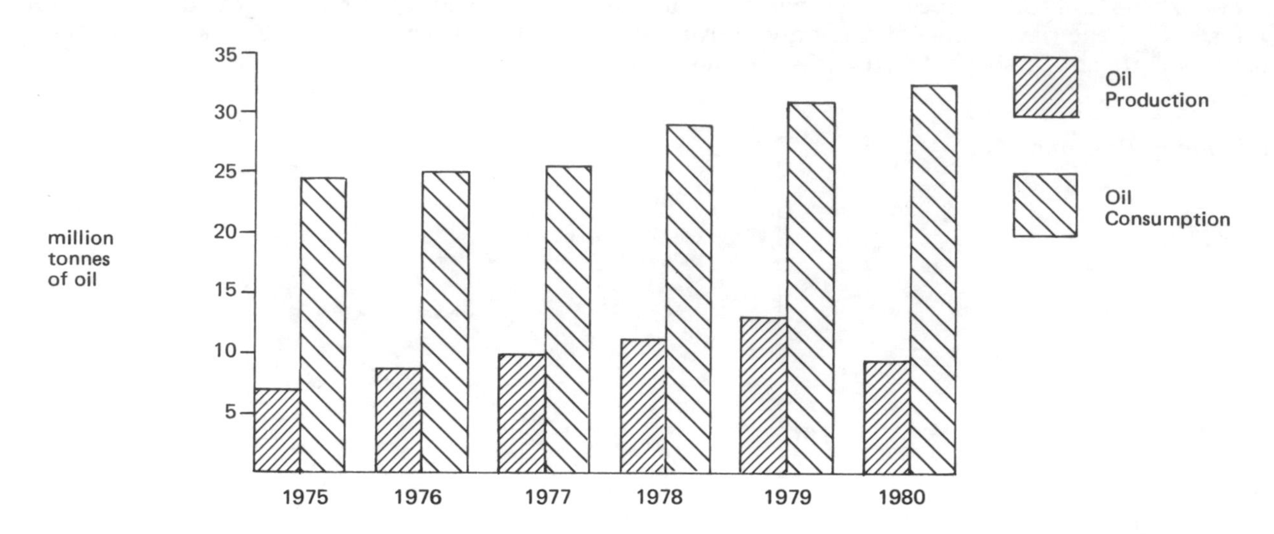

crude oil, production in old established areas such as Assam accounting for only a few million tonnes per annum. But extensive exploration work has shown substantial reserves to exist both onshore and offshore.

Most significant is production from the Bombay High area off the west coast. Here production commenced only in 1976, and output has now reached eight million tonnes per annum. There is scope for this level to be more than doubled, which, combined with other offshore discoveries to the south-east in the Palk Strait, and onshore production from Assam and Gujarat should enable basic self-sufficiency to be attained.

The main burden of exploration and production work has been undertaken by the Oil and Natural Gas Commission, a statutory body and the principal state agency for these purposes. Another state-owned company, Oil India Ltd, which until recently was a partnership with Burmah Oil Company, is responsible for exploration and production in north-east India. The Assam Oil Company also produces a very small amount of oil in Assam. International oil companies may begin to play a larger part in exploration now that more extensive areas have been proved to contain oil, and in 1981 the government opened areas, both onshore and offshore, to contract bids. Offshore reserves of oil are currently estimated to be around 250 million tonnes.

The state has gradually assumed almost complete control of the refining industry. The plants formerly owned by Burmah-Shell, Exxon and Caltex are now entirely state owned. The Bombay refinery of Burmah-Shell now operates as Bharat Petroleum Corporation and the former Exxon and Caltex plants, at Bombay and Vizagapatnam respectively, are operated as Hindustan Petroleum Corporation. These refineries have a combined throughput capacity of over 10 million tonnes per annum. In addition refineries at Cochin and Madras, with a total capacity of six million tonnes per annum, continue to be operated as joint ventures between Indian Oil Company (IOC) and foreign firms. IOC itself has become the largest refining company and is investing in the additional capacity required to keep pace with market demand. IOC has four main refineries, at Koyali, Haldia, Barauni and Mathura, with total capacity of some 19 million tonnes per annum. International majors continue to be involved in product distribution and marketing, but IOC now dominates the market with a 60 per cent market share.

Coal

India is one of the world's major coal producing countries with production now well over 100 million tonnes per annum. This level of output is enough to make India virtually self-sufficient, and provide a firm energy base for the economy, particularly as a major fuel for electricity generation purposes. Of the total less than five million tonnes was of lignite in 1980. More than 20 million tonnes of coking coal is produced annually, the rest is of thermal quality.

The mining industry has been largely state operated since the early 1970s when almost all of the mines not producing coking coal were nationalised. The state also produces most of the coking coal. The public sector industry is organised under Coal India Ltd, which is a holding company with four principal subsidiaries operating coalfields on a regional basis. These are Eastern Coalfields Ltd, Western Coalfields Ltd, Central Coalfields and Bharat Coking Coal Ltd, which operates mines in Bihar. Coal India also administers directly mines in the north-east of the country, including those in Assam.

Coal Production

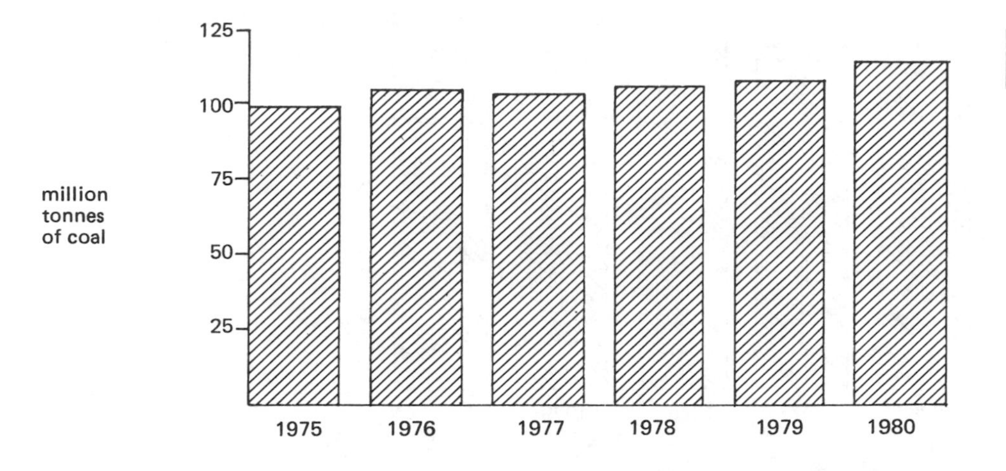

Total production of the Coal India group in 1979-80 exceeded 91 million tonnes. The balance of national output came mainly from three sources. The Tata Industries group operates mines to provide steam coal and coking coal for its steel and power generation subsidiaries. Singareni Collieries Company, owned by the Indian government and the state of Andhra Pradesh, produced over nine million tonnes of coal. Near Madras the Neyveli Lignite Corporation produced around three million tonnes in 1979-80, for an associated power station. Output at the Neyveli project is planned to increase substantially during the 1980s to fuel additional power station units and provide input for a briquetting plant. Current estimates of lignite reserves in India are put at 2,000-4,000 million tonnes.

Gas

Gas Production

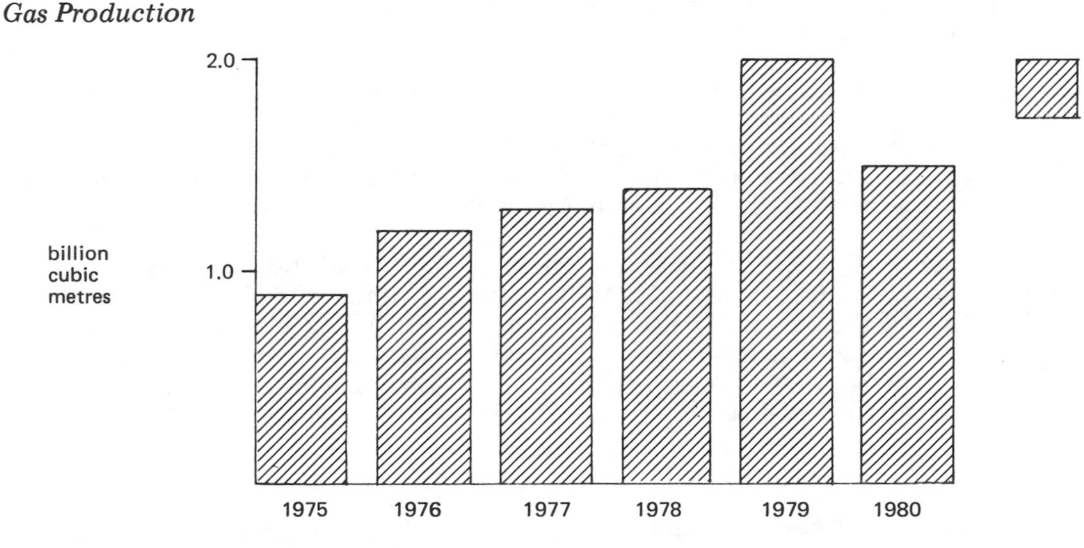

Natural gas makes only a small contribution to total energy supplies, but production has generally been on a rising trend and is expected to be stepped up significantly as offshore fields, particularly off Bombay, are exploited. Production rose from 920 million cubic metres in 1975 to nearly 2,000 million in 1979, equivalent to just under 1.8 million tonnes of oil. Output fell back sharply in 1980. Main use of the gas is in industrial bulk energy systems and in electricity generation.

The Oil and Natural Gas Commission, the statutory body responsible for exploration and for development of gas has established important reserves of natural gas, both onshore and offshore. It is estimated that reserves already proved in the western offshore areas amount to 270,000 million cubic metres. Including onshore areas the ONGC assesses national reserves of recoverable gas at 350,000 million cubic metres.

Electricity

Demand for electricity has been rising rapidly as a result of the process of industrialisation of the economy. Total demand is now well over 100 TWh per annum. The 1980 level of supply in the public system was

Electricity Production

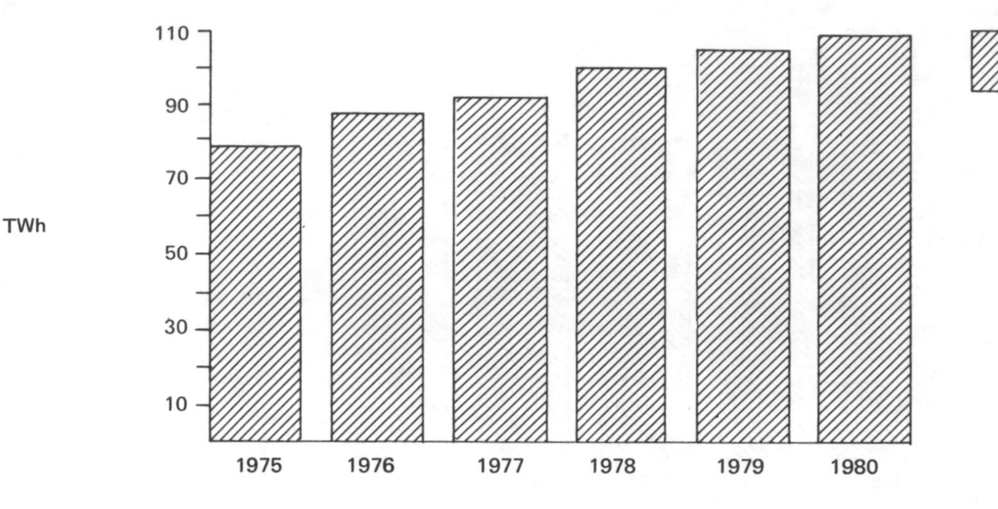

108.5 TWh, some 37 per cent higher than in 1975. This increase was much sharper than the 25 per cent increase in energy consumption as a whole. The availability and use of electricity would almost certainly have been higher but for operational problems, which had an impact on industrial and coal mining operations in 1980.

The bulk of total electricity production and distribution is undertaken by individual state authorities, but the central government is moving towards a much more important position in an attempt to provide efficient and economic production and transmission throughout the country. Through the National Thermal Power Corporation the central government has taken responsibility for constructing so-called 'super' thermal power stations, which are designed to provide power on a regional basis from large-scale stations. The states themselves are grouped into five regions in order to promote optimal development of grids. The programme of 'super' power station construction should raise the capacity operated by the National Thermal Power Corporation from its current level of 5,700 MW to around 8,000 MW by 1990.

Other agencies of the central government which are becoming increasingly important at national and regional level include the National Hydro-Electric Power Corporation and the Damodar Valley Corporation. The NHPC has wide powers and responsibilities in the field of developing hydro-electric resources. The DVC has a responsibility for all aspects of water resource development and management in the Damodar Valley, and operates generating capacity of 1,300 MW, mainly thermal.

A substantial proportion of total electricity production is based on hydro-electric plant. In 1980 this proportion was 43 per cent and some states, including Himachal Pradesh, Jammu and Kashmir, Karnataka, Kerala and Meghalaya, rely solely on this source for electricity. But more than half is produced by thermal plant and this proportion is likely to rise. The main fuel used is coal, more than 30 million tonnes in 1980, aside from three million tonnes of lignite. Other fuels included 175 million cubic metres of natural gas and around two million tonnes of oil products. Three per cent of electricity was produced from nuclear plant.

Uranium

India possesses a useful resource in the form of uranium, reserves of which are estimated to be capable of providing 34,000 tonnes of uranium oxide. Of this total some 15,000 tonnes are considered to be economically exploitable. This is considered sufficient basis for a national nuclear power generating capacity of 8-10,000 MW. At present there are two power stations in the north of the country, at Tarapur and Kota (Rajasthan) with a total capacity around 800 MW. A third station is under construction at Narora.

ENERGY TRADE

Net Imports/(Exports) 1974/75-1979/80

	1974/75	1975/76	1976/77	1977/78	1978/79	1979/80
Coal (million tonnes)	0.4	0.6	0.6	0.6	0.6	0.6
Crude Oil (million tonnes)	14.0	13.6	14.0	14.5	14.7	16.1
Oil Products (million tonnes)	2.6	2.2	2.6	2.9	3.9	4.5

The principal energy commodity in India's foreign trade is oil. Pending full exploitation of the Bombay High oil field a major proportion of India's requirements has had to be met from overseas sources. Rising availability from indigenous sources has however contained the upward trend and from the early 1980s the total level of oil imports is expected to decline. Foreign trade in coal and coal products is negligible in comparison with the scale of total production and consumption in the economy.

Sources of Imported Crude Oil

(percentage)	
Iraq	37
Saudi Arabia	21
Iran	20
USSR	11
United Arab Emirates	9
Other	2
TOTAL	100

Given its geographical location in relation to the Persian Gulf it is not surprising that Middle East exporting countries account for a high proportion of total imports. The only notable exception is the USSR, which supplies around 11 per cent, or approximately 1.5 million tonnes per annum, under general bilateral trading agreements. Iraq has been the leading source of crude oil, ahead of Saudi Arabia and Iran, with 21 per cent and 20 per cent respectively. The only other source of significance has been the United Arab Emirates.

ENERGY POLICIES

The central government is using key state agencies and development corporations to meet the needs of its industrialisation plans and at the same time ensure self-sufficiency. Economic development will lead to a continuing rapid rise in demand for energy, and in particular for electricity, which needs to be supplied more efficiently than in the past, when up to 20 per cent of electricity has been lost in transmission and distribution and disruption in supply has been a constant hazard for industrial activity. Main aspects of government policy are to provide a secure and economic base from large regional 'super' thermal power stations, coupled with a construction programme for high voltage transmission lines.

All forms of indigenous energy are to be exploited more intensively. Many projects by Coal India Ltd have been sanctioned and many hydro-electric plants constructed. Narora nuclear plant, due to be commissioned in the mid 1980s, may well be further expanded and two new plants constructed in the south and the west. But the most rapid increase in activity would be in the production of oil and gas. With the Bombay High area now a proven oil bearing province and promising discoveries elsewhere, the government is encouraging greater participation by foreign companies.

Indonesia

KEY ENERGY INDICATORS

Energy Consumption	
—million tonnes oil equivalent	31.8
Consumption Per Head	
—tonnes oil equivalent	0.21
—percentage of world average	13%
Net Energy Imports	nil
Oil Import-Dependence	nil

Indonesia is a country rich in energy resources which it is still in the process of evaluating and developing Not only is it a leading exporter of oil, but it has established large-scale LNG facilities in partnership with Japanese utilities and bulk energy consumers, epitomising the very close inter-dependence which has developed between the two countries. There remain still to be developed the largely uncharted reserves of coal which are likely to become the main source of fuel in electricity generation during the 1990s. Indonesia is also considered to have significant potential for further hydro-electric and geothermal energy exploitation, but the latent demand for energy amongst the country's 150 million plus population indicates that nuclear power is likely to be considered before long, partly to safeguard the revenue earning capability of oil and gas exports.

ENERGY MARKET TRENDS

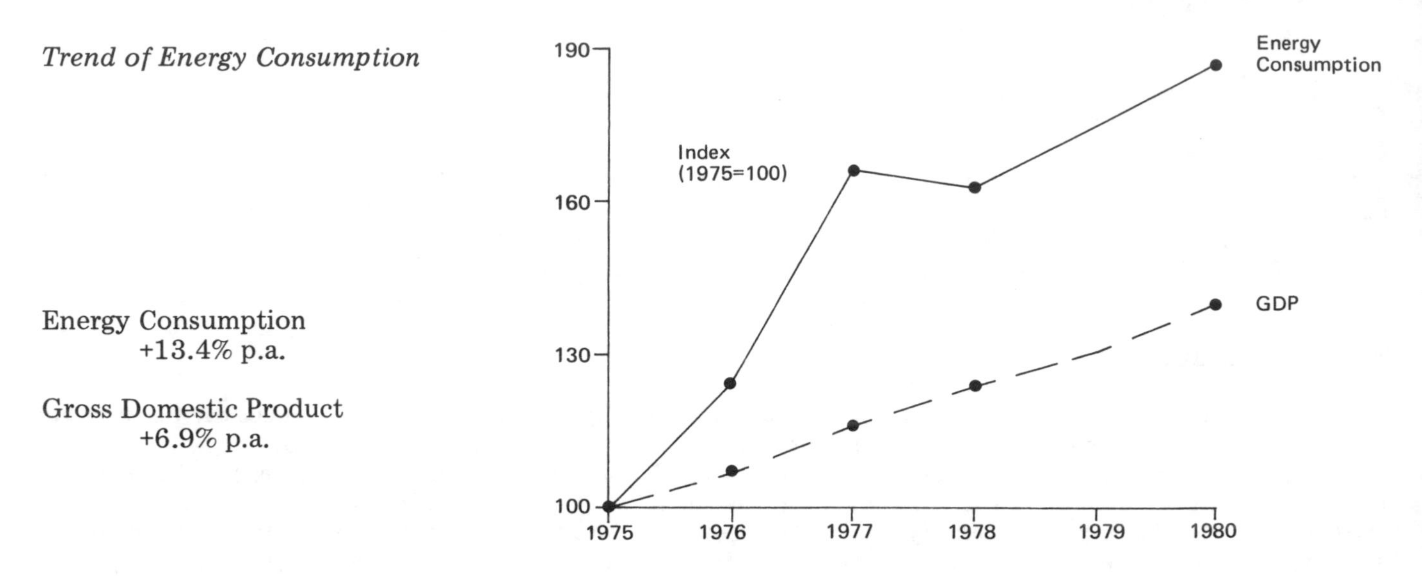

The Indonesian economy benefited from the upsurge in oil prices during the 1970s and maintained a good rate of economic growth, varying between five and nine per cent per annum between 1975 and 1980. This growth has been associated with a rapid escalation in the amount of energy consumed. In the same five-year period total energy consumption rose by almost 90 per cent. Consumption is likely to continue to rise rapidly as new industrial projects, including aluminium smelting, are implemented and the electrification programme is extended.

Pattern of Energy Supply and Consumption

Primary Fuel Supply (percentage)	
Oil	85
Solid Fuel	1
Natural Gas	12
Primary Energy	2
TOTAL	100

Consumption of Oil (percentage)	
Industry	24
Residential	33
Transport	32
Other	11
TOTAL	100

Oil dominates the pattern of primary energy consumption, with natural gas accounting for much of the balance. At present natural gas is used primarily in the energy sector and in fertiliser manufacture. Large-scale industrial consumers are few and the population distribution and state of development of the country limit the scope for mains distribution. Both coal and primary energy, whether hydro-electric or geothermal, have been neglected in favour of oil and gas. Both are expected to assume a larger share in energy supply, particularly coal, which now appears to be the most economic source for generating electricity on a large scale.

The under-developed state of the economy is emphasised by the low proportion of oil used for industrial purposes. Both residential and transport sectors consume much higher proportions. Other uses include electricity generation, which provides the main source of energy consumed in the commercial sector.

ENERGY SUPPLY INDUSTRIES

State undertakings occupy central positions in each of the energy sectors. Foremost among these is Pertamina, which controls exploration and production of oil and gas, the two energy resources most fully exploited to date. The activities and success of Pertamina are central to the health of the whole Indonesian economy. Other state bodies are responsible for coal production, gas distribution and electricity generation and supply. Those concerned with coal and electricity are likely to become increasingly important as the nation's demands for electricity increase. Foreign-based companies remain much in evidence in the search for energy resources and international oil companies retain interests in refining and marketing of oil products.

Oil

Oil Production and Consumption

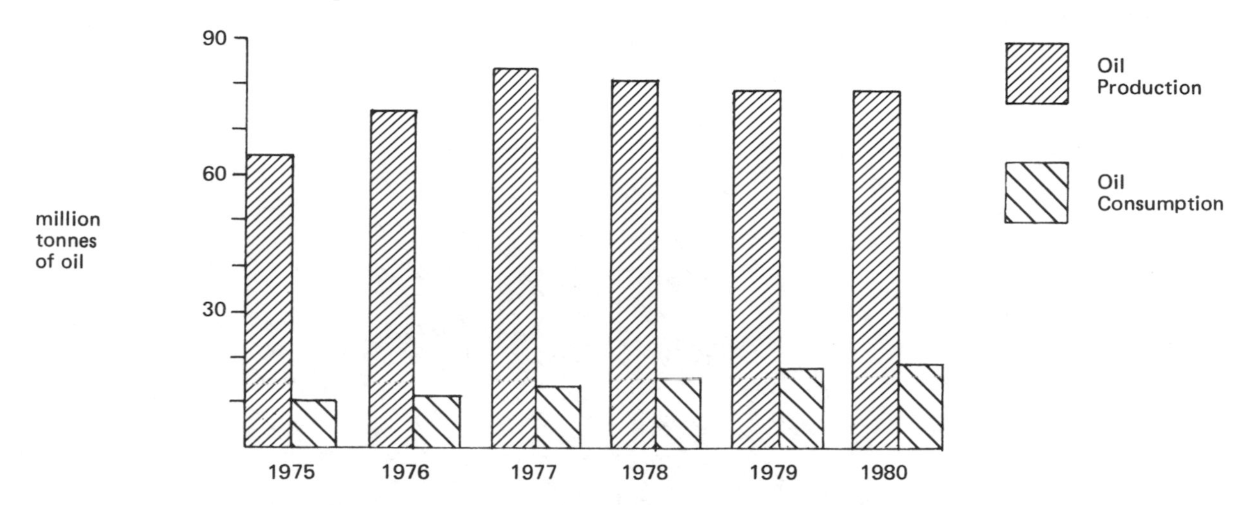

Indonesia is a major oil-producing country and the largest oil exporter in Asia. Production in both 1979 and 1980 was about 580 million barrels, equivalent to nearly 80 million tonnes per annum, although this level represented a decline from the 83 million tonnes produced in 1977. The decline was partly due to the impact of recession on the country's export markets, but also a waning of interest during that period on the part of exploration companies may have been a factor.

A state company, Pertamina, has responsibility for oil and gas exploration and production as well as a major role in oil refining and products supply. Several of the major international oil companies are involved, notably PT Caltex Pacific Indonesia, an affiliate of Standard Oil Company of California and Texaco, which accounts for around 45 per cent of total production. Standard Oil also has some production through a wholly owned subsidiary PT Calasiatic. Most of the other major oil companies, including Shell, Gulf, Mobil and Exxon, the last two of which operate through a joint affiliate, PT Stanvac Indonesia, hold contracts from Pertamina for exploration and/or production. A number of the smaller international oil companies also have interests in acreages, both onshore and offshore.

Pertamina controls virtually all of the country's refining capacity. At present this consists of six major refineries, at Dumai, Plaju, Sungei Gerong and Sungei Pakning, which are all on Sumatra, at Cilacap (Java) and Balikpapan (Kalimantan). The combined capacity of these refineries is around 24 million tonnes per annum. This is considered by Pertamina to be well short of what is needed to meet rapidly rising demand and the state company has plans to add up to 20 million tonnes per annum of capacity.

Domestic market sales in 1980 amounted to 140 million barrels, equivalent to around 19 million tonnes. Consumption has been rising rapidly, with an overall increase of 78 per cent between 1975 and 1980. Given that there has been no significant increase in basic productive capacity of the oil industry this has meant that internal demand already accounts for more than one-quarter of production. The share will continue to rise until substantial new reserves of oil are added to the current level of 14 billion barrels.

Coal

With the abundance of oil and gas discovered over many years the coal industry has not developed to any extent so far. Production of coal in 1979 amounted to only 280,000 tonnes, although this represented a significant increase on the level of around 200,000 tonnes in 1975 and 1976. Coal production is undertaken by the state owned company PT Batubara.

The use of coal is planned to increase dramatically during the 1980s and 1990s, principally to feed the new power stations of PLN. The Bukit Assam coalfield in southern Sumatra currently produces more than half of national output and has enormous reserves of brown coal which are suited to local electricity generation. In the initial stage of the coal development programme production is to be stepped up to 2.5 million tonnes per annum, in parallel with the construction of an 800 MW power station on Java. Eventually production capacity could be raised to 10 million tonnes per annum, specifically to fuel this power station.

In the long run power development plans envisage as much as 12,000 MW of coal-fired capacity, exploiting not only the Bukit Assam reserves, but also other known deposits in West Sumatra and East Kalimantan. Total reserves are currently thought to be at least 25,000 million tonnes and several international companies have shown an interest in undertaking exploration work. BP, in partnership with CRA, the Australian associate of RTZ, and Atlantic Richfield, partnered by Utah Exploration, have negotiated with PT Batubara to carry out exploration on East Kalimantan.

Gas

Gas Production

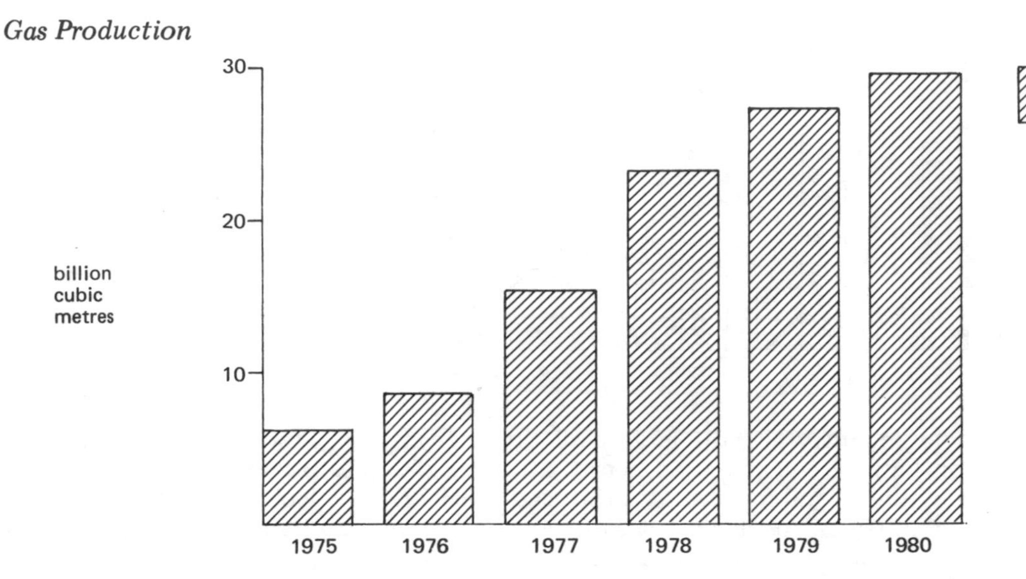

Indonesia is a major producer of natural gas. Not only are large quantities of gas produced in association with crude oil, but there are also gas fields which are being exploited for export markets in the form of liquefied natural gas. Total gas production in 1980 was nearly 30,000 million cubic metres and approximately one-third is produced specifically to feed the LNG plants at Arun, North Sumatra, and Badak, East Kalimantan. Of the associated gas produced by Pertamina, the state oil and gas enterprise, and its production partners, a substantial proportion is used in the oil fields and in refineries and processing plants, but around 50 per cent is still lost through flaring.

Attempts are being made to reduce the amount of gas lost through flaring. Four fertiliser plants have been established using gas as a feedstock. In 1980 Pertamina supplied these plants with 1,500 million cubic metres of gas. Use of gas as a fuel for mains distribution has developed to only a limited extent, partly because the densest areas of population are in Java. A state owned corporation, Perusahaan Gas Negara (PGN), supplies natural gas in Jakarta and Cirebon, and produces gas from oil in six other towns.

Liquefied natural gas is exported from the two plants at Arun and Badak to Japanese utilities and steelmakers. These consumers also participate in the LNG plant, along with the production companies, Mobil, Huffco, Union Oil and the Japanese exploration company Inpex. In 1980 these plants handled some 440,000 million cubic feet of gas and this volume may be doubled by the mid 1980s as gas reserves are further exploited. Currently proven reserves are estimated to be 30 million million cubic feet and a new field, which alone may hold 20 million million cubic feet is being evaluated.

Electricity

Demand for electricity has been rising rapidly as a result both of the operations of major resource development and processing plants and of further industrialisation of the economy and progress with electrification. Between 1975 and 1978 production by the public supply system increased by 63 per cent, an average of 15-20 per cent per annum. The country's development plants are framed to keep pace with increasing demand for electricity of this order.

The public supply system is operated by a state corporation Perusahaan Umum Listrik Negara (PLN) which has a system capacity of 2,700 MW. Plant is largely based on oil or gas as fuel, with thermal power stations and gas turbines each accounting for approximately one-third of total capacity. Some 20 per cent of capacity is in diesel and gas engines. Only 13 per cent of capacity is based on hydro-electric resources. Industrial establishments with their own generating plant operate around 1,500 MW of additional capacity, some of which is effectively supplying the PLN system.

Forward plans envisage making much greater use of non-oil resources to fuel the substantial additional capacity required under the government's economic development programme. Most important of these is the availability of low grade coal from Bukit Assam, in Sumatra, which is to be used to fuel a 3,000 MW power station on Java. This station will form the basis for a major step forward in the electrification of Java, with a 500 KV extra high tension line running the length of the country.

ENERGY TRADE

Net Imports/(Exports) 1975-80

	1975	1976	1977	1978	1979	1980
Crude Oil and Oil Products (million tonnes)	(54.3)	(59.9)	(67.3)	(69.7)	(60.4)	(49.6)
Natural Gas (000 million M^3)			(1.9)	(5.7)	(9.0)	(12.1)

Net exports of oil in 1980 were just under 50 million tonnes, substantially lower than in previous years as a result of the general decline in the international oil trade. Only two years earlier net exports were close to 70 million tonnes, and the capacity exists for exports to be increased above the 1980 level when demand increases, although the proportion of oil output needed to satisfy domestic market requirements is rising rapidly. Total exports of crude oil, part-refined oil and finished products are up to 10 per cent higher in some years, but there is some import trade in crude oil, in order to meet the technical specifications of refineries and recently also of finished products because of limitations on refinery capacity.

Natural gas, as LNG, has rapidly come to assume an important role in Indonesian exports. Exports of LNG in 1980 exceeded 12,000 million cubic metres of gas only four years after the start up of the first LNG plant. All of the LNG is purchased by Japanese companies, which foresee a continuing rise in their demands for this pollution-free fuel. The expectation is that exports to Japan could double by the mid 1980s to around 25,000 million cubic metres per annum.

Destination of Crude Oil Exports

(percentage)	
Japan	55
USA	30
Caribbean	8
Australia	3
Philippines	2
Other	2
TOTAL	100

Indonesia is particularly dependent on Japan as the main export market for its crude oil. This is partly because it is the nearest major oil-exporting country to Japan, but also because Indonesian oil is low in sulphur content. For similar reasons, Indonesian crude oil is attractive to refining companies in the United States. In addition to the oil directly exported to the USA a further significant percentage is shipped to the Caribbean, where it can be processed in export-orientated refineries.

Iran

KEY ENERGY INDICATORS

Energy Consumption	
—million tonnes oil equivalent	37.8
Consumption Per Head	
—tonnes oil equivalent	1.01
—percentage of world average	62%
Net Energy Imports	nil
Oil Import-Dependence	nil

Iran has abundant supplies of oil and associated gas, which have provided the basis of past development programmes and massive exports of energy. Internal consumption has nevertheless still not approached the world average per capita level. Following the downfall of the Shah in 1978 it has remained uncertain what lines economic development will follow and whether Iran will ever attempt to regain its former leading position as an energy exporter. However, oil and gas exports are crucial for foreign exchange earnings and a substantial increase on 1980-82 levels is to be expected. Iran was a founder member of the Organisation of Petroleum Exporting Countries and has always sought to maximise earnings from oil exports.

ENERGY MARKET TRENDS

In the period up to 1978 the country's energy resources were being exploited intensively in order to fuel industrial development and fund social programmes. Since that time the economy has suffered severe disruption as a result of internal conflict and, latterly, the war with Iraq. Production and consumption of energy have followed similarly erratic paths, both generally being at much lower levels than prior to 1978. Even when a resolution of the Iran-Iraq conflict has been arrived at, the progress of the economy and of energy consumption will remain uncertain, in view of the internal political situation.

Pattern of Energy Supply and Consumption

Primary Fuel Supply (percentage)	
Oil	63
Solid Fuel	2
Natural Gas	32
Primary Electricity	3
TOTAL	100

Oil is the cornerstone of the Iranian economy, accounting for approximately two-thirds of energy input. It is extensively used in industry and electricity generation, and as the basis for petrochemical production. There is a large refining industry, which provides products for the domestic market and for export, though the scale of this activity was severely curtailed as a result of war damage to the Abadan refinery. Gas is produced in very large quantities in association with crude oil. This has been utilised as the second main fuel and as a feedstock. Other fuels are of little significance so far. Only limited hydro-electric potential has been developed and earlier plans for a major nuclear power station construction programme have been abandoned.

ENERGY SUPPLY INDUSTRIES

The Shah of Iran was one of the most vehement of oil producer country leaders in opposing the influence of foreign oil companies and was in the forefront of moves to assume national control of energy resources. This attitude has, if anything, strengthened under the new regime, with state organisations now having full control of production and disposal of oil and gas.

Oil

Oil Production and Consumption;

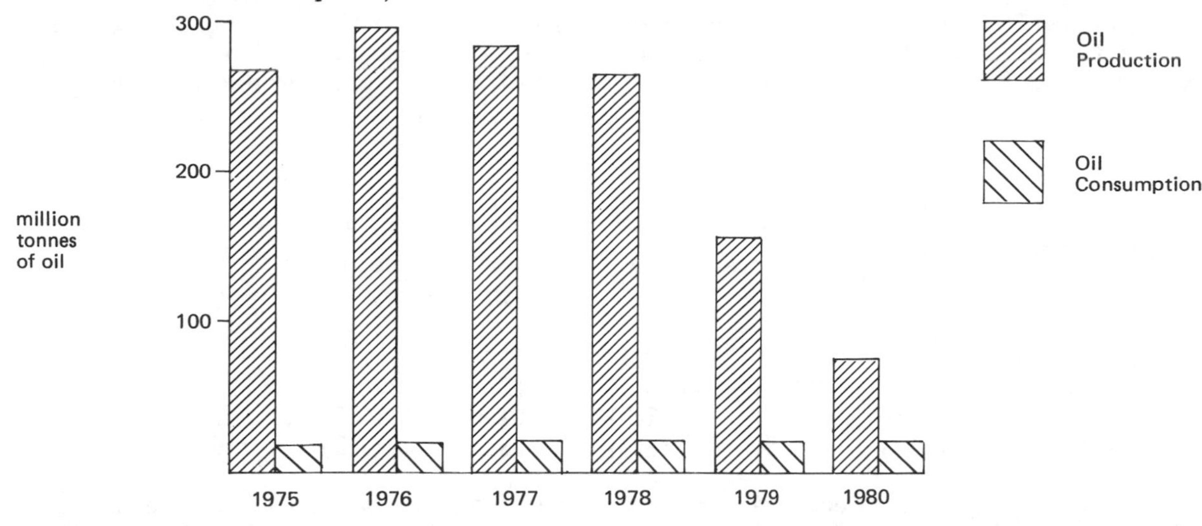

Up to 1978 Iran was second only to Saudi Arabia as producer and exporter of crude oil in the Middle East. In 1976 production had returned towards the 300 million tonnes per annum mark, following the recession in oil demand of 1974 and 1975. Output levels remained high until the fall of the Shah in 1978, since which time they have been cut back on policy grounds by the new regime and sales have been limited by the international market situation. Internal demand has stagnated or fallen. Production was only 74 million in 1980 and an estimated 65 million tonnes in 1981.

Production of crude oil is now entirely in the hands of the state oil company, National Iranian Oil Company (NIOC). NIOC operates some 39 producing fields, mainly onshore in areas formerly developed by the Iranian Consortium, led by British Petroleum, which had its origins in this area. More recently joint-ventures between NIOC and French or American Companies have been terminated and ownership assumed by NIOC. These ventures, which were mainly in offshore areas, including the Sassan and Darius fields, accounted for some 10 per cent of Iranian output. Total proven reserves of crude oil are estimated to be 60-70 billion barrels, or 8-9.5 billion tonnes. It is believed that this figure could be increased substantially if a gas re-injection programme is undertaken to raise the recovery factor.

NIOC has now assumed responsibility for operating all fields and undertaking exploration as well as operating the country's refineries and supplying finished products. Modern refineries have been built at Tehran, Isfahan, Tabriz, and Shiraz with a total processing capacity of more than 26 million tonnes per annum. NIOC formerly had at least this amount of capacity at the Abadan refinery, which was one of the largest in the world, but it has been severely damaged in the conflict with Iraq and is currently out of operation.

Gas

Production of gas is very large, resulting from the level of crude oil output. Even in 1979 when crude oil production was only around one-half of the 1976 peak, output of gas was nearly 40,000 million cubic metres. Half of this quantity was re-injected into the oil reservoirs or flared for lack of alternative use. The balance was utilised in Iranian power plant, fertiliser works, petrochemical production and other industrial uses, with a significant proportion being exported to the Soviet Union.

In the expectation that production of associated gas would rise to very high levels Iran contracted to supply the Soviet Union with up to 10,000 million cubic metres of gas per annum. The Iranian Gas Trunkline (IGAT) came into operation in 1972, fed from the Bid Boland treatment plant. A second line, IGAT II, has been under consideration for a number of years, but this project appears now to have been shelved, as have projects to develop for export as LNG other gas resources in the southern part of the country. Production and disposal of natural gas is handled by National Iranian Gas Corporation (NIGC). NIGC is effectively a subsidiary of NIOC, which is able to co-ordinate the development of oil and gas.

Electricity

Total production of electricity is around 19 TWh per annum. This is only 20 per cent higher than it was in 1975 and demand has stagnated since 1978. Hydro-electric plant produces over 20 per cent of the electricity, but the bulk is generated in thermal stations. These rely mainly on oil as fuel, but three recently constructed power stations are designed for natural gas firing.

The state organisation Tavanir is responsible for the public supply of electricity. Tavanir operates all of the hydro-electric plant, which totals 850 MW. But private producers of electricity, which include some of the key energy and industrial operations, account for a quarter of all electricity produced. Capacity of the private sector is some 1,450 MW, all using oil or gas as fuel.

ENERGY TRADE

Net Imports/(Exports) 1975-80

	1975	1976	1977	1978	1979	1980
Coal (million tonnes)	0.1	0.1	0.1	0.1	0.1	0.1
Crude Oil (million tonnes)	(233.7)	(260.9)	(242.9)	(220.7)	(119.2)	(55.0)
Oil Products (million tonnes)	(14.1)	(11.5)	(14.8)	(13.2)	(11.9)	
Natural Gas (000 million M^3)	(8.5)	(8.2)	(8.3)	(7.2)	(6.8)	

With the insignificant exception of very small imports of solid fuel, Iran has had substantial exports of oil products and gas as well as its traditionally large crude oil export operations. Exports in 1976 were approaching former peak levels, but since that time have fallen back as production has been either disrupted or cut back intentionally. Exports were at a relatively low level in 1980 and 1981 and are expected to return at least part way towards the levels of the mid 1970s.

Exports of oil products are less likely to be as significant as in the past. The Abadan refinery, one of the main sources of finished products for markets in Africa, the Middle East and Asia, is now out of operation and if ever completely rehabilitated may be orientated more towards the internal market. Doubts also hang over the scale of natural gas exports. Supplies through the IGAT pipeline have been interrupted on several occasions and the Iranian government may not be keen to export large quantities of gas in the future.

Destination of Crude Oil Exports

(percentage)	
Western Europe	40
Asia	30
Latin America	15
North America	13
Other	2
TOTAL	100

Iran has been one of the principal sources of crude oil supplying the international oil market, and oil is exported to all of the industrial importing countries. Foremost destination is Western Europe, in line with that area's position as leading oil importer. Oil is exported to many Asian countries, particularly to Japan. Latin America and North America are of approximately equal significance in the overall pattern of exports.

Iraq

Iraq's economic development, which was still in an early stage prior to hostilities with Iran, has at best stagnated in recent years. Energy consumption per head of the population is only around one quarter of the world average and amongst the lowest in the Middle East. The country is, however, one of the most abundantly endowed with oil and gas. There is particularly scope for utilising the large volumes of gas

KEY ENERGY INDICATORS

Energy Consumption	
—million tonnes oil equivalent	5.8
Consumption Per Head	
—tonnes oil equivalent	0.44
—percentage of world average	27%
Net Energy Imports	nil
Oil Import-Dependence	nil

which are currently flared. Iraq is also relatively well off for water resources compared with many of the other Arab states. Iraq was a founder member of the Organisation of Petroleum Exporting Countries, of which it has become a key member, and is part of the Organisation of Arab Petroleum Exporting Countries.

ENERGY MARKET TRENDS

Energy consumption has followed a rather erratic path in recent years. Industrial and commercial development has been limited in scale and energy consumption has been affected by the level of activity in the energy industries themselves. Overriding all of these factors in the last two years has been the state of war existing in parts of the country and the consequences on some activities of actual disruption in the eastern border areas, particularly near to the head of the Persian Gulf.

Pattern of Energy Supply

Primary Fuel Supply (percentage)	
Oil	70
Solid Fuel	–
Natural Gas	27
Primary Electricity	3
TOTAL	100

Stemming from the abundance of oil resources available, a high proportion of primary energy supply is in the form of oil. But some natural gas is utilised and this forms a second element in the supply pattern. At present as much as 85 per cent of gas production is flared, so that there is scope for increasing the share of this fuel. Iraq also has hydro-electric power potential, which has been only partially exploited so far.

ENERGY SUPPLY INDUSTRIES

Energy supply is entirely the responsibility of state organisations. The national oil company INOC deals with all aspects of crude oil and natural gas production, with foreign companies, whether private or state-owned, engaged only on a contract basis. The public supply of electricity is under state control, with limited production of electricity by auto-producers.

Oil

Iraq is the second largest oil producer in the Middle East. In the mid 1970s the country's known reserves were more intensively developed and production built up to 171 million tonnes in 1979. Since that time production has been disrupted as a result of the war with Iran, but the country has the capacity to resume an upward production path whenever the Gulf region returns to a degree of normality. Total proven reserves are estimated to be more than 30 billion barrels, with important oil fields in several parts of the country.

Oil industry operations are now entirely in the hands of state organisations. The Iraq National Oil Company was set up in 1964 and handles all aspects of exploration, production and disposal of crude oil, although foreign companies are involved in some activities on a contract basis. INOC took over the existing trunk pipelines which had been built by the Iraq Petroleum Company to transport crude oil to the Mediterranean, but it has also developed additional lines from Kirkuk, the centre of the northern oil-producing region, to Dortyol in Turkey, and between Kirkuk and Fao, on the Persian Gulf. This line serves as primary export route from the Rumeila oil fields, but also as a strategic interconnector between oil fields, refineries and export channels in the north and south of the country.

Oil Production

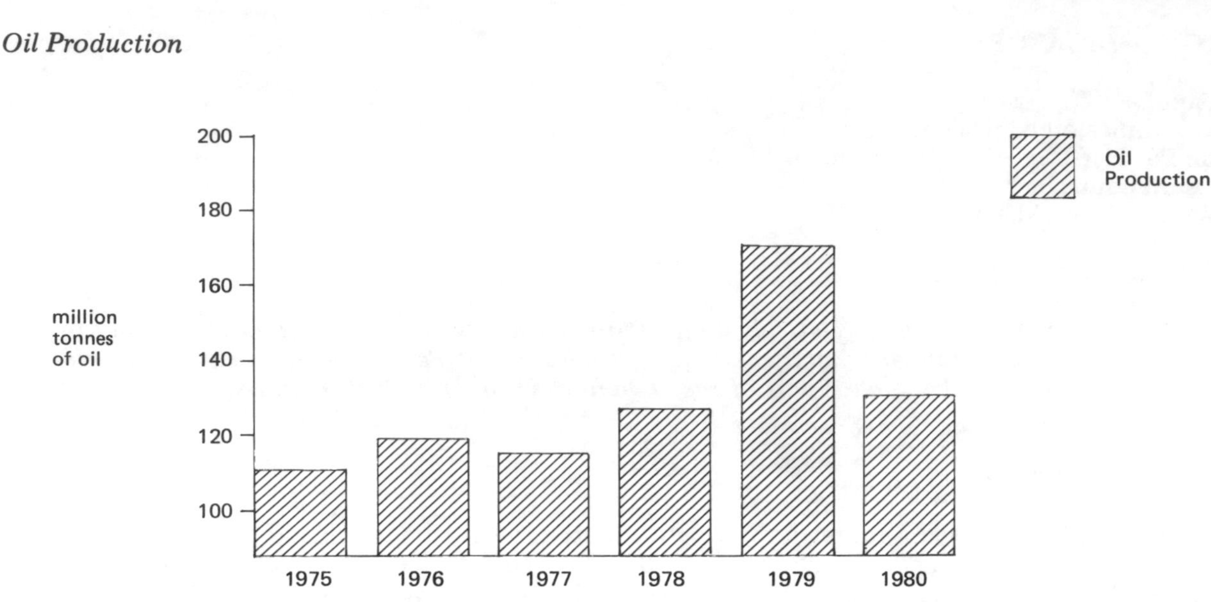

Oil refineries are all state-owned, and operated by the State Establishment for Oil Refining and Gas Processing. The two principal refineries are at Daura in the north and Basra in the south. The Basra refinery has a crude oil throughput capacity of seven million tonnes per annum, and that at Daura four million tonnes per annum. There are several very small refineries located in production areas, with a total capacity of little more than 1.5 million tonnes per annum. Domestic market consumption of refined products is around four million tonnes per annum, but has been rising at only a low rate in recent years, and there is thus the capacity to process crude oil for export markets.

Electricity

Demand for electricity increased rapidly up to 1978, when it reached 7,000 GWh, but since that time has stagnated, owing mainly to the economic effects of war conditions in the country. The public sector accounts for 85 per cent of generating capacity and major new power stations are planned, which would more than double existing capacity. At present only 10 per cent of capacity is hydro-electric. Planned capacity increases include oil-fired, hydro-electric and gas turbine plant. The country's economic development plans entail the creation of a national grid and an extension in the electrification network to cover most parts of the country.

ENERGY TRADE

Net Imports/(Exports) 1975-80

	1975	1976	1977	1978	1979	1980
Crude Oil (million tonnes)	(101.9)	(110.3)	(106.3)	(116.9)	(160.1)	(120.0)
Oil Products (million tonnes)	(0.5)	(0.7)	(0.8)	(1.3)	(1.4)	

Iraq is an important exporter of crude oil. In 1979 exports were 160 million tonnes, compared with only 102 million tonnes in 1975. In 1980 the level fell back sharply, as a result of the disruptions of the war with Iran and the weak demand for oil worldwide. Exports have subsequently been cut further, with additional problems arising from Syrian disruption of movements to Banias and Tripoli (Lebanon). Western Europe is the main destination for Iraq's exports. India and Brazil also have special contracts on a government-to-government basis for importing crude oil. Small quantities of finished products are also exported from Iraq, either across land frontiers to neighbouring countries or from ocean terminals to markets in Africa and Asia.

Israel

KEY ENERGY INDICATORS

Energy Consumption	
—million tonnes oil equivalent	7.9
Consumption Per Head	
—tonnes oil equivalent	2.03
—percentage of world average	125%
Net Energy Imports	99%
Oil Import-Dependence	99%+

Israel has the most developed economy in the Middle East with a wide range of industrial activities. Energy consumption per head of the population is well above the world average. This has been achieved despite the absence of indigenous energy resources. In the case of oil local production is negligible, and supplies have been difficult to maintain on occasions in the face of outside pressures. Israel is entitled to buy crude oil from Egypt from the Sinai oil fields for a limited period of time following its withdrawal from the area. Efforts continue to find oil and gas within the country, but the main policy is to switch to use of coal on a large scale.

ENERGY MARKET TRENDS

Trend of Energy Consumption

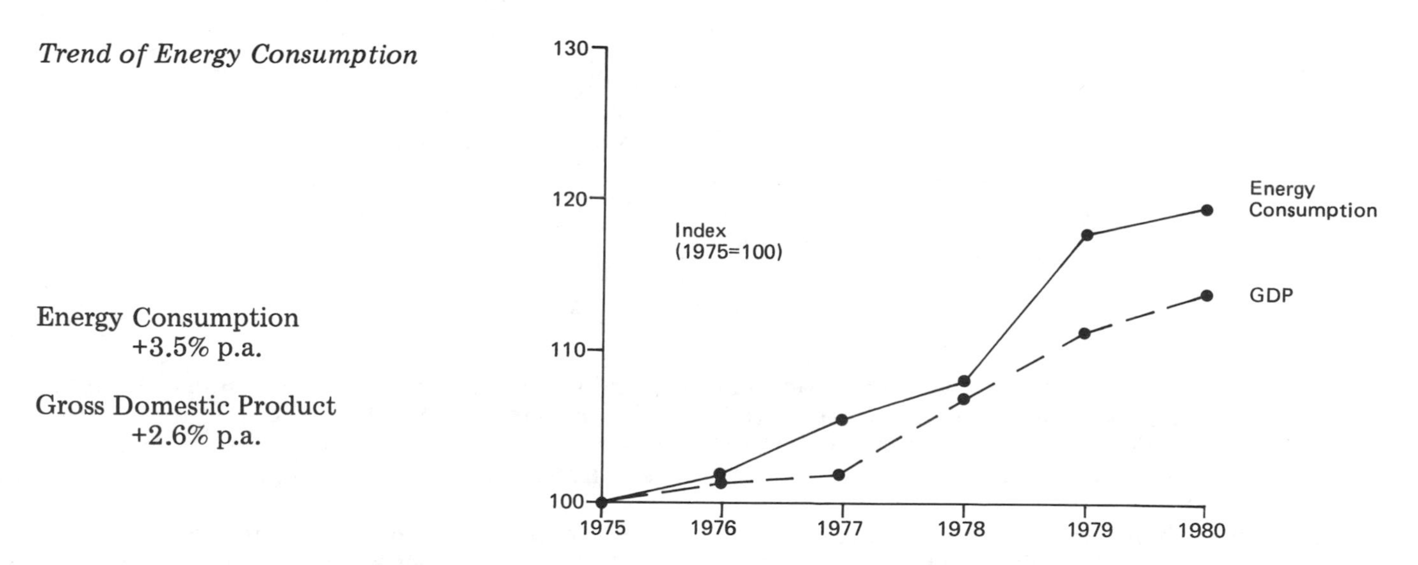

Energy Consumption
+3.5% p.a.

Gross Domestic Product
+2.6% p.a.

Despite its very high dependence on external sources of energy, and a continuingly difficult political environment, Israel has achieved solid economic growth. Between 1975 and 1980 Gross Domestic Product increased by 14 per cent, with most of the increase appearing after 1977. Energy consumption increased at a slightly greater rate, almost 19 per cent during that same period.

Pattern of Energy Supply and Consumption

Primary Fuel Supply (percentage)	
Oil	97
Solid Fuel	–
Natural Gas	2
Primary Energy	1
TOTAL	100

Final Consumption (percentage)	
Industry	32
Transport	26
Other	42
TOTAL	100

At present oil provides an overwhelmingly high proportion of energy in the economy. Local production of oil and natural gas is very small, though with some prospects of increasing output of the latter. A small contribution is obtained from solar power, which may also be developed further. Solid fuel is almost entirely absent, though this situation will change significantly, as coal-fired power stations are constructed.

Although Israel is a relatively well developed economy, the proportion of final energy consumption by the industrial sector is only 32 per cent. Energy-intensive operations are limited in view of the dependence on imported energy. Consumption by the government, commercial and residential sectors absorbs 42 per cent of the total.

ENERGY SUPPLY INDUSTRIES

Although energy is of strategic concern to the government, the state has not assumed complete control and private sector companies are involved in exploration for oil and gas and the supply of oil products. Electricity supply is publicly controlled and the state has established an agency for importing coal, for use on a large scale in power generation.

Oil

Oil Consumption

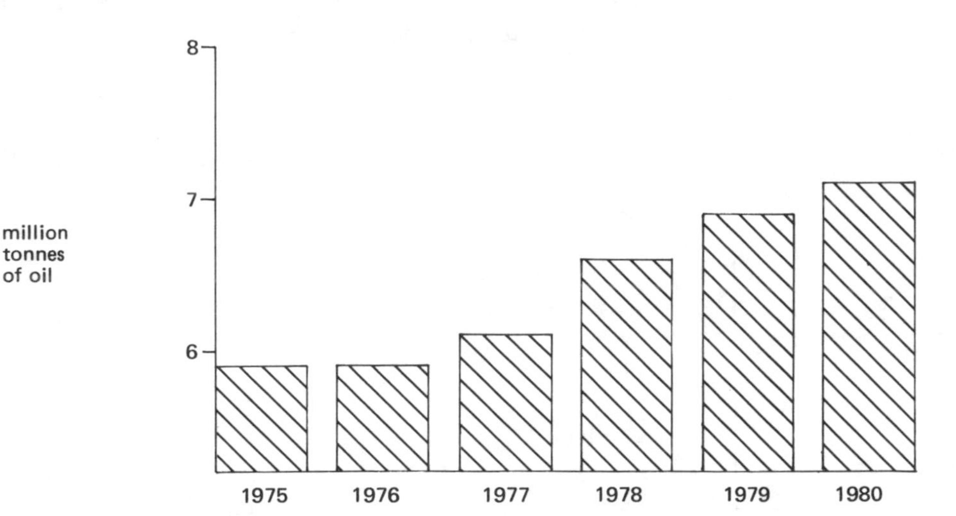

Oil is the mainstay of the economy and the consumption trend reflects the state of the economy. When the economy was progressing sluggishly in the period 1975-77 oil consumption rose very little. But in the period 1977-80 consumption rose by more than 16 per cent. Of the total some three million tonnes, or more than 40 per cent, is accounted for by the oil-fired power stations of Israel Electric Corporation.

Sufficient oil refining capacity has been established to meet market requirements for finished products. Main refinery is at Haifa, with a throughput capacity of six million tonnes per annum. A second refinery is located at Ashdod, also on the Mediterranean. Capacity of Ashdod refinery is 3.5 million tonnes per annum. Distribution of oil products from the refineries is undertaken by local private companies and by Israel Fuel Corporation (Delek), in which the state has an important interest.

Gas

A very limited amount of natural gas is produced in Israel. Output has generally been around 55 million cubic metres per annum, which has been used by local industries as a fuel. Output increased sharply in 1981 to some 130 million cubic metres. There is some expectation that more gas can be found. This is currently one of the main pre-occupations of the Israel National Oil Company.

Electricity

Consumption of electricity has been increasing much more quickly than consumption of energy as a whole. Between 1975 and 1980 it rose from 8.5 TWh to 11.1 TWh, an increase of more than 30 per cent. A high level of electrification exists in Israel, and electricity is a key energy form in many sectors.

The Israel Electric Corporation, a state-owned organisation, is responsible for the generation, transmission and distribution of electricity throughout Israel. The Corporation has 2,800 MW of generating capacity, using around three million tonnes per annum of fuel oil. A large shift to use of coal will, however, take place when work is completed on new equipment at the Hadera power station. Initial contracts have already been completed to import coal from Australia and the United Kingdom for a proportion of the station's requirements. Coal is also likely to be used in a new power station to be constructed near Ashkelon.

ENERGY TRADE

Net Imports/(Exports) 1975-80

	1975	1976	1977	1978	1979	1980
Crude Oil (million tonnes)	7.7	7.4	7.5	8.0	7.9	7.4
Oil Products (million tonnes)	(0.2)	(0.4)	(0.5)	(0.5)	0.3	1.0

Total net imports of crude oil and petroleum products have been on a rising trend since 1975. From a total of 7.5 million tonnes in that year there was a gradual increase to 8.4 million tonnes in 1980. Within the total there has been a noticeable shift in pattern, with a small net export of products being turned into a substantial net import. This has been partly offset by a reduction in crude oil imports and reflects the scale of fuel oil requirements of Israel Electric Corporation in the overall pattern of product demand.

Ivory Coast

KEY ENERGY INDICATORS

Energy Consumption	
—million tonnes oil equivalent	1.3
Consumption Per Head	
—tonnes oil equivalent	0.16
—percentage of world average	10%
Net Energy Imports	50%
Oil Import-Dependence	65%

The Ivory Coast economy has developed rapidly since independence in 1960, despite a high degree of dependence on external sources of energy. It has therefore come as a welcome additional benefit to the economy that significant discoveries of oil and gas have been made, which should make the country a net exporter of oil within the next few years. Further development of hydro-electric potential has also increased the proportion of electricity from this source. The basic energy situation of the country is thus undergoing dramatic change, underlining the sound prospects of the Ivory Coast economy.

ENERGY MARKET TRENDS

The Ivory Coast economy has been one of the more successful in developing a range of industrial and commercial activities. But energy consumption is still at a low level, reflecting the continuing significance of the agricultural sector of the economy. A diverse range of manufacturing activities has been established, but few that require much energy. Plans to exploit iron ore deposits have been delayed, but the extraction of oil and gas will itself increase the level of energy consumption. Consumption of electricity, aided by investment in hydro-electric plant, has risen much more quickly than energy consumption as a whole.

Pattern of Energy Supply

Primary Fuel Supply (percentage)	
Oil	77
Solid Fuel	–
Natural Gas	–
Primary Electricity	23
TOTAL	100

Oil accounts for a high proportion of the commercial energy consumed, with a small contribution from hydro-electricity. Constraints on the use of oil have been greatly eased by the discovery of offshore resources, which should make Ivory Coast self-sufficient within the next few years. These discoveries also include associated gas, which could become a significant energy carrier. In addition to commercial fuels a considerable amount of firewood is used, estimated to be 1.0-1.5 million tonnes per annum.

ENERGY SUPPLY INDUSTRIES

The state has a long established interest in the public company supplying electricity throughout the economy. To this has been added the establishment of a state oil company, which will become increasingly involved in the development of recently discovered offshore oil and gas resources. Most of the leading international oil companies are involved in exploration and development or in refining and marketing.

Oil

Oil is the principal energy form used in the Ivory Coast. Annual consumption is around one million tonnes per annum and tending to fall in the last few years as electricity production has been transferred to the new hydro-electric capacity at Kossou on the Bandama River. Refined products are produced at the Abidjan refinery of Société Ivoirienne de Raffinage, which also supplies products to neighbouring up-country markets in Mali and Upper Volta.

The state oil company PETROCI is largest shareholder in SIR, but is not involved in marketing. This is in the hands of the foreign oil companies which also participate in the refinery. These include British Petroleum, CFP, Elf-Aquitaine, Exxon, Mobil, Shell and Texaco. Upper Volta also has a shareholding.

PETROCI is concerned primarily with state interests in recent oil and gas discoveries. Discoveries have been made by Shell and Esso and by a group including AGIP and Phillips. PETROCI automatically participates in production from these fields, which commenced in 1979. Production capability of the fields is estimated to be eventually at least 200,000 barrels per day, equivalent to 10 million tonnes per annum.

Electricity

Consumption of electricity has risen rapidly with development and diversification of the economy. From 960 GWh in 1975 production rose to 1,720 GWh in 1980, an increase of almost 80 per cent. In 1980 almost three-quarters was hydro-electricity, whereas in 1975 60 per cent was from thermal plant. New capacity has been developed on the country's main rivers to keep pace with demand.

Electricity supply is undertaken by the state controlled Société Energie Electrique de la Côte d'Ivoire (EECI). EECI's principal hydro-electric stations have been those on the Bandama at Kossou (230 MW) and Taabo (200 MW). To these have been added the Buyo dam on the Sassandra River, with capacity of 165 MW and an even larger barrage at Soubre capable of generating 1,500 GWh per annum. Also under consideration is a joint scheme with Liberia on the Cavally River, which would produce 1,500 GWh per annum.

ENERGY TRADE

Net Imports/(Exports) 1975-80

	1975	1976	1977	1978	1979	1980
Crude Oil (million tonnes)	1.5	1.7	1.7	1.6	1.8	1.5

Imported energy is mainly in the form of crude oil for the Abidjan refinery. The trend of imports was to increase in the period up to 1979, since which time the impact of rising indigenous production has been to cut back imports. In the next few years these are expected to dwindle to very low levels and a net export position to be built up.

Sources of Imported Crude Oil

(percentage)	
Venezuela	36
Nigeria	26
Saudi Arabia	19
Iraq	10
Iran	8
Other	1
TOTAL	100

Ivory Coast has imported crude oil from a variety of sources, reflecting geography, the interests of companies participating in the Abidjan refinery and quality considerations. Nigeria has been an important source of crude oil. But larger volumes have been brought in from the Middle East. Largest individual source, however, is Venezuela, which is not only relatively well located in relation to Ivory Coast, but is also the source for heavy crude oils for use in the bitumen plant at Abidjan.

Japan

KEY ENERGY INDICATORS

Energy Consumption	
—million tonnes oil equivalent	378.0
Consumption Per Head	
—tonnes oil equivalent	3.24
—percentage of world average	200%
Net Energy Imports	90%
Oil Import-Dependence	99%

Japan is a highly industrialised economy with sizeable operations in a number of energy-intensive industries. Energy consumption per head of population is twice the world average and with limited indigenous resources this has made Japan one of the largest markets for imported energy. The country's reliance on imported energy has meant that the key elements of policy have been to increase the efficiency of energy utilisation and diversify sources of imports. Japan already imports oil from a great many countries, avoiding excessive reliance on any single one, and has readily tapped available sources of natural gas for import in liquefied form. In the future imported coal is expected to become much more important, and the large-scale programme of nuclear power station construction will continue.

ENERGY MARKET TRENDS

In the period since 1974 strenuous efforts have been made to restrain the growth of energy consumption while still meeting the needs of an expanding economy. Large-scale users of energy have sought higher levels of energy efficiency, spurred on by extensive programmes instigated by the government, combined with financial assistance in many areas. Gross Domestic Product maintained a reasonably consistent rate of

Trend of Energy Consumption

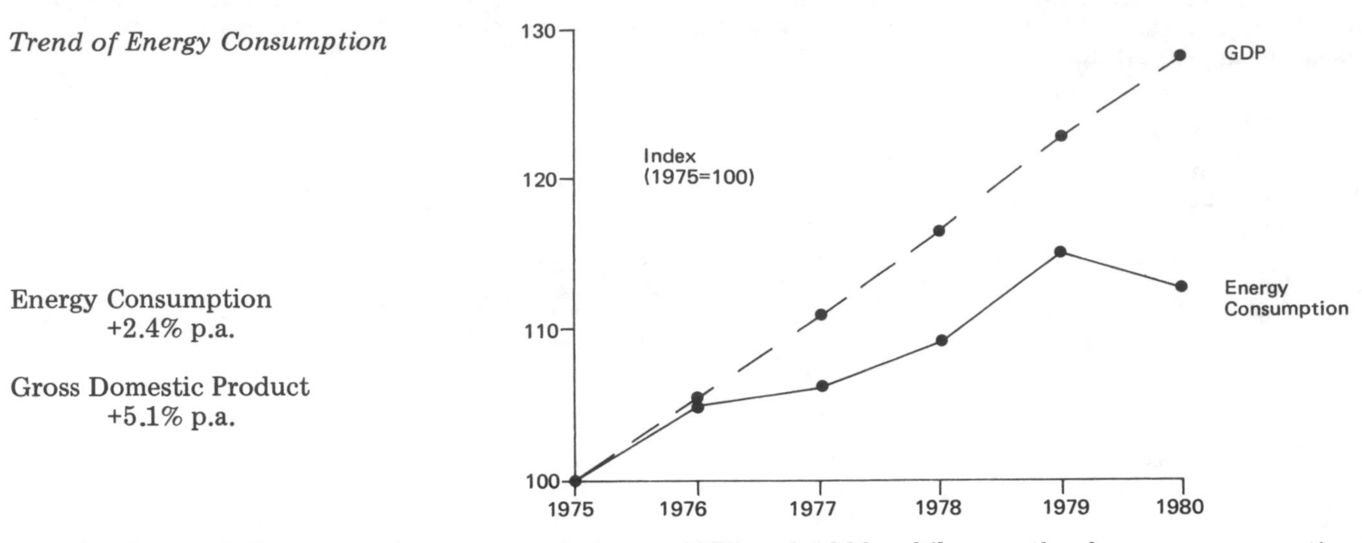

Energy Consumption
+2.4% p.a.

Gross Domestic Product
+5.1% p.a.

growth of around five per cent per annum between 1975 and 1980, while growth of energy consumption was kept to only three per cent per annum. In fact in 1980 economic growth, while below the average for previous years, was associated with a small decline in energy consumption.

ENERGY SUPPLY INDUSTRIES

Private sector companies play a more prominent role in energy supply in Japan than in any other major industrial country except the United States. The production and supply of both gas and electricity is carried out by specialised companies, involving high technology operations in running nuclear power plants and organising international LNG supply. In the oil sector foreign oil companies participate alongside Japanese concerns in refining and marketing. Coal mining is also carried out by a small group of leading Japanese industrial companies. Several ministries are involved in one way or another with the activities of the supply industries, particularly as there are implications for foreign policy and trading relationships.

Pattern of Energy Supply and Consumption

Primary Fuel Supply (percentage)		Final Consumption (percentage)	
Oil	70	Industry	53
Solid Fuel	15	Residential	19
Natural Gas	5	Transport	20
Primary Energy	10	Other	8
TOTAL	100	TOTAL	100

Oil is by far the most important source of energy for the Japanese economy, providing 70 per cent of primary energy input. There is no significant energy form within Japan and alternatives to oil are being painstakingly developed from overseas sources in the case of coal and natural gas, and at home in the form of a large-scale programme of nuclear power station construction. Hydro-electric schemes and geothermal energy resources at present constitute key elements of primary energy availability but the potential for further energy production from these sources is strictly limited and their contribution will rapidly become minor by comparison with that of nuclear energy. Output of nuclear power is planned to double within approximately a decade and imports of liquefied natural gas, which have already been rising rapidly, will double by 1985 and treble by 1990.

The Japanese economy has been developing on the basis of extensive steel making capacity, heavy engineering and energy-intensive process industries, such as petrochemical manufacture. The industrial sector thus absorbs more than half of all final energy consumption. The transport sector is less significant than in many other industrialised countries, partly due to the concentration of population in some regions and the wide use of mass transit systems.

Oil

Oil Consumption

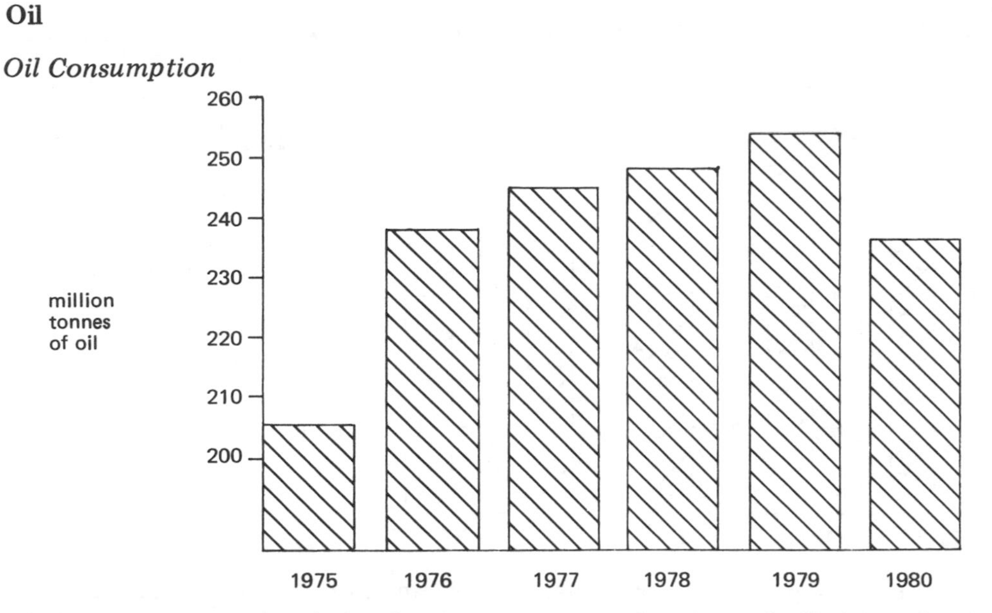

The post-war growth of the Japanese economy has been fuelled by the ready availability of oil. Total consumption of oil is of the order of 230 million tonnes per annum. This represents a substantial reduction on the pre-1973 trend, consumption having reached some 270 million tonnes in that year. Since 1975 consumption levels have risen again, although the rate of increase between 1976 and 1979 was very low by historical standards as efforts were made to conserve energy and diversify away from the use of oil. In 1980 consumption fell back slightly as a result of economic conditions.

Indigenous resources of oil in Japan are negligible, with production now down to less than half a million tonnes per annum. Japan has therefore become an important market for crude oil and finished products for both the international oil companies and, latterly, the state agencies of oil-exporting countries. During recent years a large shift has taken place away from the international oil companies towards government selling organisations as suppliers of crude oil, in line with the loss of control over oil production by the companies since the early 1970s, although these companies still retain an important role in organising the supply of a major part of the country's oil needs. Several of the major international oil companies are involved in refining and marketing activities in Japan, including Shell, Exxon, Mobil and Caltex, the joint-affiliate of Standard Oil Company of California and Texaco, on the basis of their crude oil supply capabilities in the Middle East and Far East. Their involvement has, however, been in partnership with Japanese companies, with shareholdings of 50 per cent or less. Shell is associated with Showa Oil Company, Showa Yokkaichi Sekiyu and Seibu Sikiyu KK, Caltex with the Nippon Oil group and Mobil and Exxon are closely linked with the Toa Nenryo group. In addition, Getty Oil Company, which also has interests in crude oil production in the Middle East, has a 50/50 partnership arrangement with Mitsubishi Oil Company.

The Nippon Oil group is the largest individual refining/marketing company, supplying around 18 per cent of the total market for oil products, from nine refineries located around the country. Companies with foreign affiliations account for approximately half of total sales, but there has always been a close control on the level of outside interests and wholly owned Japanese companies retain an important role. The two principal independent groups are Idemitsu Kosan and the Kyodo Oil group, which rely on their own oil supply arrangements with producer-country governments or international trading companies.

Several Japanese companies are involved, either individually or on a collective basis, in exploration for overseas oil resources. These efforts have been attended by some success in Indonesia, Kuwait and the United Arab Emirates. Under government auspices the Japan Petroleum Exploration Company was set up in 1955 and more recently the Japan National Oil Company was established to help provide financial and other support for oil and gas exploration companies.

Coal

Prior to the commencement of the rapid growth of the Japanese economy and its demand for energy in the 1960s and 1970s the coal industry was the main provider of energy. But the industry continued to run down until the mid 1970s when prospects for selling its output improved significantly. However, the importance of the industry in Japan is currently very limited with the main demand arising from coking applications and the bulk of supplies for this purpose is imported. Domestic production has, however, been maintained at around 18-19 million tonnes per annum, although the government's objective is to produce at least 20 million tonnes per annum in the coming years.

Coal Production and Consumption

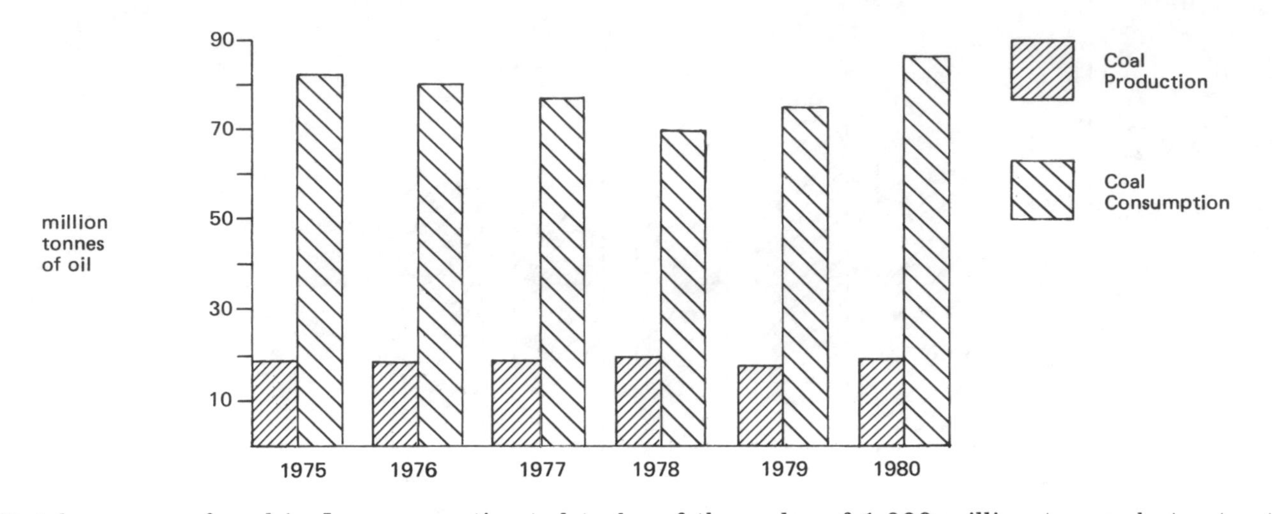

Total reserves of coal in Japan are estimated to be of the order of 1,000 million tonnes but extraction at rates of 20 million tonnes per annum or more will necessitate heavy investment programmes in deep-mining operations which will have to compete against low-cost production in other countries such as Australia. Production at Japan's principal coal mines is in the hands of several leading industrial enterprises, including the Mitsubishi and Sumitomo groups as well as other companies primarily engaged in mining, such as Hokkaido Colliery and Steamship Company. Most mines are located on Hokkaido, with some production also taking place on Kyushu.

Gas

Gas Production and Consumption

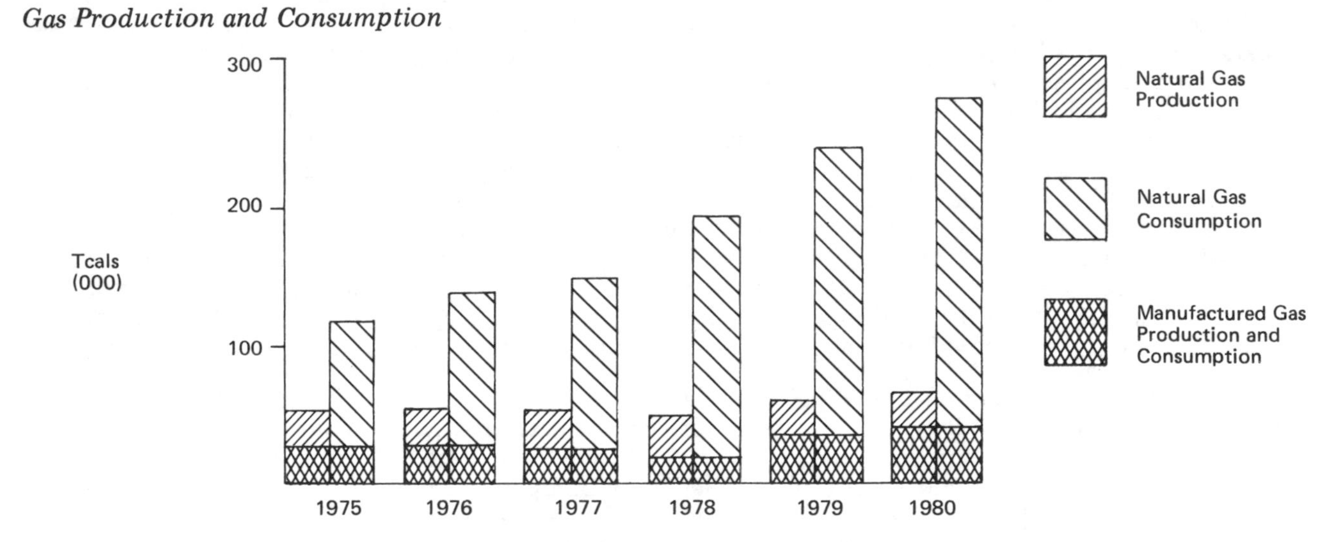

There is only limited production of natural gas in Japan. Production in 1980 was some 24,000 teracalories, equivalent to less than 2½ million tonnes of oil. This was significantly lower too than the peak level reached in 1977 when output was over 29,000 teracalories. However, total availability and use of natural gas has been rising rapidly on the basis of major international development projects in which Japanese gas and electricity companies are involved. This natural gas is imported in liquefied form from the United States, Brunei, Indonesia and the United Arab Emirates, and in total is ten times greater than the amount of gas produced indigenously.

The larger part of natural gas supplies is being used as a pollution-free fuel by electricity companies, although it is also already by far the most important form of gas handled by the gas companies. In addition oil products and to a lesser extent coal, are used to manufacture towns gas. As a result, the types of gas available and the supply industry itself have become highly complex, with production and supply on a localised basis with very little inter-connection.

There are around 255 companies involved in the supply of gas in the urban areas of the country, of which total some 75 are public corporations. But the industry is dominated by the activities of three major companies, Tokyo Gas, Osaka Gas and Toho Gas. These three companies account for three-quarters of all gas supplies and a high proportion of consumers. The remaining suppliers are often very small undertakings serving fewer than 10,000 consumers.

Electricity

Electricity Production

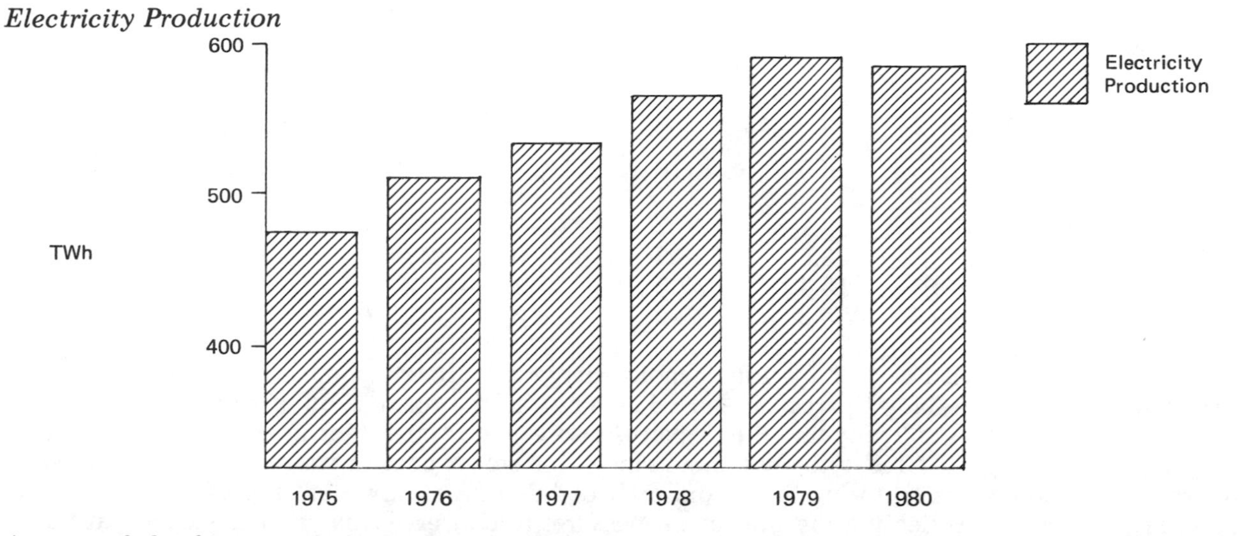

As one of the foremost industrial economies of the world with a large base of heavy industry Japan has a massive demand for electricity. Consumption is currently of the order of 600 TWh per annum. After a pause in the general upward path of consumption in 1974, consumption increased by 24 per cent between 1975 and 1979, a more rapid rate than for energy consumption as a whole, but this growth was again arrested in 1980 as a result of general economic recession.

The supply of electricity is based on nine electric power districts in each of which one private company has exclusive control over generation and supply. The three largest are Tokyo Electric Power Company, Kansai Electric Power Company and Chubu Electric Power Company, which account for nearly 60 per cent of all power supplied. Tokyo Electric Power itself had sales of 169 TWh in 1979 and is the largest private power company in the world. The remaining six companies are the Hokkaido, Tohoku, Hokuriku, Chugoku, Shikoku and Kyushu Electric Power Companies. In addition there is a separate company responsible for generation and transmission on Okinawa.

The power companies also handle electricity generated by the Electric Power Development Company and Japan Atomic Power Company. Prefectoral and local authorities are also involved in electricity distribution. Major industrial power consumers are often involved with the power companies in joint-venture operations producing electricity for both industrial bulk use and the public supply system.

The overall electricity supply system is subject to supervision and control by the government and there is inter-company co-ordination to cope with regional imbalances in demand and to optimise operations. State and industry organisations are involved in co-ordination of power station construction, exploration for, and development of, overseas fuel supplies and control of the development of nuclear power. The Electric Power Development Co-ordination Council is responsible for authorising the construction of new power stations. The Ministry of International Trade and Industry and the Science and Technology Agency are both involved in research and development work and the administration and control of nuclear facilities.

There are more than 20 nuclear power stations in operation in Japan, with a total capacity of 15.5 GW, and the country is thus second only to the United States in the scale of its nuclear generating capacity. Most of the larger power companies operate at least one station and numerous additional plants are under construction or sanctioned which will double nuclear capacity during the 1980s. Nevertheless, oil remains the main fuel input and accounts for 55 per cent of electricity generated. In an effort to reduce its share power companies are increasing their commitments to import liquefied natural gas and coal.

Uranium

Japan has no known significant source of uranium to provide the basic fuel for its nuclear power stations, but it has constructed enrichment facilities in order to reduce its dependence on other countries for nuclear fuel. Started as a pilot plant in 1979, the facility is located in Okayama Prefecture, about 500 km south-west of Tokyo and is run by the Power Reactor and Nuclear Fuel Development Corporation. The Okayama plant will be used as the basis for prototype and commercial processing plants to be built later in the 1980s. Also in order to limit the country's dependence on external agencies for operating its nuclear capacity, spent fuel processing facilities have been built by the Power Reactor and Nuclear Fuel Development Corporation and an even larger scale privately owned plant is expected to be built in due course.

ENERGY TRADE

Net Imports/(Exports) 1975-80

	1975	1976	1977	1978	1979	1980
Coal (million tonnes)	62.1	60.1	60.2	51.6	56.5	65.8
Crude Oil (million tonnes)	223.6	228.7	236.5	230.2	239.2	217.0
Oil Products (million tonnes)	15.4	22.8	23.1	23.3	27.8	24.2
Natural Gas (million tonnes)	5.0	5.9	8.3	11.7	14.9	17.0

As a major industrial nation with very limited indigenous energy resources Japan has high levels of imports of all forms of raw energy. The larger part of imported energy is in the form of oil of which Japan imports around 220-240 million tonnes per annum. But Japan is also a major customer in the international coal trade, principally for coking coal hitherto, but increasingly for steam coal as well in the future. Despite continuing growth of the economy during the second half of the 1970s, however, imports of coal and oil remained broadly unchanged, reflecting the impact of measures to conserve energy and a switch out of oil.

Japan has been in the forefront of the development of the international trade in liquefied natural gas. From one source initially, around one million tonnes per annum of LNG from Alaska, trade has increased rapidly as additional projects in other countries have been completed, so that by 1980 imports rose to 17 million tonnes, equivalent to around 10 per cent of total oil imports and representing gas supplies of the order of 20,000 million cubic metres.

Sources of Imported Crude Oil

(percentage)	
Saudi Arabia	32
Indonesia	15
United Arab Emirates	13
Iraq	7
Iran	6
China	4
Kuwait	3
Qatar	3
Other	17
TOTAL	100

Saudi Arabia is by far the largest individual source of imported crude oil, reflecting its predominant position amongst oil-exporting countries, particularly those in the Persian Gulf. But at 32 per cent of the total this share is not disproportionate and Japanese importers draw from very many sources, extending from North Africa to the Americas. African crude oil, which has an attractively low sulphur content for the stringent pollution requirements of Japan, can be transported economically, making use of otherwise empty tankers returning from Europe to the Middle East for at least part of the journey.

Japan has particularly close trading links with Indonesia and provides a natural market for the latter's low sulphur crude oil. As a result Indonesia is the second largest source of crude oil, supplying over 30 million tonnes per annum. Other sources represent a mixture of those providing low sulphur crude oils, such as Qatar and United Arab Emirates, and others which produce heavy crude oils, with a high yield of fuel oil, to meet the pattern of petroleum product requirements in Japan.

Japan also imports finished products from a wide variety of sources. Middle Eastern countries with large export-orientated refineries are important sources, eg. Saudi Arabia, Kuwait and Bahrain. Other sources are the refining centre of Singapore and Australia.

Sources of Imported Coal

(percentage)	
Australia	46
United States	28
Canada	14
South Africa	6
China	3
USSR	2
Other	1
TOTAL	100

Three-quarters of Japan's import requirements for coal are supplied by Australia and the United States, with the former being the pre-dominant source of supply. Both countries supply coking coal and steam coal roughly in proportion to the pattern of Japanese demand. Canada is the only other source of importance currently, supplying almost entirely coal for coking purposes. South Africa, which is emerging as a large-scale exporter of coal, currently provides some six per cent, largely of steam coal. Other Far Eastern sources make up the balance of the Japanese import total.

Sources of Imported LNG

(percentage)	
Indonesia	51
Brunei	33
United Arab Emirates	11
United States	5
TOTAL	100

Japan has been in the forefront of the developing international trade in liquefied natural gas (LNG). Since the early 1970s projects have been developed in Alaska, Brunei, Indonesia and the United Arab Emirates on the strength of Japan's great need for fuels with low pollution characteristics. Alaskan LNG now accounts for only a small proportion of total imports. Attention has been concentrated on developing the nearest major sources of natural gas in Indonesia and Brunei. Most recent development has been the utilisation of gas available in the United Arab Emirates. Although only accounting for 11 per cent of imports in 1980 this source is expected to become more important as imports of LNG continue to be expanded.

ENERGY POLICIES

There is no one ministry responsible for energy policy. Responsibility for different aspects of energy supply and utilisation fall under the Ministry of International Trade and Industry, the Ministry of Science and Technology or the Ministry of Finance. The State has not moved to take a significant direct involvement in the energy sector, but has relied on indirect means of supporting the efforts of private companies towards general energy objectives. Since there is no practical alternative in the near and medium term to the country's substantial dependence on imported oil to run its economy, policy is geared towards ensuring an adequate supply of oil to maintain continuing economic growth, with diversification of supply sources, while at the same time encouraging efficient use of energy and greater levels of conservation.

Energy Supply

In the absence of Japanese based integrated international oil companies the state supports indirectly the efforts of private sector companies to develop Japanese owned or part-owned energy resources overseas. This has hitherto been aimed primarily at exploration and development of oil, with support via the Japan Petroleum Development Corporation and the Japan National Oil Corporation. More recently attention has been focused on participating in overseas coal projects, particularly in Australia.

Japanese energy policy is geared as much to developing alternatives to oil as obtaining interests in oil production. No new oil-fired power stations are being allowed and existing ones are being considered for conversion to coal or to LPG/LNG. Preferential loans are available to power generation companies to convert to coal and the state will help to finance coal handling and transport facilities. Funds are also being made available under the Sunshine Project, a programme of projects to diversify Japan's energy inputs, to assist installation of solar energy systems. The programme also encompasses large scale exploitation of Japan's geothermal resources and the scope for using alcohol as a fuel.

Conservation

Japanese governments have shown reliance on the price mechanism as an important spur to energy conservation. The high degree of dependence on imported energy has meant that prices have followed international levels upwards, with a consequent response from private companies in energy supply and energy-intensive industries to improve efficiency. Normal price pressures have been further emphasised by the imposition of taxes to encourage conservation and switching of fuels.

However, governments have provided tax incentives and financial assistance, eg, through the National Finance Corporation, to influence private sector actions along the desired lines of energy policy. The government has also placed emphasis on information and advice to all sectors, and is keeping a close eye on the pace of investment in energy conservation. Under the 1979 Energy Conservation Act, companies accounting for a high proportion of national industrial energy consumption are involved in a process of formulating conservation targets, energy auditing and meeting official standards for fuel efficiency.

Jordan

KEY ENERGY INDICATORS

Energy Consumption	
—million tonnes oil equivalent	1.8
Consumption Per Head	
—tonnes oil equivalent	0.56
—percentage of world average	35%
Net Energy Imports	100%
Oil Import-Dependence	100%

Energy consumption in Jordan is well under two million tonnes of oil equivalent per annum for a population of 3.2 million. Consumption per head is thus only one-third of the world average. No conventional energy resources have been discovered so far, but attention is turning towards the extensive reserves of oil shale which could be used for fuel in power stations. The economy is almost entirely based on imported oil, which is drawn from the trunk pipeline from Saudi Arabia to the Mediterranean. A small contribution comes from hydro-electric plant. Solar energy applications also hold out some promise.

ENERGY MARKET TRENDS

Development of the economy is leading to increasing use of energy, and because of the country's location and lack of energy resources this has led to a rapid rise in oil consumption. Between 1975 and 1980 consumption increased by 90 per cent, fuelling power stations and industrial plant as well as general commercial, transport and residential uses. Jordan possesses useful mineral deposits, particularly of potash and phosphates, the extraction and treatment of which involve large scale use of energy.

ENERGY SUPPLY INDUSTRIES

State agencies and undertakings play an important part in the supply of energy. The state is involved directly in the supply of oil products through participation in the Jordan Petroleum Refinery Company, as well as being intimately involved in agreements with Saudi Arabian crude oil suppliers. Electricity generation and distribution is a state concern and other agencies are becoming of increasing importance in developing hydro-electric power and oil shale resources. Private sector companies are involved in oil distribution and foreign companies are also encouraged to explore for oil and gas.

Oil

Despite the apparent absence of oil and gas deposits, Jordan's location next to major oil producing countries and between Saudi Arabia and the Mediterranean has meant that oil is inevitably the principal energy form in use. Consumption of oil is rising towards two million tonnes per annum, having been less than one million tonnes in 1975. Crude oil is drawn from a spur line from the Trans Arabia Pipeline (TAPLINE) feeding the Zarqa refinery of Jordan Petroleum Refinery Company.

Refinery throughput in 1980 was 1.8 million tonnes, virtually at the limit of capacity. An expansion programme is underway which should double capacity. A full range of products is made, with a high proportion of kerosine and diesel oil to meet market requirements.

Electricity

The development of the national electricity supply system is the responsibility of the state-owned Jordan Electricity Authority. Production exceeds 700 GWh per annum, mainly from oil-fired plant with some hydro-electric output. One of the Authority's main power stations, at Zarqa, burns crude oil directly. Consideration is being given to constructing a power station of 300-400 MW at Qatrana, in central Jordan, which would utilise the oil shale reserves in the Lajjoun area. These are estimated to be of the order of 1.4 billion tonnes, and even so represent only a small part of total national oil shale reserves. Hydro-electric power will also increase in importance as water development projects combining irrigation and desalination are progressed.

ENERGY TRADE

Jordan is virtually completely dependent on imports of energy. For the most part these comprise crude oil for processing at the Zarqa refinery or direct use as a fuel. Additional small quantities of finished products, such as LPG, are imported from Iraq, Lebanon and Kuwait. Imports of oil currently total some 1.8 million tonnes per annum. Crude oil is imported entirely from Saudi Arabia via a spur from the TAPLINE to Sidon. Electricity has also been imported from time to time from neighbouring grids.

Kenya

KEY ENERGY INDICATORS

Energy Consumption	
—million tonnes oil equivalent	2.1
Consumption Per Head	
—tonnes oil equivalent	0.13
—percentage of world average	8%
Net Energy Imports	85%
Oil Import-Dependence	100%

Kenya has one of the best records in Africa for economic development over the last two decades, although this success has been qualified by the relatively rapid rate of increase in population. Thus consumption of commercial energy is still only eight per cent of the world average. Kenya is handicapped by the absence of indigenous oil, gas or coal resources, but it does have hydro-electric potential, which is only partially exploited at present, and geothermal energy resources which could be providing up to 500 MW in the 1990s.

ENERGY MARKET TRENDS

The Kenyan economy has continued to make progress despite the impact of high oil prices since 1974. Gross Domestic Product rose by 25 per cent between 1975 and 1980. This was accompanied by a somewhat higher rate of increase in consumption of energy, but this amounted to only 30 per cent and within the total the use of imported oil was kept at a much lower rate. Measures are being taken to increase output of indigenous energy and improve the level of energy conservation in order to sustain economic progress in the future.

Pattern of Energy Supply and Consumption

Primary Fuel Supply (percentage)	
Oil	83
Solid Fuel	..
Natural Gas	–
Primary Electricity	17
TOTAL	100

Consumption of Oil (percentage)	
Industry/Commerce	26
Transport	62
Electricity Generation	9
Other	3
TOTAL	100

Of the commercial fuels oil is of dominant importance. Negligible amounts of coal and coke are used. At present hydro-electricity is the only significant indigenous source of energy. Hydro-electric power is being further developed, and to this will be added an increasing amount of geothermal energy, which will tend to increase the proportion of energy produced from indigenous resources. Outside the modern sector of the economy fuelwood and charcoal continue to be of importance. Annual consumption of wood and charcoal is estimated to be over 10 million tonnes per annum, making it as important as oil in raw energy terms.

Transport demands account for a high proportion of oil consumption. Although this includes aviation fuel and marine bunkers, the government is also making attempts to contain the local demand for fuels. Industry and commerce account for only one-quarter of consumption. Use of oil in electricity generation fluctuates. In normal years this proportion is less than 10 per cent, but in 1980 the amount of oil used doubled to compensate for low hydro-electricity production.

ENERGY SUPPLY INDUSTRIES

The supply of electricity and associated needs for investment in generating capacity, transmission equipment and rural electrification programmes inevitably involve the government. The supply of oil products is, however, largely left to local and international oil companies, which operate the national refinery and distribute products.

Oil

Consumption of petroleum products is on a rising trend as development of the economy takes place. However, the rate of increase has been kept to a modest level, rising significantly in 1980 only as a result of the need to resort to oil-fired generating plant to make up for the shortfall in hydro-electric production. The overall increase between 1976 and 1980 was only 13 per cent, and omitting the special factors in operation in 1980, the average rate of increase has been only two per cent per annum.

The supply of products is obtained from the Mombasa oil refinery, in which all of the leading marketing companies participate. The refinery has a crude oil throughput capacity of around four million tonnes per annum and supplies products for other inland markets, including Uganda, and the substantial trade in bunkers. Even so the refinery is only 60 per cent utilised.

Electricity

Production of electricity is around 1,800 GWh per annum and on a strongly rising trend. Between 1976 and 1980 the increase was 34 per cent and would have been greater but for restrictions on the use of electricity in 1980. Two-thirds of consumption is by large commercial and industrial operations, whereas residential and small commercial consumption accounts for only one-quarter.

Generating capacity is 485 MW, of which 314 MW is hydro-electric. In 1980 hydro-electric output was constrained to a little over 1,000 GWh, compared with 1,300 GWh in 1979. Total capacity is expected to reach 555 MW by 1983 with the addition of 40 MW of hydro-electric and the first 30 MW of geothermal capacity. Production and importation of electricity from Uganda are the responsibility of Kenya Power Corporation, which sells power to East African Power and Lighting Company, the sole distributor of electricity in Kenya.

ENERGY TRADE

Net Imports/(Exports) 1976-80

	1976	1977	1978	1979	1980
Crude Oil (million tonnes)	2.5	2.6	2.4	2.5	3.1
Oil Products (million tonnes)	(1.3)	(1.3)	(0.9)	(0.7)	(1.4)
Electricity (TWh)	0.2	0.3	0.2	0.2	0.3

Kenya depends on imported energy to a large extent, and particularly on oil, which is entirely imported. Crude oil for the Mombasa refinery is imported from the Middle East. The refinery also produces products for export markets, including Uganda, and for marine bunkers. The balance of trade in oil has shown an increase in net imports as consumption has risen. There is also a net inflow of electricity from Uganda. In 1979 this had fallen to less than 10 per cent of electricity supplies, but almost doubled in 1980, as drought conditions reduced the potential available within Kenya.

ENERGY POLICIES

A full range of policy measures is being developed by the government to improve the local supply of energy and improve the level of energy conservation. In the absence of oil, gas or coal resources attention is being concentrated on exploiting hydro-electric and geothermal energy, which could be supplying 1,100 MW by the 1990s. The government is also assessing the viability of the project to install additional refining plant in order to meet the pattern of domestic oil demand more closely.

Because of the importance of wood and charcoal in the overall energy supply pattern, the government is concerned to raise the level of reafforestation. In an attempt to supplement the available gasoline supplies molasses are being utilised as an extender. On the conservation side attention is being paid to reducing the use of fuel in the transport sector through a variety of measures to encourage the use of private transport more efficiently and to increase the role of public transport.

Korea (Republic of)

KEY ENERGY INDICATORS

Energy Consumption	
—million tonnes oil equivalent	44.0
Consumption Per Head	
—tonnes oil equivalent	1.15
—percentage of world average	71%
Net Energy Imports	70%
Oil Import-Dependence	100%

Korea has a relatively developed industrialising economy built up on imported oil during the last two decades. The impact of oil supply problems during the 1970s has not brought the process of economic growth to a halt, but strenuous efforts are being made to assure adequate and reliable overseas sources of supply and to scrutinise the use to which energy is put. The government has accordingly developed a range of policies to utilise as far as possible the country's limited indigenous resources and to become more active in overseas exploration and development ventures.

ENERGY MARKET TRENDS

Trend of Energy Consumption

Energy Consumption
+8.6% p.a.

Gross Domestic Product
+7.3% p.a.

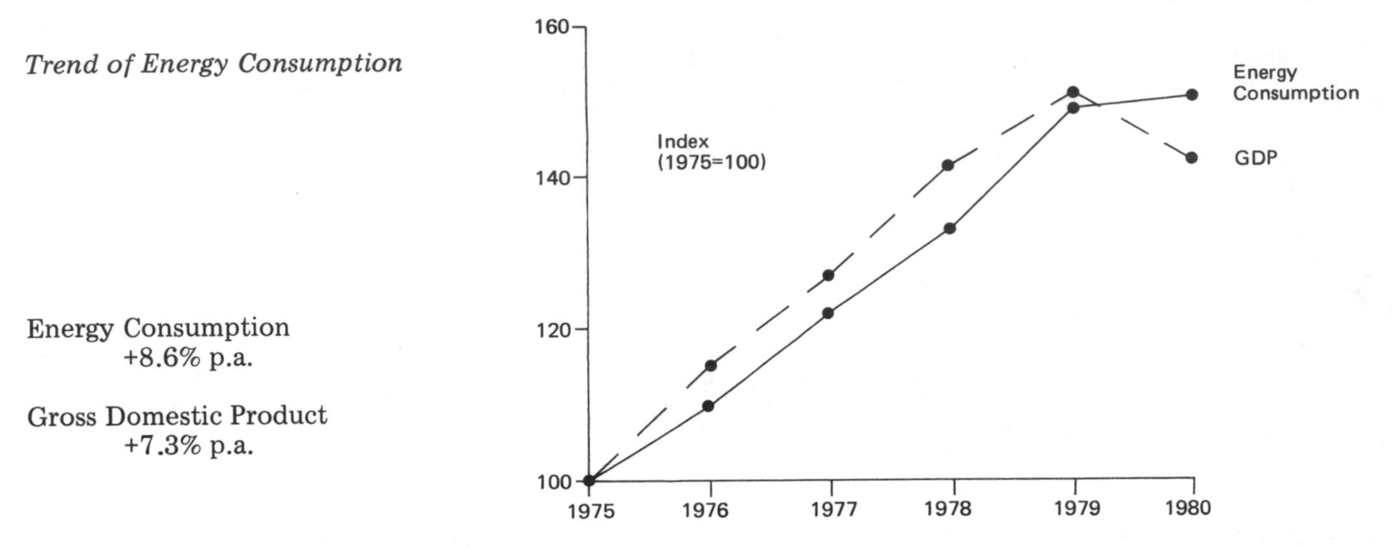

Successive plans for the Korean economy have stressed the development of heavy industries with consequent demands for energy. In the four years 1975-79 Gross Domestic Product rose by more than 50 per cent, but the trend was generally downward from 1976 and with the impact of international recession in 1980 GDP actually declined. Despite the energy-intensive nature of many of the growth industries, the rate of growth of energy consumption has been broadly similar to that of GDP, with a rise of just over 50 per cent in the period 1975-80, reflecting the effects of improved energy efficiency and conservation measures.

Pattern of Energy Supply

Primary Fuel Supply (percentage)	
Oil	60
Solid Fuel	37
Natural Gas	–
Primary Electricity	3
TOTAL	100

Development of Korea's industrial economy has involved a rapid increase in the use of oil, which now accounts for 60 per cent of primary energy. Almost all of the balance is obtained from solid fuel, including anthracite, coking coal, steam coal and locally available firewood, which is estimated to account for six per cent of energy supply. Natural gas has not been found in Korea and production of hydro-electricity and nuclear power is still small.

The pattern of fuel supply is undergoing considerable change, however, as oil consumption is held down and use of nuclear power and coal is expanded. It is also anticipated that imported LNG will be introduced by about 1985 for use as a fuel in power stations and for distribution in Seoul and other cities. Importation of LPG on a large scale is also envisaged.

Oil

Consumption of oil increased rapidly in the period to 1979, but in 1980 it was held at a level of around 28 million tonnes. This reflected the impact of conservation measures and attempts to use indigenous and imported coal on a wider scale. The 1980 level of consumption was nevertheless more than 60 per cent higher than in 1975.

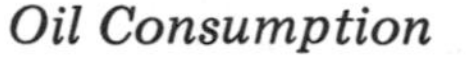

Oil Consumption

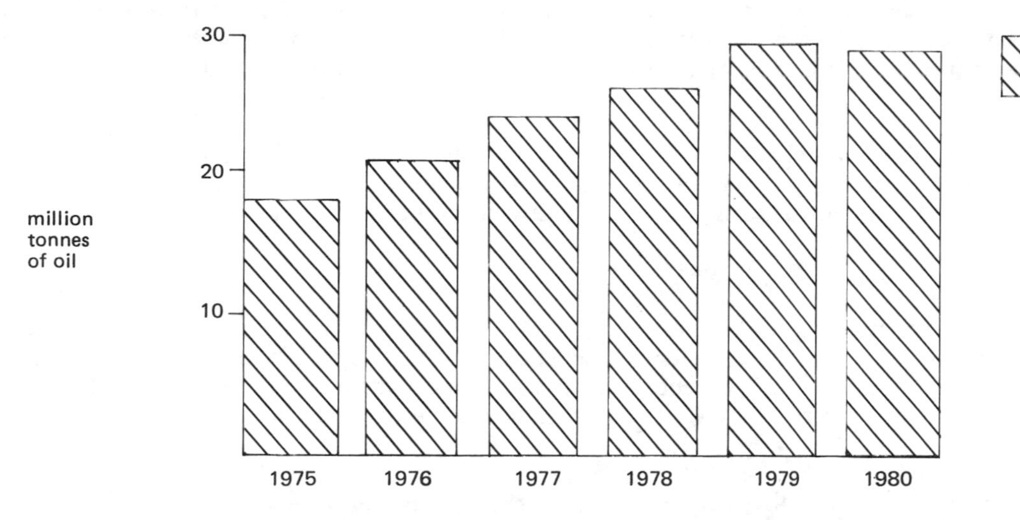

The supply of oil is undertaken by the state-owned and private corporations involved in oil refining. Oil refineries have usually been established as joint ventures with international oil companies or producer country-interests, in view of the lack of Korean involvement in oil production.

Caltex Petroleum is a partner in the country's largest refinery at Yosu, which has been expanded to almost 20 million tonnes per annum capacity, and Union Oil of California partners the Korea Explosives Group in the Kyung In Energy Company, which operates a three million tonnes per annum refinery at Inchon.

Gulf Oil was involved for many years in the operations of Korea Oil Corporation, which owns the country's second main refinery, at Ulsan, but has recently withdrawn, leaving the company wholly state owned. The National Iranian Oil Company has also withdrawn from the recently completed refinery venture with the Ssanyong Group.

Oil refiners sell products wholesale to dealers or direct to industrial consumers. Prices and marketing arrangements are subject to state control. The state owned Korea Oil Corporation was responsible for importing 50 per cent of all crude oil in 1980. Government agencies are also involved in overseas exploration, whilst foreign companies continue to explore in Korean waters.

Coal

Coal Production and Consumption

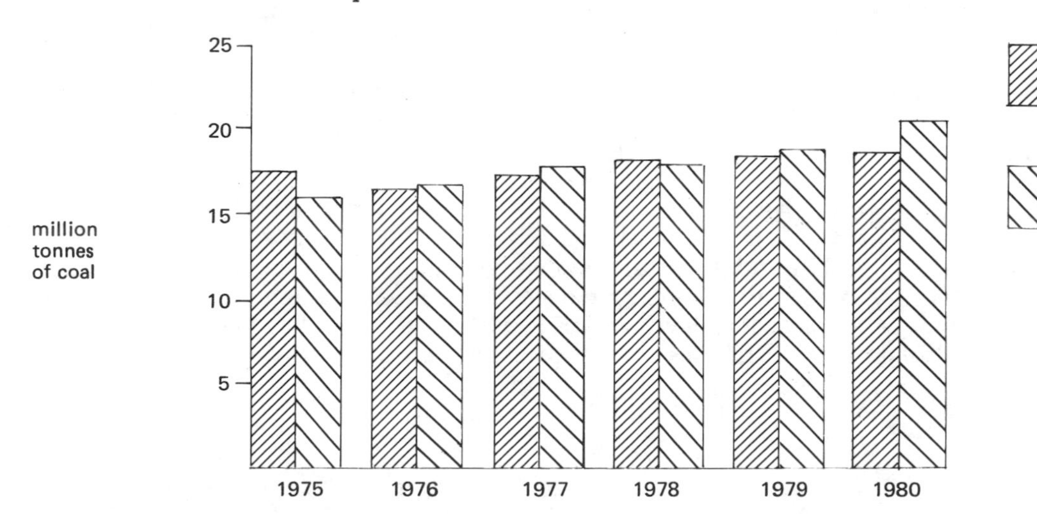

Korea has a substantial coal mining industry, although reserves are basically of anthracite, which limits the range of end uses to which it may be economically put. There is some lignite, but most steam and metallurgical coal must be imported. Proven reserves of anthracite are put at 630 million tonnes. Reserves are located at various places in a belt running from south-west to north-east across the country. Some 45 per cent of reserves are in the Samcheok area with other main locations Chungnam and Jeongseon.

Coal mining is undertaken by a state corporation and by a large private sector composed of almost 200 companies, often operating very small mines. The state-owned Dai-Han Coal Corporation produces almost five million tonnes or approximately 20 per cent of industry output. Private mines are encouraged to consolidate in order to raise efficiency and increase investment in new plant and capacity. An important role in this respect is being played by the Korea Mining Promotion Corporation, which provides advice, technical support and finance.

Indigenous production is consumed mainly in the residential sector, with very limited consumption by power stations and industry. Use in power generation will increase somewhat, but imported bituminous coal is to be the fuel for 4,000 MW of new thermal capacity being completed in the next few years.

Electricity

The pace of industrialisation and the scale of certain energy-intensive activities have placed great demands on the electricity supply industry. Electricity production in 1980 exceeded 37 TWh, almost twice the level of 1975. This increase is much greater than for energy as a whole, and though the increase in 1980 was less than in previous years, it was nevertheless well above four per cent, even though total energy consumption stagnated.

The supply of electricity is the responsibility of a state undertaking, Korea Electric Corporation, which operates the national grid and distributes electricity throughout the country. The Kyung In Energy Company, which operates an oil refinery at Inchon, feeds some of its output to the grid. The Industrial Sites and Water Development Corporation also provides some power to Korea Electric from its 290 MW of hydro-electric plant.

Indigenous anthracite and hydro-electric resources provided only 22 per cent of electricity output in 1980. A small contribution of seven per cent came from the 590 MW nuclear station at Kori. But the major proportion, around 70 per cent, was produced in oil-fired stations. These include large installations at Ulsan (1,400 MW), Incheon (1,150 MW), Pyong Taek (700 MW), Honam (560 MW) and Seoul (360 MW).

Korea Electric has a large programme of construction of nuclear power stations and additional plans to diversify fuel sources to increase use of coal considerably and introduce LNG as a fuel. Plans are for construction of some 11,000 MW of nuclear capacity at four or five locations by 1991, completion of 4,000 MW of coal-fired plant over the same period and commissioning of LNG-fired plant by about 1985. Hydro-electric capacity is to be increased by 500 MW. Much of the Corporation's existing oil-fired capacity is to be examined with a view to possible conversion to coal.

ENERGY TRADE

Net Imports/(Exports) 1975-80

	1975	1976	1977	1978	1979	1980
Coal (million tonnes)	0.8	1.6	2.1	2.5	6.6	7.4
Crude Oil (million tonnes)	15.4	18.3	20.7	23.0	25.0	24.8
Oil Products (million tonnes)	(0.3)	(1.0)	(2.5)	0.6	1.8	2.0

Energy trade is very significant in Korea. All oil supplies are imported and increasing quantities of steam coal are being added to existing imports of metallurgical coal and coke. Coal imports tripled between 1975 and 1978 and tripled again between 1978 and 1980 to become an important element in energy imports. Imports of oil have been contained, as a result of energy conservation measures and diversification of fuel supplies. Oil imports in 1980 were at the same level as 1979, although there has been a trend towards imports of finished products. This should be offset as a result of expansion of refining capacity.

Sources of Imported Crude Oil

(percentage)	
Saudi Arabia	61
Kuwait	25
Iran	8
Neutral Zone	3
Other	3
TOTAL	100

Considering the importance of oil in the Korean energy picture and the country's reliance on foreign sources, crude oil has been drawn from a limited number of sources. This has been partly the consequence of arrangements with particular companies such as Gulf and Caltex, and partly reflects the need for heavier grades of crude oil to meet demands for industrial fuels. Saudi Arabia holds a dominant position currently, with Kuwait and the Kuwait/Saudi Arabia Neutral Zone accounting for most of the balance.

ENERGY POLICIES

Korea has been evolving radical programmes to increase the availability and security of its energy base and to control the rapidly rising demand for energy. Given the limited indigenous potential this is taking the form of increasing the number of supply sources for crude oil, expanding greatly the use of nuclear power and coal, following the route taken by Japan in establishing long-term supplies of LNG and participating in overseas resource development projects. A full range of measures in the field of conservation is also being implemented. This has involved the setting up of a special agency, the Korea Energy Management Corporation, to promote improved awareness, management, technology, training and research.

Kuwait

KEY ENERGY INDICATORS

Energy Consumption	
—million tonnes oil equivalent	5.8
Consumption Per Head	
—tonnes oil equivalent	4.14
—percentage of world average	256%
Net Energy Imports	nil
Oil Import-Dependence	nil

Kuwait is a major crude oil producer with limited energy consumption owing to its population of less than 1.5 million. Oil consumption is only two per cent of annual output, but the country is reliant on oil exports to generate foreign exchange earnings. This, allied to high economic expectations within the state, maintain pressure to keep up oil output. A technical consideration having a similar impact is that difficulties arise in gas-using industries if crude oil output falls below a certain level. Kuwait is an active member of OPEC, concerned to raise average revenue, but is concentrating on investment to increase the proportion of refined products in total exports and to improve the yield of high value products.

ENERGY MARKET TRENDS

The abundance of local energy, including the availability of gas which would otherwise be flared, has led to a rapidly rising trend of energy consumption. Already high levels of consumption of electricity and water (which must be expensively produced from desalination plant) continue to rise further. New processes to utilise oil or gas and add value to raw energy also tend to involve intensive use of energy. On the other hand, demand for transport fuels is likely to be less dynamic, in view of the existing high level of car ownership and the compact nature of the state.

The energy supply situation is dominated by the abundant availability of oil and gas. As the gas is associated with crude oil, but is inherently more costly to export, it has been used as the principal form of fuel in refining, petrochemicals and power generation and as a feedstock for fertiliser manufacture. Oil products, principally for transport fuels, constitute the balance of energy supply.

Pattern of Energy Supply

Primary Fuel Supply (percentage)	
Oil	36
Solid Fuel	-
Natural Gas	64
Primary Electricity	-
TOTAL	100

ENERGY SUPPLY INDUSTRIES

Foreign companies now have a negligible role in energy supply in Kuwait, although they continue to be important in oil production, refining and export marketing in the Neutral Zone. Kuwait Petroleum Corporation, the state holding company, has subsidiary companies in all sectors of the oil and gas industries and state organisations are also important in related activities, such as fertiliser manufacture, and in extending Kuwaiti involvement overseas.

Oil

Oil Production

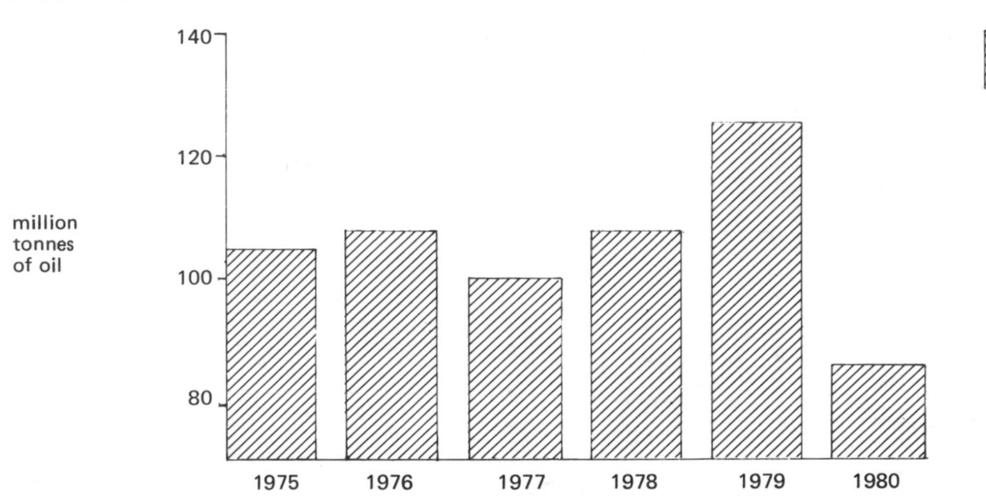

Kuwait is one of the key oil producing states of the Middle East. Peak production was achieved in 1972 at a total of 153 million tonnes, to which was added a share of Neutral Zone output of some 14 million tonnes. Production has never reached these levels subsequently and in general is subject to a ceiling. Kuwait nevertheless has very large reserves, already proven at over 70 billion barrels, equivalent to around 100 years supply at recent rates of extraction. Neutral Zone reserves are additional to this.

With Neutral Zone production remaining firm at around 14 million tonnes per annum, the full impact of fluctuations in the international oil market has been felt by production in Kuwait itself. Production in 1980 was only 77 million tonnes, half the level of 1972 and sharply lower than in 1979. Almost 80 per cent of crude oil output is exported, with a mere two million tonnes per annum satisfying local market demand. But substantial quantities are processed in export-orientated refineries, from which refined products are shipped to Africa, Asia and Europe.

The oil industry in Kuwait is now almost entirely in Kuwaiti hands, with state corporations predominant. Kuwait Petroleum Corporation is holding company for subsidiaries involved in exploration and production, refining, transportation and marketing, both at home and abroad. Kuwait Oil Company formerly the vehicle for important crude oil producing concessions of British Petroleum and Gulf Oil, is now KPC's subsidiary responsible for exploration and production. Kuwait National Petroleum Company operates large refineries at Shuaiba, Mina Al Ahmadi and Mena Abdullah, which have a total processing capacity of more than 25 million tonnes per annum. Both Mina Al Ahmadi and Mena Abdullah refineries are being modernised in order to produce higher value products instead of fuel oil from the country's relatively heavy crude oil.

Kuwait has taken a particular interest in downstream possibilities. Apart from increasing the proportion of refined products in its exports it has sought to develop refining capacity and marketing facilities abroad. This has been accompanied by involvement in other countries as a form of investment as well as aid. KPC itself participates in the International Energy Development Corporation, which is exploring in Oman, Africa, Canada, Australia and elsewhere.

State involvement in the Neutral Zone is more limited. KPC participates with foreign companies in oil exploration and production, but refining and marketing of crude oil and products is left to the private companies. These include Getty Oil and a consortium of Japanese companies which participate in offshore production.

Gas

Production of natural gas is substantial and is associated with the production of crude oil. Production is more than 4,000 million cubic metres per annum. Close attention has been paid to utilising this gas in fertiliser manufacture, petrochemical plant and hydrogenation equipment at the Shuaiba refinery, as well as in power generation. It is estimated that three-quarters of the gas is now utilised. A field of non-associated gas has been found in the Neutral Zone and may be exploited in the near future, particularly if crude oil output remains at a low level.

The large quantities of associated gas contain significant proportions of liquids, which are now being extracted. A large plant has been built at Mina Al Ahmadi capable of extracting 4.4 million tonnes per annum of LPG. Additional schemes to gather gas from the southern part of Kuwait and from the Neutral Zone are being considered.

Electricity

Demand for electricity is very high in relation to the small population of Kuwait. This arises from the loads used in energy production and processing, the consumption of power in the large scale desalination plants being built and a high level of demand arising from the standard of living of the average Kuwaiti. The indications are that recent rapid rates of growth will continue, requiring substantial investment in new power stations. A 1,050 MW station was completed in 1980 at Doha and a 2,400 MW station is to be built nearby for electricity supply and desalination processes. Shuaiba is the site of two other key desalination plants.

ENERGY TRADE

Net Imports/(Exports) 1975-80

	1975	1976	1977	1978	1979	1980
Crude Oil (million tonnes)	(90.9)	(90.1)	(80.7)	(88.7)	(107.0)	(67.0)
Oil Products (million tonnes)		(18.1)	(17.5)	(18.9)	(18.3)	(14.5)

Kuwait is one of the world's leading crude oil exporters with export levels generally of the order of 90-100 million tonnes per annum. Exports fell back considerably in 1980 to only 67 million tonnes in the wake of the international recession. In addition there is a significant export trade in refined products amounting usually to 18 million tonnes per annum and further quantities of LPG from the recently completed extraction plant at Mina Al Ahmadi. Exports from this plant are expected to be in the range of two to four million tonnes per annum.

Kuwait crude oil is exported to many countries, including most of the industrial countries of Western Europe and Japan. Japan absorbs almost all of the output from the Neutral Zone, and takes more than 30 million tonnes per annum in total. EEC countries imported 23 million tonnes in 1980, almost 20 million tonnes less than in the previous year.

ENERGY POLICIES

Kuwait is a leading member of the Organisation of Petroleum Exporting Countries and of the Organisation of Arab Petroleum Exporting Countries. Internally it has pursued a path of gradual takeover of foreign assets, although allowing private Kuwaiti involvement in some of the oil-related activities. It has been an important element in the country's policy for many years to be closely involved in downstream operations, leading to joint-ventures or even own operations in refining and marketing. Kuwait state corporations and agencies are active abroad in energy exploration and development.

Policy towards oil and gas production is to minimise the flaring of associated gas and maximise the value of export sales. This has meant both the consideration of a limit on crude oil exports and massive investment in gas-based industries and in up-grading plant to handle heavy crude oil.

Lebanon

KEY ENERGY INDICATORS

Energy Consumption	
—million tonnes oil equivalent	2.1
Consumption Per Head	
—tonnes oil equivalent	0.66
—percentage of world average	41%
Net Energy Imports	91%
Oil Import-Dependence	100%

The Lebanese economy and the growth in demand for energy have both been disrupted in recent years as a result of political turmoil and economic disruption. Energy consumption per head of the population is only two-fifths of the world average, with industrial development only on a limited scale. Lebanon is one of the least well endowed countries of the Middle East as far as energy resources are concerned. It has some hydro-electric potential, but it remains to be seen whether oil shows onshore and offshore are of any significance. Lebanon is, however, strategically located on the Mediterranean and terminals for exporting Iraqi and Saudi Arabian crude oil are located there, providing a ready supply of crude oil, and revenue in the form of transit dues.

ENERGY MARKET TRENDS

Energy consumption in Lebanon has remained largely stagnant in recent years because of the difficult political situation in the country following civil war in the mid 1970s and consequent disruption to industrial and commercial life. By 1980 energy consumption had scarcely attained the level of 1975.

Pattern of Energy Supply

Primary Fuel Supply (percentage)	
Oil	91
Solid Fuel	..
Natural Gas	–
Primary Electricity	9
TOTAL	100

Lebanon has limited indigenous energy production and depends almost entirely on imported oil. Use of solid fuel has so far been negligible. Oil has been discovered, but not so far developed, and the only indigenous energy is derived from a number of hydro-electric plants.

ENERGY SUPPLY INDUSTRIES

The central government has only a limited involvement in the energy supply industries. The electricity company Electricité du Liban operates autonomously and the country's two oil refineries are owned and managed by independent companies.

Oil

Oil refineries are located at Sidon and Tripoli, the terminals for crude oil pipelines from Saudi Arabia and Iraq respectively. The Trans Arabian Pipeline has remained operational for most of the time, but the Iraqi pipeline, which also crosses Syria, has been more subject to interruption. This has given rise to the idea of linking the two refineries by pipeline, so as to reduce the risk of disruption.

Both refineries are of relatively small size. The Tripoli refinery has a capacity of 1.5 million tonnes per annum. The Sidon refinery is not much more than half that size. Total capacity is therefore adequate for the current level of consumption, which is only around two million tonnes per annum. Consumption fell sharply in 1977 and has been recovering only slowly. It is still short of the level reached in 1975.

Electricity

The responsibility for ensuring the public supply of electricity lies with Electricité du Liban (EDL) which operates as an autonomous organisation under general supervision of the government. EDL supplies most of the country, although a number of local franchises still exist. On expiry of these franchises EDL is expected to assume responsibility for supply. There is a very high degree of electrification in Lebanon, put at 99 per cent.

EDL operates seven hydro-electric stations and eight thermal power stations, based on oil. Production of electricity was some 1,850 GWh in 1975, but declined sharply in 1976 as a result of the civil war. The 1975 level of production was not passed until 1980, when consumption in the EDL network jumped to 2,300 GWh. During the next few years two new power stations are being constructed, at Zhoug and Jiyyeh.

Liberia

KEY ENERGY INDICATORS

Energy Consumption	
—million tonnes oil equivalent	0.5
Consumption Per Head	
—tonnes oil equivalent	0.26
—percentage of world average	16%
Net Energy Imports	85%
Oil Import-Dependence	100%

Liberia, which has a population of less than two million, has considerable natural resources. The agricultural sector is important, but above all the economy is dominated by iron ore mining. This sector is partly responsible for raising energy consumption per head to as much as 26 per cent of the world average. Other industrial activities are limited. Hydro-electric potential is being increasingly exploited and uranium prospecting is being undertaken, but no oil or gas has been found, consequently around 85 per cent of energy is imported.

ENERGY MARKET TRENDS

The growth of Gross Domestic Product is dominated by the fortunes of the iron ore mining industry, which accounts for a significant share of the total. Liberia possesses 20 per cent of iron ore reserves evaluated in Africa. On average the quality of the ore is high, so that output has not suffered unduly in the general recession affecting steel in recent years, and output has been relatively stable. Consumption of energy in the economy has reflected this situation, with very little growth since 1975.

Pattern of Energy Supply

Primary Fuel Supply (percentage)	
Oil	85
Solid Fuel	-
Natural Gas	-
Primary Electricity	15
TOTAL	100

Oil is the dominant source of energy used in Liberia, for transport purposes and power generation as well as other commercial and residential uses. There is quite a high level of consumption of electricity, of which one-third is generated from hydro-electric plant. Of the non-commercial fuel forms wood makes a useful contribution, as almost 40 per cent of the land area is forested.

ENERGY SUPPLY INDUSTRIES

The state plays little part in the supply of energy. It exercises general control of the national oil refinery at Monrovia and the public sector role in electricity is increasing as the activities of Liberia Electricity Corporation raise the level of electrification. But iron ore mining companies remain of great importance as producers and consumers of energy.

Oil

Consumption of oil has fluctuated from year to year, reflecting activity in iron ore mining. The capabilities of hydro-electric plant and the general trend of the economy. Consumption is in the range 400-450,000 tonnes per annum, with little upward movement in the trend. Product supplies are obtained mainly from the Monrovia refinery of Liberia Refinery Company. The refinery has a capacity to process around 750,000 tonnes per annum of crude oil. Output is sold to local marketing companies for distribution.

Electricity

Production of electricity is of the order of 900 GWh per annum from both private and public sectors. Approximately half of this is produced by the mining companies for their own internal consumption. They also produce some electricity for the publicly owned Liberia Electricity Corporation. The Corporation's operations are based on the hydro-electric plant at Mount Coffee, which provides the larger part of its requirements. Hydro-electric capacity represents only 25 per cent of total capacity, but produces 35 per cent of the electricity. The Electricity Corporation has under consideration important additional projects in conjunction with Sierra Leone and the Ivory Coast.

Libya

Prior to the development of oil and gas resources, which were discovered only as recently as the late 1950s, Libya was counted amongst the poorest of nations. The abundance of oil and gas transformed this situation in many respects, such that energy consumption per head of population is now close to the world average. However, this arises mainly from the operations of certain energy intensive activities, amongst which must

KEY ENERGY INDICATORS

Energy Consumption	
—million tonnes oil equivalent	4.5
Consumption Per Head	
—tonnes oil equivalent	1.50
—percentage of world average	93%
Net Energy Imports	nil
Oil Import-Dependence	nil

be included the oil and gas industries themselves, in a country of only three million people. Energy resources are thus ample for the needs of economic development and the full extent of the country's resources has still not been fully assessed.

ENERGY MARKET TRENDS

The level of energy consumption tends to mask the state of the rest of the economy. This remains patchy in its development, partly because of the very low base from which the country started and the limited population. The oil and gas industry still accounts for more than half of Gross Domestic Product and almost all of the country's export earnings. This seems likely to continue to be largely the case as additional oil-and gas-related projects are undertaken, although there is a project to produce steel from indigenous iron ore by the second half of the 1980s. It remains a matter of policy to establish operations which add value to oil production, but attempts to establish a sound basis of manufacturing activity are progressing only slowly.

Pattern of Energy Supply

Primary Fuel Supply (percentage)	
Oil	78
Solid Fuel	-
Natural Gas	22
Primary Electricity	-
TOTAL	100

The oil and associated gas produced from Libya's large oil fields supply the needs of all sectors of the economy. Oil is of primary significance both as a fuel and as a feedstock. Efforts are also made to utilise gas which might otherwise be flared. Libya is in any case without hydro-electric potential or known reserves of other fuels.

ENERGY SUPPLY INDUSTRIES

The state is intimately concerned with all energy sector activities. It has gained control through nationalisation, strict legislation on exploration and production and the presence in all aspects of the state oil company. However, foreign companies are encouraged to participate with the state company in oil and gas development.

Oil

Production of crude oil was at its peak in 1970 when it reached 160 million tonnes. Since that time the government has reined back production from such high levels, and sought to increase or maintain revenue through price rises. In 1979 production reached 100 million tonnes after the general fall in oil demand in the mid 1970s, but fell back again subsequently as a result of weak demand. In 1981 production was cut substantially to only 55 million tonnes. Internal demand is only 3.5 million tonnes, although increasing quantities of crude oil are being processed for export markets.

Production of crude oil is closely controlled by the state through the Libyan National Oil Company (LINOCO) and its subsidiaries. But foreign companies remain important, both in exploration and production, and include Amerada Hess, Conoco and Marathon, which form the Oasis Group, and Occidental Oil. Companies operate in partnership with the National Oil Company under various contract arrangements,

Oil Production

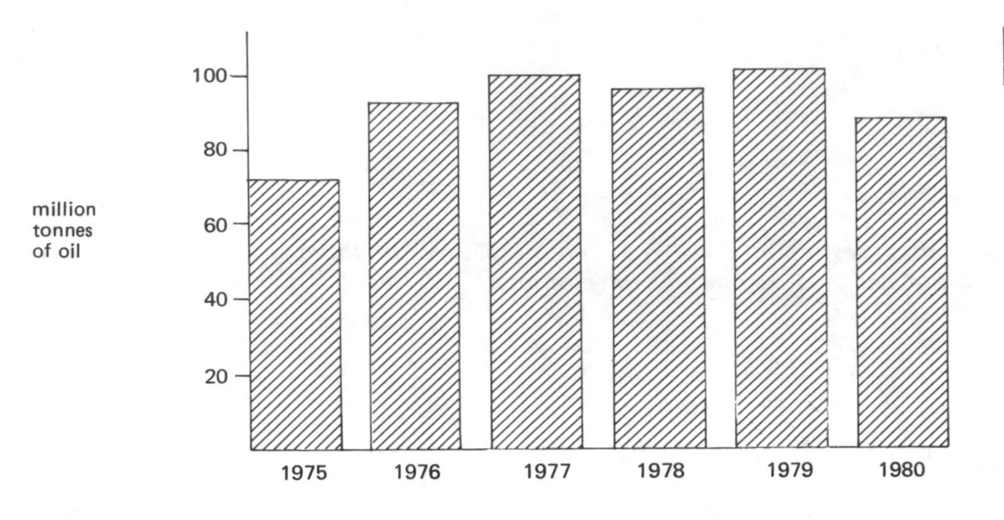

undertaking to lift oil attributable to their interest and to part of the state's. The Oasis group is the largest producing group, accounting for 35 per cent of production. Arabian Gulf Oil Company, a subsidiary of LINOCO, is second largest producer, with over 20 per cent, and Occidental the third, with 13 per cent of production.

LINOCO is the only organisation involved in oil refining in Libya. The main refinery is at Azzawiya, with a throughput capacity of six million tonnes per annum. There is also a small plant at Marsa El Brega of half a million tonnes per annum capacity. A much larger and more complex refinery is under construction at Ras Lanuf, with a capacity of 10-11 million tonnes per annum, which is linked to petrochemical production plant and will also produce refined products for export.

Gas

Large quantities of gas are produced in association with crude oil. Only a fraction of this gas is utilised in the domestic economy and a proportion either re-injected into the oil fields or flared. Libya was one of the first countries to establish an export trade in LNG, which has been exported since the early 1970s to Italy and Spain. Total maximum contractual quantities amount to almost 4,000 million cubic metres per annum.

Electricity

Production of electricity is over 1,600 GWh per annum. The rate of increase has been between six and seven per cent per annum. This is much lower than the rate of increase for energy as a whole and reflects the predominance of oil-and gas-using activities and a lack of development in other manufacturing and commercial sectors. The supply of electricity is entirely in the hands of state agencies, operating some 900 MW of thermal power plant.

ENERGY TRADE

Net Imports/(Exports) 1975-80

	1975	1976	1977	1978	1979	1980
Crude Oil (million tonnes)	(69.2)	(89.3)	(93.6)	(89.7)	(92.8)	(80.1)
Natural Gas (000 million M3)	(3.3)	(3.8)	(3.9)	(3.7)	(3.9)	

A high proportion of Libya's crude oil output is exported. In 1977 and 1979 the level exceeded 90 million tonnes, but in 1980 sales fell back to 80 million tonnes and a further substantial fall, to around 50 million tonnes was recorded in 1981, as demand for oil on the international market fell. Exports from Libya were particularly badly affected as LINOCO consistently pressed for high prices. Exports of LNG to Spain and Italy have been maintained at 3.7-3.9 billion cubic metres per annum, after earlier disruptions in supplies following problems over price renegotiations with importing countries.

Destination of Oil Exports

(percentage)	
Western Europe	48
North America	35
Latin America	8
Other	9
TOTAL	100

Destinations of Libyan crude oil reflect the geographical proximity of the country to Western Europe and the United States, in both of which markets low sulphur crude oil is valued. Large volumes are shipped to northern Europe or across the Mediterranean to refineries and trunk pipeline terminals.

Malagasy Republic

KEY ENERGY INDICATORS

Energy Consumption	
—million tonnes oil equivalent	0.5
Consumption Per Head	
—tonnes oil equivalent	0.06
—percentage of world average	4%
Net Energy Imports	88%
Oil Import-Dependence	100%

Economic development in the Malagasy Republic has been taking place only slowly and energy consumption per head of population is still at very low levels. The extent of the country's natural resources is still largely uncharted. Numerous mineral deposits, some on a substantial scale, have been proved, but energy resources other than hydro-electric potential, are apparently lacking. All major activities are undertaken by state organisations as part of the marxist-socialist ideology adopted by recent governments.

ENERGY MARKET TRENDS

Consumption of energy has risen in an erratic fashion since 1975 with an insignificant upward trend. Consumption of commercial energy has remained around 500,000 tonnes of oil equivalent per annum. Within this total consumption of electricity has risen at a steady rate since 1975, but averaging less than three per cent per annum.

Pattern of Energy Supply

Primary Fuel Supply (percentage)	
Oil	84
Solid Fuel	4
Natural Gas	-
Primary Electricity	12
TOTAL	100

Of the commercial energy forms available oil accounts for the larger part. Small amounts of coal/coke are imported too. The only significant indigenous energy resource currently being exploited is hydro-electricity, which provides some 12 per cent of energy supply. Non-commercial fuel forms, principally wood, remain important in many areas.

ENERGY SUPPLY INDUSTRIES

Since the institution of programmes of socialisation of the economy, the control of basic commodities such as energy has been entirely in the hands of state organisations. Supply of both petroleum products and electricity is undertaken by integrated organisations under the control of the government.

Oil

Consumption of petroleum products is around 450,000 tonnes per annum. Consumption of transport fuels has been rising, but use of fuel oil has fluctuated, partly owing to variable availability of hydro-electric power. Most of the market's requirements are met from the national refinery at Tamatave. The state organisation SOLIMA has sole responsibility for operating the refinery and supplying crude oil and refined products. SOLIMA also handles exports of oil products as surpluses arise.

Electricity

Production of electricity is 380-400 GWh per annum, and is rising at a rate of about three per cent per annum. More than 50 per cent of production is from hydro-electric plant. Public supply of electricity is undertaken by the state organisation JIRAMA, which has a generating capacity of 65 MW. A further 30 MW is operated by larger industrial enterprises for their internal consumption.

Malawi

KEY ENERGY INDICATORS

Energy Consumption	
—million tonnes oil equivalent	0.2
Consumption Per Head	
—tonnes oil equivalent	0.03
—percentage of world average	2%
Net Energy Imports	65%
Oil Import-Dependence	100%

Malawi is still a mainly agriculture-based economy, with limited manufacturing activities. The growth of such industry is retarded by the lack of indigenous energy as well as of development capital. Use of commercial energy is small, consumption per head being only two per cent of the world average. Hydro-electric potential exists to provide electricity, but other forms of energy are lacking and have to be imported from, or through, neighbouring countries, as Malawi is entirely landlocked.

ENERGY MARKET TRENDS

During the late 1970s the economy experienced growth rates of 5-10 per cent per annum, but towards the end of the period there was a tailing off in this process and the economy has suffered from poorer returns from a number of key agricultural sectors. Consumption of energy rose at a lower rate than Gross Domestic Product, of the order of three per cent per annum, reflecting the relatively low degree of energy-intensiveness of the economy.

Pattern of Energy Supply

Primary Fuel Supply (percentage)	
Oil	57
Solid Fuel	14
Natural Gas	-
Primary Electricity	29
TOTAL	100

Oil products provide over half of the input of commercial energy to the economy. A useful contribution is made by indigenous hydro-electric production, which continues to be further exploited. Coal is also used to some extent.

ENERGY SUPPLY INDUSTRIES

The state occupies a key role in the provision of electricity and is responsible for almost all electricity production. The supply of other fuels is left in the hands of private companies. Affiliates of several of the major international oil companies are present, engaged in the importation and marketing of petroleum products.

Oil

Oil is the principal form of fuel used for other than electricity generation, with annual consumption in the range of 150-200,000 tonnes. There is no oil refinery in Malawi and all finished products are imported, although a programme has been started using products from the sugar industry to extend gasoline in an effort to reduce the burden of imports. The supply of oil products is handled by local marketing affiliates of several of the international oil companies.

Electricity

Electricity is the only indigenous form of commercial energy of any significance. Production is around 350 GWh per annum, of which almost 90 per cent is obtained from hydro-electric power stations at Nkula and Tedzani on the Middle Shire River. Standby capacity of some 40 MW of thermal capacity is available for public and private sector use. The state-owned Electricity Supply Commission is responsible for public supply of electricity and accounts for 90 per cent of output. The Commission has established a grid system covering the southern part and interconnection with Dedza, Salima and Lilongwe in Central Malawi.

ENERGY TRADE

Imports of energy consist of the entirety of national requirements of petroleum products and coal. Imports of the former exceed 150,000 tonnes per annum. In addition around 50,000 tonnes per annum of coal is imported from coal fields in Mozambique.

Malaysia

The energy situation of Malaysia has improved radically since the mid 1970s with increasing production of oil and the establishment of an international LNG export trade starting from 1983. Malaysia is already exporting a large part of its energy production and by the mid 1980s exports should exceed domestic

KEY ENERGY INDICATORS

Energy Consumption	
—million tonnes oil equivalent	8.1
Consumption Per Head	
—tonnes oil equivalent	0.60
—percentage of world average	37%
Net Energy Imports	nil
Oil Import-Dependence	nil

consumption. Energy consumption per head of population is currently only 37 per cent of the world average and will continue to rise, but greater emphasis is to be placed on greater utilisation of natural gas and more exploitation of hydro-electric potential in order to reduce over-reliance on oil products.

ENERGY MARKET TRENDS

Trend of Energy Consumption

Energy Consumption
+7.8% p.a.

Gross Domestic Product
+8.5% p.a.

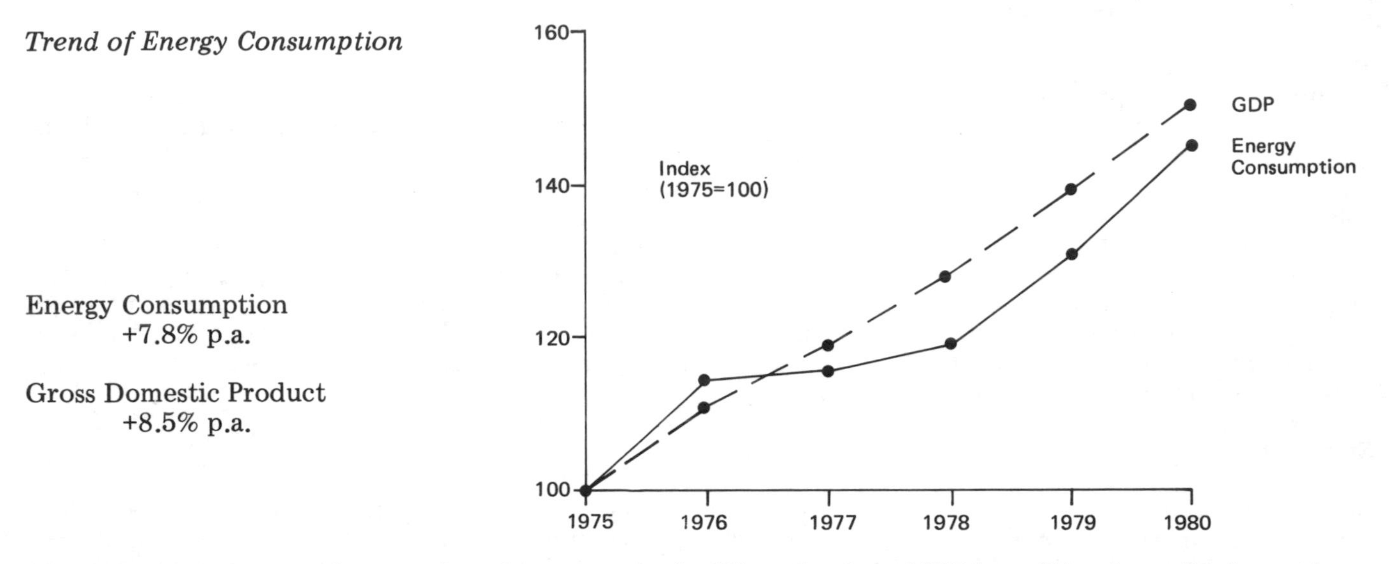

The Malaysian economy has continued to grow at a healthy rate since 1975 benefiting from the increasing availability of oil. This has ensured an adequate supply of energy without significant repercussions on the country's balance of payments. Between 1975 and 1980 Gross Domestic Product increased by more than 50 per cent in real terms, whereas energy consumption is estimated to have risen by only 45 per cent. So far indigenous energy resources have been used to earn revenue for the economy rather than fuel energy-intensive industrial developments.

Pattern of Energy Supply

Primary Fuel Supply (percentage)	
Oil	95
Solid Fuel	..
Natural Gas	1
Primary Electricity	4
TOTAL	100

The Malaysian economy has developed a particularly heavy reliance of oil, which at present accounts for as much as 95 per cent of commercial fuel supplies. The only other source of any significance has been hydro-electricity. Very limited use is made of commercial solid fuel, although firewood and other non-commercial fuels are used.

The pattern of fuel supply is expected to change significantly during the 1980s. A much increased contribution is expected from natural gas which is to be used as a main fuel in new thermal generating plant and also as feedstock in fertiliser production. It is also likely to be a preferred fuel for some other major energy-consuming projects in view of the high level of reserves which have been proved in Malaysian fields. The role of hydro-electricity will also increase, though probably only marginally.

ENERGY SUPPLY INDUSTRIES

Since the establishment in 1975 of a state corporation with wide responsibilities for the supply of oil and gas, the government has become an increasingly dominant factor in the energy supply situation. The existence of the state company coincided with the increase in oil production which turned Malaysia into a net exporter of oil during the late 1970s and an increasing level of interest in the prospects for further discoveries. Refining and marketing remain largely the responsibility of affiliates of international oil companies, although the state company is likely to assume the role of an integrated company. Electricity production and supply are undertaken by three public authorities, each operating systems of varying sophistication on Peninsular Malaysia, Sarawak and Sabah.

Oil

Oil Production and Consumption

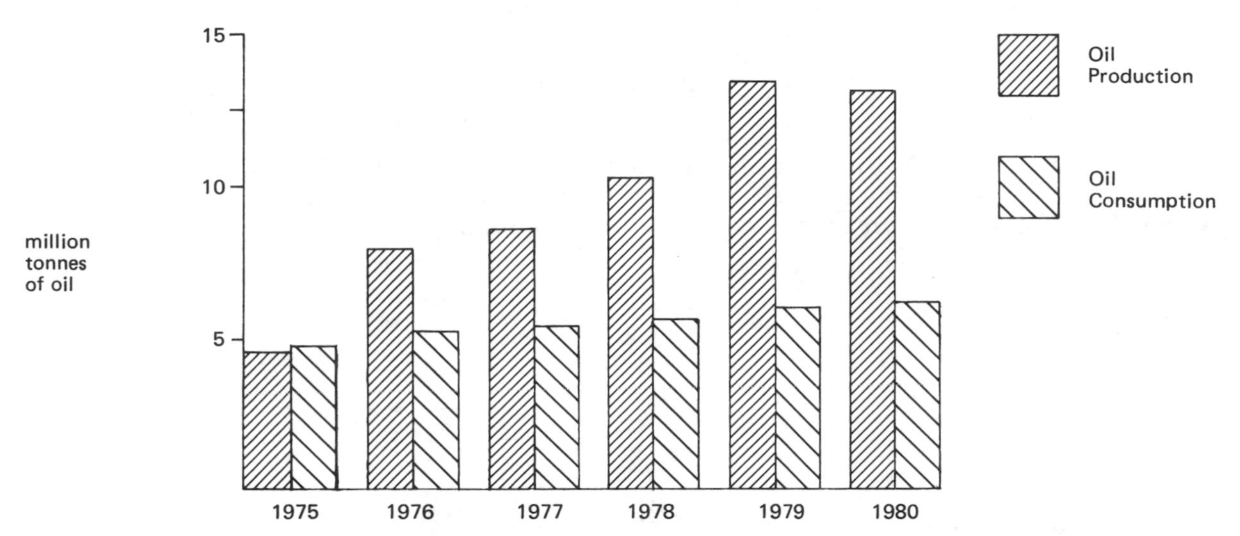

Since the mid 1970s Malaysia has become a net exporter of oil. Production is at present mainly from Sarawak and Sabah, but a significant output from offshore fields near to Peninsular Malaysia is now being achieved. Exploration work continues throughout Malaysia. Consumption of oil is concentrated in Peninsular Malaysia, where the majority of the population are to be found, and it is here that the main refineries have been built. Largest is that of Shell, which has a fully owned refinery at Port Dickson with a crude throughput capacity of around 4.5 million tonnes per annum. The local affiliate of the Exxon group also has a refinery at Port Dickson, the capacity of which is two million tonnes per annum. Shell has a small refinery at Lutong, Sarawak, processing local crude oil for meeting market requirements in Sarawak and Sabah, but also exporting some products to Peninsular Malaysia.

In 1975 a state oil company, Petronas, was established, principally to handle the state's participation in oil and gas production and to control exploration and development. Petronas is also engaged in distribution and marketing of finished products, but this sector is dominated by the affiliates of the international oil groups, British Petroleum, Exxon, Mobil and Shell. Shell itself holds around 40 per cent of the market and also supplies products to Mobil and Petronas. BP has refinery capacity at Singapore, which is the main source for additional products needed to meet market demand. Malaysia's current level of refining capacity is insufficient to meet total domestic consumption and Petronas is likely to initiate the construction of additional capacity.

Gas

At present gas is produced only in Sarawak and at a limited level of total output, reaching a maximum of 108 million cubic metres in 1980, equivalent to around 100,000 tonnes of oil. Output is consumed locally for electricity generation by the Sarawak Electricity Supply Corporation. Gas has also been discovered off the east coast of Trengannu in Peninsular Malaysia and this is to be exploited to provide fuel for electricity generation. The full extent of Malaysia's gas reserves is still uncertain, but is estimated to exceed 30 million million cubic feet.

Natural gas production is set to increase sharply from 1983 with the coming into operation of a large scale liquefaction plant exploiting a major gas discovery made by Shell. The gas is being acquired by Petronas, the state oil and gas company, which is main participant in the liquefaction plant, along with Japanese

Gas Production

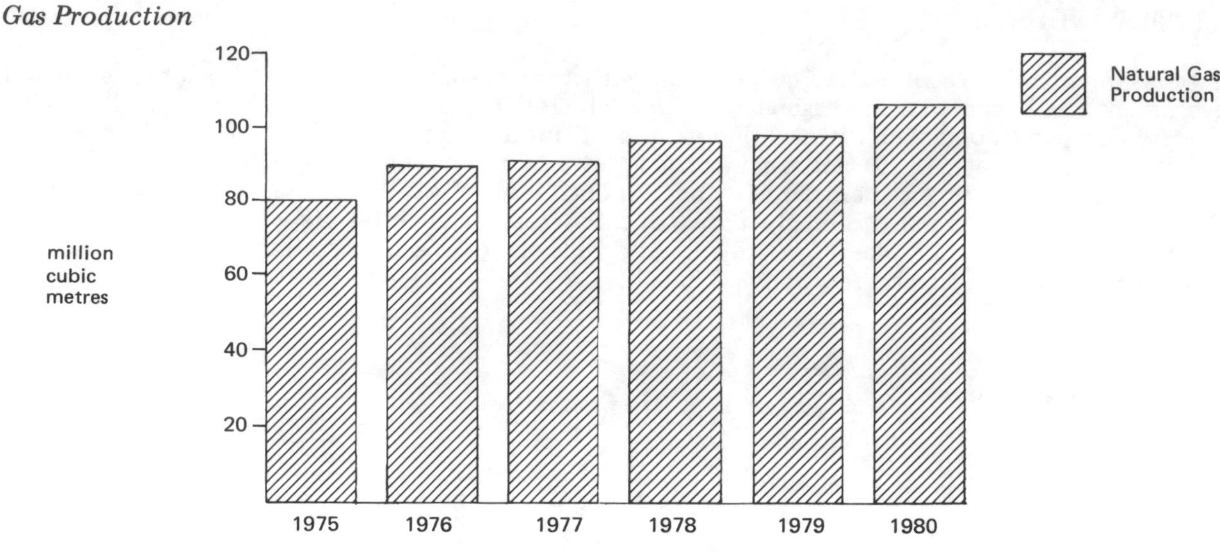

utilities which will import the LNG into Japan. The liquefaction plant is expected to handle up to 1,000 million cubic feet per day, exporting some six million tonnes of LNG per annum. The gas resource will also be used to provide feedstock for a urea plant to be located at Bintulu.

Electricity

Electricity Production

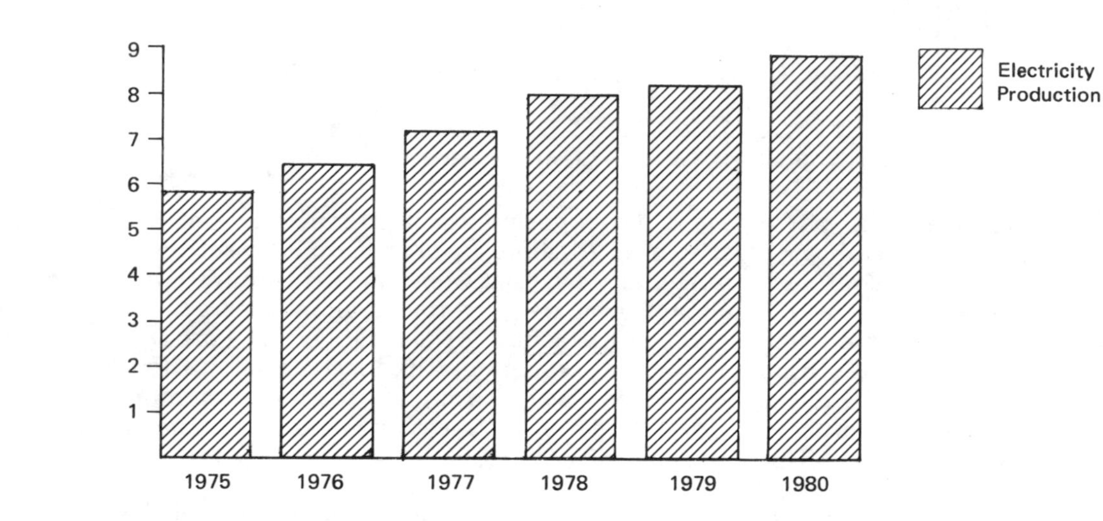

Demand for electricity rose at almost 13 per cent per annum during the 1970s, as economic activity rose and industrialisation continued. Between 1975 and 1980 consumption increased by more than 50 per cent. Consumption is predominantly within Peninsular Malaysia, which includes most of the population and economic activity.

Responsibility for generation, transmission and distribution of electricity lies almost entirely in the hands of three state undertakings, covering Peninsular Malaysia, Sarawak and Sabah respectively. In Peninsular Malaysia the National Electricity Board operates a national grid based on more than 2,100 MW of generating capacity. The Perak River Hydro-electric Power Company has a further 94 MW. NEB has a large hydro-electric power station at Temengor, but 66 per cent of the Board's capacity is oil-fired and this plant is responsible for an even higher proportion of electricity production. The NEB's main stations at Port Dickson, Prai and Temengor have a total capacity of 1,210 MW.

Both the Sarawak Electricity Supply Corporation (SESCO) and the Sabah Electricity Board (SEB) operate systems supplying some 3-4 per cent of total Malaysian electricity supplies, of the order of 300 GWH per annum. SESCO and the SEB supply electricity from numerous small diesel driven sets for localised distribution. Largest power station is SESCO's expanded station at Kuching, with 77 MW, representing 55 per cent of SESCO's total capacity. Both SESCO and the SEB are developing hydro-electric projects. The SEB has a 44 MW plant under construction on the Padas River. SESCO plans a 92 MW scheme at Kuching for commissioning in the mid 1980s.

ENERGY TRADE

Net Imports/(Exports) 1975-80

	1975	1976	1977	1978	1979	1980
Coal (million tonnes)	..	0.1	0.1	0.1	0.1	
Crude Oil (million tonnes)	(0.4)	(3.2)	(3.6)	(5.0)	(7.6)	(7.3)
Oil Products (million tonnes)	0.9	0.8	0.9	0.9	1.3	

Since 1976 Malaysia has achieved a substantial level of net exports of crude oil, rising to 7.6 million tonnes in 1979. The level fell back slightly in 1980 as a result of rising internal demand and slack conditions in international oil markets. In the last few years exports of oil have been partly offset by the need to make up deficiencies in the supply of some finished products, which has prompted Petronas to embark on an expansion of refinery capacity.

While the level of net oil exports may not rise significantly until major new discoveries are made, Malaysia will add greatly to its energy exports from 1983 when the Bintulu LNG plant is commissioned. Exports from Bintulu are planned to rise to six million tonnes per annum, equivalent to around six and a half million tonnes of oil.

Destination of Crude Oil Exports

(percentage)	
Japan	43
USA	28
Singapore	20
Other	9
TOTAL	100

Exports of Malaysian crude oil have been going largely to three particular destinations. The main market is Japan, for which Malaysian crude oil is particularly suited. The second largest market is the United States, for which Malaysia is one of the most economic sources. Some 20 per cent of exports is fed into the export-orientated refineries at Singapore, whence products are supplied to numerous countries in the Far East and Australasia. Smaller quantities of crude oil are shipped to other local deficit markets such as Thailand and the Philippines.

ENERGY POLICIES

The state oil and gas undertaking, Petronas, is set to occupy an ever increasing role in energy supply to the Malaysian economy. It will be the primary vehicle of energy policy under the Fourth Malaysian Plan for the period 1981-85, and the subsequent economic development plan. Key points of policy are to implement a depletion policy in respect of oil, combined with a much fuller utilisation in the domestic market of the huge reserves of gas now believed to be present.

Originally the 1981-85 plan envisaged a continuing rise in oil output. But this may not be justified on the basis of known reserves currently and Petronas is seeking to encourage continuing exploration for oil. To the present time natural gas, on the other hand, has not been much utilised. This will now be used, not only as a revenue earner via the Bintulu LNG project, but also as fuel and feedstock for major energy consuming plants. First such projects are expected to be a 450 MW power station in Trengannu and a fertiliser plant at Bintulu.

Mali

KEY ENERGY INDICATORS

Energy Consumption	
—million tonnes oil equivalent	0.13
Consumption Per Head	
—tonnes oil equivalent	0.02
—percentage of world average	1%
Net Energy Imports	90%
Oil Import-Dependence	100%

Mali is at a very early stage of development with few natural resources and only limited hydro-electric potential as an indigenous source of energy. The country is land-locked and dependent on the transportation of oil products from refineries in Senegal and the Ivory Coast. Energy consumption per head of population is only one per cent of the world average.

ENERGY MARKET TRENDS

The Mali economy has been developing only slowly, with an increase in real Gross National Product per head of population of less than two per cent per annum. The lack of indigenous energy is one of the handicaps to economic development. Energy demands are relatively simple. Oil meets commercial, transportation and other fuel markets. Electricity from hydro-electric plant contributes around half of electricity supplies, equivalent to 10 per cent of total energy consumption. Consumption of electricity has grown slowly, reflecting the pace of economic development.

ENERGY SUPPLY INDUSTRIES

Private sector companies are closely involved in the supply of energy, and the state plays only a small part. Oil products are brought in through the supply networks of international oil companies. The state is involved in the provision of electricity, in which it receives technical and managerial advice from the French state electricity authority.

Oil

Consumption of oil in Mali is of the order of 100,000 tonnes per annum of all products, meeting demands for all purposes. Supplies are entirely of finished products which are brought in from refineries in neighbouring countries. Several international oil companies are involved, including British Petroleum, CFP, Mobil, Shell and Texaco, using their availability of products from the refineries at Abidjan and Dakar.

Electricity

Annual consumption of electricity is slightly over 100 GWh, but is rising at less than two per cent per annum. Supply of electricity is by Société Energie du Mali, which is publicly controlled, and accounts for the major part of production. The company's main power station is at Bamako, but a new station is under construction at Manantali as a joint project with Senegal.

Of the existing 42 MW of capacity only 6 MW is hydro-electric, but this accounts for almost half of production and oil-fired plant is kept largely for standby purposes. The role of hydro-electricity will increase with the commissioning of the Manantali project, which is of 45 MW.

Mauritania

KEY ENERGY INDICATORS

Energy Consumption	
—million tonnes oil equivalent	0.2
Consumption Per Head	
—tonnes oil equivalent	0.12
—percentage of world average	7%
Net Energy Imports	100%
Oil Import-Dependence	100%

Mauritania is a large and sparsely populated country, with a population of little more than one and a half million throughout its 400,000 square miles. Consumption of energy is very limited and even including the important mineral operations amounts to little more than seven per cent of the world average. There are no indigenous resources of commercial energy forms so far identified, although exploration is taking place.

ENERGY MARKET TRENDS

Use of energy in the Mauritanian economy is dominated by the operations of the iron ore and copper mining industries. Mauritania is a leading producer of iron ore, with peak output of nearly 12 million tonnes in 1974. A new iron ore project is under way for completion during the mid 1980s. Virtually all mining output is exported, necessitating large scale transport facilities. Industrial activity independent of the mining operations is insignificant. Consumption of energy has fluctuated correspondingly.

ENERGY SUPPLY INDUSTRIES

Supply of energy is fragmented and geared to the mining operations of COMINOR, the iron ore producing company and other mining operators. A state-controlled company supplies electricity in the capital and a few other urban centres.

Oil

The Mauritanian economy is dependent on oil for all purposes. Upwards of 200,000 tonnes per annum is imported to fuel power plant, maintain transport systems and meet other commercial and domestic demands as exist, plus marine and aviation bunkers. The supply of oil products is undertaken by several of the international oil companies.

Electricity

Annual consumption of electricity is of the order of 100 GWh per annum, of which a substantial proportion is used in the mining extraction and processing industries. Generating capacity totals 53 MW, which is operated by a state-controlled company SONELEC, the Société Nationale d'Eau et d'Electricité and by COMINOR. COMINOR has a generating capacity of over 30 MW at Nouadhibou itself. SONELEC operates remaining plant, principally at Nouakchott and is planning additional capacity for Nouakchott and Nouadhibou.

Morocco

KEY ENERGY INDICATORS

Energy Consumption	
—million tonnes oil equivalent	4.3
Consumption Per Head	
—tonnes oil equivalent	0.22
—percentage of world average	14%
Net Energy Imports	79%
Oil Import-Dependence	99%

Morocco has a variety of energy resources, including coal, lignite, oil shale and, prospectively, uranium as well as hydro-electric potential deriving from the Atlas mountains. But commercial deposits of oil and gas are negligible, and the country is highly import-dependent. Per head of population energy consumption is only 14 per cent of the world average.

ENERGY MARKET TRENDS

Progress in the economy has been maintained despite the impact of higher oil prices and Morocco's dependence on imported supplies. Between 1975 and 1980 there was a steady increase in Gross Domestic Product averaging six per cent per annum. On the other hand the rise in energy consumption was contained at slightly less than this rate. Emphasis has been on developing a range of manufacturing industries, but this has generally avoided energy-intensive activities. Energy consumption will, however, increase if projects to extract oil from shale and construct a steel-making plant are implemented.

Pattern of Energy Supply

Primary Fuel Supply (percentage)	
Oil	79
Solid Fuel	11
Natural Gas	1
Primary Electricity	9
TOTAL	100

Oil is the principal form of commercial energy being used in Morocco. Other forms of energy contribute 20 per cent of supply in total. The proportion seems unlikely to increase as no significant oil or natural gas resources have yet been found and extraction of oil from shale, or the use of shale directly as a fuel in electricity generation, will build up only slowly. Oil shale does, however, present the best opportunity for Morocco to reduce the level of energy imports.

ENERGY SUPPLY INDUSTRIES

Morocco has a relatively complex energy supply system incorporating all of the main commercial fuel forms and with the possibility of introducing unconventional energy resources. In all of these activities state organisations play a key role, although involvement of foreign companies is encouraged in exploration and development.

Oil

Consumption of oil is around 3.5 million tonnes per annum. Fuel oil accounts for around half of total consumption, as it is a major fuel in electricity generation. Gas oil demand is also very high. Products are supplied from the refineries at Mohammedia and Sidi Kacem, which have a total throughput capacity of more then 4.5 million tonnes.

The oil industry is largely controlled by the state through agencies of the Ministry of Energy. The Mohammedia refinery has majority state participation and a national oil distribution company has been established. The state agency ONAREP is partnered by Occidental Petroleum Corporation in assessing the feasibility of extracting oil from shale reserves at Timhadit. Oil shale reserves, which are generally of high oil content, are estimated to be 10 billion tonnes.

Coal

Consumption of coal is running at around 700,000 tonnes per annum Production is of anthracite, and is consumed mainly in the Jerada electricity generating station. The state agency BRPM has discovered reserves of lignite in the Oued Nja area, which have been estimated at over 40 million tonnes.

Electricity

Gross production of electricity has been rising at a rate of around 10 per cent per annum on average, since 1975, reaching a total of 4.7 TWh in 1980. Production is based on a range of fuels, including oil, gas and coal as well as a substantial element of hydro-electricity. The contribution from hydro-electric plant fluctuates, depending on water resources, but is on average around 45 per cent, from numerous locations in the northern part of the country. At Jerada the power station burns over half a million tonnes per annum of coal.

The generation and distribution of electricity is undertaken by the state organisation ONEL, which operates 85 per cent of installed capacity. ONEL operates the national grid and retail distribution network, which extends to over 400,000 consumers. To meet growing demands for electricity and cut down on the use of imported fuels ONEL is planning to construct four or five 250 MW plants which would burn oil shale directly. Also under consideration is a nuclear plant, based on use of indigenous uranium which could be extracted from phosphoric acid.

ENERGY TRADE

Almost all of crude oil run in Moroccan refineries is imported. Total imports of crude oil in 1980 were four million tonnes, to meet internal demand and the bunker trade. There are both imports and exports of refined products. In addition small quantities of coal are exported.

Mozambique

KEY ENERGY INDICATORS

Energy Consumption	
—million tonnes oil equivalent	0.9
Consumption Per Head	
—tonnes oil equivalent	0.09
—percentage of world average	6%
Net Energy Imports	nil
Oil Import-Dependence	100%

The Mozambique government is struggling to revitalise an economy which has experienced many difficulties since independence. There is undoubted potential in terms of mineral, energy and other resources, but the development process is slow. Consumption of commercial energy per head of population is only six per cent of the world average. Massive amounts of hydro-electric energy are actually and potentially available and known coal resources are very large, but no commercial deposits of oil, and only a limited amount of natural gas, have been found.

ENERGY MARKET TRENDS

The Mozambique economy is only slowly developing and restoring stable activities following periods of internal unrest, involvement in guerrilla activity in Zimbabwe and attempts to restructure fundamentally the organisation of the economy. Economic development is being generally guided within a long term plan, which involves the exploitation of mineral resources, adding value by processing and establishing a range of manufacturing activities. In recent years recovery and development have led to a rising trend in energy consumption, although the rate has averaged less than two per cent per annum.

Pattern of Energy Supply

Primary Fuel Supply (percentage)	
Oil	45
Solid Fuel	37
Natural Gas	–
Primary Electricity	18
TOTAL	100

The availability of hydro-electric power from the Cabora Bassa dam and established coal mining activities in Tete Province enable Mozambique to meet more than half of total energy consumption from these indigenous resources. Consumption of oil is limited and is likely to be restrained in favour of indigenous fuels. Further exploitation of coal and hydro-electricity will tend to reduce the share of oil and exploitation of natural gas will introduce a new element into the energy balance.

ENERGY SUPPLY INDUSTRIES

Since the establishment of the marxist-socialist government after independence the activities of key sectors of the economy have been brought under state control. Public enterprises PETROMOC and CARBOMOC have full responsibility for the oil and coal industries respectively. Control of the important Cabora Bassa dam operation is being transferred to the state over a period of time.

Oil

Consumption of oil has fluctuated around an average level of 430,000 tonnes per annum. Finished products are obtained from the national refinery at Maputo. The refinery has a throughput capacity of over 800,000 tonnes per annum and is owned by PETROMOC, the state organisation responsible for importing crude oil, refining and disposal of finished products.

Coal

Coal production is in excess of 0.5 million tonnes per annum, having recovered from a level of only 370,000 tonnes in 1976. Production has been subject to disruption, mainly as a result of explosions in the mines. There is some movement of both exports and imports, which are in approximate balance. The main mining area is around Moatize, in Tete Province. Reserves are estimated to be at least 400 million tonnes and the state coal producing company CARBOMOC is planning to increase capacity to five million tonnes per annum during the 1980s in order to supply major new domestic development projects and to build up an export trade.

Gas

Prior to political upheavals in the mid 1970s, exploration by several international companies had established the presence of three gas fields, at Pande, Temane and Buzi. Exploration work has recommenced only recently and a priority will be to assess the extent of reserves in the Pande field, which had been put tentatively at the order of one million million cubic feet. Exploration is under the control of a state organisation, Empresa Nacional de Hidrocarbonetos, which was established in 1980.

Electricity

Mozambique is an important producer of hydro-electric power at the Cabora Bassa dam on the Zambezi River. The project came into operation in the mid 1970s in order to supply power to South Africa and

Electricity Production and Consumption

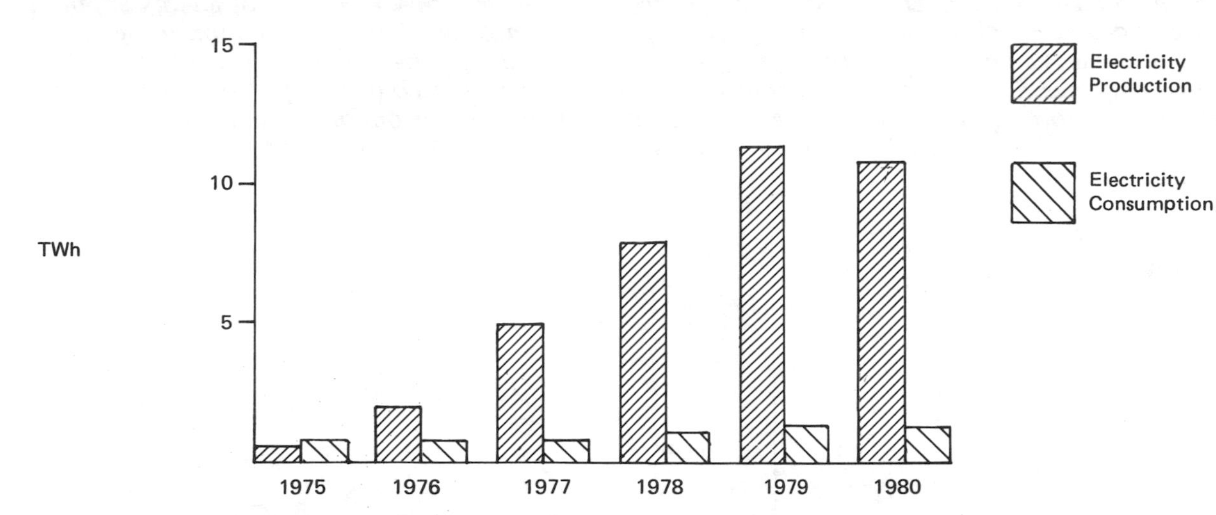

Mozambique. Approximately 90 per cent of output is exported to South Africa. Internal consumption of electricity in Mozambique is around 1,400 GWh of which 1,200 GWh is hydro-electric. Distribution of electricity is carried out by the public enterprise Electricidade de Moçambique, which takes power from Hidroelectrica de Cabora Bassa and operates some supplementary thermal power plant.

ENERGY TRADE

Net Imports/(Exports) 1975-80

	1975	1976	1977	1978	1979	1980
Crude Oil (million tonnes)	0.4	0.4	0.5	0.5	0.4	
Electricity (TWh)	0.2	(1.0)	(4.1)	(6.8)	(10.2)	(9.6)

Imports of crude oil are in the range of 0.4-0.5 million tonnes per annum, sufficient to meet all of the country's requirements. Since 1976 the major component of energy trade has been the transfer of electricity from the Cabora Bassa dam to South Africa. In 1979 this exceeded 10 TWh, but in 1980 the output was reduced owing to sabotage of the connector to South Africa. There is also some trade in coal, with imports offsetting the movement of coal to Malawi and other neigbouring countries.

New Zealand

KEY ENERGY INDICATORS

Energy Consumption	
—million tonnes oil equivalent	10.8
Consumption Per Head	
—tonnes oil equivalent	3.48
—percentage of world average	215%
Net Energy Imports	34%
Oil Import-Dependence	91%

New Zealand suffered considerably from the impact of higher oil prices during the 1970s as a result of its dependence on foreign sources. The use of coal had declined and the only firm basis of energy supply was the country's hydro-electricity production, which has enabled a high level of electrification to be achieved. North Island also has valuable geothermal energy resources producing heat and electrical power. Recent exploration activity has however brought about an improvement in the country's fortunes. Discovery and development of offshore gas and condensate has not only introduced an important new element into the energy supply equation, but also is taken to indicate that there may be further commercial fields to be discovered in the future.

ENERGY MARKET TRENDS

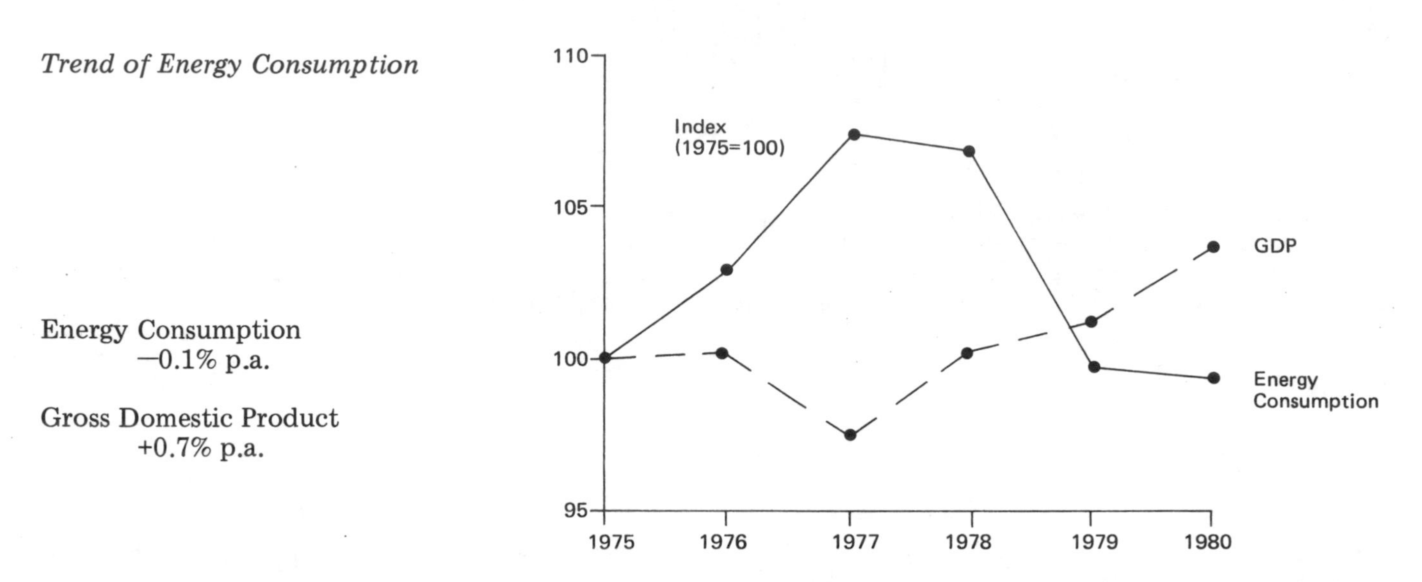

The economic problems of the late 1970s have meant that Gross Domestic Product has followed an uncertain path. Overall, between 1975 and 1980, GDP grew by only 3.7 per cent. Consumption of energy has been constrained by economic circumstances and by conservation measures. In fact primary energy consumption was lower in 1980 than in 1975, although this was largely as a result of a change in the pattern of fuel supply as use of oil, coal and gas for electricity generation was cut back. Consumption of energy by final consumers showed an increase similar to the rise in GDP.

Pattern of Energy Supply and Consumption

Primary Fuel Supply (percentage)		Final Consumption (percentage)	
Oil	37	Industry	37
Solid Fuel	11	Residential	13
Natural Gas	8	Transport	35
Primary Energy	44	Other	15
TOTAL	100	TOTAL	100

New Zealand is unusual in being an advanced economy in which primary energy plays a larger role than any of the fossil fuels. Thanks to the country's suitability for hydro-electric development and the existence of geothermal energy sources, the use of oil has been kept to below 40 per cent despite the limited role in recent years of coal, which exists in considerable quantities, and natural gas, although this has been exploited only relatively recently.

The intention is to make greater use of indigenous coal resources for electricity generation by NZ Electricity and industrial consumers in the future as well as continuing to exploit further hydro-electric and geothermal potential. At the same time increasing availability of gas and condensate will enable oil to be supplanted to some extent in all of its main existing markets. Current plans are that the share of gas will rise to more than 20 per cent in the 1990s and coal should then meet around 17 per cent of total primary energy requirements. The importance of imported oil will then be more than halved, to well under 20 per cent.

The share of final energy consumption attributable to industry is about average for the industrialised countries of the OECD, as energy intensive projects have not been encouraged, because of the country's dependence on imported oil. However, the availability of gas from the Maui field is to be exploited as fuel and feedstock in several major energy projects over the next decade. These include plants for producing methanol and ammonia/urea and use of gas liquids. LPG and CNG (compressed natural gas) are expected to make a significant contribution to meeting the need for transport fuels, which is particularly high, given the sparse distribution of population over much of the country and the distances between key cities.

ENERGY SUPPLY INDUSTRIES

The Ministry of Energy has a direct involvement in the production and supply of electricity and a major role in the production of coal. This role is expected to become more predominant as use of coal is expanded for power generation and, in the longer term, other transformation processes. A close surveillance is kept on the supply of oil products, although refining and distribution is left to private companies. The recently established state oil and gas company is pre-occupied with exploitation of gas from the Maui field and the further exploration for oil and gas, both onshore and offshore.

Oil

Oil Consumption

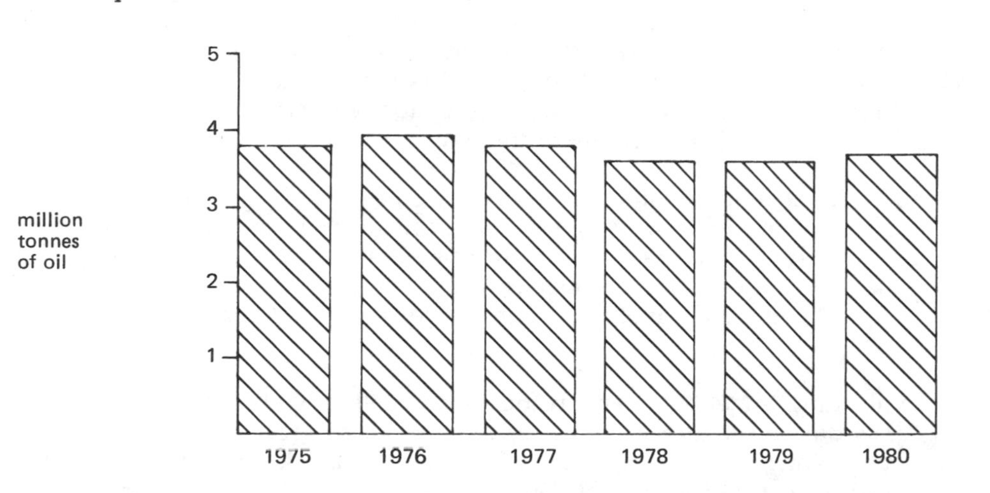

The supply of oil to the New Zealand economy lies largely in the hands of private sector companies. International oil companies have always played the major part in crude oil supply, oil refining and distribution of oil products. These companies, which include Shell, BP, Mobil and Caltex, the joint operating company of Texaco and Standard Oil Company of California, all hold interests in the industry refinery at Marsden Point, Whangarei, in North Island, approximately 70 miles from Auckland. The Marsden Point refinery has a capacity of only three million tonnes per annum, which is not sufficient to meet the country's full requirement for petroleum products, and one-third of the market is supplied by direct imports. An expansion of the refinery is being undertaken, which should not only increase the capacity of the refinery, but more importantly reduce the need to import higher value products.

The state maintains a close supervision of the activities of the oil industry, particularly through its control of market prices. But it has also come to have an increasingly direct role as a result of the establishment of the Petroleum Corporation of New Zealand (Petrocorp). Petrocorp's principal concerns in the oil sector are the management of the State's 50 per cent interest in the Maui field development, participation and promotion of oil exploration, particularly offshore, through Offshore Mining Company, and guiding the development of a new industry company Liquigas, which has been formed to distribute liquefied petroleum gas in bulk.

Crude oil for the refinery is predominantly supplied by imports from the Middle East and Far East countries. Since 1974, some indigenous oil has been contributing to the overall balance, from the Kapuni and Maui fields. The Kapuni field, in Taranaki Province, was first to be developed and this was joined in 1979 by output from the Maui field, which is about 50 miles offshore between North Island and South Island. Both fields produce condensate rather than normal full range crude oil and the Maui field is considered primarily as a gas field. Proven and probable reserves in these two fields are currently put at no more than some 12 million tonnes. However, a recent offshore crude oil discovery is expected to come onstream in 1982 and holds promise of further discoveries.

Coal

Coal Production

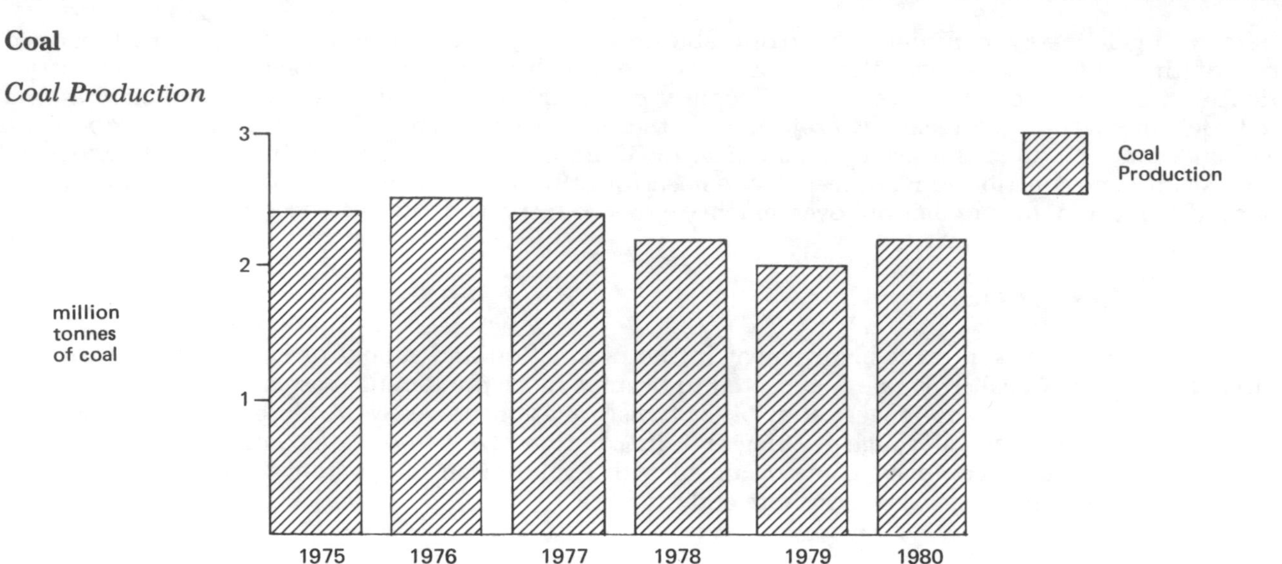

Indigenous sources of coal were for a long time a mainstay of the New Zealand energy economy. The country is well endowed with coal deposits. The most clearly defined category of 'measured' recoverable reserves is estimated at over 200 million tonnes, equivalent to around 100 years supply at current extraction rates. But total recoverable reserves could be an order of magnitude greater at well over 3,000 million tonnes, although two-thirds of this total is in the form of lignite. A more intensive programme of exploration is now under way to determine the country's reserves more precisely.

The state plays an important part in the coal-mining industry and state-controlled mines produce more than half of the country's output. In 1979-80 production by State Coal Mines was 1.3 million tonnes out of a total of 2.1 million tonnes, with 0.7 million tonnes coming from open-cast operations. On the other hand open-cast mining, which is inherently a lower cost operation is less significant for the privately owned mines. These mines are owned by a variety of companies, including some major industrial energy consumers, such as NZ Paper Mills.

As greater emphasis is placed on exploiting indigenous energy resources the role of state-run mines is expected to increase. To a large extent this is due to the increasing use of coal for electricity generation by state-run power stations and by the rising requirement for coking coal by NZ steel. Constructionof a major coal-fired power station is under way at Huntly, the first of four 250 MW units coming into service in 1980. It is also expected that eventually the Marsden Point power station will be converted from oil-firing to coal. As a result the proportion of coal mined by the private sector is expected to fall from 30 per cent currently towards 15 per cent by the mid 1990s.

Gas

Gas Production

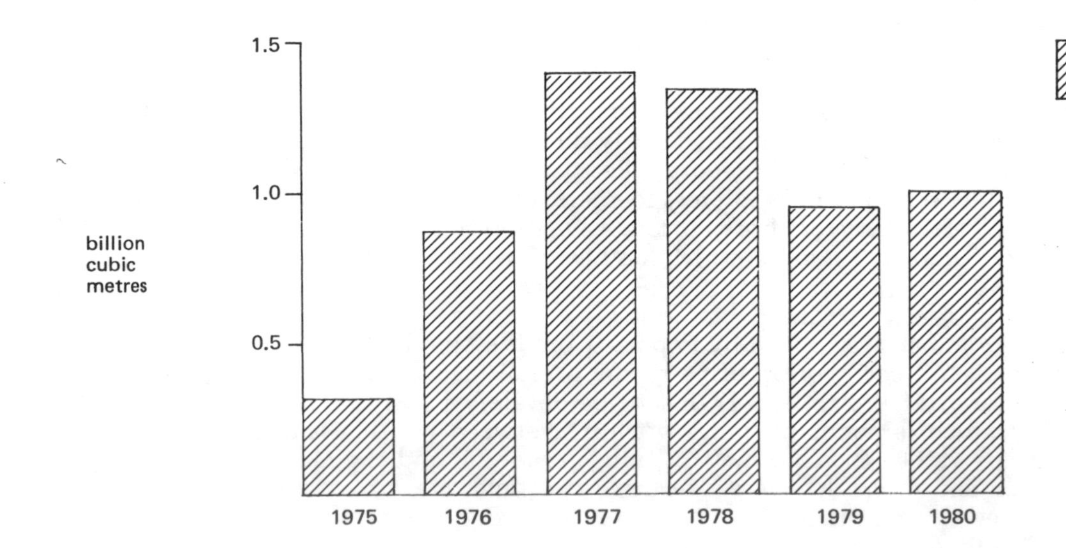

The situation of the gas industry, which had for many years been based on manufactured towns gas, was fundamentally changed as a result of the discovery of natural gas in association with condensate at Kapuni and subsequently offshore in the Maui field. Since 1970 natural gas has been supplied to an increasing number of local distribution undertakings in North Island as well as to some industrial consumers and to power stations at Stratford and New Plymouth. Control of the main transmission of natural gas lies with the Natural Gas Corporation, a subsidiary of the state oil and gas company Petrocorp. The National Gas Corporation operates the gas processing terminal at Kapuni and is also involved in the distribution and sale of liquid by-products in the form of liquefied petroleum gas and natural gasoline, which is used as a feedstock for towns gas by some municipalities.

The remaining gas supply undertakings, which are owned or controlled by municipalities under the supervision of the Ministry of Energy, continue to receive substantial financial support to offset the high cost of feedstocks. Subsidies are in the process of being phased out by 1984 and some undertakings may close down. Others, however, may undergo a revival on the basis of indigenous LPG, which will become available in increasing quantities as the Maui field is developed.

Indigenous gas resources are seen as an important element in the country's drive to reduce dependence on imported energy, and also as a valuable resource for upgrading and transformation to other products. Accordingly, proportions of the available natural gas from the Maui field are being allocated for use as compressed natural gas (CNG), as feedstock for synthetic fuels, methanol production and other possible petrochemical activities. Total gas reserves are at present estimated to be around 260,000 million cubic metres.

Electricity

Electricity Production

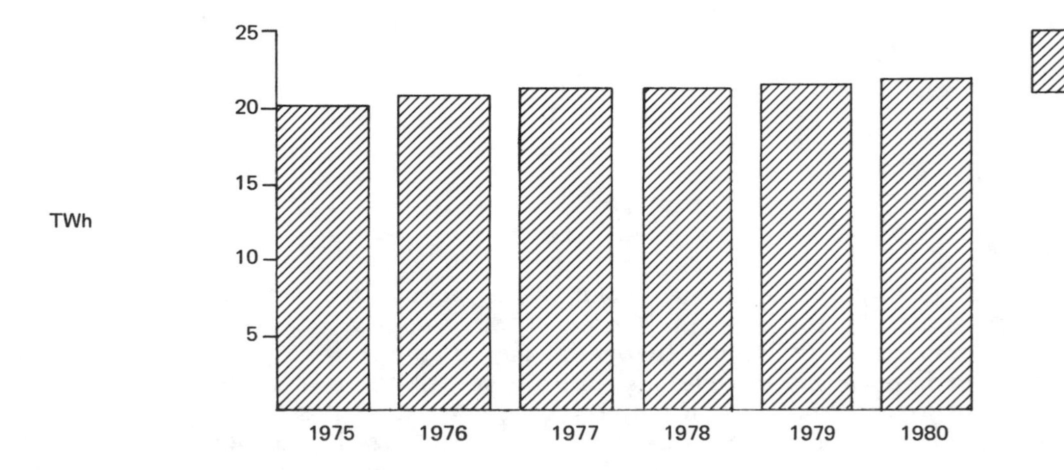

State agencies and undertakings are responsible for almost all generation and distribution in the public supply system. The system consists of electric power boards, which are in charge of designated electric power districts, and electric supply authorities, together with New Zealand Electricity, the operations side of the Electricity Division of the Ministry of Energy. New Zealand Electricity also acts as distributor for Southland Electric Power Supply to channel the bulk power which generally flows from South Island to North Island. But the 59 electric power boards and supply authorities have a generating capacity of only 146 MW out of the total of 5,623 MW in the public supply system. NZ Electricity thus accounts for almost 98 per cent of electricity produced for the public system.

Historically electricity has been produced to a large extent from natural energy. Even today around 75 per cent comes from hydro-electric schemes throughout the country, while a further five per cent has been derived from geothermal energy, including an element of natural steam drive. There remains considerable potential for continuing development of both hydro-electric and geothermal resources and new projects are under construction or in the country's energy plan. The current capacity of hydro-electric plant to produce 22 TWh per annum is expected to be increased by 50 per cent.

Thermal power stations have been making only a limited contribution. One power station has been supplied with fuel oil through a pipeline link to the oil refinery at Whangarei, but this is to be converted to using indigenous coal. A major thermal power station now under construction at Huntly is starting up using Maui gas, but is intended to be entirely coal-fired in due course.

ENERGY TRADE

Net Imports/(Exports) 1975-80

	1975	1976	1977	1978	1979	1980
Crude Oil (million tonnes)	3.0	3.0	2.7	2.4	2.6	2.6
Oil Products (million tonnes)	1.1	0.9	1.2	1.2	1.3	1.3

New Zealand became progressively dependent on imported energy during the post-war years, up until the mid 1970s. The many purposes to which oil could be put increased the country's dependence on overseas sources, while the coal industry was in a period of stagnation. But the result of the 1973-74 turnabout in international oil markets combined with the discovery of significant oil and gas reserves began to reverse the trend of the previous period. Thus the country's basic energy position improved at the same time as high energy prices and economic recession cut into the demand for energy.

Between 1975 and 1980 total oil imports stabilised at around four million tonnes per annum. But finished products have been claiming an increasing proportion, rising from one-quarter in 1975 to one-third in 1980, as crude oil runs at the refinery were cut back. This has had the effect of limiting the need to export surplus fuel oil, which has been the only product to be exported in any quantity. Other energy forms have not figured in the import/export balance, apart from small quantities of special coals, but the greater efforts to utilise the country's coal reserves is expected to show up in the form of a substantial level of exports. The objective is to try to establish markets for up to 500,000 tonnes annually.

Sources of Imported Crude Oil

(percentage)	
Saudi Arabia	44
Kuwait	17
Singapore	13
Indonesia	12
Qatar	5
Iran	3
Other	6
TOTAL	100

Imports of crude oil and feedstocks into the Marsden Point refinery come to a large extent from Saudi Arabia, and Middle East sources account for a high proportion of total imports. Kuwait, the second largest source of oil, accounts for a much smaller amount. Indonesia, the oil exporting country nearest to New Zealand, is a natural source for imports and provides a significant contribution to the total. Singapore is officially recorded as source for some 13 per cent of imports. To some extent this reflects its position as an entrepôt port, but mainly Singapore is a source of other partially processed feedstocks which enable the New Zealand refinery to provide a balance of finished products closer to the requirements of the market.

The New Zealand refinery is however unable to meet market requirements either in total or in relation to the market pattern and accordingly finished products are imported on a substantial scale. The large export orientated refineries of Singapore are the main source for these, accounting for just over 50 per cent in 1980. Australian refineries supply around 30 per cent of New Zealand imports, helping to fill the country's deficit in gasoline and diesel fuel.

ENERGY POLICIES

Government influence is exerted on all aspects of energy policy. Most energy prices are controlled directly or indirectly and state organisations are central to the supply of coal, gas and electricity. Active policies are being pursued to encourage exploration for, and evaluation of, indigenous onshore and offshore resources. Energy conservation and the optimal utilisation of energy resources are important features of the current long term energy plan, developed in 1980.

Energy Supply

The Ministry of Energy was formed only in 1978 and this move represented a decision to consolidate the State's involvement in the energy sector and to provide a framework for the development of energy policy.

The country's first overall long-term energy plan was formulated in 1980, in the context of rising world energy prices, a dynamic offshore situation in New Zealand and the need to focus attention on the full range of indigenous energy resources. The state-owned Petrocorp is the vehicle for development of the Maui field and a subsequent discovery, the McKee field, onshore North Island, will also be exploited. Further exploration is being encouraged in offshore areas.

Greater attention is now being focused on the other more familiar indigenous resources, hydro-electricity, geothermal energy and coal. Programmes are under way to evaluate reserves both as to usability and cost of development. In the case of geothermal energy the conflict with environmental policy is acknowledged. In the case of coal relatively little is known about some of the main deposits, such as those of bituminous coal in Taranaki, and the very large lignite resources in Southland and Otago. Also of importance is the need to assess the value of each coal type for its quality as steaming or coking coal. Coal will be favoured as a source of fuel for rising electricity requirements, particularly if it can be extracted cheaply, although policy is moving towards restricting the over-exploitation of open-cast availability in the short term.

Conservation

An important aspect of policy is that the general level of energy prices should not be too low, encouraging excessive demand for energy, and prices have to reflect the costs of supplying additional energy. The state has substantial control of energy prices through its control of many mines and of electricity generation. Private sector coal prices are not controlled but usually reflect prices for state coal. In the electricity sector some variation in distributors' tariffs exists, but in the case of natural gas the Natural Gas Corporation sets prices to end users. Price controls are most evident in the markets for oil products. Price levels for each product are set and these, together with varying levels of taxation, have been used to effect conservation and change the relative attractiveness of fuels.

At times of supply crisis the government has readily stepped in to curtail consumption of oil products. But it has also introduced a wide range of directives and incentives to encourage conservation, including mandatory insulation of new housing, taxes on energy intensive appliances, finance and tax incentives for investment in energy-saving plant and the encouragement of use of LPG in place of conventional transport fuels. In this respect a prominent role is being played by the Liquid Fuels Trust Board, which is supporting various schemes and studies for the application of natural gas, gas liquids, biomass, alcohol and coal-derived products as substitutes for petroleum.

Niger

KEY ENERGY INDICATORS

Energy Consumption	
—million tonnes oil equivalent	0.2
Consumption Per Head	
—tonnes oil equivalent	0.04
—percentage of world average	2%
Net Energy Imports	100%
Oil Import-Dependence	100%

Niger has a very low level of energy consumption per head of the population, only two per cent of the world average, and a significant proportion of that is attributable to uranium mining operations. Niger is fortunate in having valuable deposits of uranium and is in the process of developing coal and hydro-electric power. Evidence of oil has been shown in the eastern part of the country which borders on Chad, but no production is taking place yet.

ENERGY MARKET TRENDS

The Niger economy is still very underdeveloped and dependent to some extent on aid flows and preferential trade arrangements. It is mainly an agriculture-based economy, with manufacturing accounting for only 10 per cent of Gross Domestic Product. Of fundamental importance to the economy and to energy consumption is the exploitation of uranium deposits. Other minerals are also present and energy consumption is likely to develop in parallel with activity in this sector.

Indigenous production of commercial energy is virtually non-existent, other than for the energy content of the uranium. Hydro-electricity is now used, but only on import from Nigeria. But indigenous production of hydro-electricity will begin when the Dyodonga barrage is completed, and a coal-fired power station is under construction.

ENERGY SUPPLY INDUSTRIES

The state has become increasingly involved in energy supply matters as a result of the growing scale of uranium production, the country's main energy-consuming activity, of the importance of that industry in itself and of the potential of indigenous energy resources. State interests are now active in building hydro-electric plant, developing coal for power station fuel and investigating the prospects for oil or gas.

Oil

Consumption of oil has been rising rapidly. Between 1975 and 1980 it increased from 100,000 tonnes to 180,000 tonnes. Oil products are used to meet almost all of the energy needs of the country, including the generation of at least 50 per cent of electricity requirements. Niger has no production of oil and no oil refinery, so that all products must be transported from refineries in neighbouring countries, particularly Nigeria.

Electricity

Electricity demand has reflected the increasing activity of the uranium mining companies, which absorb 60-65 per cent of the total electricity demand of 150-160 GWh per annum. The companies themselves operate generating plant and in the case of SOMAIR the company undertakes to provide electricity for the town of Arlit. The state-controlled electricity company NIGELEC is concerned primarily with supplying electricity in the Niamey area, where a large proportion of manufacturing industry is located. NIGELEC operates some thermal plant, but currently relies on importation of electricity from the Kaindji dam power station in Nigeria.

NIGELEC will eventually have additional hydro-electric capacity of its own when construction of the Dyodonga barrage, 30 km from the confluence with the Niger River, is completed. An additional coal-fired power station is to be built at Anou Araren, to utilise coal deposits and generate power for the uranium mines, replacing oil fired plant.

Uranium

Niger is an important uranium producing country with mines in operation at Arlit and Akouta. The Arlit mine, operated by SOMAIR, includes French, Italian and German interests representing purchasers or contractors for uranium output. The Akouta operation involves French, Spanish and Japanese interests. In both mines the state has a large interest, through ONAREM in the case of SOMAIR and through Uraniger in the case of the Akouta mine. Both mines are capable of producing 1,500-2,000 tonnes per annum of uranium oxide.

ENERGY TRADE

At present Niger imports all of its energy, in the form of either oil products or hydro-electricity. In value terms, however, this is more than offset by export flows of uranium. The import bill will be significantly reduced by the start up of hydro-electric and coal-fired plant within the country.

Nigeria

KEY ENERGY INDICATORS

Energy Consumption	
—million tonnes oil equivalent	10.9
Consumption Per Head	
—tonnes oil equivalent	0.13
—percentage of world average	8%
Net Energy Imports	nil
Oil Import-Dependence	nil

Nigeria is the most populous country in Africa with a population now exceeding 80 million. It is fortunate in being one of the world's key oil-producing countries, so that there is little constraint on the availability of energy to the economy and export revenues can provide the basis for economic growth in all sectors. Currently, the level of energy consumption per head of population is only eight per cent of the world average, but is rising at a rapid rate. Accordingly, part of the country's revenues from energy exports will be used in energy-related projects, such as utilisation of natural gas, construction of hydro-electric dams and power stations and a possible liquefaction terminal and related facilities to enable gas to be exported.

ENERGY MARKET TRENDS

The ready availability of oil at relatively low prices has relieved the Nigerian economy of one of the key constraints on many other countries. Between 1975 and 1980 total consumption of energy increased by more than 80 per cent, allowing expansion in the use of energy in all sectors. The relationship of increases in energy consumption to growth of the economy has been erratic and at least in the period 1975-78 was not dissimilar to the rate of growth in the economy. But in future years the rate of consumption of energy will be pushed up further by the development of energy-intensive projects as well as by the natural population increase of the country.

Pattern of Energy Supply

Primary Fuel Supply (percentage)	
Oil	77
Solid Fuel	1
Natural Gas	12
Primary Electricity	10
TOTAL	100

Oil has come to assume as important a role in the pattern of fuel supply in Nigeria as it has in many of the developed industrial oil-importing countries. The rapidly expanding economy has been absorbing increasing quantities of oil such that it now accounts for more than three-quarters of all primary energy input. Natural gas, which is theoretically available in plenty, meets 12 per cent of requirements but it has not so far been practicable to harness more than a small proportion of the energy available from this source. Primary energy, in the form of hydro-electricity provides around 10 per cent. Of the commercial forms of fuel, coal is of very small significance, although there is in addition an unassessed level of consumption of wood and other non-commercial fuels.

ENERGY SUPPLY INDUSTRIES

State-owned undertakings play the leading part in the supply of all forms of energy. The Nigerian National Petroleum Corporation (NNPC) has the major interest in crude oil production and sales and dominates refining, although private companies are still involved in much of the marketing of finished products.

Foreign companies are heavily involved in the search for new energy resources. Electricity supply and the small-scale coal industry are controlled by state corporations.

Oil

Oil Production

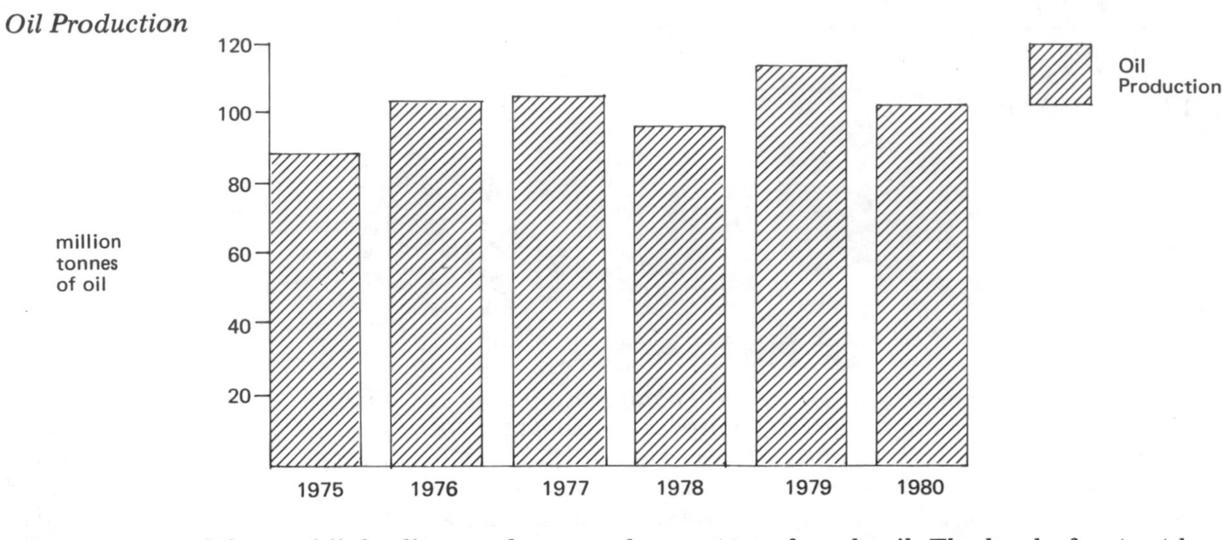

Nigeria is one of the world's leading producers and exporters of crude oil. The level of output has fluctuated considerably from year to year as exports have reflected the changing situation in the international oil market. After the cutbacks of 1974 production rose from 89 million tonnes in 1975 to nearly 115 million tonnes in 1979, dropping back to just over 100 million tonnes in 1980. By contrast internal consumption accounts for only a small proportion of total production, less than 10 per cent currently, though rising rapidly. The proportion was less than four per cent in 1975.

The state oil company Nigerian National Petroleum Corporation (NNPC) holds generally a 60 per cent share in production facilities with a corresponding interest in disposal of crude oil and refined products. However, as a result of the expropriation of the Nigerian assets of British Petroleum Company in 1979 NNPC interest in the former Shell-BP production areas rose to 80 per cent. Shell Nigeria currently is the largest oil producing company, accounting for more than half of the country's output. Other international oil companies are also prominent in production, particularly offshore. Gulf, Mobil, AGIP, Phillips and Elf-Aquitaine have fields in production producing between four and 15 million tonnes per annum. Other companies involved in production include Standard Oil Company of California, Texaco and Ashland.

To meet the rapidly rising demand for oil products oil refinery capacity is being increased substantially by NNPC. To the established refinery at Port Harcourt NNPC has added new refineries at Warri and Kaduna, both with capacities of five million tonnes per annum. The Warri refinery remains close to the oil producing areas of the south-east, although nearer than Port Harcourt to the largest area of consumption around Lagos. The Kaduna refinery will meet market requirements in the northern inland areas, including parts of neighbouring countries such as Niger and Benin.

The state has built up a scale of involvement in oil distribution and marketing comparable to its interests in production and refining. NNPC holds 60 per cent shares in most marketing companies, but the former affiliate of BP has been taken over and now operates as African Petroleum Ltd, with NNPC holding a majority interest. Shell and other international oil companies distribute the bulk of finished products.

Coal

Significant resources of coal have not so far been discovered in Nigeria, but some coal has been produced for many years. Principal mines, producing hard coal from deep mining operations, are located in the Enugu area of Anambra state in the south-eastern part of the country. Output was at a peak of 283,000 tonnes in 1976 but has been on a downward trend since that year, falling to only 155,000 tonnes in 1980. A resurgence of activity in the coal sector should result from current programmes of investment in the industry by the state undertaking, the Nigerian Coal Corporation.

Gas

Nigeria has substantial resources of natural gas, only a part of which has so far been evaluated. Production, at present of the order of 500 million cubic metres per annum, is mainly of associated gas, which is largely burnt off and wasted. Other fields, essentially of gas, are known to exist, but cannot be developed until home and overseas markets can justify it. Little gas is used within the country, consumption being confined to a few larger scale industrial users and the government proposes as part of its economic development programme to channel more gas to industrial plants and electricity generating stations rather than waste it.

The government wishes to avoid large-scale flaring of gas, but the prospects of achieving this, while still maintaining crude oil output levels, depends on bringing to fruition the project to liquefy the gas for exporting it to North America or Western Europe. This project would probably involve utilising some 2,000 million cubic feet per day but so far it has not proved possible to make firm arrangements either for project participation or for purchasing contracts.

Electricity

Electricity Production

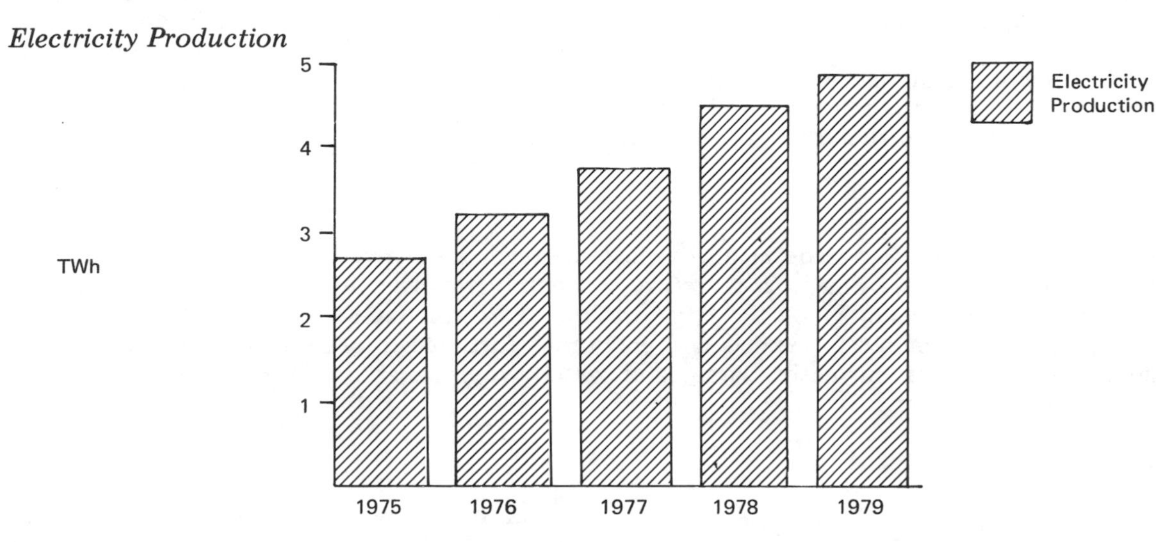

The demand for electricity has been rising very rapidly, increasing from approximately 2,700 GWh in 1975 to over 4,900 GWh in 1979. The rate of increase in installed capacity has been even greater. From a level of 811 MW in 1975 capacity was increased to 1,823 MW in 1979 and it is envisaged that a further 3,000 MW should be added by 1985.

Nigeria has considerable resources for hydro-electric power generation and has a large scheme in operation at Kaindji in the west of the country, on the Niger River. In 1979 hydro-electric plant still accounted for the major part of total electricity production but the proportion is steadily decreasing, despite projects for new schemes, at Jebba, also on the Niger River, and Shirora, which would give a further 1,100 MW. Thermal power stations totalling 2,200 MW are to be built, of which 1,500 MW is intended to utilise natural gas from oil fields in the delta region. In addition coal and oil fired stations are to be built.

The production of electricity and its transmission throughout the country is the responsibility of the state undertaking, the National Electric Power Authority (NEPA), although some industrial companies, such as African Timber and Plywood Company are also involved in operating thermal power stations. In addition to building power stations NEPA has a programme of construction of high voltage transmission lines.

Uranium

There is at present no production of uranium in Nigeria, but prospects have been considered sufficiently promising for a state company, Nigerian Uranium Mining Company (NUMCO), to be organised. NUMCO will have a 60 per cent interest in a joint venture exploration programme with the French state company Minatome. The NUMCO-Minatome programme is likely to concentrate initially on prospects in the Gombe area in the east of the country.

ENERGY TRADE

Net Imports/(Exports) 1975-80

	1975	1976	1977	1978	1979	1980
Crude Oil (million tonnes)	(85.5)	(95.9)	(100.2)	(81.6)	(108.5)	(98.4)

Nigeria's energy trade is overwhelmingly in terms of crude oil with exports now usually exceeding 100 million tonnes per annum. Destinations are primarily the United States, where the low sulphur quality of the oil is attractive, but also many of the oil-importing countries of Western Europe. Japan imports Nigerian oil to some extent as a result of its need to diversify the sources of its enormous import requirement. The exact pattern of trade however varies considerably from time to time as individual contracts with the state-owned NNPC are arranged. Small quantities of refined products and electricity from Kaindji are exported to neighbouring countries.

ENERGY POLICIES

The general stance of Nigerian policy towards the energy sector reflects the desire to maintain control of key industrial activities and a substantial level of Nigerianisation in the economy. State involvement is also felt through the budgeting of oil-derived revenues towards major development projects in energy supply and energy-intensive consuming industries. But, apart from the isolated instance of the nationalisation of BP's assets in 1979, private sector companies are looked to for a large part of the activity in exploration for oil, gas and minerals and in the distribution and marketing sectors.

Energy Supply

Although Nigeria has a massive export trade in oil the level of revenues derived from it is critical for the country's whole economy. The government is therefore intimately involved in controlling the price and volume of exports. Nigeria is a member of the Organisation of Petroleum Exporting Countries, but is aware of the overriding need to maintain exports and encourage further exploration. It is also concerned to utilise associated gas, which is currently flared, and which would enable domestic consumption to increase without diverting oil exports.

Energy Conservation

It is hardly surprising that energy conservation policies have not developed far in a country such as Nigeria, which is basically an energy supplier. Nevertheless, the pace at which internal demand is rising, combined with the forward development programme, which includes energy-intensive projects, has lead to an awareness of the need for efficiency in the use of energy and the blame attaching to low energy prices, which have in the past been held down to well below international market levels.

Oman

KEY ENERGY INDICATORS

Energy Consumption	
—million tonnes oil equivalent	1.2
Consumption Per Head	
—tonnes oil equivalent	0.8
—percentage of world average	49%
Net Energy Imports	nil
Oil Import-Dependence	nil

Oman is well endowed with oil and gas resources and has a small population of around 1.5 million. However, energy consumption per head is still only half the world average, reflecting the late movement of the country towards economic development. Energy consumption is rising very quickly and Oman is relatively free of the balance of payments problems now being faced by some of the more populous oil producers with their ambitious development programmes. Oman is not a member of either the Organisation of Petroleum Exporting Countries or the Organisation of Arab Petroleum Exporting Countries.

ENERGY MARKET TRENDS

The Omani economy is in a process of rapid development in all sectors. Revenues are being generated from oil exports to finance major programmes of economic and social infrastructure and investment in new industrial operations. Energy production operations themselves nevertheless account for a significant proportion of energy demand. This will increase with commissioning of the national oil refinery. The need for desalination plant will also increase energy demand and the new Sohar copper extraction and smelting complex will be energy-intensive. The rapid rise in energy consumption is therefore expected to continue.

Pattern of Energy Supply

Primary Fuel Supply (percentage)	
Oil	75
Solid Fuel	–
Natural Gas	25
Primary Electricity	–
TOTAL	100

Oil is the mainstay of the Oman economy, as it is abundantly available in relation to the level of national demand. But gas is also present, both associated with crude oil or in separate gas fields. Gas is being increasingly used for some major projects and accounts for a significant proportion of total energy supply. Energy consumption is dominated by these projects, which include electricity generation, desalination and mineral extraction and processing.

ENERGY SUPPLY INDUSTRIES

The state holds a central role in the energy supply industries. It has not yet established a state oil company as such. Main participation is through the 60 per cent shareholding in Petroleum Development Oman, but the state is financing directly the construction of a national oil refinery. Electricity production and supply is controlled and co-ordinated by the state, which is allocating funds for generating capacity and development of a grid.

Oil

Oil Production

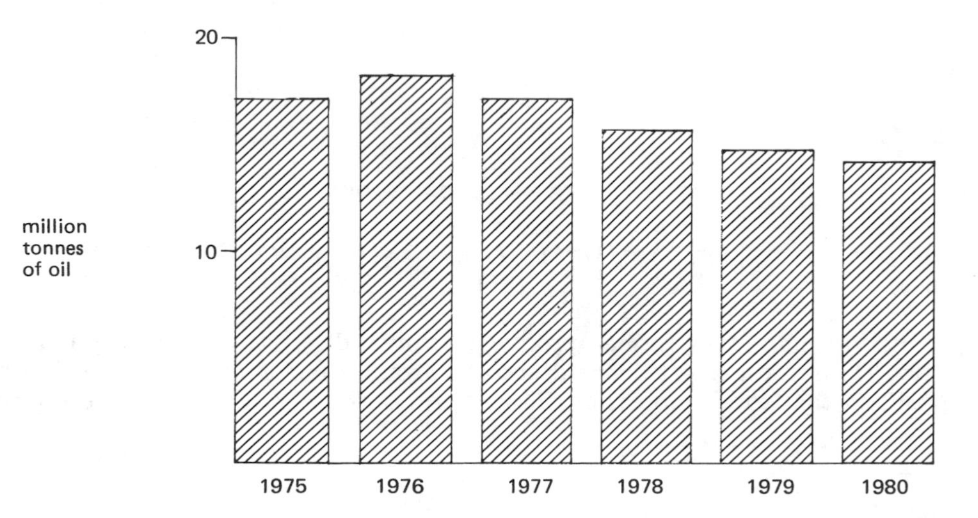

Oil production has declined in successive years since 1976, when it exceeded 18 million tonnes. This has been partly due to technical considerations as well as reflecting the general conditions of the international oil market. With the bringing on stream of new oil fields the production potential has been increased, so that output of well over 15 million tonnes per annum can be maintained.

The principal oil producing group is Petroleum Development Oman (PDO), in which the government has a 60 per cent interest. Operator for the group is Shell, with the additional participation of the French group CFP. PDO produces around 90 per cent of Oman crude oil. New fields developed by PDO in the southern part of the country have added capacity of 70,000 barrels per day (approximately 3.5 million tonnes per annum) to existing capacity. Other exploration groups are also active, including the French state company Elf-Aquitaine, Wintershall AG of West Germany and consortia containing a number of Japanese production, trading and consuming companies. AGIP, British Petroleum, Cluff Oil, Deminex, Gulf Oil and Hispanoil are also involved in exploration.

Internal consumption of oil products now exceeds half a million tonnes per annum. In the past products have been imported, but the government is building a 2.5 million tonne per annum refinery at Mina Al Fahal, which is also the export terminal for crude oil.

Gas

Oman's first natural gas project was completed in 1978, with the completion of a 330 km pipeline from Yibal to the Ghubrah desalination and electricity generating complex in the capital area. This represented a substantial shift towards gas from oil in the total energy picture. Gas will also be used in new power stations being built or planned as well as in other major mineral and industrial projects. These include the new copper smelting operations starting up in Sohar. Total proven recoverable reserves of gas have been increased considerably, so that by 1985 it is expected that gas consumption will rise to around 130 million cubic feet per day.

Electricity

Electricity demand has been rising very rapidly as a result of continuing mineral development programmes, commissioning of desalination plant, a build-up of manufacturing industry and general improvement in the level of electrification. Total production of electricity in 1980 was around 960 GWh compared with little more than 300 GWh in 1975.

The bulk of electricity is produced in government-owned power stations, of which the most important is at Ghubrah in the capital city area. In 1980 the Ghubrah station generated 611 GWh. The government-owned station at Salalah produced a further 122 GWh, with the remainder coming from several small plants. The leading oil and gas development group PDO generates its own electricity requirements, amounting to 193 GWh in 1980.

Generating capacity of the government power stations is 320 MW, of which 235 MW is at Ghubrah. PDO stations have a capacity of around 60 MW. Most power stations are fuelled with oil, but the Ghubrah power station has been converted to using gas. New power stations are to be built in the capital area, in the north at Sohar and between Sohar and the capital with a total capacity of 300 MW, mainly using gas as fuel. This will strengthen the establishment of a national grid.

ENERGY TRADE

Net Imports/(Exports) 1975-80

	1975	1976	1977	1978	1979	1980
Crude Oil (million tonnes)	(17.1)	(18.4)	(16.7)	(15.8)	(14.7)	(14.1)
Oil Products (million tonnes)	1.1	1.2	1.2	1.2	1.3	1.3

Oman exports all of its oil production, importing refined products to meet domestic market requirements pending completion of a refinery. Domestic market demand absorbs around half a million tonnes, but the market for bunkers is greater still, so that total imports of finished products are around 1.3 million tonnes per annum.

Destination of Crude Oil Exports

(percentage)	
Japan	51
Western Europe	25
Singapore	11
Other	13
TOTAL	100

Omani crude oil is exported to many countries, but Japan and Western Europe account for three-quarters of all exports. Japan is a particularly important export market, partly as a consequence of the substantial involvement of Japanese companies in consortia operating some of the oil fields. The only other destination of significance is Singapore, where Shell, which is a leading participant in Petroleum Development Oman, has a major international refinery. No other destination takes a significant percentage of exports.

Pakistan

KEY ENERGY INDICATORS

Energy Consumption	
—million tonnes oil equivalent	12.5
Consumption Per Head	
—tonnes oil equivalent	0.16
—percentage of world average	10%
Net Energy Imports	43%
Oil Import-Dependence	89%

The level of energy consumption in Pakistan remains very low by world standards, reflecting limited industrialisation to date and the inhibiting effect of dependence on foreign sources for most of the oil used. There exists a firm base of hydro-electric power generation, which continues to be expanded. But the main platform on which the economy is now developing is the additional natural gas reserves which have been discovered, together with more promising indications of oil. The amount of natural gas will make further inroads into oil demand in the future and may be sufficient for the government to contemplate exporting some to neighbouring countries.

ENERGY MARKET TRENDS

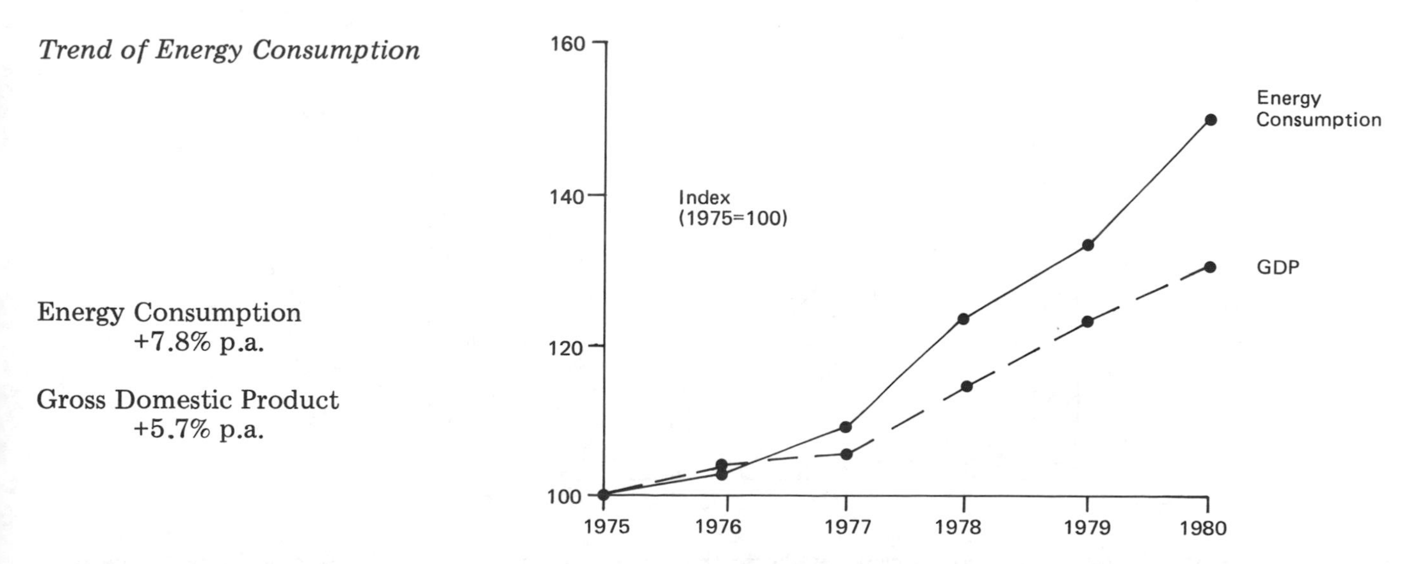

The economy has been growing at a reasonable rate during the last few years, by approximately 25 per cent between 1976-77 and 1979-80, after low rates of growth in the previous years. Overall during the five years to 1980 Gross Domestic Product increased by more than 30 per cent. However, after an increase of nearly 12 per cent in 1977-78 the rate of increase has been much lower subsequently. Consumption of energy rose at a significantly higher rate, by more than 45 per cent during the five-year period, and natural gas has been used in increasing quantities as a feedstock and as a bulk heat-raising fuel by certain industries.

Pattern of Energy Supply and Consumption

Primary Fuel Supply (percentage)	
Oil	37
Solid Fuel	6
Natural Gas	36
Primary Electricity	21
TOTAL	100

Consumption of Natural Gas (percentage)	
Electricity Generation	30
Fertiliser Manufacture	22
Other Industries	39
Domestic/Commercial	9
TOTAL	100

Oil, natural gas and primary electricity each make an important contribution to total energy supply. The position as between oil and natural gas is undergoing constant change as indigenous gas is further exploited, and in particular, substituted for oil. Until recently oil was the major energy carrier in the economy. Hydro-electric resources are also being exploited, although in an expanding economy their share is not likely to increase significantly.

Consumption of oil products is largely of kerosine and diesel oil for domestic sector and transport needs. Natural gas has assumed a key role in providing the fuel input for electricity generation, industrial energy needs and as feedstock for fertiliser manufacture. Experimental projects are under way to evaluate the scope for use of compressed natural gas as a substitute fuel for the transport sector. The small contribution from coal mining is almost entirely used in brick kilns.

ENERGY SUPPLY INDUSTRIES

Government controlled enterprises and agencies now play an important part in all energy supply industries. This includes the electricity supply industry, where the position of the two public agencies has been consolidated, in coal exploitation and in oil and gas exploration. Nevertheless, foreign-owned private companies are still important, usually in partnership with state companies, in oil and gas production, crude oil refining and distribution and marketing of petroleum products.

Oil

Oil Production and Consumption

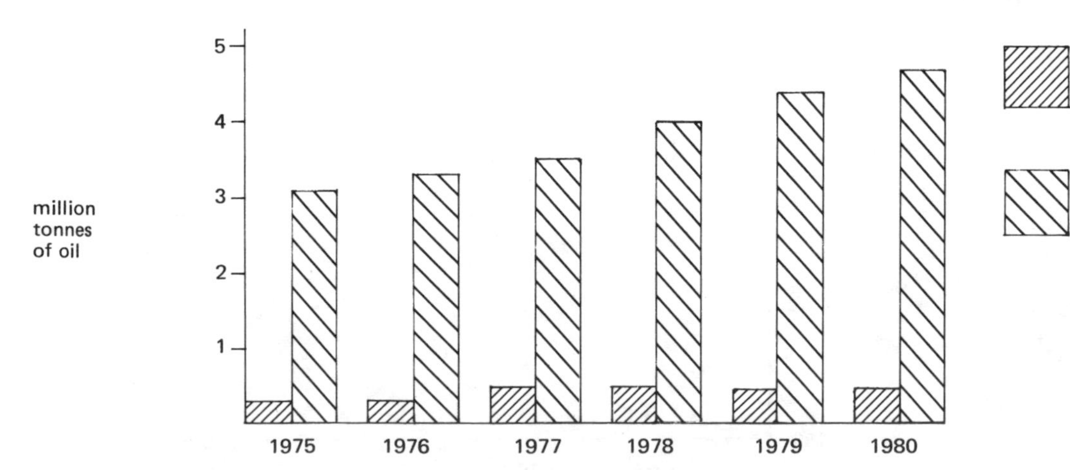

Consumption of oil has continued to grow apace in the last few years, despite the increasing use of natural gas. In 1979-80 consumption rose to 4.7 million tonnes, compared with only 3.1 million tonnes in 1974-75. Indigenous production accounts for only a small proportion of total needs and is static at the level of half a million tonnes per annum. Oil production is currently confined to the northern part of the country with the Meyal field accounting for a large part of total production.

There are three refineries. The two main refineries, at Karachi, are both government controlled and have a total capacity of around 4.5 million tonnes per annum. Attock Refinery is privately owned and is located in the oil fields of the north. This refinery's capacity is only 0.5 million tonnes per annum, but it may be expanded to 1.5 million tonnes during the current Five Year Plan. Marketing and distribution is carried out by both state and private companies. Pakistan State Oil is now the major marketing company, supplying approximately 60 per cent of the market. A further 25 per cent of the market is supplied by Pakistan Burmah-Shell in which Burmah Oil Company and the Royal Dutch/Shell group each hold a 24.5 per cent interest. A smaller share of the market is held by an affiliate of Caltex Petroleum. Pakistan State Oil is building a pipeline from Karachi to Mahmood Kot, in central Punjab, for transporting products in bulk to northern markets.

Exploration for oil is the responsibility of a state company, the Oil and Gas Development Corporation (OGDC), which is currently developing the Toot field to at least half a million tonnes per annum output. The expectation of the current Five Year Plan is that oil production can be increased to 1.7 million tonnes per annum by 1982-83. The OGDC has agreements with a number of international oil companies for further programmes of exploration, both onshore and offshore, including Burmah Oil, Shell, Amoco, Gulf, Occidental and Husky.

Coal

Coal production in 1979-80 amounted to 1.5 million tonnes, representing a steady increase since 1975-76, when output stood at only 1.1 million tonnes. This is small in relation to the country's known reserves, which exceed 440 million tonnes. Coal deposits are mainly in the Punjab, Sind and North West Frontier Provinces and at several locations in Baluchistan. Reserves in the Lakhra area of Sind are estimated to be at least 250 million tonnes.

But production capacity is old and inefficient and the quality of the coal is poor. Of total production 90 per cent is used in the brick kiln industry. Other consumers include the railways and small industries. The coal industry falls under the responsibility of the Pakistan Mineral Development Corporation, which is developing new capacity at Sharigh, Makerwal (Punjab), and Jhimpir-Meting (Sind). Output from Sharigh will be used entirely by Pakistan Steel Mills at Karachi.

Gas

Gas Production

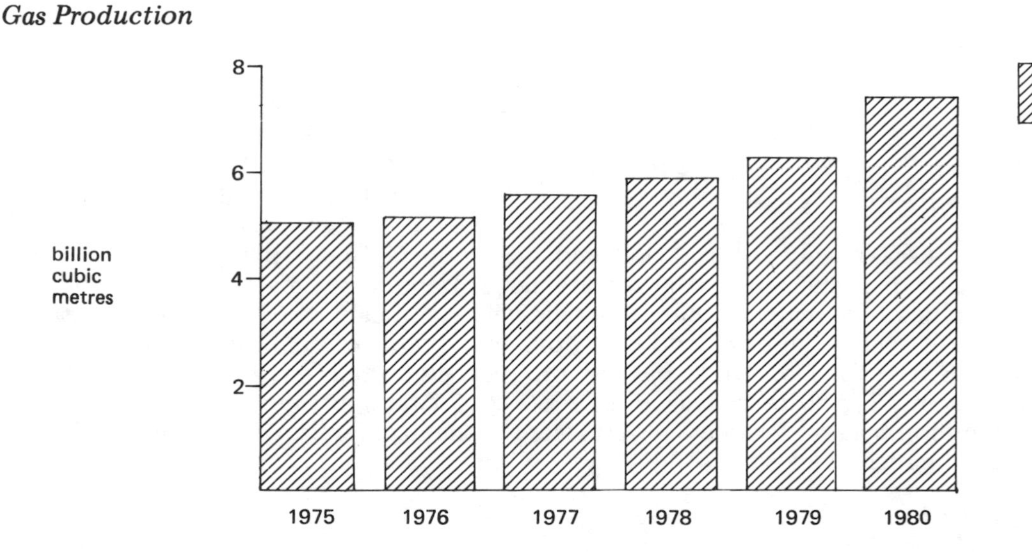

Production of gas in 1979-80 reached 7,500 million cubic metres, a 50 per cent increase since 1974-75. Use of gas is increasing relatively rapidly as a result of the proved reserves available in the country, and its attraction as an alternative to oil, which has to be largely imported. Output from the Sui gas field in Baluchistan has been making an important contribution to the country's energy supplies for many years and new fields have added to total reserves. These are estimated to stand at 500,000 million cubic metres, excluding two major new finds at Pirkoh and Dhodak, Baluchistan, which are still being evaluated.

The Sui gas field is operated by Pakistan Petroleum Ltd, in which Burmah Oil Company has a 70 per cent holding. The Pakistan government holds nearly all of the remaining shares. The Sui field has been producing gas since the mid 1950s, but is now supplemented by other supplies from fields more recently developed, particularly by the state undertaking, the Oil and Gas Development Corporation. Gas from these fields now accounts for 20 per cent of total output. The amount of gas contained in new discoveries is as yet uncertain, but the Dhodak field may have reserves of more than 110 million cubic metres of gas, and this is associated with useful reserves of condensate.

Continuing success in exploration for gas means that natural gas can be considered for a wider range of end users. At present a high proportion is being used as a fuel in power stations, as feedstock for fertiliser manufacture and fuel for industrial bulk energy users. Small-scale consumption by the commercial and residential sectors is limited. But projects have been approved for the extension of pipelines to a further 10 major towns, which will lead to an expansion of consumption. In addition, the extent of reserves which are being proved may allow some exporting of gas to take place.

Electricity

Electricity production is currently around 13 TWh per annum, having risen from 10.5 TWh in 1975-76, a slower rise than the increase in energy consumption overall. Principal sources of power are hydro-electric dams and natural gas-fired stations. Both of these indigenous fuel forms will be further exploited to meet rising power demands. In addition, there is a project to build a coal-fired power station in the Lakhra area in Sind Province, based on the extensive reserves of lower grade coal in that area.

The generation and transmission of electricity is largely the responsibility of two public authorities, the Water and Power Development Authority (WAPDA) and the Karachi Electric Supply Company (KESC). WAPDA has generating capacity of around 3,000 MW including a large amount of hydro-electric generation which is capable of producing more electricity than the Authority's thermal power stations. KESC has capacity of 730 MW, including access to the power produced by the small 137 MW nuclear plant near Karachi, although this has been supplying only a small proportion of energy to the public supply system.

Apart from the project for a coal-fired power station at Lakhra construction has been concentrated on additional hydro-electric capacity at Mangla, where capacity is now 800 MW, Warsak and Tarbela, in the north of the country. Four units of 175 MW are currently being added, and four further units, also of 175 MW, may be built by the mid 1980s, making the Tarbela scheme by far the largest centre of electricity production.

ENERGY TRADE

Net Imports/(Exports) 1975-76—1979-80

	1975-76	1976-77	1977-78	1978-79	1979-80
Crude Oil (million tonnes)	2.8	2.9	3.4	3.6	3.9
Oil Products (million tonnes)	..	0.9	0.5	1.0	0.8

Pakistan's imports of energy consist mainly of crude oil and oil products. Small quantities of coke and special coals are also imported for particular industrial requirements. Imports are mainly of crude oil, but because of the pattern of product requirements there is a relatively significant import and export trade, with imports of middle distillates such as kerosine and diesel oil, and exports of naphtha, and particularly, fuel oil. In the period 1975-78, the total of crude oil and products imported was of the order of four million tonnes per annum. But in 1979-80 there was an increase to 4.7 million tonnes in total, of which crude oil accounted for 3.9 million tonnes. Product imports were at the high level of 1.8 million, offset by exports of one million tonnes, the highest level for some years.

Sources of Imported Crude Oil

(percentage)	
Saudi Arabia	60
Abu Dhabi	35
Other	5
TOTAL	100

Middle East countries supply all of the imports of crude oil into Pakistan. The overwhelming proportion of oil is obtained from two countries at present. Saudi Arabia has been supplying as much as 60 per cent of all crude oil imported. Almost all of the balance has come from Abu Dhabi. Minor quantities are also imported from Iraq, Qatar and other Persian Gulf states.

ENERGY POLICIES

State agencies are pressing ahead in the exploration for commercial deposits of oil and gas and the exploitation of water and coal resources. The intention is to use the substantial reserves of gas for a variety of purposes, including electricity generation, as feedstocks and possibly as compressed natural gas for transport fuel. Hydro-electric potential is being exploited, involving very high voltage transmission lines from Tarbela to major cities in the centre and south. During the Fifth Five Year Plan all main power markets should be inter-connected forming a national grid. At the same time the rural electrification programme aims to reach 1,000 new villages each year.

In the drive to develop the country's resources state agencies and corporations are involving private sector and foreign companies a number of which are involved in exploration work. Eastern European teams are also supplying expertise in exploration and development.

Papua New Guinea/Pacific Islands

KEY ENERGY INDICATORS

Energy Consumption	
—million tonnes oil equivalent	3.5
Consumption Per Head	
—tonnes oil equivalent	0.70
—percentage of world average	43%
Net Energy Imports	95%
Oil Import-Dependence	100%

The islands of Australasia and the Pacific have a total population of some five million, of which more than three million are in Papua New Guinea. Although their economies are largely undeveloped, energy consumption per head is relatively high. However, this includes a significant volume of aviation fuel, and exaggerates the actual use of commercial energy. Primary electricity is still undeveloped, and, outside Papua New Guinea, is not considered to have much potential. Geothermal energy is a promising source of energy in some areas and this will supplement existing consumption of indigenous non-commercial fuel.

ENERGY MARKET TRENDS

The development of energy consumption is dependent primarily on external factors, such as demand for primary products and the level of international transport demand, and on individual mining and processing operations. Such activities are of dominant importance in several countries, such as New Caledonia and Papua New Guinea. However, each country is evaluating the scope for substituting indigenous resources of hydro-electricity or geothermal energy for imported oil.

Pattern of Energy Supply

Primary Fuel Supply (percentage)	
Oil	92
Solid Fuel	3
Natural Gas	–
Primary Electricity	5
TOTAL	100

Oil is the predominant commercial fuel used in the islands. Transport sector demands are high and almost all electricity generating plant is in small oil-fired diesel sets. Hydro-electric power is produced in Papua New Guinea and several smaller islands. Non-commercial fuels play a significant part in the total energy economy of some areas, notably Fiji, where wood and bagasse provide as much energy as the commercial fuels.

ENERGY SUPPLY INDUSTRIES

With very limited markets for energy state governments play a minor role in overseeing the activities of private sector companies. Public sector involvement is usual in the supply of electricity for general consumption. In the supply of oil products the international oil companies meet market demands from various refineries outside the area.

Oil

Consumption of oil in the region is more than three million tonnes per annum, excluding the demand for marine bunkers. Several of the major international oil companies have distribution operations, drawing supplies from Australia, New Zealand and Singapore. There is only one refinery, located on Guam, which is owned by the independent Guam Oil Refining Company. The refinery has a throughput capacity of over two million tonnes per annum. Expectations of finding oil or gas deposits in the area are low. Exploration activity has taken place in Fiji and Tonga, but without success so far.

Electricity

Total consumption of electricity is in excess of 600 GWh per annum. Production is mainly from oil-fired plant, with localised distribution from diesel generating sets. In Papua New Guinea some 80 GWh per annum is generated by hydro-electric plant and it is estimated that the potential for hydro-electric generation is as high as 25,000 MW. A scheme is currently being developed at Monasavu, in Fiji, which will meet virtually all of the country's electricity demand. Small scale hydro-electric schemes are being examined for many locations, as is geothermal energy. There is known to be exploitable potential in the Solomon Islands, and geothermal resources are also being evaluated in Fiji.

Philippines

KEY ENERGY INDICATORS

Energy Consumption	
—million tonnes oil equivalent	12.0
Consumption Per Head	
—tonnes oil equivalent	0.25
—percentage of world average	15%
Net Energy Imports	83%
Oil Import-Dependence	95%

The Philippines has a low average consumption of energy per head of population. This is as low as 15 per cent of the world average, although somewhat higher if account is taken of the non-commercial fuels which represent around 30 per cent of total energy consumption. Economic growth has until recently been based on increasing quantities of imported oil, there being very little indigenous production, while coal resources have been neglected. But the government has taken determined action to increase the proportion of energy obtained from indigenous resources and has formulated a comprehensive plan of action for the next decade or more. Particular attention is being paid to the geothermal energy potential of the country and the Philippines expects to become the world's leading producer of geothermal energy during the 1980s.

ENERGY MARKET TRENDS

Trend of Energy Consumption

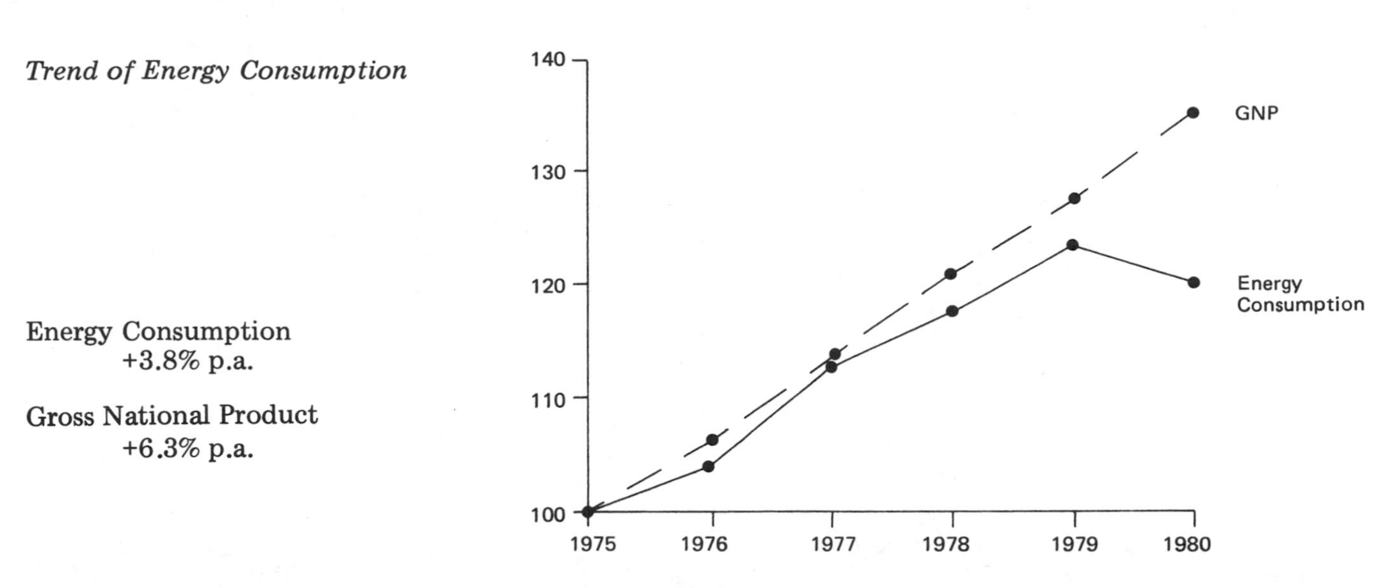

The Philippine economy has maintained a relatively steady rate of growth in recent years and has been adapting to high energy costs while continuing to meet rapidly rising demand for energy. Between 1975 and 1980 Gross National Product expanded by over 35 per cent in real terms. On the other hand consumption of commercial energy fell back by more than three per cent, while GNP nevertheless continued to rise. Overall between 1975 and 1980 energy consumption rose by only 20 per cent.

Pattern of Energy Supply and Consumption

Primary Fuel Supply (percentage)		Final Consumption (percentage)	
Oil	87	Industry	43
Solid Fuel	1	Residential	13
Natural Gas	–	Transport	33
Primary Energy	12	Other	11
TOTAL	100	TOTAL	100

Oil represents a very high proportion of total commercial energy supplies to the Philippine economy. The balance is derived from indigenous hydro-electric and geothermal sources. Natural gas and coal output make only a negligible contribution. However, non-commercial solid fuel, consisting of wood, bagasse and other wastes, is an important additional component in primary energy supply. The contribution from these fuels is estimated to equate to around 5.5 million tonnes of oil equivalent annually, or 30 per cent of total energy supply. This percentage is expected to increase during the 1980s.

Emphasis on potential indigenous resources is reflected in plans to meet the energy requirements of the two main sectors of energy consumption, industry and transport, which together account for three-quarters of final energy consumption. Industry is being encouraged to use coal and bagasse for its increasing demand for energy, while in transportation the use of alcogas, a mixture of 15 per cent anhydrous alcohol from sugarcane and cassava and 85 per cent gasoline, will help to contain rising demand for transport fuels.

ENERGY SUPPLY INDUSTRIES

In recent years the state has moved to take up a central role in energy supply. A state company, PNOC, has been set up with wide-ranging responsibilities for ensuring crude oil supply and it has taken a majority share in the country's largest refinery. PNOC subsidiaries are also taking a leading role in the development of coal production and distribution and in exploiting geothermal energy. The National Power Corporation was set up to provide the nationally co-ordinated production and transmission of power necessary for a developing economy. However, in all energy sectors private sector involvement is also encouraged and private or non-state organisations are mainly responsible for supplying oil, coal, gas and electricity at retail level.

Oil

Oil Consumption

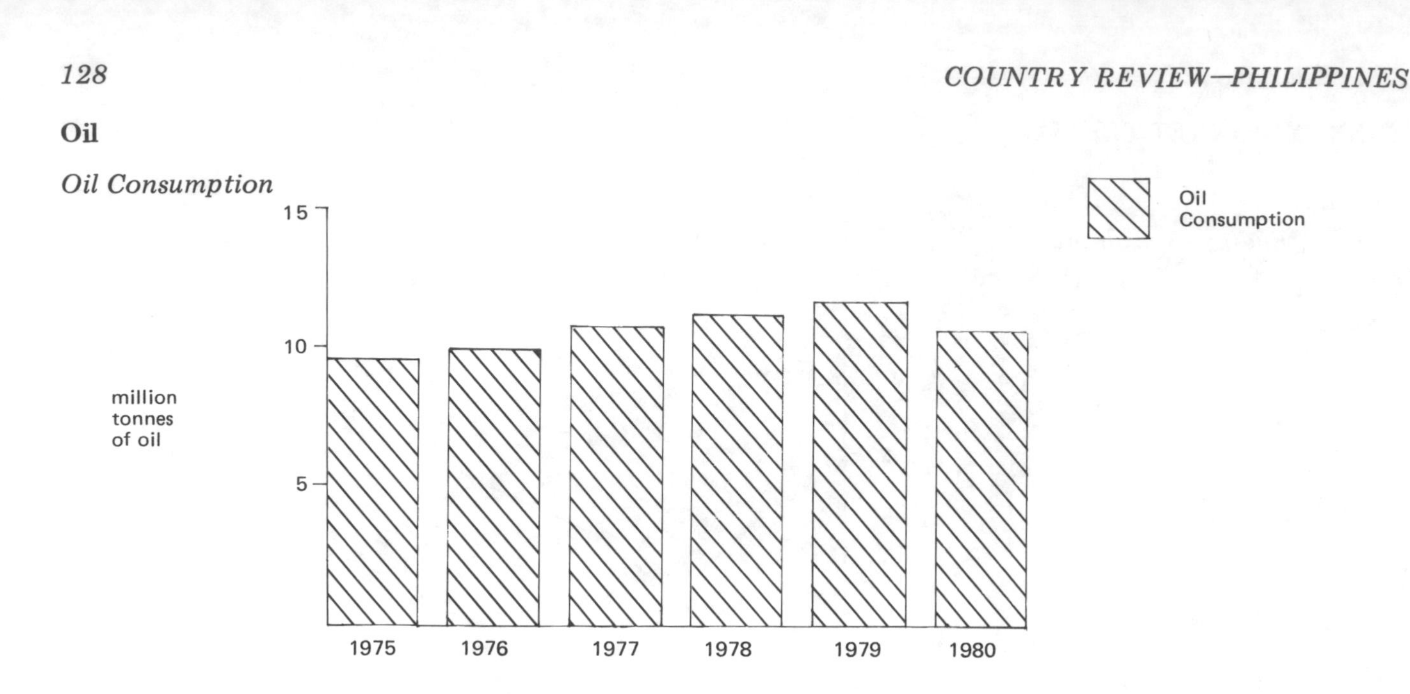

Oil is the single most important source of energy for industry, transport and other sectors currently. Consumption reached a peak of 11.4 million tonnes in 1979, falling back to 10.4 million tonnes in 1980. Domestic production of oil is still small, although new fields have been brought into production since 1979. These are the Nido field, off Palawan, and the Cadlao field, which have a combined production of 12,000 barrels per day, and in 1982 the level of indigenous production is being more than doubled with exploitation of discoveries in the Matinloc-Pandan area. Exploration under service contracts continues to be encouraged by the government.

Until the formation of Philippine National Oil Company (PNOC) the supply of oil products was entirely in the hands of affiliates of the major international oil companies. This group includes Mobil, Caltex (the partnership of Standard Oil Company of California and Texaco), and Shell. All continue to have a major involvement in exploration, refining, crude oil supply and product distribution and marketing, although PNOC's involvement is increasing. PNOC's market share is now approaching 30 per cent and the company is investing in distribution facilities for all products.

There are three oil refineries. Largest is the Limay refinery of 155,000 barrels per day throughput capacity, which is operated by Bataan Refining Corporation, and owned 60 per cent by PNOC and 40 per cent by Mobil. Smaller refineries are at San Pascual (Caltex) and at Tabangao, in which Shell has a 50 per cent interest. Total refining capacity is over 260,000 barrels per day, equivalent to 13 million tonnes per annum, sufficient to meet national market requirements.

Coal

Coal has hitherto been of little significance in the energy picture, but output is on a strongly rising trend with even faster growth envisaged during the 1980s. Between 1975 and 1980 production, which is of hard coal, tripled to just under 330,000 tonnes. During the period up to 1986 consumption is planned to rise to an annual level of over five million tonnes, as the result of a determined switch from oil to coal in power generation and energy-intensive industries, such as sugar processing, mining, cement and metal trades.

In 1980 80 per cent of output was derived from mines on Cebu. The balance came from Malangas, Polillo, Batan and Semirara. Production from all areas will be stepped up, with the expectation of meeting at least a large part of market requirements from indigenous mines. There has been a considerable expansion in exploration and development activity since 1979, involving both private companies and the state, through subsidiaries of PNOC. At the end of 1980 proven reserves stood at 186 million tonnes, while potential reserves were estimated to approach 1.7 billion tonnes.

Because of the scale of the increase in consumption which is envisaged, and the problem of co-ordinating the coal supply chain, the National Coal Authority (NCA) has been established. The NCA has a general responsibility to oversee the sale, handling and distribution of coal, and if necessary its importation.

Gas

Indications of the presence of natural gas have been found in many places throughout the Philippines, but to date no significant discovery sufficient to justify commercial exploitation has been found. Reserves of the order of 2,500 million cubic feet have been delineated in Southern Isabela and smaller gas reserves have been located in Northern Cebu, estimated in total to be around 400 million cubic feet. Exploration continues in the expectation that more sizeable fields will be discovered in due course.

Electricity

Electricity Production

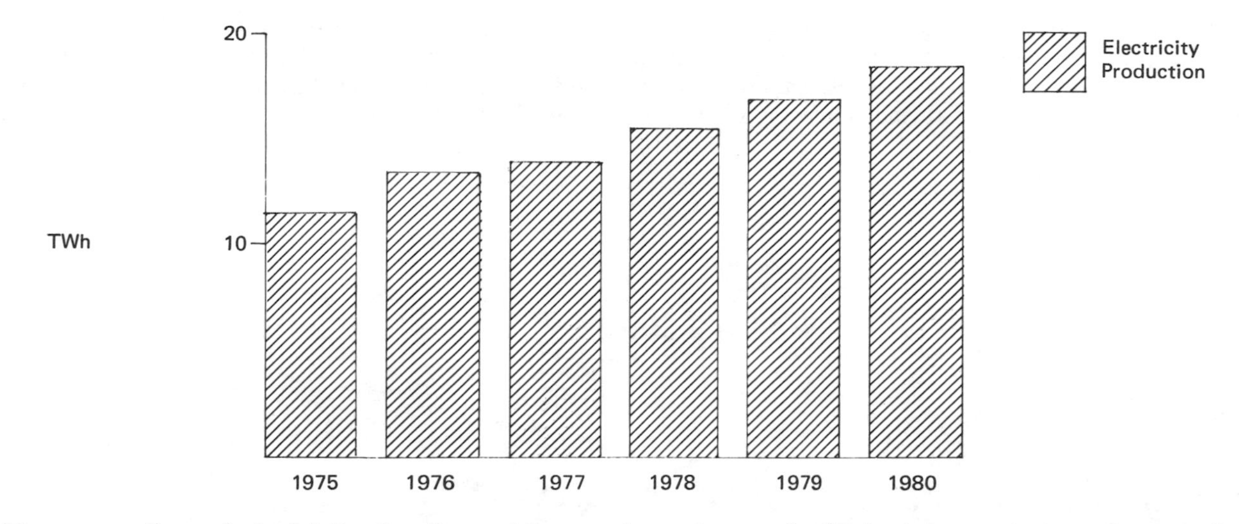

The generation of electricity for the public supply system and efficient transmission of power throughout the country is the responsibility of the National Power Corporation (NPC), a state undertaking responsible to the Ministry of Energy. Since its formation in 1973 NPC has had the task of ensuring adequate power supplies for the rapidly developing economy. Distribution at retail level remains the responsibility of local municipal authorities and companies which are franchised by the state authorities. Largest of these is Meralco, the electricity supplier in Manila.

Between 1975 and 1980 electricity consumption increased by 50 per cent, with power handled by NPC rising from 2.7 TWh in 1975 to 15.5 TWh in 1980, largely as a result of integrating the output of Meralco into the Luzon grid. NPC accounts for 85 per cent of national electricity supplies. A small percentage is produced by local co-operatives, and the rest, around 11 per cent, by other companies, including industrial energy consumers.

The electricity industry is still highly dependent on oil-fired plant, with 64 per cent of all electricity being generated from this fuel in 1980. This, however, represents a significant decline from only two years previously, when the proportion was 75 per cent. The main reason for this sharp reduction has been the growth in exploitation of geothermal energy, which provided 2.3 TWh in 1980, or 13 per cent of total supply. The development of geothermal resources is being pursued vigorously with the objective of producing 10 TWh from this source by 1990.

The responsibility for this development programme lies primarily with NPC. It has two main projects under way at present. In one of them NPC is partnered by the Energy Development Corporation, a subsidiary of PNOC, and in the other by Philippine Geothermal Inc (PGI), a subsidiary of Union Oil Company of California. PGI has already constructed a 440 MW plant on Luzon. PNOC-EDC is developing resources at Tongonan and Palimpinon and is evaluating several other areas in addition.

The exploitation of geothermal resources, which will lead to the Philippines becoming the world's largest producer of this form of energy, combined with further exploitation of hydro-electric potential and the commissioning of the country's first nuclear power station in the mid 1980s, is expected to virtually eliminate the use of oil in large-scale power generation. By that stage it is planned that hydro-electricity and geothermal energy could each be meeting 30-35 per cent of electricity supplies, coal around 20 per cent and nuclear power at least 10 per cent.

ENERGY TRADE

With only a small proportion of oil requirements being met from indigenous sources the Philippines has had to import a high proportion of total commercial energy requirements. Between 1975 and 1979 net imports of crude oil and finished products rose from 9.3 million tonnes to 10.9 million tonnes, an increase of some

Net Imports/(Exports) 1975-80

	1975	1976	1977	1978	1979	1980
Coal (million tonnes)	..	..	0.2	0.3	0.3	..
Crude Oil (million tonnes)	9.2	9.6	9.2	9.5	9.1	9.2
Oil Products (million tonnes)	0.1	0.8	1.2	0.8	1.8	

17 per cent. This was however rather less than the rate of increase in total energy consumption. Coal has begun to figure in the country's foreign trade. Imports prior to 1977 were negligible, but in both 1978 and 1979 approached 300,000 tonnes as efforts were made by industrial bulk energy users to move away from fuel oil. Imports in 1980 fell back, however, as domestic production increased.

Sources of Imported Crude Oil

(percentage)	
Saudi Arabia	34
Kuwait	16
Iraq	15
Indonesia	14
China	11
Brunei/Malaysia/Sarawak	7
Other	3
TOTAL	100

Saudi Arabia is the largest individual source of imported crude oil, accounting for around one-third of the total. Middle East countries together account for more than half, with Kuwait and Iraq both supplying well over one million tonnes per annum. But Far Eastern sources are also significant. Not only do Philippine refinery companies import substantial quantities from the main oil-exporting country, Indonesia, but they also draw on several other export sources. These include Brunei, Malaysia and Sarawak, and there is an important trade with the People's Republic of China.

ENERGY POLICIES

Since the time of the international oil crisis of 1973-74 the government has moved decisively towards altering the pattern of energy use within the economy, encouraging the development of all forms of indigenous energy, while restraining consumption of key energy products such as gasoline. Institutions and agencies have been developed, giving the state a strong influence in energy supply, research and development, backed up by controls on prices of oil products, gas and electricity. A detailed programme is under way to implement government priorities within the context of an energy policy for the period up to 1990.

Energy Supply

Although PNOC and NPC both have the power to engage in developing and distributing energy products to the Philippine market, it is a point of policy to continue to encourage private sector involvement in all aspects of energy supply. Service contracts are widely used to bring in foreign and domestic private finance, expertise and operating skills in exploration for oil, gas, coal and uranium.

Conservation

The government has used price and taxation measures to help contain rising demand for certain forms of energy and influence consumption along the lines established in the country's energy plans to 1990. Price has been used particularly to restrain consumption of motor gasoline. Tax incentives and finance have been available to encourage conversion from fuel oil to coal and the installation of more energy efficient equipment in industry.

Of particular note is the progress being made in developing unconventional and renewable forms of energy. Most striking is the exploitation of geothermal resources by the state-owned NPC, which already has a programme to increase its use of hydro-electric potential throughout the country. Of critical importance in reducing dependence on oil is the possibility of introducing alternative fuels for transportation. Two programmes are being developed here. The Alcogas programme involves the introduction of a 15 per cent alcohol component into motor gasoline, utilising the country's capacity to produce sugar cane. A second programme is seeking to use a 5 per cent element of coconut oil as an extender in diesel oil.

Qatar

KEY ENERGY INDICATORS

Energy Consumption	
—million tonnes oil equivalent	1.8
Consumption Per Head	
—tonnes oil equivalent	9.0
—percentage of world average	555%
Net Energy Imports	nil
Oil Import-Dependence	nil

Qatar's small population (only 200,000) possesses extensive oil and gas resources. These ensure ample availability of energy for economic development and, at least for the medium term, for export earnings. In the longer term the country may become increasingly dependent, from both points of view, on gas. Qatar is a member of the Organisation of Petroleum Exporting Countries and of the Organisation of Arab Petroleum Exporting countries.

ENERGY MARKET TRENDS

With only a limited population and high standards of living, the growth of energy consumption is determined mainly by the impact of individual industrial developments, as well as by the level of activity of the oil and gas industries themselves. Total energy consumption therefore appears to have stabilised at around 1.8 million tonnes of oil equivalent pending the next set of development projects. One such will be the start up of oil refining operations, expected in 1983. But emphasis is now on less energy intensive activities which will help to broaden the range of industries in Qatar and reduce dependence on imports for a whole range of manufactured goods.

Pattern of Energy Supply

Primary Fuel Supply (percentage)	
Oil	23
Solid Fuel	–
Natural Gas	77
Primary Electricity	–
TOTAL	100

Qatar has a substantial base of oil resources which it exploits as main source of foreign exchange. In its internal economy, however, gas has been, and will continue to be, the principal fuel used. Not only has it made sense to utilise associated gas as far as possible, but Qatar in any case possesses very large reserves of non-associated gas. These are being tapped at present, though only to a small extent, and are of such a scale that a large LNG export project can be envisaged as well.

Oil

Crude oil production has been maintained at well over 20 million tonnes per annum, despite adverse conditions in the international oil market. Production in 1980 was some 23 million tonnes, down on the previous year's level, but in line with that of 1978. However, the generally slack demand for oil was felt in 1981, when production was cut back to under 20 million tonnes. Internal consumption of oil is less than two million tonnes per annum. Reserves are estimated to be less than 500 million tonnes, split approximately equally between onshore and offshore fields.

Oil industry activities are controlled by the state oil company Qatar General Petroleum Company (QGPC). QGPC is holding company for state interests in exploration, production, refining and products distribution.

Oil Production

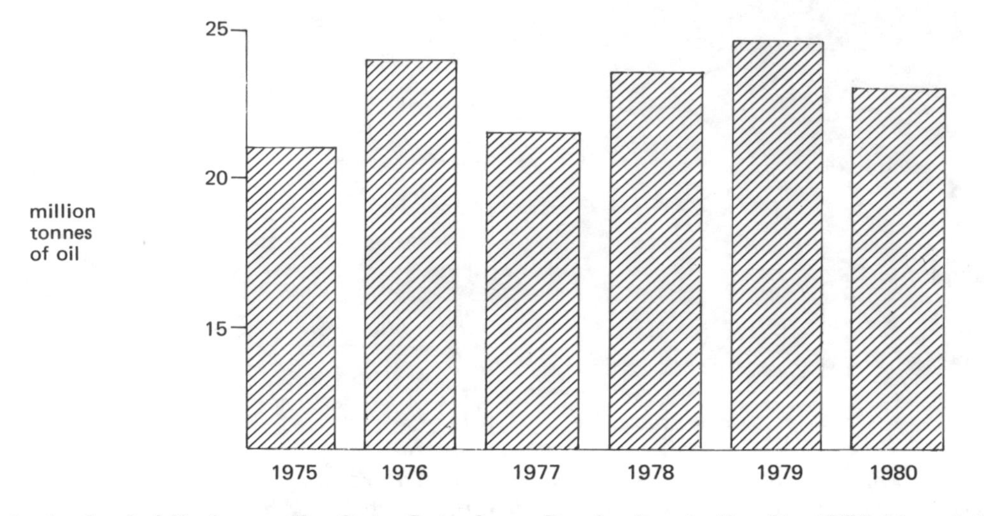

Principal subsidiaries are the Qatar Petroleum Producing Authority (QPPA) and the National Oil Distribution Company (NODCO). British Petroleum Company and Shell are involved as management companies for onshore and offshore production respectively.

At present there is only one small refinery in Qatar, at Umm Said. The refinery has a capacity of less than one million tonnes per annum. A new refinery of 2.5 million tonnes per annum is under construction. This will provide all of NODCO's requirements.

Gas

There is production of associated and non-associated gas by QGPC. Total production in 1980 was 3,000 million cubic metres, of which only 30 per cent was associated gas. Gas liquids are extracted, leaving gas for industrial fuel, power generation and fertiliser manufacture. The increasing quantities of non-associated gas from the Khuff zone, which underlies the Dukhan oil field, have been needed to compensate for shortfalls in associated gas. A major project to develop an offshore gas field to produce LNG is under consideration. This field, known as the North-West Dome, is believed to hold vast reserves, capable of meeting Qatar's medium and longer term requirements as well as 6-7 million tonnes per annum of LNG exports.

Electricity

The growth of the Qatar economy and the commissioning of industrial and energy-related projects has brought about a substantial rise in demand for electricity. Total production in 1980 was 1,400 GWh, more than double the level of five years before. Almost all electricity is supplied from the state-run power stations. Capacity is 520 MW, excluding the Ras Abu Fontas power station and desalination complex being built near the capital.

ENERGY TRADE

Net Imports/(Exports) 1975-80

	1975	1976	1977	1978	1979	1980
Crude Oil (million tonnes)	(20.7)	(23.5)	(20.7)	(23.4)	(23.8)	(22.0)
Oil Products (million tonnes)	(0.1)	(0.2)	..	0.1	0.1	

Almost all of Qatar's crude oil production is exported, with only 0.5-1.0 million tonnes per annum being retained as feedstock for the small refinery at Umm Said. Small quantities of finished products are usually imported to make up for local market deficits. With the completion of the new refinery, expected in 1983, product imports are expected to be eliminated and a net export of products achieved.

Saudi Arabia

KEY ENERGY INDICATORS

Energy Consumption	
—million tonnes oil equivalent	11.8
Consumption Per Head	
—tonnes oil equivalent	1.40
—percentage of world average	86%
Net Energy Imports	nil
Oil Import-Dependence	nil

Saudi Arabia is the world's largest producer of crude oil, apart from the United States and the Soviet Union, and, as a result of its small internal market, a key force in the international oil market. Production has never approached productive capacity and the extent of the country's reserves is still uncharted. Very few oil importing countries do not rely on Saudi Arabia to some extent, and in Asia the proportion is often high. Saudi Arabia is actively trying to establish large scale downstream operations, in order to add value to oil and gas production and to reduce dependence on the export of raw energy.

ENERGY MARKET TRENDS

Saudi Arabia's economic development plans involve massive investment in social infrastructure and the industrial sector. Growth of Gross Domestic Product has been at a rate of up to 10 per cent per annum on average. State agencies are promoting a wide range of manufacturing activities and a reasonable level of self-sufficiency in construction materials has been achieved. These, together with major expansion of oil- and gas-related industries, such as petrochemicals production, export oil refining and fertiliser manufacture, have led to energy consumption rising at almost 15 per cent per annum.

Pattern of Energy Supply and Consumption

Primary Fuel Supply (percentage)		Final Consumption (percentage)	
Oil	95	Utilities	29
Solid Fuel	–	Energy Industries	15
Natural Gas	5	Transport	32
Primary Electricity	–	Other	24
TOTAL	100	TOTAL	100

Given the super-abundance of oil resources it is not surprising that oil accounts for a high percentage of energy supply. Gas provides the balance, but this is still only five per cent, and reflects the difficulties in developing the infrastructure for use of gas on a wide scale. Saudi Arabia has no commercial exploitation of solid fuel and lacks water resources for hydro-electricity generation.

ENERGY SUPPLY INDUSTRIES

The government has gradually assumed full control of energy supply industries. In the oil sector the state oil company is responsible for all disposals of crude oil, oil refining for the domestic market and sale of products. Foreign participation is now minor, with the role of Aramco now essentially that of operations management. This expertise is being drawn on in the process of rationalising and developing regional electricity grids and operating companies.

Oil

Since 1980 production of crude oil has been at record levels of nearly 500 million tonnes per annum, nearly 40 per cent higher than in 1975. This level has reflected the basic use of crude oil output by Saudi

Oil Production

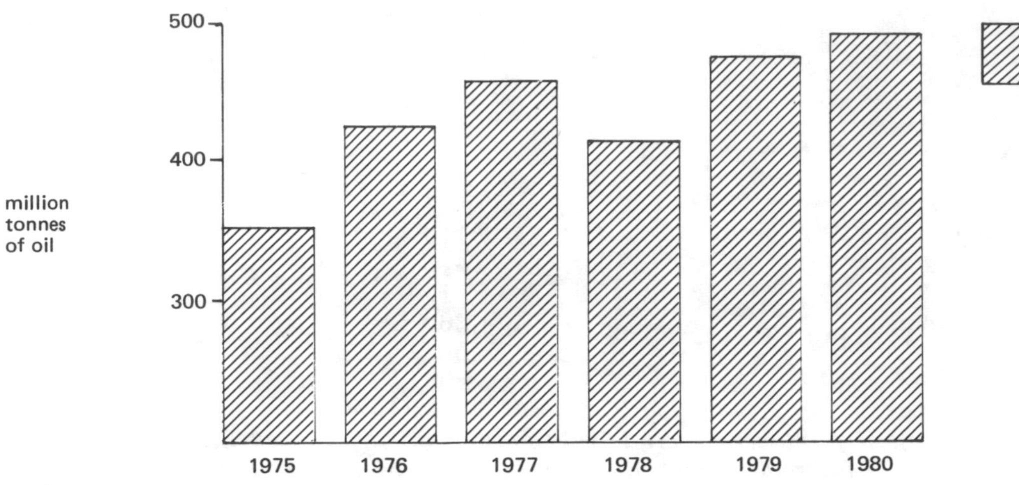

Arabia to influence price levels in the international oil market, running counter to export trends for most other OPEC countries. A production ceiling of 8.5 million barrels per day (425 million tonnes per annum) was set officially in 1978, but was waived in the interests of broader policy objectives.

Almost all of Saudi Arabian production takes place in the Eastern Province, where the four American major oil companies, Exxon, Mobil, Standard Oil of California and Texaco, participated in the development of Aramco's concessions. These exploited the massive Ghawar field, which is estimated to contain at least 60 billion barrels and the offshore Safaniya field, of some 13 billion barrels. Total proven reserves of crude oil, essentially within the Aramco operating area, are put at 110 billion barrels, but the full extent of potential reserves remains to be evaluated. In the Neutral Zone, which is shared by Saudi Arabia and Kuwait, Getty Oil and the Japanese group Arabian Oil Company retain production interests, in partnership with the Kuwait state oil company.

Internal consumption of oil is less than 10 million tonnes per annum, with a further 10 million tonnes for the marine bunker market. The overwhelming proportion of crude oil is thus exported, mainly from the Persian Gulf. But in 1981 a new strategic pipeline to Yanbu on the Red Sea was opened. This pipeline is capable of transporting almost 100 million tonnes per annum. The pipeline may be either expanded or doubled in order to reduce the strategic risk of the Straits of Hormuz.

Yanbu is the site of one of the major oil-based projects currently being developed by Saudi Arabian state agencies in partnership with foreign oil companies. At Yanbu Mobil is partner in a 12 million tonnes per annum petrochemical refinery. Shell is partner in a potentially larger project at Jubail on the Persian Gulf. This export capability will be added to the existing major international refinery at Ras Tanura, which is owned by Aramco. The refinery has a capacity of over 20 million tonnes per annum, orientated to export markets. Refineries supplying the inland market are at Jeddah and Riyadh, with a total capacity of four million tonnes per annum. Petromin, the state oil company, plans to build a major new refinery to keep pace with market demand.

Gas

Very large volumes of gas are produced in association with crude oil, which until the last few years have been mostly flared. Investment in gas-gathering systems and treatment plant, to separate gas liquids and utilise gas for feedstock and fuel, has formed one of the state's principal programmes. A gas-gathering system has been established covering 75 per cent of the output of the Aramco producing area, with three key separation plants at Berri, Shedgum and Uthmaniyah. Annual output of gas is 40-50 billion cubic metres and reserves are estimated at some 2,500 billion cubic metres. Little exploration has been undertaken to assess the resources of non-associated gas.

Gas processing plants produce natural gas liquids, LPG and dry gas. Gas liquids and LPG are exported, but a large part of the output will be used in the petrochemical plants being constructed at Yanbu and Jubail. Dry gas is piped to the Jubail industrial area on the Persian Gulf.

Electricity

The electricity sector is one of the most dynamic in the Saudi Arabian economy with demand for electricity rising at a very high rate in all sectors. Residential and commercial premises use air conditioning on a large scale, new industrial plants are being added each year and water preparation and supply is energy-intensive. Capacity is being increased in all areas. The structure of the supply industry is being rationalised and up-graded to meet demand and cope with large new plant. The process of electrification is expected to extend to virtually all of the country by 1985.

The level of generating capacity in 1980 stood at just under 4,000 MW, of which more than half was in the area of the Eastern Province, where oil industry operations are concentrated. Current construction programmes will take generating capacity in the region up to 4,700 MW, with a further 2,000 MW at the Jubail and Aziziyah desalination plants. The scale of construction and investment has necessitated the setting up of a national agency, the General Organisation for Electricity, and regional operating companies to manage power supply and transmission. The management of Aramco has been crucial in establishing the first of these groupings, the Saudi Consolidated Electricity Company (Eastern Region), which consolidated the power system of Aramco itself with those of numerous local companies.

ENERGY TRADE

Net Imports/(Exports) 1975-80

	1975	1976	1977	1978	1979	1980
Crude Oil (million tonnes)	(328.2)	(397.9)	(428.1)	(385.3)	(443.0)	(463.0)
Oil Products (million tonnes)	(10.0)	(11.8)	(9.6)	(10.7)	(10.6)	

Crude oil forms the overwhelming proportion of Saudi Arabian energy exports. Exports continued to increase to over 460 million tonnes in 1980 as Saudi Arabia used its production capacity to gain leverage over international crude oil prices. By contrast, exports of finished products are relatively small, of the order 10 million tonnes per annum. This level will rise as new export refining capacity comes on stream. LPG is also becoming more important, although volumes remain small by comparison with crude oil.

ENERGY POLICIES

The scale of Saudi Arabia's oil resources in relation to its population of around 8.5 million is such that the country is not dependent on maintaining levels as high as have been the practice in the past, in order to maintain investment and economic progress. Thus the idea of an official ceiling on production has been adopted. This has been combined with a policy of limiting the ratio of light crude oil, which is more in demand, but relatively limited in terms of reserves, to heavier oil. This is to guard against becoming excessively dependent on selling heavy crude oil in a period when fuel oil is in excess supply.

Construction of gas-gathering pipelines was initiated in order to reduce to an acceptable level the amount of gas being flared and to utilise existing production more usefully. An important point of this approach is the extraction of gas liquids, which are easier to transport than dry gas itself. Gas liquids are being exported, but some are earmarked as petrochemical feedstock. The construction of extensive capacity to produce petrochemicals as well as the normal range of refined products for overseas markets is one of the main ways in which value is added to oil and gas resources within the country. This is perhaps more important in Saudi Arabian policy towards oil and gas than trying to establish major new unrelated activities.

Senegal

Senegal has limited natural resources to form the base for economic development and is heavily dependent on production of ground nuts and phosphate mining. No commercial energy forms are available. The level of oil consumption per head of population, at 18 per cent of the world average, is higher than in a number

KEY ENERGY INDICATORS

Energy Consumption	
—million tonnes oil equivalent	1.0
Consumption Per Head	
—tonnes oil equivalent	0.18
—percentage of world average	11%
Net Energy Imports	100%
Oil Import-Dependence	100%

of other under-developed African countries, but is exaggerated owing to the proportion of oil products used as marine bunkers. At present Senegal relies on oil for all of its commercial energy, supplementing the use of firewood and charcoal, but hydro-electric power is to be established, utilising the Senegal River. Development of an offshore oil discovery remains problematic.

ENERGY MARKET TRENDS

Total consumption of commercial energy is around one million tonnes of oil equivalent per annum, compared with little more than 800,000 tonnes in 1975. The increase, of nearly four per cent per annum on average indicates a very moderate rate of growth in the economy. Consumption of electricity has increased at a higher rate, of between six and seven per cent per annum. This is in part the cause of rising oil demand, as all electricity generation is based on oil-fired plant. Outside the commercial fuel sector indigenous production of firewood and charcoal is significant. Production of firewood is estimated to be over 100,000 cubic metres per annum and production of charcoal more than 100,000 tonnes per annum.

ENERGY SUPPLY INDUSTRIES

Private sector companies play the major part in ensuring supplies of energy to the Senegal economy. Importation, refining and marketing of oil is handled by several international oil companies. The electricity supply company is controlled by the state, with participation of local authorities, but the supply network is not well developed as yet.

Oil

Oil industry operations are centred on the national refinery at Dakar. The refinery has a throughput capacity of 1.2 million tonnes per annum and supplies finished products for the Senegal market and the international marine bunker trade and is a source of products for inland markets in Mali and Mauritania. The Dakar refinery is owned principally by the marketing companies involved in Senegal. Largest shareholder is the French state oil company Elf-Aquitaine. Other industry participants are BP, CFP, Esso, Mobil and Shell. Oil has been discovered offshore, but there are currently no firm plans for its exploitation owing to the technical problems involved.

Electricity

Consumption of electricity has risen to over 500 GWh from a level of less than 400 GWh in 1975. The rate of increase has averaged six to seven per cent per annum. Electricity supply is undertaken by SENELEC, which is state controlled. SENELEC's production is almost entirely at Dakar, where it has 155 MW of generating plant. Several other towns are linked in a limited grid system. All power stations are run on oil, but hydro-electric power is to be introduced from plant on the Senegal and Gambia rivers. The Senegal River project at Manantali is being undertaken jointly with Mali.

Sierra Leone

KEY ENERGY INDICATORS

Energy Consumption	
—million tonnes oil equivalent	0.3
Consumption Per Head	
—tonnes oil equivalent	0.09
—percentage of world average	6%
Net Energy Imports	99%
Oil Import-Dependence	100%

Sierra Leone has an under-developed economy and a consequently low level of consumption of commercial energy per head of the population. The lack of indigenous energy resources is hampering development and only a limited increase will be attained from the prospective Mano River scheme. Growth in energy consumption may increase significantly if iron ore mining is recommenced and other mining and food-based industries expanded.

ENERGY MARKET TRENDS

Growth of the economy has been very slow, with only limited success in establishing manufacturing activities, which remain on a small scale. Mining is a mainstay of the economy, but this sector contracted after 1975 following the cessation of iron ore mining. Consumption of energy shows a minimal rise, with oil consumption virtually flat and use of electricity rising at a bare one per cent per annum.

ENERGY SUPPLY INDUSTRIES

The state has a close involvement in energy supply. The Sierra Leone Electricity Corporation is controlled by the Ministry of Energy and will increase in importance if major hydro-electric projects proceed. The state also has a controlling interest in the national oil refinery.

Oil

Consumption of oil is relatively static, with imports of crude oil for processing at the refinery close to 360,000 tonnes per annum. The refinery, which is located at Freetown, has a throughput capacity of 500,000 tonnes per annum and meets most requirements for finished products. The state has a controlling interest in the refinery company. Other participants are the companies involved in distribution and marketing of products and include BP and Shell.

Electricity

Electricity consumption is a little over 200 GWh per annum and rising at only one per cent per annum. Supply of electricity is from thermal power plant totalling 95 MW. The publicly owned Sierra Leone Electricity Corporation is responsible for electrification of the economy, but 50 per cent of generating capacity is still operated by industrial auto-producers. A significant boost to local production of electricity and the process of electrification will take place if the proposed Mano River Project, to be undertaken jointly with Liberia, is achieved.

Singapore

KEY ENERGY INDICATORS

Energy Consumption	
—million tonnes oil equivalent	3.9
Consumption Per Head	
—tonnes oil equivalent	1.62
—percentage of world average	100%
Net Energy Imports	100%
Oil Import-Dependence	100%

Singapore has a relatively complex and developed economy, with active commercial and light industrial sectors. Heavy industry is not significant, except that Singapore is location for several major international export refineries. Energy consumption in refining operations is significant and raises total energy consumption per head to world average levels. The country lacks indigenous energy resources and energy supply is influenced by the presence of these refineries.

ENERGY MARKET TRENDS

Despite a complete dependence on foreign sources for its energy supplies the Singapore economy has not been adversely affected by problems of availability or cost. Gross Domestic Product continued to rise during the latter part of the 1970s, increasing by more than 50 per cent between 1975 and 1980, the rate not slackening in 1979 and 1980 when many other industrially based economies were suffering from international recession. The resilience of the economy stems from its low energy intensity.

ENERGY SUPPLY INDUSTRIES

Oil

The location of Singapore in relation to Far East and Pacific region markets and to sources of crude oil in the Middle East and ASEAN countries led to it becoming a focus for export-orientated refineries. These combined the economies of scale in crude oil supply and refining with the small-scale requirements of many marketing operations in the area.

The Royal Dutch/Shell group established one of its principal refineries at Pulau Bukom, the capacity of which is well over 20 million tonnes per annum. The Exxon and Mobil groups have refineries of 12 and nine million tonnes per annum capacity respectively. Singapore Refining Company, a joint-venture of BP, Caltex and Singapore Petroleum Company, has eight million tonnes of capacity.

Gas

Gas is supplied in most of the urban areas and is the responsibility of the Public Utilities Board (PUB). Sales of gas in 1980 were 614 million cubic metres, equivalent to around 550 teracalories. Consumption has continued to increase steadily, the increase in the period of 1975-80 being more than 40 per cent. The PUB operates a modern naphtha reforming plant at Kallang as main source of supply to its network. Supplementary production is from LPG plants located at various points in the distribution system. The Board's sales are mainly to public housing areas and to industry. Demand from private housing has been declining, with bottled gas the main form in use.

Electricity

The Public Utilities Board is the sole supplier of electricity for public supply in Singapore and operates capacity of 2,000 MW to meet the high level of demand. Production in 1980 was just under 7,000 GWh, an increase of 64 per cent over the five-year period from 1975. Generating plant is located principally at Senoko (1,100 MW) and Jurong (600 MW).

South Africa

KEY ENERGY INDICATORS

Energy Consumption	
—million tonnes oil equivalent	61.0
Consumption Per Head	
—tonnes oil equivalent	2.14
—percentage of world average	132%
Net Energy Imports	20%
Oil Import-Dependence	100%

South Africa has a unique energy supply situation resulting from its policy of self-sufficiency based on coal, which the country has in abundance, but which is also the only form of energy available in quantity. Efforts to find commercial deposits of oil and gas have been very unrewarding and hydro-electric potential is limited. Uranium is produced in large quantities, but nuclear power has only recently been considered and policy is in any case to rely on foreign sources for the enriched fuel necessary for any reactors which will be built.

Energy consumption in South Africa is well above the world average on a per capita basis, although the actual level of energy consumption since 1978 is a matter of estimation in view of the shroud of secrecy surrounding the use of conventional crude oil and refined products. South Africa is in the forefront of producing oil from coal and is building large-scale plants to increase this capability. But at present imported crude oil remains an important element in the energy picture.

ENERGY MARKET TRENDS

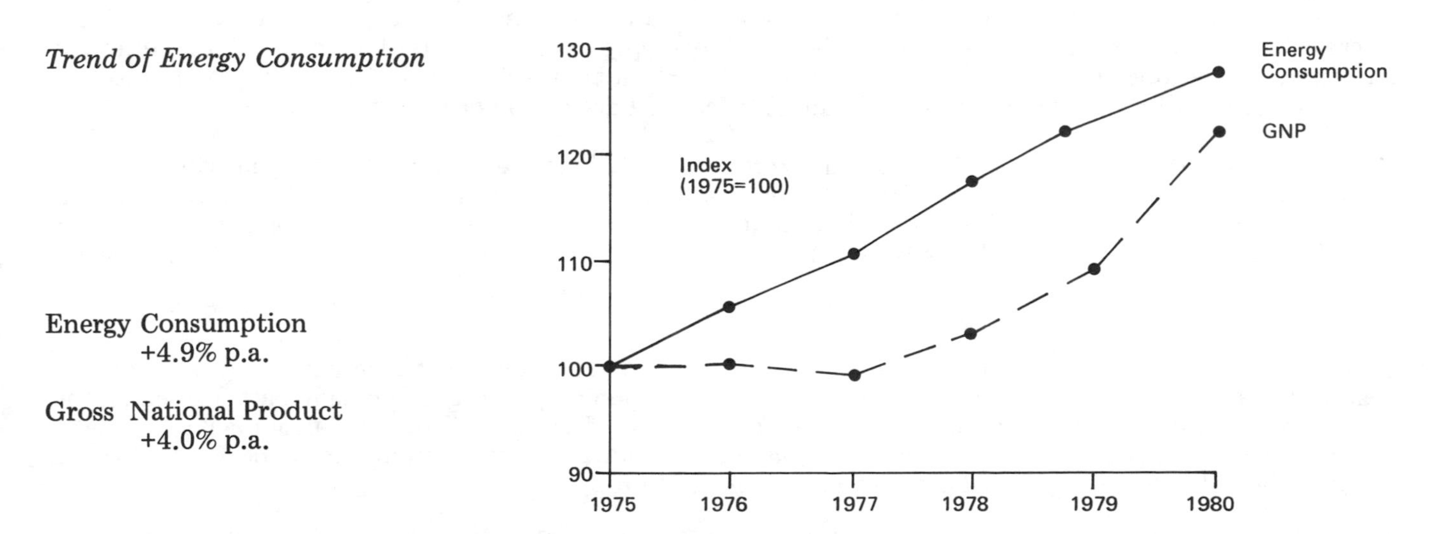

The South African economy showed a strong resurgence in activity during the years 1978-80 following a period of stagnation. Gross National Product for the period 1975-80 rose by 22 per cent. Energy consumption appears to have followed a more steady upward path, although the level of crude oil processing in each year is not known. Consumption of energy other than oil (coal, together with primary electricity imported from Mozambique) has increased by at least four per cent in each year with an overall increase between 1975 and 1980 of 25-30 per cent. This steady growth reflects the healthy state of the energy-intensive mining sector and related processing industries, which account for such a large proportion of total energy consumption.

Pattern of Energy Supply and Consumption

Primary Fuel Supply (percentage)		Final Consumption (percentage)	
Oil	15	Industry	48
Solid Fuel	85	Mining	9
Natural Gas	–	Transport	28
Primary Electricity	..	Other	15
TOTAL	100	TOTAL	100

Coal is the dominant form of fuel in the economy, providing around 85 per cent of all primary energy supply. It is used for virtually all electricity generation and is used as feedstock for conversion plant producing oil and gas. Conventional crude oil makes up the balance of energy supply and, pending development of the large new plants at Sekunda, still accounts for the major part of liquid fuels. Primary electricity will also make a contribution with the commissioning of the Koeberg nuclear reactor in the mid 1980s.

The South African economy makes heavy demands on energy. The mining industry itself takes a significant percentage and other industries, including iron and steel production and a wide range of manufacturing activities, absorb nearly half of final energy. There is also a high demand for energy in transportation, reflecting distances between main centres and the well developed state of transport systems.

ENERGY SUPPLY INDUSTRIES

Oil

South Africa has not so far discovered any commercial deposits of oil, and the few shows off the west coast have not been adequate to generate private sector interest. A programme of exploration continues to be undertaken by SOEKOR, the state-owned oil and gas exploration company. Emphasis in the drive towards reducing dependence on imported oil continues to be on increasing the scale of operations of SASOL, which runs complex plant at Sasolburg, Transvaal, producing oil from coal. The SASOL plant, now known as SASOL I is of a scale an order of magnitude larger than anywhere else and uses some 14 million tonnes of coal per annum to produce gasoline and chemical feedstocks.

SASOL has in the past been entirely state financed, but the even larger plants planned as SASOL II and SASOL III are intended to be more self-financing with public loan and share capital. Local South African interests are not otherwise significantly involved in the supply of oil products. The major oil refineries and marketing operations are run by affiliates of international oil companies. Shell, BP, Mobil and Caltex, the joint operation of Standard Oil of California and Texaco, all have refining capacity.

Linked to SASOL's activities is the Sasolburg refinery of National Petroleum Refiners, in which SASOL Limited has a majority interest. This refinery processes oil from SASOL's plant as well as conventional crude oil. The French group CFP has a shareholding as base for its marketing. During the last 10 years Trek Petroleum has grown as the only South African based oil distributor and marketer of significance.

Coal

Coal is one of the cornerstones of the South African economy. Not only is it used directly as a fuel, but it is almost the sole source of electricity produced in the Republic, plays an important part in fuelling the railway system and is used on an increasing scale as feedstock for SASOL's oil-from-coal plants. Its role in the economy is being further enhanced as mining and trading companies exploit the scope for exports to energy deficient countries, particularly Japan and Western Europe.

Production of coal, which is largely of sub-bituminous type, with a small proportion of anthracite, is undertaken by the major mining companies operating in South Africa—GENCOR, Anglo-American Corporation, Rand Mines, Lonrho, Goldfields of South Africa and Johannesburg Consolidated Investment Company. SASOL and ISCOR, the South African Iron and Steel Corporation, also operate mines dedicated to their own electricity requirements. In addition there are more than twenty smaller mine operators.

More than half of South African production is derived from mines operated by Anglo-American Corporation (AAC) and GENCOR. AAC production was some 37 million tonnes in 1980. GENCOR's production also exceeded 30 million tonnes. AAC and GENCOR also provide a large part of the coal exported by the

Coal Production and Consumption

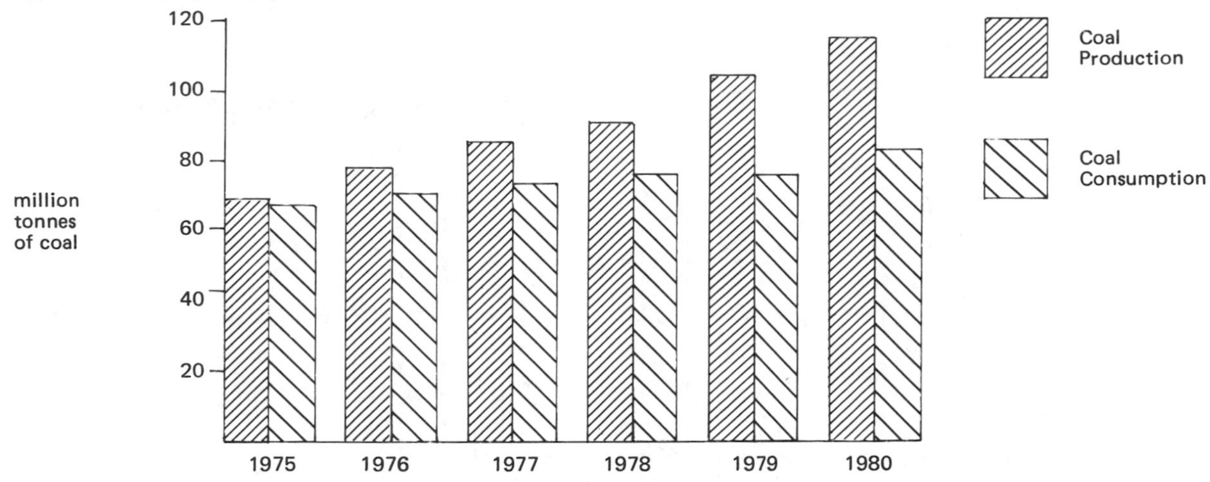

Transvaal Coal Owners' Association (TCOA), which is the main vehicle for South African coal exports through the terminal at Richards Bay. Coal export licenses are also held by mining companies on an individual basis.

Inland sales of coal are mainly handled by producers' associations. For the 20 per cent of production not tied to individual consumers under long-term arrangements such as with ESCOM, ISCOR or SASOL, distribution and marketing is carried out through the TCOA, the Anthracite Producers Association, Natal Associates Collieries and the recently formed Independent Coal Producers Association, which represents a number of smaller producers.

TCOA, and the other producers' organisations to a lesser extent, perform a co-ordinated marketing function on behalf of mining companies, the main concern of which is to operate individual mines or groups of mines. This type of role is even more explicitly undertaken by the Chamber of Mines on behalf of the industry in the fields of industrial relations, recruitment and technical development. The state retains only limited functions, principally in licensing mines on grounds of safety etc. Export allocations are also strictly controlled by the state.

Gas

Gas plays a very small part in the energy economy of South Africa. No significant deposits of natural gas have been found onshore and those discovered offshore have not so far been considered commercially viable. SOEKOR, the state owned oil and gas exploration company has a continuing programme of exploration work, but prospects do not appear to be promising.

Towns gas systems have developed in some of the major centres of population, based on the use of coal. The Johannesburg city authority operates coal gasification plant with a capacity of around 100,000 tonnes of coal per annum, and smaller scale operations are undertaken by Cape Gas Company in Cape Town and by Port Elizabeth Municipality.

The largest source of gas currently is the SASOL oil-from-coal complex at Sasolburg. The processes used here generate significant volumes of gas which cannot be used within the plant. A separate company, GASCOR, a subsidiary of SASOL, distributes this gas to bulk consumers in the Witwatersrand/Vereeniging area, both to industrial plants and to Johannesburg Municipality.

Electricity

The state-owned Electricity Supply Commission (ESCOM) is responsible for the generation and transmission of electricity for the public supply system. Its sales represent around 93 per cent of all electricity consumption, with the balance being produced by mining and industrial companies for their own internal use. ESCOM sells very large amounts of electricity to bulk energy-consuming industrial and mining establishments, and these groups account for well over 60 per cent of ESCOM's sales. ESCOM also supplies power direct to South African Railways.

Retail sales to the commercial and residential sectors is mostly handled by local municipal undertakings, which buy in bulk from ESCOM. Additional areas being supplied under similar arrangements as separate

Electricity Production and Consumption

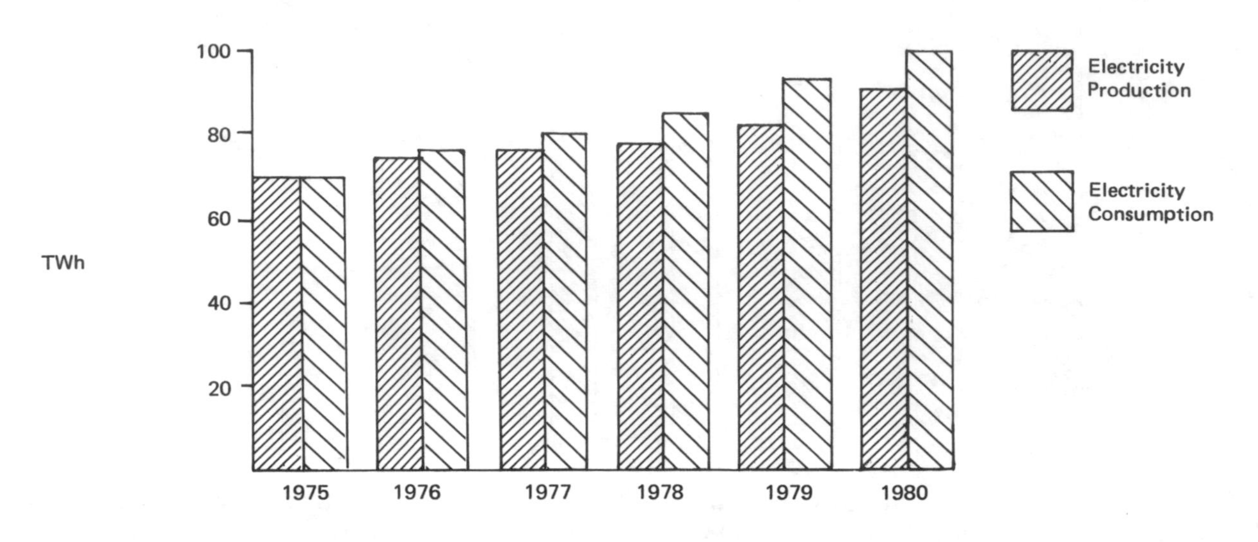

distribution operations include the Transkei and other 'homeland' areas. ESCOM itself is organised on a regional basis as far as generation and supply operations are concerned. Most significant of these is the Rand/Orange Free State region which accounts for over 55 per cent of total consumption, owing to the great concentration in that area of mining and industrial activity. Other regions cover Natal, Eastern Transvaal and Cape Province, which is divided into Northern, Eastern and Western divisions.

ESCOM is unusual in that its electricity is generated virtually entirely from coal-fired stations, which are concentrated in the coal mining areas of the Rand and Northern Natal. Some 83 per cent of ESCOM's capacity is located in Transvaal and Orange Free State. Outside this region other forms of generation are emerging. In addition to the supply from Cabora Bassa dam in Mozambique, which meets almost 10 per cent of ESCOM's requirements, nuclear capacity and a pumped storage scheme are being developed.

At Koeberg, 30km north of Cape Town, the country's first nuclear power station is being built. Work commenced in 1976 on this 1,844 MW station and commissioning is expected in 1982-83. Cape Town is remote from the coal fields and Koeberg station would avoid long distance transmission of electricity from the Rand area. A pumped storage scheme in Natal province is also designed to utilise low-cost base-load coal-fired capacity in order to help meet peak load requirements in the Durban area.

Uranium

The raw material for uranium processing has for many years been produced as a by-product of gold mining operations. It is a characteristic of the Rand area that gold and uranium tend to occur together, but until relatively recently there has been little requirement to extract uranium. The economic viability of uranium production therefore depends at least in part on the valuation applicable to gold as well as to the market value for uranium itself.

Production of uranium metal is currently of the order of 5,000 tonnes per annum. Capacity is expected to be increased during the 1980s in order to have a production capability of 8,200 tonnes by 1984. However, it is believed that this figure could be as high as 11,000 tonnes if the demand arose. There are some 18 processing plants in operation, fed by more than 20 mines. Uranium production therefore is being carried out by the main gold mining companies, such as Anglo-American Corporation, GENCOR and Goldfields of South Africa.

In 1979 uranium production in South Africa amounted to 11.4 per cent of Western output. There is the potential to raise this level in view of the reserves held by the country. Total resources economically recoverable at less than $50 per pound of uranium oxide were estimated at 1st January 1979 to be over half a million tonnes.

In addition to the reserves and production position of South Africa itself is the development at Rössing, in Namibia. The Rössing mine, which is one of the largest uranium extraction operations in the world, is located some 70 km from Swakopmund, and has been in commercial production since 1976. This mine produces around 5,000 tonnes of uranium oxide annually.

ENERGY TRADE

Net Imports/(Exports) 1975-80

	1975	1976	1977	1978	1979	1980
Coal (million tonnes)	(2.3)	(6.3)	(12.1)	(15.6)	(23.3)	(29.2)
Electricity (TWh)	..	1.2	4.2	6.9	10.4	9.6

Since 1975 energy trade has grown to substantial proportions. On the export side movements of coal from South African terminals rose from around two million tonnes to over 29 million tonnes in 1980. In net energy terms, however, this has been to some extent offset by the import of electricity from the Cabora Bassa dam in Mozambique, which came into service in 1975. Imports built up rapidly from 1.2 TWh in 1976 to 10.4 TWh in 1979. The quantity in 1980 was less, owing to disruption of the overland links. South Africa will remain a net importer of energy for some time as oil must be wholly imported. Firm data on movements of oil since 1978 have been kept undisclosed for security reasons.

Destination of Coal Exports

(percentage)	
France	32
Japan	11
Denmark	11
Italy	11
Belgium/Luxembourg	7
Taiwan	6
West Germany	5
Korea and others	17
TOTAL	100

The South African coal export trade has built-up during the last six years as industrial energy-importing countries have switched from oil to coal. France is a leading destination for coal exports and several other West European countries, which have switched electricity generation towards coal, are also prominent. Other main export markets are in the Far East, with Japan, Taiwan and Korea all involved.

ENERGY POLICY

Energy policy is dominated by the over-riding need to keep dependence on external sources to a minimum. At the same time energy-intensive mining and treatment activities are an integral part of the economy and self-sufficiency in basic manufactures, which involves other energy-intensive operations, such as steel-making, is also an objective. Energy policy therefore has to reconcile the limitations of indigenous energy resources with inevitably rising demand for energy.

Energy Supply

Since 1973 government control over the importation of crude oil and petroleum products and over the exportation of coal has increased. The government retains close supervision of all aspects of the supply of oil, including operation of a strategic stockpile. The continuing dependence on imported oil is expected to be reduced markedly with the commissioning of SASOL's new plants during the early 1980s. Coal exports are licensed by the government and maximum levels have been raised consistent with maintenance of an adequate reserves:production ratio.

Diversification away from coal is difficult within the policy parameter of limiting the risk of dependence on overseas supplies, as oil and gas exploration results have been very disappointing. The construction of the country's first nuclear power station introduces a new element into the equation, but no plans have been announced for a programme of nuclear power plant construction.

Conservation

Apart from seeking further ways of utilising coal for energy and feedstock purposes, attention is also being paid to the scope for producing fuel components from biomass and increasing the applications of solar equipment. An alcohol/gasoline blend is already being marketed in the Witwatersrand area. A regulatory approach to energy conservation is being largely avoided so far, with emphasis on information and advisory services to the public, to industry and commerce. Increasing the efficiency of fuel usage will also become important in view of the policy of increasing the degree of beneficiation in mineral processing industries.

South Yemen

KEY ENERGY INDICATORS

Energy Consumption	
—million tonnes oil equivalent	0.7
Consumption Per Head	
—tonnes oil equivalent	0.35
—percentage of world average	22%
Net Energy Imports	100%
Oil Import-Dependence	100%

South Yemen, the Yemen Democratic Republic, remains at a very low level of economic development. Energy consumption per head of population is little more than a fifth of the world average. The country is entirely dependent on imports of oil for its energy requirements. The availability of oil for South Yemen and for processing in the Aden refinery has sometimes also been affected by political relations between the state and other Arab countries.

ENERGY SUPPLY INDUSTRIES

Reflecting the political ideology of the government the state is of predominant importance in industrial activity. In the oil sector it has established the Yemen National Oil Company. The state also has full responsibility for public supplies of electricity.

Oil

The supply of oil products in South Yemen is the responsibility of the Yemen National Oil Company (YNOC). YNOC either imports refined products or processes crude oil at the Aden refinery. Internal consumption of oil products is around 700,000 tonnes per annum, with additional quantities required for bunkers. The Aden refinery has a nominal capacity of 8-9 million tonnes per annum, having been built as an international refinery to meet the large bunkering market in that region. Closure of the Suez Canal in 1967 and the subsequent trend towards use of VLCCs by-passing the area reduced this demand considerably. The refinery is therefore used mainly to meet local demand, and has a large amount of spare capacity, which is partially used for export-processing.

Electricity

Electricity demand has been rising at a low rate of less than four per cent per annum. Total consumption is around 250 GWh per annum, of which 75 GWh is produced within industrial operations for own consumption. Generating capacity, which is all oil-fired, is only 100 MW.

ENERGY TRADE

Net Imports/(Exports) 1975-79

	1975	1976	1977	1978	1979
Crude Oil (million tonnes)	1.5	1.6	1.9	1.9	2.0
Oil Products (million tonnes)	(0.7)	(0.4)	(0.5)	(0.5)	(0.5)

South Yemen's energy trade consists of oil imported for its own requirements and crude oil which is processed at the Aden refinery for other markets. Several Arab countries use the refinery in this way. Around half a million tonnes per annum of finished products are exported to countries, mainly in the Indian Ocean area, by the crude oil supply companies.

Sri Lanka

KEY ENERGY INDICATORS

Energy Consumption	
—million tonnes oil equivalent	1.3
Consumption Per Head	
—tonnes oil equivalent	0.09
—percentage of world average	6%
Net Energy Imports	75%
Oil Import-Dependence	100%

Energy consumption per head of population in Sri Lanka is amongst the lowest in Asia at no more than six per cent of the world average, with total annual primary energy consumption of less than one and a half million tonnes of oil equivalent. Oil is the mainstay of the economy, meeting three-quarters of total energy requirements, though Sri Lanka is entirely dependent on imports. There is useful hydro-electric potential which is being exploited. Other indigenous energy resources are lacking, but recent discoveries of oil and gas in Indian areas of the Palk Strait have introduced a new element of hope into the Sri Lankan supply picture.

ENERGY MARKET TRENDS

Despite a high degree of dependence on imports of energy the Sri Lankan economy has continued to grow, the increase in real Gross Domestic Product between 1975 and 1980 being more than 30 per cent. Total energy consumption has risen more rapidly. Sales of principal oil products increased by 40 per cent and primary electricity production rose by around 34 per cent.

Pattern of Energy Supply

Primary Fuel Supply (percentage)	
Oil	75
Solid Fuel	..
Natural Gas	–
Primary Electricity	25
TOTAL	100

Sri Lanka uses oil to meet three-quarters of its energy requirements. Oil has maintained this predominance, despite the need to rely wholly on foreign sources. The balance of energy supplies is derived almost entirely from hydro-electric plant, which also continues to be expanded as advantage is taken of the potential of rivers in the central mountain region. Limited quantities of solid fuel are imported.

ENERGY SUPPLY INDUSTRIES

Energy supply is controlled entirely by state undertakings. Importation of crude oil, oil refining and sale of petroleum products is the responsibility of Ceylon Petroleum Corporation. Electricity generation, transmission and distribution is carried out by the Ceylon Electricity Board. Towns gas in the Colombo area is also supplied by a state-owned company.

Oil

Sales of petroleum products rose steadily from 1976 to reach more than one million tonnes in 1980. Overall increase in the four-year period was 41 per cent, although it was not until 1980 that sales regained the level of 1973, prior to the oil supply difficulties and economic recession. The increase in sales would have been greater, but for price increases introduced in 1979 and 1980 to curb consumption of gasoline, kerosene and automotive diesel. But 1980 sales of heavy fuel oil were particularly high as thermal power plant was used to compensate for poor hydro-electric production conditions.

Supply of oil products to the domestic market is in the hands of the state owned Ceylon Petroleum Corporation (CPC). CPC operates the country's only oil refinery, at Sapugaskanda, which has a crude oil processing capacity of 2.5 million tonnes per annum. CPC purchases crude oil from Middle East countries and exports some products to countries in the Indian Ocean area and further afield.

Electricity

Electricity production in 1980 reached 1,670 GWh, an increase of 45 per cent over the level of 1975. This is greater than the increase in consumption of oil, which is the country's main source of energy, but this was in fact due to the need to increase the use of thermal generation considerably during 1980 as there was inadequate water supply for hydro-electric plant. Output of hydro-electricity in the period 1975-80 rose by only 34 per cent.

Electricity production, transmission and distribution are the responsibility of the Ceylon Electricity Board. The Board operates 330 MW of hydro-electric plant and maintains 90 MW of thermal plant for peak demand and standby purposes. New hydro-electric capacity utilising the Mahaveli River potential is expected to provide a further 450 MW by 1984.

ENERGY TRADE

Net Imports/(Exports) 1980

Crude Oil (million tonnes)	1.6
Oil Products (million tonnes)	(0.5)

Ceylon Petroleum Corporation imports some 1.6 million tonnes per annum of crude oil meeting the bulk of domestic market requirements and producing substantial quantities for export to Aden, India and other countries, including Egypt and the Philippines. Additional products are imported from Kuwait and Singapore.

Sources of Imported Crude Oil

(percentage)	
Saudi Arabia	51
Iraq	25
Iran	24
TOTAL	100

Saudi Arabia is the principal source of crude oil supplying just over half of Ceylon Petroleum Corporation's imports in 1980. The balance was evenly divided between Iran and Iraq.

Sudan

KEY ENERGY INDICATORS

Energy Consumption	
—million tonnes oil equivalent	1.6
Consumption Per Head	
—tonnes oil equivalent	0.09
—percentage of world average	6%
Net Energy Imports	92%
Oil Import-Dependence	100%

Sudan is the largest country in Africa and one with considerable potential. Economic development is patchy and subject to continuing foreign exchange and internal political problems. The agricultural resources of the country are well known, but evaluation of mineral wealth and energy resources is still at an early stage. The recent discovery of oil in southern Sudan has generated increased exploration activity, both onshore and offshore, and has illuminated the development prospects of the country.

ENERGY MARKET TRENDS

Economic development progressed well in the mid 1970s but was brought to a halt in 1979 when a balance of payments crisis led to an austerity programme. This was in part caused by the failure to gain full benefit from the agricultural resources which represent the main strength of the economy. Lack of foreign exchange prevented demand for oil being met and consequently energy consumption fell in 1979 to below the 1975 level.

Pattern of Energy Supply

Primary Fuel Supply (percentage)	
Oil	92
Solid Fuel	–
Natural Gas	–
Primary Electricity	8
TOTAL	100

Until recently Sudan's indigenous resources of conventional energy were thought to lie only in hydro-electric potential, the value of which has been affected by the highly seasonal nature of the flow of water. However, a significant level of crude oil production is now anticipated following discoveries of crude oil. These may be large enough to achieve self-sufficiency eventually.

ENERGY SUPPLY INDUSTRIES

The state controls all activities in the energy supply industries. A state oil company has been established to give effect to this. There is a public electricity supply system in Khartoum and main centres, but the larger industrial establishments produce their own electricity. Private companies are encouraged to participate in exploration for minerals and energy and a number of American and European oil companies are engaged in exploration and development projects.

Oil

Oil is the predominant form of commercial energy being used in Sudan. Total consumption is around 1.5 million tonnes per annum. As yet there is no local production of oil and imports of crude oil and finished products are brought in through Port Sudan, where there is a refinery. The refinery is a joint-venture between the state and two leading marketing companies, Shell and BP, and is linked by a products pipeline to Khartoum.

Several oil fields have been identified during the last two years in the southern part of the country. Standard Oil of California has been leading the exploration campaign, which seems certain to lead to early exploitation of the deposits and involving construction of a new oil refinery, probably at Kosti, a regional centre south of Khartoum.

Electricity

Consumption of electricity is around 900 GWh per annum and is met by a combination of hydro-electric and thermal plant. The public electricity supply undertaking operates hydro-electric plant at the Roseires and Sennar dams on the Blue Nile. However, in the dry season most electricity has to be generated by thermal plant and private industrial generation also uses oil-fired plant. Current projects include the expansion of the Roseires station to 250 MW and the installation of new thermal plant totalling 100 MW at Khartoum and Burri.

Syria

KEY ENERGY INDICATORS

Energy Consumption	
—million tonnes oil equivalent	5.6
Consumption Per Head	
—tonnes oil equivalent	0.62
—percentage of world average	38%
Net Energy Imports	nil
Oil-Import-Dependence	nil

Energy consumption in Syria remains low by world standards and relative to most Middle East countries. Consumption has tended to increase at around five per cent per annum, but is unlikely to rise more quickly until further industrial development takes place. The country has a firm base of oil production and useful natural gas resources, which are being little utilised currently. Syria is a member of the Organisation of Arab Petroleum Exporting Countries.

ENERGY MARKET TRENDS

The Syrian economy has progressed erratically in the period since 1975, with an overall increase in Gross Domestic Product of the order of 45 per cent for the five-year period to 1980. Energy consumption rose at a slightly higher rate, with the demand for electricity particularly strong. However, industrial and commercial development remains limited and energy consumption trends reflect at least in part the level of activity in the oil refining sector.

Pattern of Energy Supply

Primary Fuel Supply (percentage)	
Oil	81
Solid Fuel	..
Natural Gas	9
Primary Electricity	10
TOTAL	100

Oil is Syria's principal energy resource and a cornerstone of the economy. Output exceeds the country's own level of demand. Oil accounts for four-fifths of energy consumed. The balance is split approximately equally between natural gas, of which there is also ready availability, and hydro-electricity, derived from plant on the Euphrates River. At present a high proportion of electricity is produced from hydro-electric plant.

ENERGY SUPPLY INDUSTRIES

The supply of energy is entirely undertaken by state-owned organisations. Separate companies are responsible for exploration and production, refining, transportation and distribution in the oil sector. The Establishment for Electricity controls operations of the public supply grid. Some encouragement has, however, been given to foreign companies, in an attempt to discover and prove additional oil and gas reserves.

Oil

Oil Production and Consumption

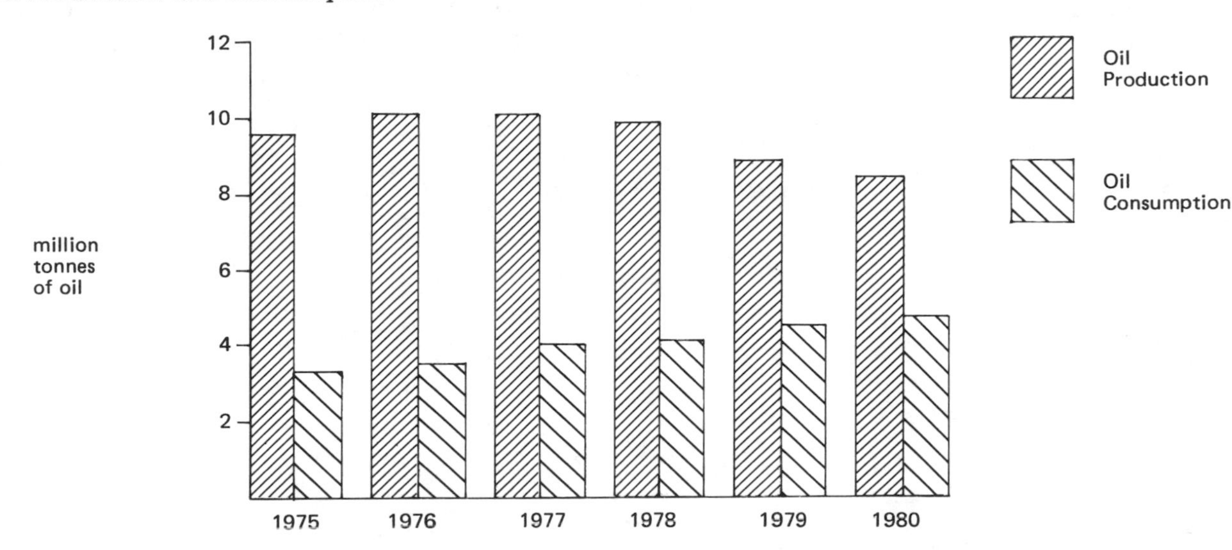

Syria has a modest level of oil production by Middle East standards, output having peaked at around 10 million tonnes per annum between 1976 and 1978. Since that time the level has fallen back to 8.5 million tonnes, although existing proven reserves are reported to be around 300 million tonnes. Annual output is nevertheless well above the requirements of the internal market, although these are now rising towards five million tonnes per annum.

Syrian crude oil is produced in the north-eastern part of the country, but refineries are located in the west at Homs and Banias, which is on the Mediterranean. Capacities of these refineries are six million tonnes and five million tonnes per annum respectively. Banias is the terminal for the Iraq export pipeline from Kirkuk to the Mediterranean and Homs is also linked to the pipeline. Their location enables them to use Iraqi crude oil in conjunction with indigenous oil to provide a better blend for refining purposes. The major part of Syrian crude oil production is thus exported.

All oil industry operations are undertaken by state companies. Syrian Petroleum Company is responsible for exploration and production. Other companies operate the Banias and Homs refineries and the Syrian Company for Oil Transport, which was formed to take-over the pipeline facilities formerly owned by the Iraq Petroleum Company, is responsible for shipment of crude oil to the refineries and Banias export terminal. SADCOP, the Syrian Storage and Distribution Company for Petroleum Products, handles the distribution and marketing of refined products.

Electricity

The main impact of economic development has been an increasing demand for electricity. This has also been a reflection of the government's programme of increasing the degree of electrification in rural areas. Production of electricity in 1980 reached 4,000 GWh, compared with less than 1,700 GWh in 1975. Significant increases in demand were recorded in each year.

Around 95 per cent of production is from the public power stations controlled by the Establishment for Electricity. More than 60 per cent of electricity is produced from hydro-electric sources, but the proportion has been decling as additional demand has been met by new thermal capacity. Since 1980 some 230 MW have been added to the country's total generating capacity of 1,500 MW and a 300 MW oil-fired station is planned for the Damascus area.

ENERGY TRADE

Net Imports/(Exports) 1975-79

	1975	1976	1977	1978	1979
Crude Oil (million tonnes)	(7.1)	(6.9)	(5.0)	(5.9)	(0.1)
Oil Products (million tonnes)	1.0	0.9	1.1	0.5	(3.2)

Syria maintains a useful net export of oil, either as crude oil or as refined products. However, the total has been declining steadily as crude oil production has tailed off and internal demand has continued to rise. In 1975 net oil exports totalled just over six million tonnes. In 1979 the figure was only 3.3 million tonnes. Compensating to a limited extent for this was a switch away from crude oil to finished products in the pattern of exports.

Taiwan

KEY ENERGY INDICATORS

Energy Consumption	
—million tonnes oil equivalent	27.0
Consumption Per Head	
—tonnes oil equivalent	1.52
—percentage of world average	94%
Net Energy Imports	78%
Oil Import-Dependence	99%

Taiwan has developed rapidly in the post-war period to become an industrialised nation with per capita consumption of energy close to the world average. But it has limited indigenous resources and is almost completely dependent on overseas sources for crude oil, the main source of energy for the economy. The country is therefore likely to expand its use of coal, albeit imported from Australia, South Africa and elsewhere, in an attempt to diversify away from the use of oil and there are plans to import liquefied natural gas from the mid 1980s if no significant indigenous resources are located. Already the state electricity authority has embarked on a major programme of nuclear power station construction.

ENERGY MARKET TRENDS

The Taiwan economy expanded rapidly during the second half of the 1970s, notwithstanding its dependence on imported energy. Gross National Product increased by 64 per cent during the period, although the rate of growth had tailed off by 1980 to a 6.6 per cent increase over the previous year. Demand for energy rose even more rapidly, by more than 80 per cent over the five-year period. However, in 1979 and 1980 the relationship of growth of energy consumption to growth of GNP indicated some improvement in the efficiency of energy use in the economy, although also a reflection of the impact of economic recession on some of the energy-intensive sectors.

Trend of Energy Consumption

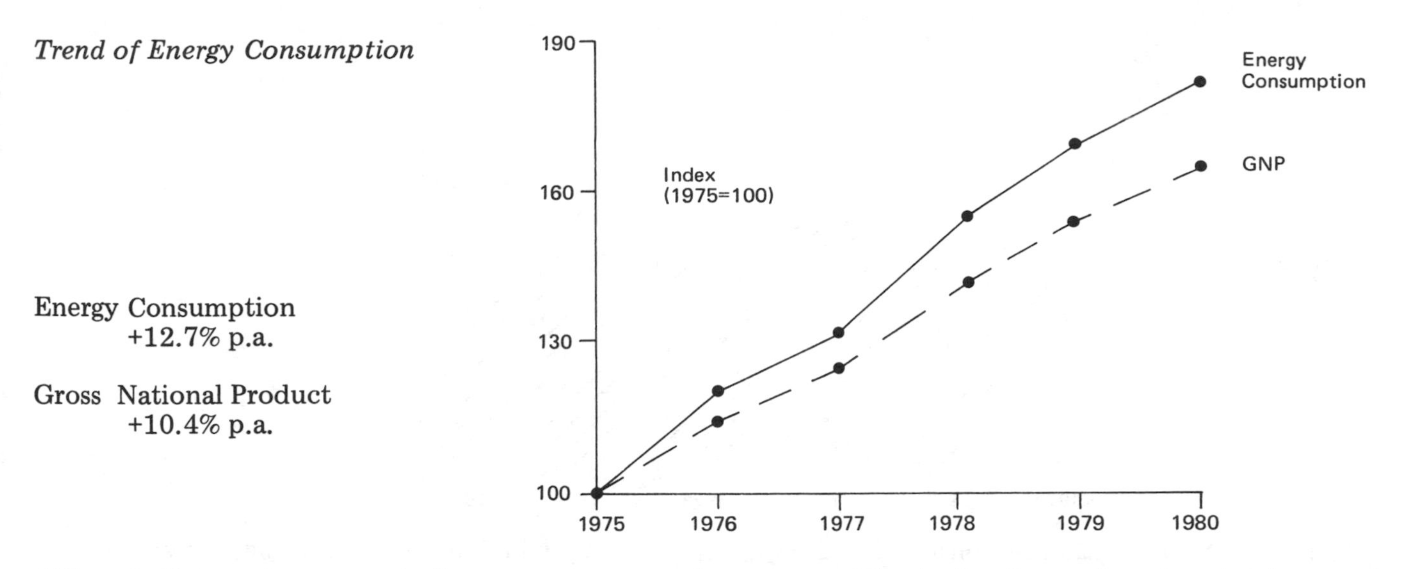

Energy Consumption
+12.7% p.a.

Gross National Product
+10.4% p.a.

Although Taiwan has some indigenous resources of energy, mainly in the form of coal, natural gas and hydro-electric potential, it is highly dependent on oil, as is usually the case with industrialised economies. Oil accounts for 69 per cent of primary energy inputs: coal only 16 per cent. The proportion of natural gas is small, but should rise, particularly if a major LNG import chain is established. Primary energy production is mainly nuclear generation, but both nuclear and hydro-electricity will be increased. As a result the role of oil may be expected to decline.

Pattern of Energy Supply and Consumption

Primary Fuel Supply (percentage)	
Oil	69
Solid Fuel	16
Natural Gas	6
Primary Electricity	9
TOTAL	100

Final Consumption (percentage)	
Industry	52
Transport	12
Other	36
TOTAL	100

The industrial sector is a dominant factor affecting energy consumption. More than half of final energy consumption is by industry. In addition a further 12 per cent is consumed for non-energy purposes such as petrochemical feedstock. The transport sector is relatively small. Residential and commercial sectors combined account for less than 20 per cent of the total.

ENERGY SUPPLY INDUSTRIES

Supply of the main forms of energy for the Taiwan economy has for many years been treated as a fundamental state responsibility. The Chinese Petroleum Corporation is responsible for all aspects of the supply of oil and gas in bulk. Taiwan Power Company (Taipower) is responsible for providing the high levels of electricity demand for the country's industries. Only in the coal industry do private firms play a significant role, but even here the state is likely to become increasingly important as efforts are made to sustain domestic production and imports rise in order to take over a major share in the production of electricity.

Oil

Production of indigenous crude oil is of negligible proportions in relation to the level of consumption. Output is only around 200,000 tonnes per annum and has been on a downward trend since 1976-77. It therefore amounts to less than one per cent of the country's needs. Large volumes of crude oil are imported by Chinese Petroleum Corporation (CPC) for processing in its refineries at Kaohsiung, in the south-west of the country, and at Taoyuan, in the north near the capital of Taipei. Kaohsiung is a major refinery with a capacity of 470,000 barrels per day, equivalent to 23.5 million tonnes per annum, and is one of the largest in Asia. The Taoyuan refinery, of capacity five million tonnes per annum, was brought on stream in 1977 to keep pace with the rising demand for oil products.

Oil Consumption

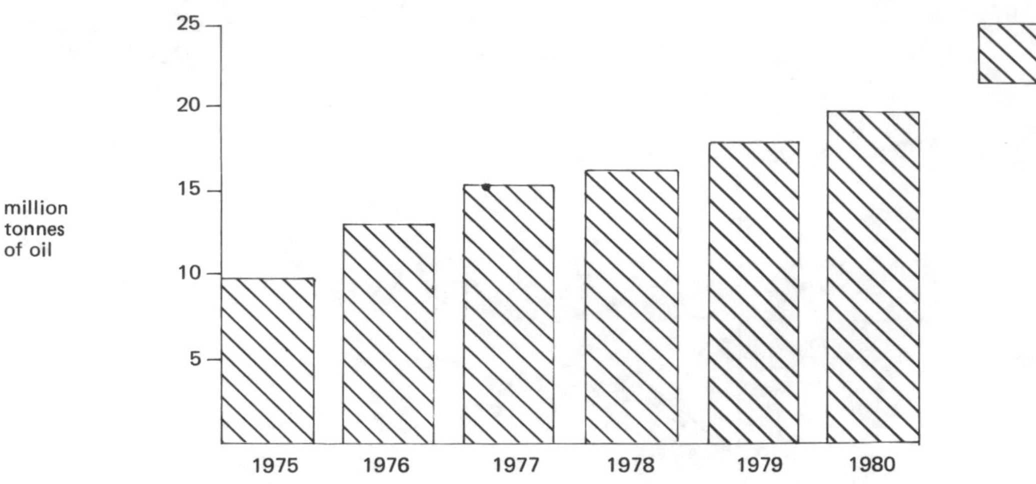

CPC is a state corporation responsible for all aspects of oil supply in Taiwan and is engaged in exploration and production activities as well as refining, distribution and marketing. In recent years CPC has also assumed responsibility for the development of the petrochemical industry. So far CPC's intensive exploration activity, both onshore and offshore, over many years has revealed some oil deposits, but these are as yet of uncertain commercial viability.

Coal

Coal Production and Consumption

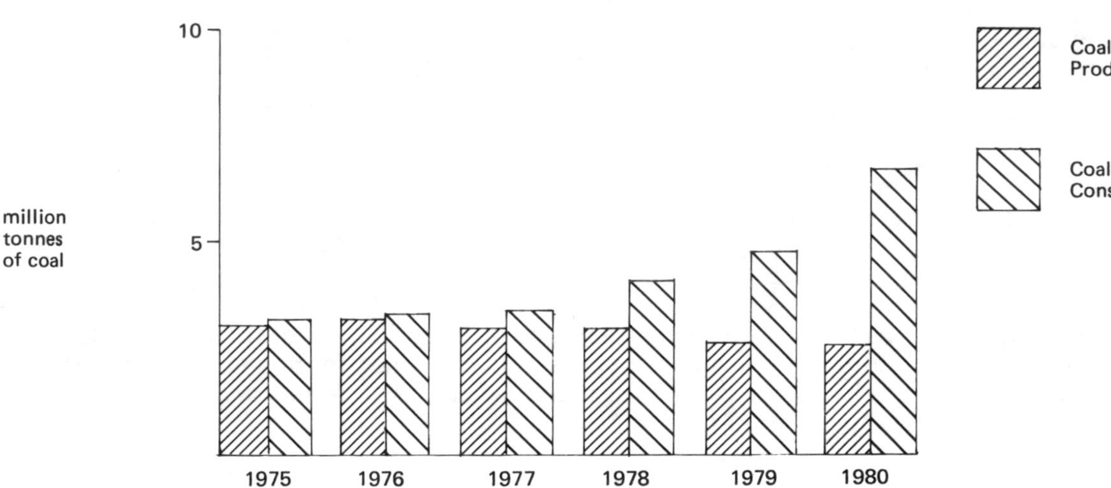

The coal industry is at present the largest individual energy producing sector in Taiwan, with output of around 2.5 million tonnes per annum. Production has been on a downward trend since the mid 1970s when output exceeded three million tonnes per annum. Total reserves in the island are estimated to be at least 200 million tonnes, and the government's long-term plans assume that the rate of output can be got back up to three million tonnes per annum.

But the industry faces difficulties in trying to reverse past trends. The industry consists of a number of private mining companies, which are experiencing difficulties common to deep-mining operations in many parts of the world. A certain degree of rationalisation is therefore envisaged and investment is required to revitalise the industry. Nevertheless, a significant expansion of the use of coal will be possible only through a rising level of imports. Already in 1980 imports exceeded domestic production and the level will continue to rise rapidly as emphasis is placed on the use of coal rather than oil in thermal power stations.

Gas

Indigenous production of natural gas was just over 1,700 million cubic metres in 1980. This was lower than in any of the previous four years, but still substantially higher than the output recorded in 1975. Exploration

Gas Production

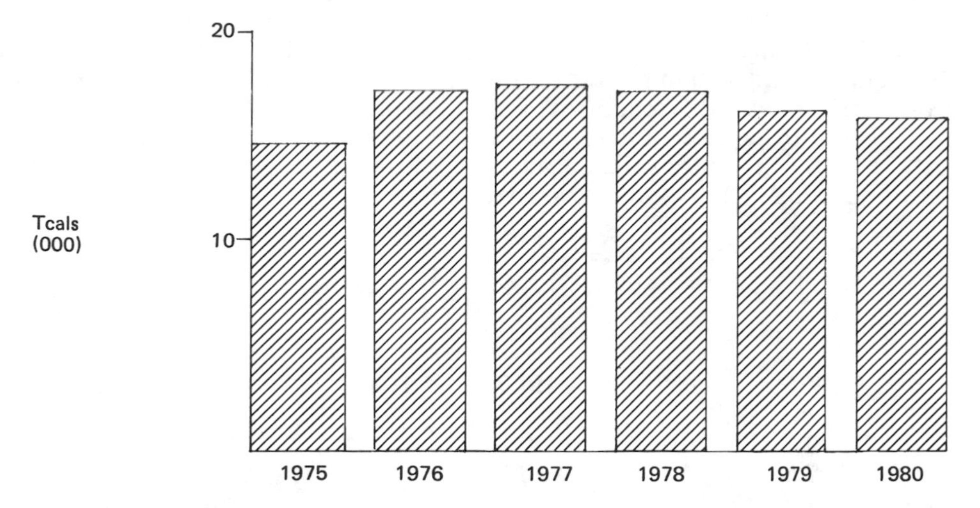

and development are continuing and the country's long-term energy plans allow for output to double by the middle of the decade. In any event the scale of the gas industry's operation will expand substantially, as liquefied natural gas will be imported to complement indigenous production.

Responsibility for the production and supply of gas lies with the long established state undertaking Chinese Petroleum Corporation. CPC also supplies an important additional supply of gas in the form of liquefied petroleum gas, total sales of which amounted to 600,000 tonnes in 1980. Local distribution and sales at retail level in the capital city region is carried out by the Great Taipei Gas Corporation.

Electricity

Electricity Production

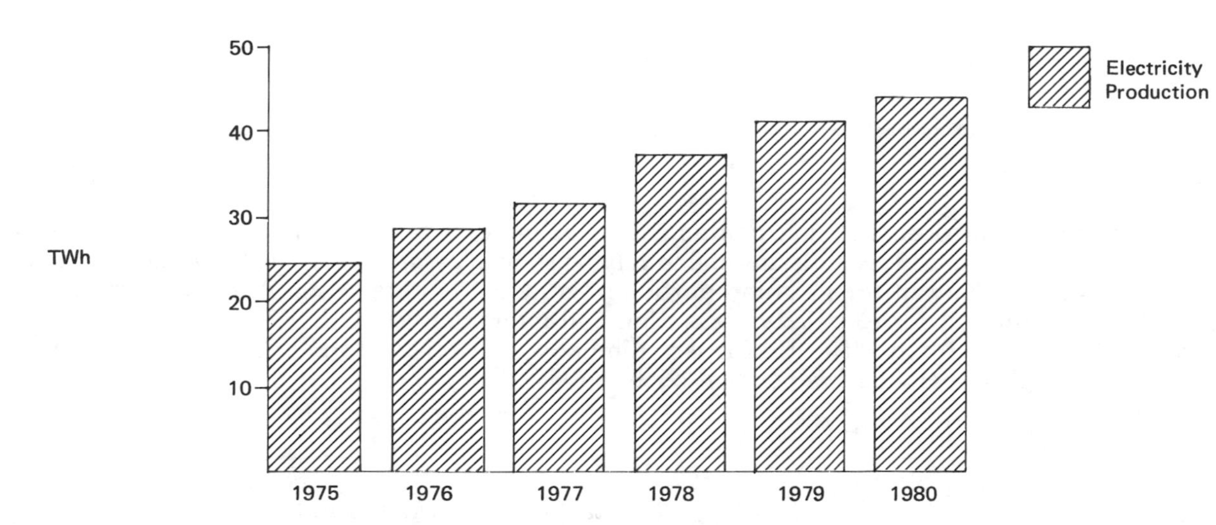

Taiwan's consumption of electricity has been rising sharply. Between 1975 and 1980 demand increased from 24.6 TWh to 43.8 TWh, an increase of 78 per cent in total energy consumption during that period. There is intensive use of electricity in Taiwan. The national grid covers almost all of the population, a total of 4.2 million customers. Even so, industrial consumers take more than 50 per cent of electricity production.

Approximately three-quarters of all electricity is generated by thermal power stations, fuelled largely on oil. In the short term this proportion is rising as new oil-fired capacity is completed. During the 1980s however, the proportion should fall substantially if construction programmes for hydro-electric plant and nuclear power stations are maintained, although more oil-fired capacity will still be required. A large increase in use of coal for electricity generation is also envisaged.

Taiwan Power Company (Taipower), a state undertaking, has sole responsibility for generation and supply of electricity. Taipower operates more than 45 power stations with a total capacity of 8.2 GW. Its first nuclear power station was commissioned in 1979, with a capacity of 1,270 MW, and a second station, of capacity just under 2,000 MW is expected to be completed by the end of 1982. At least two more nuclear stations are planned for the 1980s taking total nuclear capacity up to 7,000 MW.

CPC also has two major hydro-electric schemes under construction and planned. These are on the Mukua River in the eastern part of the country and a pumped storage project of 500 MW in Nantou county, central Taiwan. By 1990 hydro-electric capacity is planned to have trebled to 4,200 MW. In terms of capacity the greatest increase is expected to be in coal-fired plant. From only 900 MW in 1979 the total of such plant is aimed to reach 7,400 MW by the end of the decade. This should include the conversion from oil-firing of the Hsinta power station (1,000 MW) now under construction.

ENERGY TRADE

Net Imports/(Exports) 1975-80

	1975	1976	1977	1978	1979	1980
Coal (million tonnes)	0.1	0.1	0.3	1.2	2.0	4.1
Crude Oil (million tonnes)	7.5	12.4	13.6	17.1	17.4	18.1
Oil Products (million tonnes)	1.7	1.7	1.8	0.2	0.3	2.0

Taiwan's energy imports consist primarily of oil, the combined total of crude oil and finished products exceeding 20 million tonnes in 1980. This represents a steady rise since 1976, when demand recovered from the economic disruptions of the previous two years. Crude oil imports were only 7.5 million tonnes in 1975 compared with 13.7 million tonnes in the previous year. Imports have largely been of crude oil as Chinese Petroleum Corporation has maintained adequate domestic oil refining capacity. Imports of coal are now increasing rapidly as Taiwan seeks to diversify its energy base. From a mere 100,000 tonnes in 1976 imports rose to two million tonnes in 1979 and doubled to over four million tonnes in 1980.

Sources of Imported Crude Oil

(percentage)	
Kuwait	46
Saudi Arabia	39
Other Middle East	6
Other	9
TOTAL	100

Figures for 1980 reflected an overwhelming predominance of imports from just two Middle East oil exporters, Kuwait and Saudi Arabia. Taiwan is unusual in relying so heavily on such a small number of supply sources and in particular on Kuwait for a high proportion of crude oil requirements. This is in part a reflection of the demand for fuel oil for electricity generation and industrial use. Other Middle East countries with light crude oils exported very little to Taiwan. Similarly, the relatively near sources of Malaysia, Indonesia and Brunei supplied less than five per cent of the total.

Sources of Imported Coal

(percentage)	
Australia	42
South Africa	41
USA	13
Other	4
TOTAL	100

During the last few years Taiwan has entered the international coal trade, importing hard coal and coke for industrial purposes. Australia and South Africa each account for more than 40 per cent of Taiwan's imports, with the balance largely coming from the United States.

ENERGY POLICIES

With state undertakings well established in the oil and electricity supply industries, the organisations are in place for implementing positive policies to provide a more secure and diversified base of energy supplies.

The government has become aware of dependence on oil to fuel its industries and of the need to draw on other potential energy sources in the same way that Japan has been doing. An important element in its approach to developing energy security is to promote co-operative ventures with mineral producing countries through broad arrangements to use Taiwan's economic and technological capacity in exchange for natural resources. Co-operation with the Philippines and Colombia is particularly likely.

Energy Supply

Although only limited levels of output of oil and gas have been achieved to date the government is planning on the possibility that both sources can increase their contribution in due course. It is hoped that more intensive exploration and development work will permit oil production to rise to 2,800 cubic metres per day (approximately 870,000 tonnes per annum) and natural gas to be maintained at over 1,800 million cubic metres per annum.

There is a strong possibiility of setting up arrangements to import liquefied natural gas during the 1980s, particularly if indigenous energy production falls short of expectations. The use of LPG will also be expanded. Coal is to be used on a wider scale, especially in electricity generation. Rationalisation of the private sector coal industry is to be encouraged in order to help domestic production to play its part. A major programme of nuclear power plant construction is already being followed.

Conservation

As an industrial country with very limited indigenous resources and a high dependence on imports for oil, government policy is to be slanted strongly towards encouraging industries which are not intensive in their use of imported resources, particularly of energy. Renewable resources of hydro-electric power are to be further exploited and greater attention is being paid to the scope for re-cycling of water for power production.

Tanzania

KEY ENERGY INDICATORS

Energy Consumption	
—million tonnes oil equivalent	0.7
Consumption Per Head	
—tonnes oil equivalent	0.04
—percentage of world average	2%
Net Energy Imports	80%
Oil Import-Dependence	100%

Tanzania ranks as one of the poorest countries in Africa, with very limited industrial and commercial development and a rural sector which has failed to achieve its potential. Energy consumption per head is only two per cent of the world average. The country has resources of coal and natural gas, as well as further hydro-electric potential, and these will need to be developed in parallel with other projects in mining and manufacturing.

ENERGY MARKET TRENDS

The poor state of the Tanzanian economy was undermined by several major events during the 1970s, which checked any upward trend. These included sharply rising oil prices, disbanding of the East African economic community and droughts. Resultant difficulties in earning sufficient foreign exchange have also restricted economic development. Consumption of energy has followed a downward trend since 1976.

Pattern of Energy Supply

Primary Fuel Supply (percentage)	
Oil	80
Solid Fuel	..
Natural Gas	–
Primary Electricity	20
TOTAL	100

Up to the present time Tanzania has relied on oil for the larger part of its energy input. Oil has had to supply all sectors of the economy as well as international aviation and marine bunker markets. Hydro-electricity has been the only indigenous source of energy of any significance. Resources of coal and natural gas will be exploited in the future, providing fuel and feedstock, and increasing both the range of fuels available and the proportion of indigenous production.

ENERGY SUPPLY INDUSTRIES

The socialisation policies of the government have concentrated most of the energy industry's operations in the hands of state agencies. Electricity generation and supply and the exploitation of coal and natural gas are entirely state controlled. Private oil companies continue to conduct oil supply operations, although under the supervision also of the government.

Oil

The oil industry supplies products to all markets in Tanzania. Supplies are obtained mainly from the industry refinery at Dar-es-Salaam, and distributed by local affiliates of the international oil companies, principally BP and Shell. Throughput at the refinery has declined from a peak of nearly one million tonnes in 1976 to only half that level as a result of balance of payments problems and disruption of the economic development programme.

Coal

At present consumption of solid fuel in the modern sector of the economy is negligible, but coal deposits have been established in the Mbeya region at Igogo. Reserves are estimated to be of the order of 50 million tonnes. Projects to utilise the coal include power generation and use in conjunction with iron ore smelting.

Gas

Gas is not currently being used in the energy supply pattern, but a substantial gas field has been discovered at Songo Songo Island in the south-east of the country. Estimates of reserves are 30 billion cubic metres, of which almost two-thirds should be recoverable.

Electricity

Total production of electricity is 700 GWh per annum of which 75 per cent is derived from hydro-electric plant. Generating capacity is 260 MW, of which 240 MW is in the public sector and operated by TANESCO the state electricity supply corporation. Future power stations are expected to rely on the introduction of coal-fired plant based on known reserves in the south-west of the country.

Thailand

KEY ENERGY INDICATORS

Energy Consumption	
—million tonnes oil equivalent	12.7
Consumption Per Head	
—tonnes oil equivalent	0.27
—percentage of world average	17%
Net Energy Imports	91%
Oil Import-Dependence	99%+

Thailand has managed to support an expanding economy, despite a high level of dependence on imported energy. Consumption per head is only 17 per cent of the world average, but likely to continue to rise as indigenous resources, including a substantial offshore gas field, are developed. Hydro-electric potential and known lignite resources are likely to be largely exploited by the end of the decade, when imports of oil, and to a significant extent coal, will be rising once more. Strenuous attempts are to be made to ensure that profligate consumption of energy as a result of low prices does not occur, and positive help is to be provided to encourage greater efficiency in industry and transportation and economy in private consumption.

ENERGY MARKET TRENDS

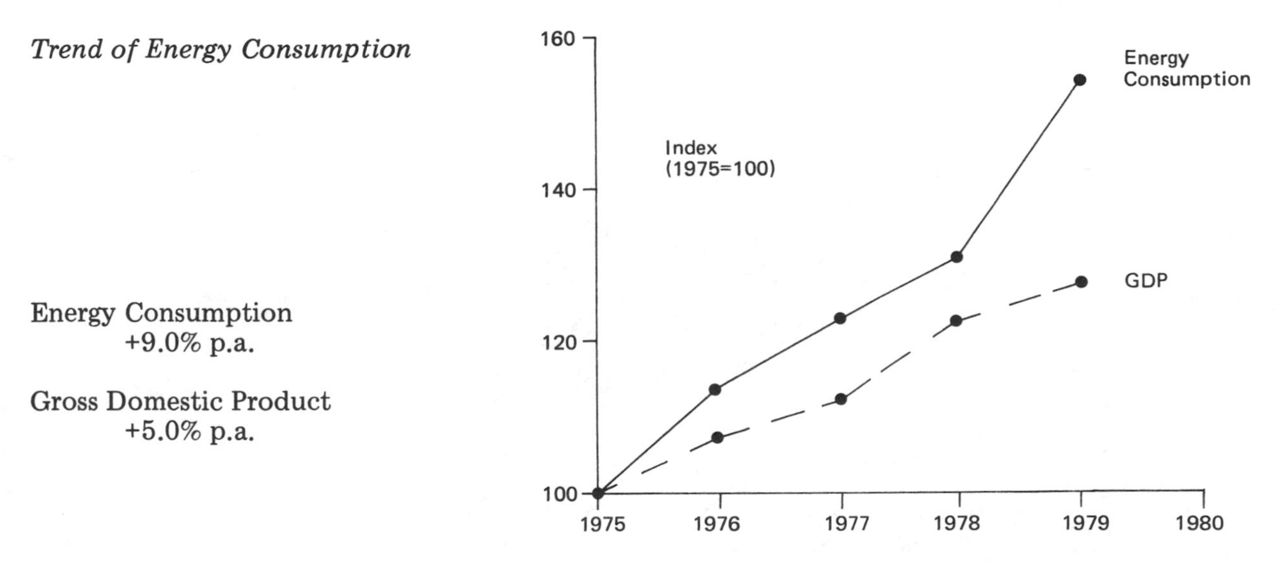

In the period since 1975 the Thai economy has been growing at a reasonably high rate despite dependence on imported energy, with the consequent escalation in balance of payments cost to the country. Overall increase in Gross Domestic Product between 1975 and 1979 was 27.5 per cent, but this was associated with an increase in energy consumption of 40 per cent. Taking into account a major expansion in use of charcoal in 1979 and 1980 the growth in total consumption of energy, including non-commercial fuels was much greater. The increase in use of charcoal meant that in 1980 consumption of commercial fuels fell by around five per cent.

The supply of commercial forms of energy is currently overwhelmingly met by oil, with hydro-electricity the only other significant source of energy. There is a substantial amount of non-commercial fuel being used, comprising firewood, charcoal, paddy husk and bagasse, which contribute approximately 15 per cent to the overall energy balance. Even taking this into account the share of oil remains as high as 75 per cent. However, there is expected to be a significant drop in the contribution from oil, as natural gas from the offshore fields in the Gulf of Thailand is fed into the economy. By 1985 natural gas may then be supplying at least 25 per cent of commercial energy supplies.

Pattern of Energy Supply and Consumption

Primary Fuel Supply (percentage)		Final Consumption (percentage)	
Oil	88	Industry	29
Solid Fuel	4	Transport	39
Natural Gas	–	Residential and Other	32
Primary Electricity	8		
TOTAL	100	TOTAL	100

Consumption of energy by final users shows that transport accounts for a relatively high proportion and has been particularly responsible for the rising demand for imported oil. Industrial development on the other hand has been held back as a result of import-dependence. This situation may ease somewhat in the future with the encouragement of more energy-intensive activities based on the indigenous natural gas.

ENERGY SUPPLY INDUSTRIES

With the growing importance of oil as a fuel in the Thailand economy and the consequent impact of imports on the balance of payments, the Thai government has assumed a key role in the industry through the state agency, Petroleum Authority of Thailand (PAT). PAT is responsible for oil supply, market pricing policies and exploration programmes. Nevertheless, affiliates of international oil companies play an important part, particularly in exploration, but also in refining and distribution of oil products. PAT is co-ordinating the development of offshore natural gas as a principal source of energy, a major proportion of which will be used by the state electricity generating authority, the Electricity Generating Authority of Thailand (EGAT). EGAT also absorbs almost all of the output of indigenous coal mines.

Oil

Oil Consumption

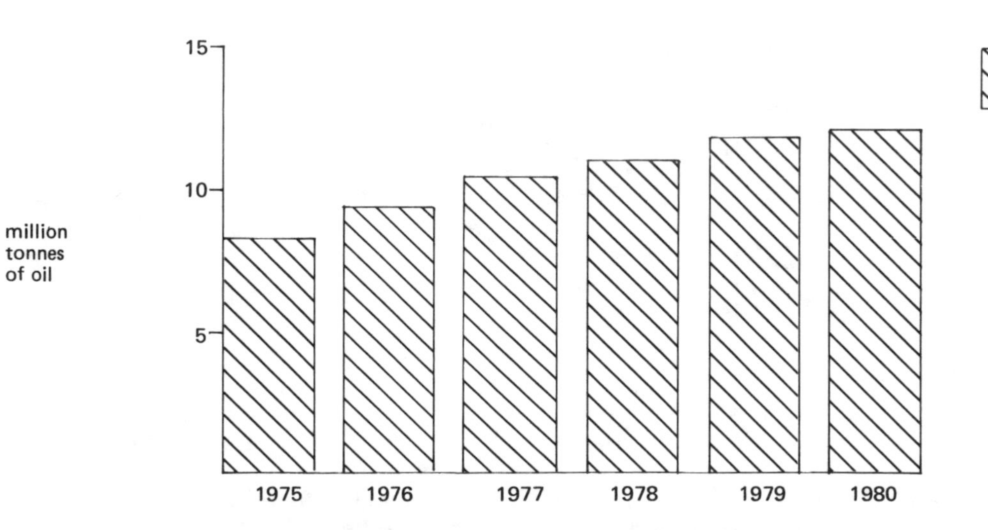

Consumption of crude oil has been rising at a rapid rate since 1975. In that year consumption of oil was 8.2 million tonnes. By 1980 consumption had risen by almost 50 per cent to 12 million tonnes. Increases in demand were across the board, for automotive and aviation fuels and for heavy fuel oil and LPG. Finished products are produced by the three refineries located in Thailand, but their total crude oil throughput capacity is less than nine million tonnes per annum and imports have become increasingly significant. The further increase of oil imports will be arrested during the next few years as natural gas displaces up to five million tonnes of fuel oil and other products.

The Petroleum Authority of Thailand has been established as a state agency involved in crude oil and products supply, as well as in exploration for oil and gas, but international oil companies, are also involved in refinery operations and marketing. Esso Standard Thailand, the local affiliate of the Exxon group operates a refinery at Cholburi and Shell has an interest in the Thai Oil Refinery at Soiacha. The third refinery is at Bangchak, operated by Summit Industrial Corporation of Panama.

Coal

There are deposits of lignite in many parts of Thailand, which are being increasingly exploited. Currently production is taking place at three locations. Largest of these is at Mae Moh, Lampang Province, in the northern part of the country. Here, lignite is being produced by open-cast extraction to fuel a 225 MW power station. Construction of a further four 150 MW units is planned during the 1982-86 period. Some lignite is also used in fertiliser manufacture. Total reserves at Mae Moh are put at 350 million tonnes.

The other two deposits currently being worked are on a smaller scale. At Li, also in Lampang Province, some 120,000 tonnes is being produced annually for industrial and commercial markets. Reserves are put at only 24 million tonnes. At Bangpudam, Krabi Province, in the south-west of the country, minable reserves of some 5-6 million tonnes are being extracted to fuel a 60 MW power station. Of the total output of the coal mining industry, which increased to 1.4 million tonnes in 1980 from only 0.6 million tonnes per annum between 1975 and 1977, 93 per cent is used in power generation. The balance is used largely in industries such as tobacco curing. Bituminous coal and coke have to be imported.

Lignite at Li is found in association with oil shale, and in the Mae Sot basin in Tak Province, to the south of Li, a large deposit of shale has been found. Total reserves of oil shale are estimated to be approximately 18,000 million tonnes, although the oil content of the shale is put at only five per cent.

Gas

Until very recently gas has not figured in the energy supply situation of Thailand, but exploration activity over a number of years has led to the discovery of an important gas field offshore in the Gulf of Thailand. Union Oil Company of California is operator for this development, which should be producing 250 million cubic feet per day in 1982. Total potential reserves of gas are estimated to be around 16 million million cubic feet, and higher annual production rates are anticipated in succeeding years. The Petroleum Authority of Thailand plans to construct a gas separation plant with capacity of 700 million cubic feet per day to extract gas liquids. This facility would also be able to accommodate other potentially commercial finds of gas in the Gulf which are currently being evaluated. The main market for the gas will initially be as a fuel in thermal power plant or as fuel or feedstock in other large scale industrial activities.

Electricity

Electricity Production

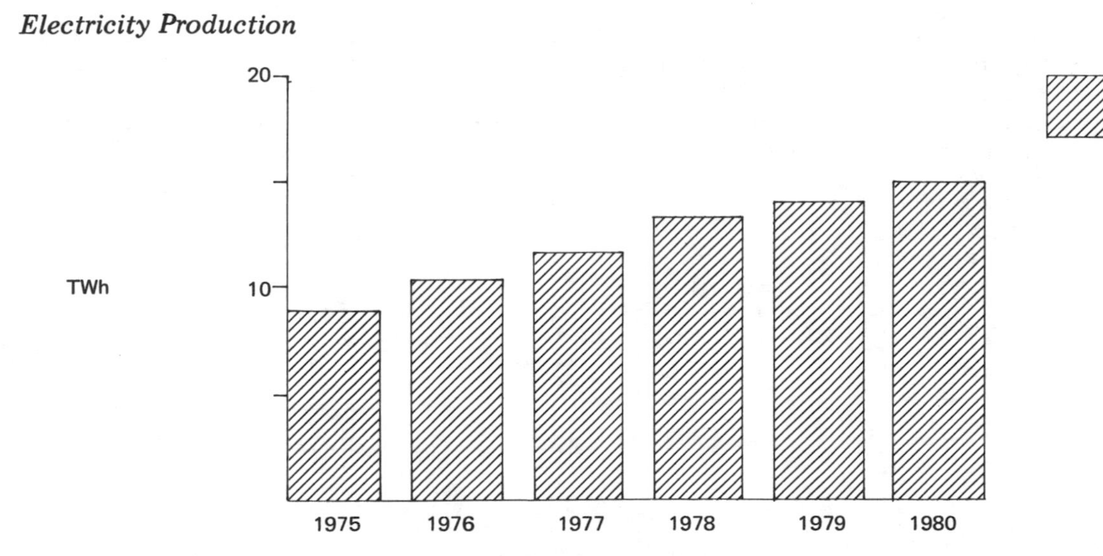

The demand for electricity has been rising quickly and this trend is expected to continue through the period of the current economic development plan. From a level of just under 9,000 GWh in 1975 consumption increased by 70 per cent to 15,000 GWh in 1980, approximately twice the rate of increase for energy consumption overall. Predicted demand for the current decade indicates that total demand for electricity will more than double by 1990.

The responsibility for generation and transmission of electricity lies almost entirely in the hands of the Electricity Generating Authority of Thailand (EGAT). EGAT accounts for over 90 per cent of production. EGAT sells electricity in bulk to the Metropolitan Electricity Authority, which distributes it mainly in the Bangkok area, and to the Provincial Electricity Authority and the National Energy Administration, which are responsible for distributing electricity in the more remote areas of the country. These organisations also maintain numerous small generating stations for localised supply. EGAT sells direct to some major industrial power users, with total requirements of some 450-500 GWh per annum, but industrial consumers also produce upwards of 700 GWh themselves.

As of 1980 aggregate generating capacity was some 3,450 MW, of which EGAT operated 3,360 MW and the PEA 90 MW of diesel generators. Thermal power plant constitutes more than half of the capacity, using both indigenous lignite and import-based fuel oil. Hydro-electric plant accounted for 36 per cent of capacity, but in some years contributes a much smaller proportion to total electricity supply, because of the dual function of reservoirs for irrigation as well as power generation purposes. Thermal power stations accordingly provide a very high proportion of electricity (80 per cent in 1980).

EGAT's construction programme includes more than 600 MW of lignite fuelled capacity at Mae Moh, being completed in stages between 1980 and 1984. Hydro-electric plant continues to be built at a number of locations, notably at the Srinagarind project. Of greater significance, however, is the conversion of existing oil-fired units to be able to run on natural gas now becoming available from the offshore field in the Gulf of Thailand. Three of the five 300 MW units at South Bangkok power station have been converted and the two remaining ones will be considered for conversion in due course. In addition a number of gas turbines are being installed.

ENERGY TRADE

Net Imports/(Exports) 1975-80

	1975	1976	1977	1978	1979	1980
Crude Oil (million tonnes)	7.3	7.7	8.5	8.4	8.6	7.6
Oil Products (million tonnes)	0.8	1.2	1.7	2.4	2.2	3.2
Natural Gas (million M^3)	..	..	2.4	39.9	53.0	89.1
Electricity (TWh)	0.5	0.5	0.6	0.8	2.4	2.3

Up to the time of the development of the country's offshore natural gas deposits in 1981-82 Thailand is highly dependent on imported energy. Principally imports have been of crude oil and finished products, the total of the two categories rising from just over eight million tonnes in 1975 to approaching 11 million tonnes in 1979 and 1980. In addition imports of coal, gas and electricity have been on a rising trend. Some coal and coke is required for particular industrial purposes, as indigenous production is of lignite. Electricity imports rose sharply in 1979 and 1980, largely as a result of a loss of hydro potential, but the trend of net imports may be reversed on completion of more gas-fired generating capacity, and cross-border movements of natural gas should become much smaller.

Sources of Imported Oil

(percentage)	
Saudi Arabia	34
Qatar	19
Singapore	16
China	9
Brunei	7
Other	15
TOTAL	100

Thailand has relied on Middle East sources for the bulk of its imported oil, with Saudi Arabia accounting for around one-third of the total. Qatar has been the only other prominent country individually, although oil is imported from several other countries in the Middle East, including Iran, Kuwait and the United Arab Emirates. Brunei, the nearest oil-exporting country, supplies seven per cent and China nine per cent. Singapore is the predominant source of imported oil products, the refineries there also being heavily dependent on Middle East sources for crude oil supplies.

ENERGY POLICIES

Faced by rising demand for energy, compensated for only partially by limited indigenous resources, the Thai government has developed a detailed plan of action for encouraging energy supply to maintain the momentum of the economy, while at the same time curbing the inefficient use of energy. The institutions already exist to enable strong action to be taken in advancing energy policies, but private sector involvement from both local and foreign companies will be encouraged, particularly in exploration and development.

Energy Supply

Plans for energy supply involve a build-up in the exploitation of natural gas from the Gulf of Thailand and increased electricity generating capacity from hydro-electric projects and lignite. Natural gas use is planned to rise from 250 million cubic feet per day in 1982 to 525 million cubic feet per day in 1986. Over the same period lignite-based power station capacity is to be expanded from 210 MW to 885 MW and hydro-electric potential from 1,300 MW to 2,000 MW. This will largely exploit the existing scope for production from these sources and the application of non-conventional energy production will also be sought, although these involve technical, economic and institutional issues and the basic potential is in any case limited. Meanwhile intensive exploration for oil and gas, in both onshore and offshore areas, will be continued.

Conservation

The government is aware of the need to involve the full range of conservation measures available in order to curb consumption, particularly in transportation. This will involve removal of subsidies and the imposition of taxes on transport fuels, linked to the promotion of mass transit schemes. Reduced taxation and the provision of financial assistance will also be used to improve energy efficiency in industry and promote conservation projects. For all sectors extensive use is to be made of information and advisory services, the establishment of standards, energy audits and mandatory controls.

Tunisia

KEY ENERGY INDICATORS

Energy Consumption	
—million tonnes oil equivalent	2.5
Consumption Per Head	
—tonnes oil equivalent	0.39
—percentage of world average	24%
Net Energy Imports	nil
Oil Import-Dependence	nil

Tunisia is still in the process of emerging from a low level of economic development, and energy consumption per head of population is low by world standards. The country is a net exporter of energy, with earnings from crude oil a vital element in the balance of payments. A wide range of fuels is used, including solid fuel, natural gas and oil as well as traditional fuels. The government has also promoted the development and use of geothermal, biomass and wind energy.

ENERGY MARKET TRENDS

Economic development was not greatly affected by the economic crisis of 1974 in the international economy. With a sound infrastructure, a net energy export position and activity in a wide range of industries the country has fared relatively well. This is reflected in the trend of energy consumption, which has risen at around 10 per cent per annum to over 2.5 million tonnes of oil equivalent per annum.

Pattern of Energy Supply

Primary Fuel Supply (percentage)	
Oil	83
Solid Fuel	3
Natural Gas	14
Primary Electricity	..
TOTAL	100

In view of the availability of crude oil in Tunisia it is not surprising that oil accounts for more than 80 per cent of energy consumption. Natural gas, produced in association with the oil, makes up almost all of the balance. Coal is little used as there are no indigenous resources. Similarly, the potential for hydro-electricity is very limited.

ENERGY SUPPLY INDUSTRIES

Private sector companies and state agencies are both involved in energy supply. The supply of gas and electricity is controlled by the state, and official agencies are particularly involved in promoting exploration and development for all forms of energy. A number of foreign oil companies have production and exploration rights, with French-based companies prominent.

Oil

Tunisia produces between five and six million tonnes per annum of crude oil. The level fell below four million tonnes in 1976 as output from the El Borma field declined, but the introduction of secondary recovery methods, and, more importantly, the beginning of production in new areas, are lifting output levels once more. The main field is now Ashtart in the Gulf of Gabès, where there are prospects for further finds. Other recent discoveries have been made in central Tunisia, between El Borma and Gabès, and at El Franig, in the south of the country.

A large proportion of Tunisian crude oil is exported, with imports of Algerian and other foreign crude oil partially offsetting these exports. Crude oil is refined at the Bizerta refinery of SOTURAF, which is state-controlled. Affiliates of several foreign companies are engaged in the distribution of refined products.

Gas

Availability of gas has increased sharply with the bringing on stream of new oil production in the Gulf of Gabès. Production in 1980 exceeded 350 million cubic metres compared with little more than 200 million cubic metres in 1975. A further increase will come with the start up of production in the Miskar field offshore, which may yield 1,800 million cubic metres per annum by 1985. Reserves of the Miskar field are estimated to be some 60 billion cubic metres, and total national reserves are put at around 170 billion cubic metres. Distribution of gas is handled by the state company STEG.

Electricity

Electricity production has risen rapidly since 1975, at an average rate of more than 10 per cent per annum, to 2,100 GWh. Generating capacity is 640 MW, of which 590 MW is operated by the state corporation STEG, which is responsible for electricity supply. There is a small amount of hydro-electric capacity, which contributes two per cent of electricity, but thermal power plant predominates.

ENERGY TRADE

Net Imports/(Exports) 1975-79

	1975	1976	1977	1978	1979
Coal (million tonnes)	0.1	0.1	0.1	0.1	0.1
Crude Oil (million tonnes)	(3.7)	(2.5)	(3.1)	(3.6)	(4.0)

Tunisia is a net exporter of energy in the form of crude oil. The level of exports recovered from a low point of 2.5 million tonnes in 1976, when production was at a minimum, to four million tonnes and may well increase in future years as production of oil and gas rises. Small quantities of solid fuel, averaging 100,000 tonnes per annum are imported.

Turkey

KEY ENERGY INDICATORS

Energy Consumption	
—million tonnes oil equivalent	27.9
Consumption Per Head	
—tonnes oil equivalent	0.62
—percentage of world average	38%
Net Energy Imports	46%
Oil Import-Dependence	85%

Turkey has reached an intermediate stage of economic development, in which agriculture and extractive industries still play an important part, but industrial, transport and other activities give rise to substantial demands for commercial energy forms. Consumption of these fuels is currently less than 40 per cent of the world average, with a continuing widespread use of certain non-commercial fuels. Indigenous energy resources are significant and future economic development may be dependent on exploiting them.

ENERGY MARKET TRENDS

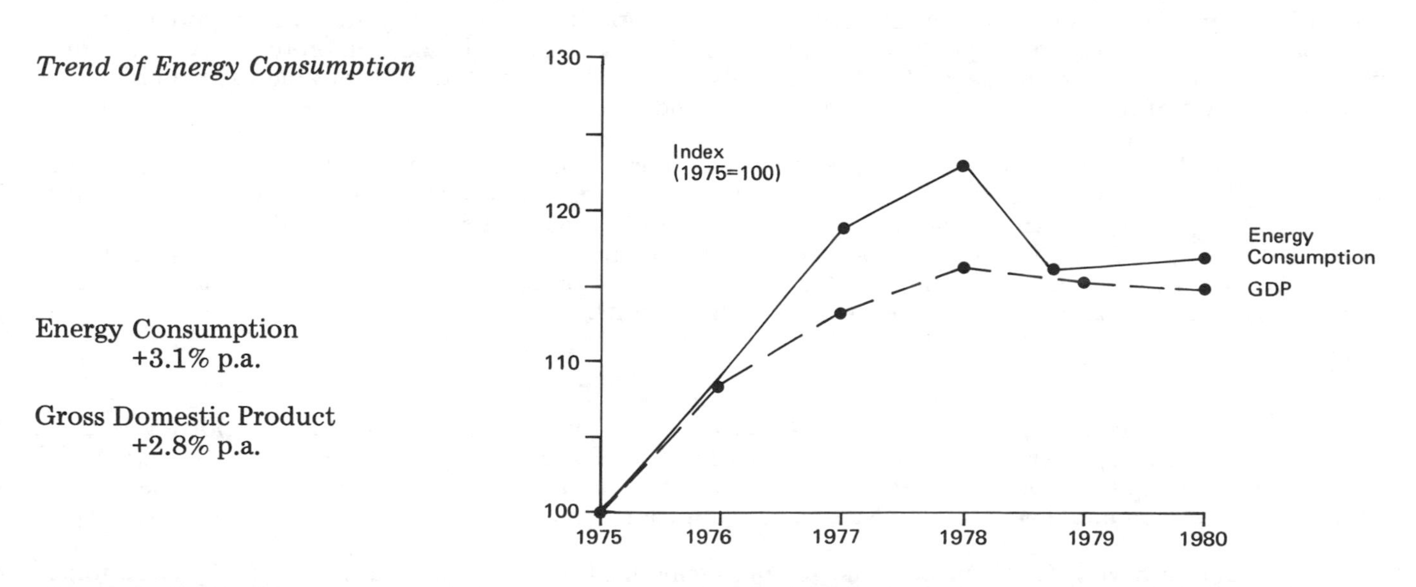

Although the Turkish economy is in the process of developing and includes some heavy industries, the rate of increase in energy consumption has been kept down to close to the rate of growth of Gross Domestic Product, which averaged less than three per cent per annum between 1975 and 1980. After running ahead of GDP in the period 1975-78, consumption was cut back sharply in 1979 and 1980, although this partly reflected economic recession.

Pattern of Energy Supply and Consumption

Primary Fuel Supply (percentage)		Final Consumption (percentage)	
Oil	52	Industry	23
Solid Fuel	38	Electricity Generation	19
Natural Gas	–	Transport	17
Primary Electricity	10	Other	41
TOTAL	100	TOTAL	100

Oil is the single most important form of primary energy, accounting for more than half of energy supply. But there is a substantial availability of hard coal, lignite and hydro-electricity, which almost equal oil in their contribution. Coal is used for more than a quarter of electricity production, and this share is rising steadily. Coal is also used widely in the domestic and industrial sectors. Non-commercial solid fuels, principally wood, are estimated to be equivalent to a third of total commercial energy inputs.

Oil

Oil Production and Consumption

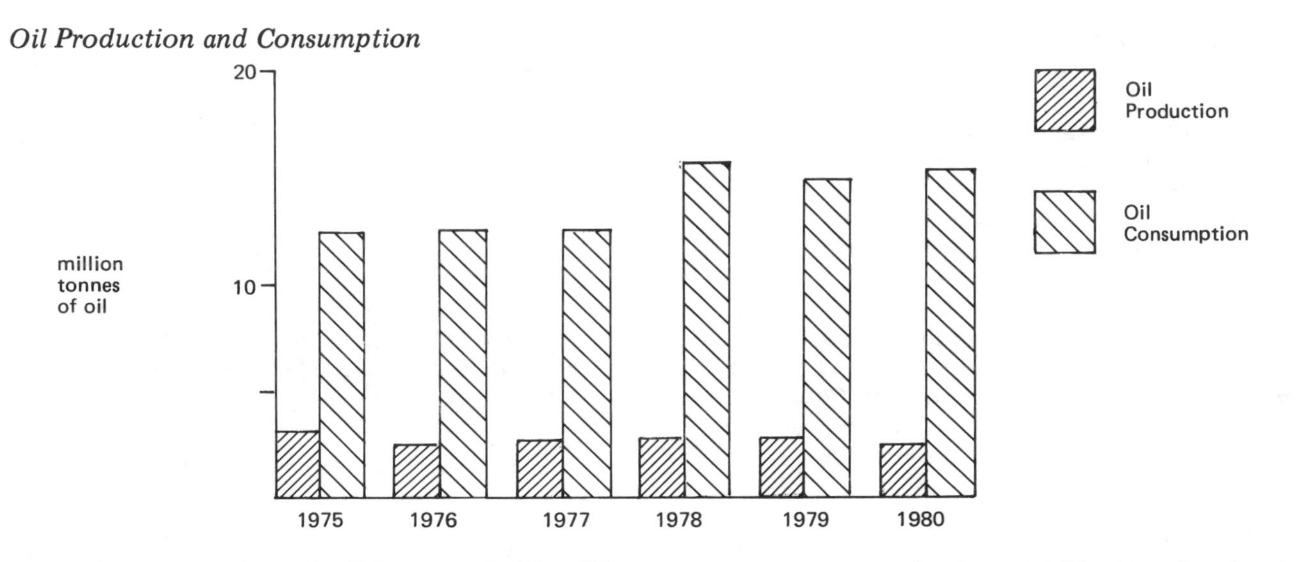

Annual consumption of oil is around 15 million tonnes per annum, having stabilised at that level after increasing from 12 million tonnes in 1975. Inland production meets only 15 per cent of requirements and the country relies heavily on imports. Production in 1980 was in fact only 2.3 million tonnes and the trend has been downward in recent years. Producing fields are in the south-east, with production mainly by the Turkish affiliate of Shell, the state oil company TPAO and Mobil.

Indigenous crude oil is processed mainly in the Batman refinery of TPAO, with small quantities transported to other refineries in the west. TPAO operates the country's largest oil refinery, of over eight million tonnes per annum capacity, at Izmit, and a second major refinery at Izmir. TPAO's total annual processing capacity is some 13-14 million tonnes. Shell and Mobil are involved in the joint-venture refinery at Mersin, along with BP and the local oil company Marmara Refining Company. Distribution and marketing of oil products is carried out by the companies with refining interests.

Coal

Coal is one of the main indigenous energy resources of Turkey, but an increasingly high proportion consists of lower grade lignite, whereas output of hard coal is on a downward trend. In 1980 lignite accounted for 70 per cent of output in tonnage terms. In 1975 output of hard coal, at 8.4 million tonnes, exceeded that of lignite. Other non-commercial fuels also form a significant input to the total energy balance.

The state coal authority TKI is responsible for producing most of the coal and lignite. The production of lignite as an indigenous alternative to oil as a fuel in electricity generation is an important reason for this trend. TKI's investment programme is centred on the development of the Afsin-Elbistan deposits of lignite to fuel a 1,200 MW power station, but investment is also taking place in lignite mines at several other locations. Afsin production is intended to rise to 20 million tonnes per annum, and a similar level of output is anticipated from other projects due to be completed by 1984.

Coal Production

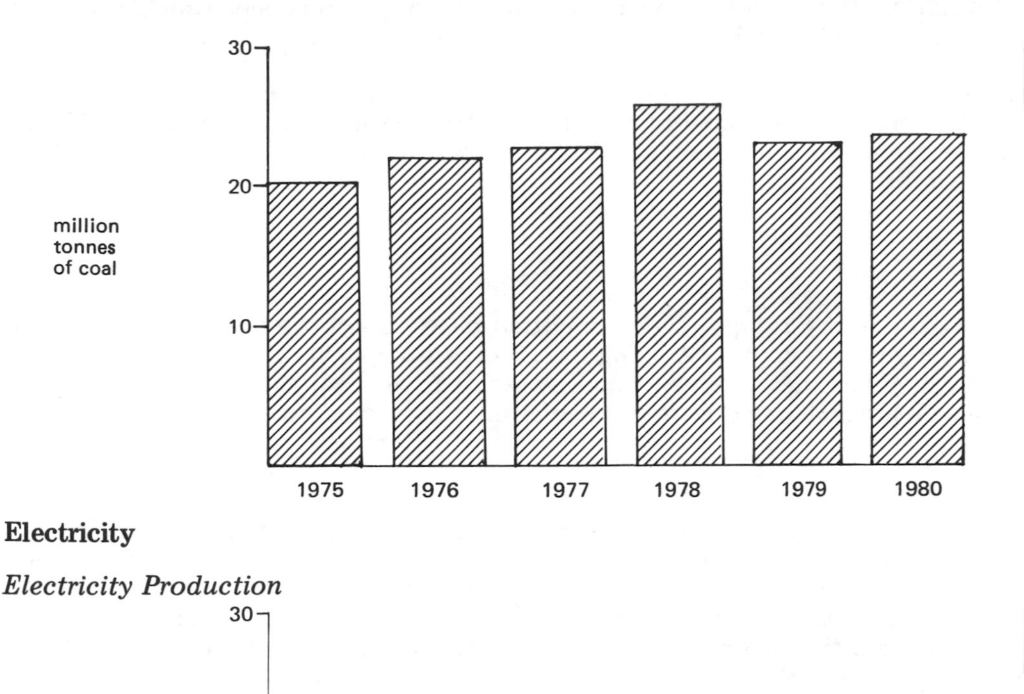

Electricity

Electricity Production

Consumption of electricity rose by about 50 per cent between 1975 and 1980. But the major part of this increase took place in the period 1975-78, after which the rate decreased. Production of electricity in 1980 was over 23 TWh. The state electricity supply organisation TEK accounts for 90 per cent of production. The balance is produced by industrial establishments for their own use.

Total generating capacity is over 5,000 MW, of which TEK operates 4,400 MW. This includes important hydro-electric plant as well as coal and oil-fired stations. Hydro-electric capacity is nearly 2,000 MW, and provides more than 40 per cent of electricity. The Keban dam on the Euphrates River has a capacity of 640 MW, and is to be doubled in size. A 1,200 MW scheme on the Tigris is due to be completed by the mid-1980s, with further exploitation of the Euphrates to follow.

ENERGY TRADE

Net Imports/(Exports) 1975-80

	1975	1976	1977	1978	1979	1980
Coal (million tonnes)	..	0.3	0.5	0.4	0.5	
Crude Oil (million tonnes)	9.6	11.2	12.1	10.8	11.6	10.5
Oil Products (million tonnes)	0.2	1.4	2.6	2.8	2.8	2.9
Electricity (TWh)	0.4	0.3	0.5	0.6	0.6	0.6

Oil is the most important component of Turkish energy trade. The total of crude oil and refined products amounts to 13-14 million tonnes per annum. Imports of products have tended to rise since 1975, but there is now an excess of refining capacity in the country. In addition to imports for Turkish consumption, there is also an important transit trade in Iraqi crude through the Kirkuk-Dortyol pipeline, which can carry 25-30 million tonnes per annum. There is a regular import of coal, and of electricity, from Bulgaria and the Soviet Union.

Iraq draws its crude oil imports from a limited number of countries. Iraqi crude is drawn from the Kirkuk-Dortyol pipeline and there are contracts with Iran and Saudi Arabia. A proportion of Turkey's total imports comes from the Soviet Union.

ENERGY POLICIES

Turkish governments have been forced by rising costs of imported oil to develop a range of measures to increase supplies of indigenous energy and contain existing levels of consumption. State organisations are central to the supply of all energy, but these are hampered by lack of technical and financial resources. Recent policy changes indicate a greater role for private companies in coal production.

Energy Supply

The crux of Turkish policy towards energy supply is to try to increase the availability of energy, particularly indigenous energy, in order to reduce the dependence on imported oil. The objectives are to maintain levels of crude oil production, which involves investment in recovery technology, to increase output of lignite, and to exploit the country's hydro-electric potential to a much greater extent. As far as exploration for additional deposits of oil is concerned, the government is likely to rely on the expertise and resources of international oil companies.

Conservation

The main instrument of conservation policy has been the use of controls on prices of all energy products to curb demand. Other elements of a conservation programme have been introduced, such as encouraging attention to energy management practices, the scope for combined heat and power plant in new projects and the installation of waste heat recovery equipment in existing ones. New plants based on oil are unlikely to be approved. In the residential sector mandatory insulation standards have been introduced and in transportation greater emphasis is being placed on public systems.

United Arab Emirates

KEY ENERGY INDICATORS

Energy Consumption	
—million tonnes oil equivalent	3.5
Consumption Per Head	
—tonnes oil equivalent	3.5
—percentage of world average	216%
Net Energy Imports	nil
Oil Import-Dependence	nil

The United Arab Emirates includes two important oil producing countries, Abu Dhabi and Dubai. A third sheikhdom, Sharjah, also has some oil reserves and production. Major gas projects have been undertaken and are being planned, to utilise associated gas and develop a substantial export trade in LNG and LPG. Energy consumption by energy industries, manufacturing and other sectors is more than twice the world average per head of population. The UAE is a member of both the Organisation of Petroleum Exporting Countries and the Organisation of Arab Petroleum Exporting Countries.

ENERGY MARKET TRENDS

The demand for energy in the UAE is relatively high, and rising as a programme of new projects unfolds. This programme includes oil refining, gas liquefaction and fertiliser production. Oil continues to meet a high proportion of total energy requirements, but gas now accounts for more than 10 per cent and this share may be expected to rise as gathering systems are established and gas treatment facilities increased.

ENERGY SUPPLY INDUSTRIES

State involvement in the oil and gas industries in Abu Dhabi is considerable, through the control and influence of the state oil company, though foreign companies remain very active and retain shareholdings in the main operations. State organisations are responsible for electricity supply. In Dubai the ruler has a substantial interest in oil and gas exploitation.

Oil

Oil Production

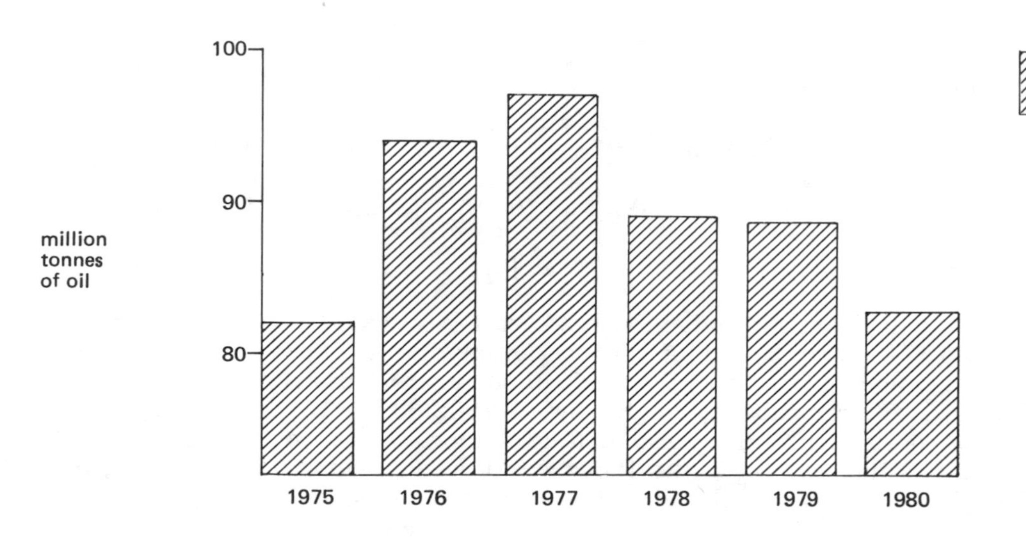

Total UAE production of crude oil in 1980 was 83 million tonnes. This was substantially below the 1977 peak level of 97 million tonnes, and in 1981 there was a further fall, owing to the very slack demand for oil in the international oil market. Abu Dhabi accounts for the major part of output, approximately 80 per cent. Production from Sharjah has declined rapidly from the level of nearly two million tonnes in 1975 to only half a million tonnes per annum currently. Dubai output has remained the most stable since 1976, at 15-18 million tonnes per annum.

Abu Dhabi has established a state oil corporation (ADNOC) which is active in all sectors of the industry. ADNOC has acquired a 60 per cent interest in the production operating companies ADMA-OPCO and ADCOPO, which are responsible for offshore and onshore operations respectively. British Petroleum, Shell and CFP are involved in both groupings. In Dubai Conoco is operator for Dubai Marine Areas Limited, in partnership with CFP and the Spanish state exploration company Hispanoil. Dubai production is almost entirely from the offshore Fateh fields. Reserves are put at 1.4 billion barrels.

Consumption of oil in the UAE amounts to some four million tonnes per annum. This total includes substantial quantities of aviation fuel and around 200,000 tonnes per annum of bunkers. ADNOC, BP and Shell are principal suppliers of products, with a smaller participation by Caltex. ADNOC owns a small refinery at Umm Al-Nar, which is currently being expanded to 3.5 million tonnes per annum capacity. A larger refinery, of more than five million tonnes per annum is being constructed by ADNOC at Ruwais. In the meantime, products are mostly imported from other Persian Gulf refineries.

Gas

Oil industry operations produce massive volumes of gas, which until 1977 were largely wasted by flaring. But Abu Dhabi has installed a liquefaction plant on Das Island which utilises almost one-third of total gas available. The plant is capable of handling at capacity more than 500 million cubic feet per day, extracting important quantities of LPG as well as producing LNG.

ADNOC, the state oil company holds a 51 per cent majority share in the Abu Dhabi Gas Liquefaction Company, which operates the plant. Other participants are BP and the French Group CFP, both of which are closely involved in oil and gas production in Abu Dhabi, and the Japanese companies Mitsui and Bridgestone, which are extensively involved in importing LPG and LNG into Japan.

Gas is also being utilised in industrial projects in Abu Dhabi. These are located in the Ruwais industrial complex, and include production of fertiliser. Additional gas-gathering and treatment plant is being completed, to utilise onshore gas which is currently flared, increasing the overall availability of natural gas liquids and gas for use as fuel or feedstock. Dubai has also established a natural gas treatment plant where gas liquids are separated for export, yielding gas for local industrial and power generation use.

Electricity

Demand for electricity rose very sharply after 1975 to reach levels of more than 4,300 GWh. This is well over three times the 1975 level of 1,330 GWh. Production is undertaken by public undertakings, which have a capacity of 1,100 MW of thermal plant utilising oil or gas as fuel. An additional 90 MW dual-fired power station is being built to service the growing demand of the Ruwais industrial area.

ENERGY TRADE

Net Imports/(Exports) 1975-80

	1975	1976	1977	1978	1979	1980
Crude Oil (million tonnes)	(82.1)	(95.4)	(96.8)	(89.2)	(89.1)	(83.0)
Oil Products (million tonnes)	1.0	1.2	1.8	1.7	1.5	1.5
Natural Gas (000 million m^3)	–	–	(0.7)	(1.8)	(1.8)	(2.7)

Up to the present time virtually all crude oil production has been exported. In 1980 this was around 83 million tonnes, well below the 1977 level of 97 million tonnes. Net imports of oil products rose toward two million tonnes in 1977, but have since fallen back as a result of the start-up of export trade in LPG. With the commissioning of additional refinery capacity in the period 1982-83 the UAE should become a net exporter of oil products. Net exports of LNG have also built up. Exports started in 1977 and by 1980 had reached capacity. Exports will rise further later in the 1980s with the construction of an even larger plant, also geared to supplying LNG to Japan, which will utilise onshore gas resources.

Upper Volta

KEY ENERGY INDICATORS

Energy Consumption	
—million tonnes oil equivalent	0.1
Consumption Per Head	
—tonnes oil equivalent	0.01
—percentage of world average	1%
Net Energy Imports	100%
Oil Import-Dependence	100%

Upper Volta is an extremely poor country, with very limited economic development. Consumption of commercial energy forms is no more than one per cent of the world average. The country is deficient in indigenous energy resources and is completely reliant on imported oil for its energy requirements.

ENERGY MARKET TRENDS

Growth of the economy has been very slow and has suffered serious setbacks through drought conditions in the Sahel region. The agricultural sector is of great importance. Manufacturing activity is very limited. Energy consumption is accordingly small and not growing very fast. However, consumption of electricity, an important energy form in service activities, has been rising at 15-20 per cent per annum. Known reserves of minerals have been little exploited so far, and energy consumption will rise if development programmes are undertaken.

ENERGY SUPPLY INDUSTRIES

Although oil is crucial to the operation and development of the economy, there is no state oil company in existence and the supply of oil is undertaken by affiliates of several of the major international oil companies. The state is, however, involved in the provision of electricity, which involves extension of the supply system and investment in hydro-electric plant.

Oil

Consumption of oil exceeds 100,000 tonnes per annum. The country has no oil refinery and all finished products are brought in from neighbouring countries. Principal source of products is the Abidjan refinery, in which the Upper Volta government is a minor participant. Products are imported and distributed by affiliates of the international companies also participating in the Abidjan refinery.

Electricity

Electricity consumption now exceeds 100 GWh per annum compared with little more than 50 GWh in 1975. Electricity is supplied by the state-owned undertaking Société Voltaique d'Electricité (VOLTELEC). VOLTELEC has capacity of 30 MW of thermal plant, producing electricity for the main centres of population, and is in the process of strengthening its generating base and connected system in order to increase the degree of electrification of the economy. At present VOLTELEC has fewer than 20,000 customers. Hydro-electric plants are to be commissioned using the waters of the Komtenga and White Volta rivers.

Yemen Arab Republic

KEY ENERGY INDICATORS

Energy Consumption	
—million tonnes oil equivalent	0.3
Consumption Per Head	
—tonnes oil equivalent	0.05
—percentage of world average	3%
Net Energy Imports	100%
Oil Import-Dependence	100%

The Yemen Arab Republic is the least developed country in the Middle East. Consumption of commercial energy is very low by world standards and development programmes have made only slow progress. The country has no energy resources under exploitation, although exploration for oil or gas is continuing. Oil is used for all types of energy requirements.

ENERGY MARKET TRENDS

Although energy consumption is at a very low level, the rate of increase is not large. Consumption of oil has been rising at less than 10 per cent per annum. Consumption of electricity, which might be expected to rise at a greater rate in a developing economy, is increasing at no more than 10 per cent annually.

ENERGY SUPPLY INDUSTRIES

The government is involved in the provision of energy to the Yemen market. A state organisation accounts for most of the electricity supplied. In the oil sector private companies play more of a part and several foreign companies are particularly involved in exploration activity.

Oil

Oil is the sole source of energy of significance in Yemen. Annual consumption of all products is less than 300,000 tonnes. The level has been rising erratically from around 200,000 tonnes in the mid 1970s. Exploration activity has so far proved unrewarding, but several companies are currently involved. These include Shell, the Japanese oil company Tonen and the United States oil and gas engineering company Santa Fe International Corporation.

Electricity

Electricity production rose to over 60 GWh in 1976, but in succeeding years has tailed off at around 70 GWh. Generating capacity amounts to some 25 MW, most of which is operated in the public sector. The Yemen General Electricity Corporation is responsible for generating, transmission and distribution through the public supply system.

ENERGY TRADE

Net Imports/(Exports) 1975-79

	1975	1976	1977	1978	1979
Oil Products (000 tonnes)	209	234	282	249	291

Imports of energy consist of refined oil products to meet the requirements of all sectors of the economy. Imports in 1979 totalled 291,000 tonnes, significantly higher than in 1978, but only three per cent above the 1977 level.

Zaire

KEY ENERGY INDICATORS

Energy Consumption	
—million tonnes oil equivalent	1.9
Consumption Per Head	
—tonnes oil equivalent	0.07
—percentage of world average	4%
Net Energy Imports	nil
Oil Import-Dependence	nil

Zaire is a net exporter of energy, owing to local production of crude oil and substantial hydro-electric development. But the margin is not large and Zaire's own energy demand is a function of its very limited state of economic development. Energy consumption per head of population is very low by world standards. Unless further oil reserves are discovered an upturn in economic fortunes is likely to lead to Zaire becoming a net importer of energy, despite projects to expand hydro-electric production.

ENERGY MARKET TRENDS

The economy has been at a low ebb since 1974. The mineral output on which Zaire relies for foreign earnings was badly hit by recession, with a consequent impact on the overall economy, in which mining constitutes one-third of Gross Domestic Product, and on energy consumption, which has increased by less than 10 per cent since 1975. Agricultural and manufacturing sectors are both considered to have deteriorated in recent years as well.

Pattern of Energy Supply

Primary Fuel Supply (percentage)	
Oil	43
Solid Fuel	9
Natural Gas	–
Primary Electricity	48
TOTAL	100

The pattern of energy input to the economy indicates the importance of hydro-electricity. But it is also in part a reflection of the state of economic development since overall energy consumption is quite small. Around 300,000 tonnes per annum of coal/coke is used, of which the majority is imported. No account is taken in this balance of the amount of traditional fuels which is used.

ENERGY SUPPLY INDUSTRIES

The state has a majority holding in the important sector of electricity production and supply, although some of the major mining and industrial companies are largely self-sufficient in electricity. The state is involved in varying degrees throughout the oil industry, but international oil companies are important in exploration for crude oil and distribution of refined products.

Oil

Production of crude oil is static at around one million tonnes per annum. Some is exported, but the crude oil is used as main feedstock for the SOZIR refinery at Moanda. The refinery has a throughput capacity of around three-quarters of a million tonnes per annum and normally meets most of local market demand. The refinery is operated by the Italian oil company AGIP. Crude oil production is expected to increase as additional fields are brought on stream.

Electricity

Zaire has massive untapped hydro-electric potential, which has been estimated at over 100 GW, representing more than 50 per cent of the continent's potential and 13 per cent of the world total. Production of electricity has remained stagnant at under 4,000 GWh per annum as a result of the state of economic activity. Hydro-electric capacity is currently 1,000 MW, and is the source of power for the state-controlled public supply company SONEL. SONEL's major station is at Shaba, on the Zaire River, but the station at Inga, also on the Zaire River, is being expanded from 350 MW to 1,630 MW. Industrial and mining companies operate almost 700 MW of thermal capacity.

ENERGY TRADE

Energy trade shows a small positive balance overall. Production of crude oil is slightly greater than current depressed levels of consumption. Exports are partially offset by imports of solid fuel, amounting to some 200,000 tonnes per annum. Cross-border transfers of electricity between Zaire and Zambia also take place.

Zambia

KEY ENERGY INDICATORS

Energy Consumption	
—million tonnes oil equivalent	3.1
Consumption Per Head	
—tonnes oil equivalent	0.53
—percentage of world average	33%
Net Energy Imports	6%
Oil Import-Dependence	100%

Zambia has an average level of consumption per head of population only one-third of the world average. This is high by African standards and is entirely a reflection of the importance of the mining industries of the Copper Belt. The exploitation of at least part of the potential of the Zambezi and other rivers for the benefit of Zambia and Zimbabwe means that Zambia is a net exporter of energy. But the country has no oil resources and all oil has to be imported.

ENERGY MARKET TRENDS

Consumption of energy is dominated by the level of activity in the mining sector. This has been adversely affected in recent years by the slack demand for raw materials, compounded by production difficulties in Zambia and continuing problems of transporting exports of bulk commodities. Gross Domestic Product thus fell back to below the 1975 level in 1979 and 1980 after an uncertain rise in the intervening period. Energy consumption has shown a stronger upward trend with a net rise over the period as a whole.

Pattern of Energy Supply and Demand

Primary Fuel Supply (percentage)	
Oil	21
Solid Fuel	12
Natural Gas	–
Primary Electricity	67
TOTAL	100

Consumption of Oil (percentage)	
Mining	36
Transport	28
Industry/Commerce	17
Other	19
TOTAL	100

The supply of energy is unusual in that a high proportion of all energy is in the form of electricity, which is produced from the country's hydro-electric resources. This is mainly for consumption in the mining industries of the Copper Belt. This sector also consumes a large part of indigenous coal production and is a significant user of oil. But use of oil is biased towards the patterns of demand of the commercial, residential and transport sectors.

ENERGY SUPPLY INDUSTRIES

The state occupies a key role in the supply of energy. Production and transmission of electricity is undertaken by state corporations, and the mining industry, which has extensive auto-production of power, is coming increasingly under direct state control. The only oil refinery is state owned as are the mines at Maamba.

Oil

Consumption of oil products is around 700,000 tonnes per annum. Products are obtained mainly from the government owned refinery at Ndola operated by Indeni Petroleum Refinery Limited. The refinery has a throughput capacity of over 1.2 million tonnes per annum and is supplied via the Tazama pipeline from Dar-es-Salaam. Because of the extensive use of coal and hydro-electric power in the key mining sector, demand is relatively high for gasoline and diesel fuel, relative to demand for heavy fuel oil, necessitating the supply of special feedstocks. Installation of fuel oil conversion equipment is being considered.

Coal

Coal is produced at the state-owned mines of Maamba Collieries Limited, near the southern border with Zimbabwe. The mines produce less than 700,000 tonnes per annum, well below capacity, which is one million tonnes per annum. Coal was formerly much used on the railways as well as in the mining sector, but total demand has declined. Coke is produced at Kabwe for the mining industry, with further quantities imported into Zambia.

Electricity

Electricity Production and Consumption

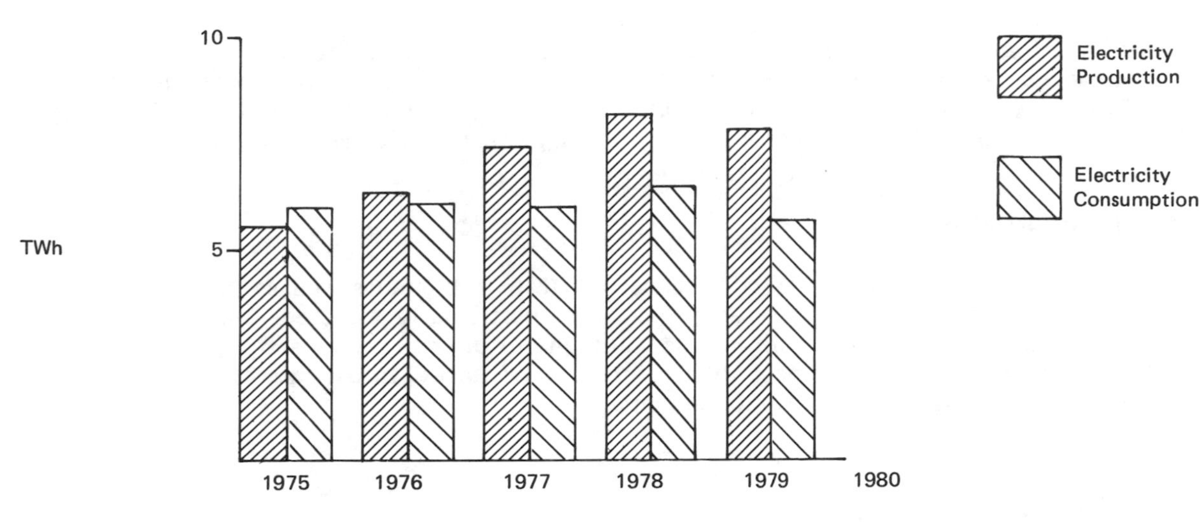

Zambia has very large hydro-electric production from the Zambezi River, at Kariba and Victoria Falls, and at Kafue Gorge on the Kafue tributary of the Zambezi. Of the total generating capacity of 1,860 MW 1,670 MW is hydro-electric, with 85 per cent of all electricity coming from these plants. Consumption of electricity is very high because of demand in the mining industry, which accounts for almost 80 per cent of national consumption.

State organisations control the production of electricity from the main hydro-electric plant on the Kafue and Zambezi Rivers, and distribute electricity throughout Zambia. The Zambia Electricity Supply Corporation also organises the mass transfer of power to the Copper Belt in the north of the country, where the major part of all power is consumed. Mining companies are themselves significant producers of electricity too.

Uranium

Amongst Zambia's mineral wealth are deposits of uranium. European and Japanese interests have identified workable reserves, which are to be brought into production. It is anticipated that output of yellow-cake could be some 1,000 tonnes per annum by the mid 1980s.

ENERGY TRADE

Zambia's foreign trade in energy is characterised by a large-scale export of electricity to Zimbabwe amounting to more than 2,000 GWh per annum. There are also exchanges with Zaire, involving a small net export. On the import side is the full requirement for crude oil and feedstock for the Ndola refinery and several hundred thousand tonnes per annum of coke.

Zimbabwe

KEY ENERGY INDICATORS

Energy Consumption	
—million tonnes oil equivalent	3.8
Consumption Per Head	
—tonnes oil equivalent	0.51
—percentage of world average	31%
Net Energy Imports	29%
Oil Import-Dependence	100%

Zimbabwe's pattern of energy usage has been influenced by the extensive indigenous resources of coal and hydro-electric power and by external circumstances, which have made oil a difficult commodity to obtain until recently. The pressure to restrict use of oil remains, however, in order to limit expenditure of foreign exchange. Accordingly, coal and hydro-electric power are being more intensively exploited.

ENERGY MARKET TRENDS

The economy has reflected the vicissitudes of recent political problems as well as the impact of general economic recession. The important mining sector has been particularly affected. Manufacturing activity has developed in the face of difficulties with imports, but agriculture still remains important. Consumption of energy fell considerably during the most difficult period of unrest in the late 1970s, but since 1979 consumption has shown a net increase of 15-20 per cent.

Pattern of Energy Supply

Primary Fuel Supply (percentage)	
Oil	16
Solid Fuel	58
Natural Gas	–
Primary Electricity	26
TOTAL	100

Coal is the major indigenous fuel available in Zimbabwe and accounts for a high proportion of total energy supplies. Oil is relatively little used and is concentrated in applications for which substitution is difficult. Hydro-electric potential has been exploited, principally through development at Kariba on the Zambezi, in conjunction with Zambia. In addition to coal and coke, there is a substantial amount of non-commercial solid fuel, principally wood, which is used.

ENERGY SUPPLY INDUSTRIES

The supply of electricity has become a primary concern of state enterprises. But private sector companies are important in the supply of coal and oil. The Wankie colliery is run by the South African based Anglo American Corporation and several of the international oil companies have maintained distribution operations within the country. The role of private companies may decline in both sectors as policies of greater state control are implemented.

Oil

Consumption of oil is of the order of half a million tonnes per annum. Use has been curtailed until recently as a result of general embargoes on oil exports to the country. Sanctions meant that the country's oil refinery at Umtali was not able to operate, owing to closure of the crude oil pipeline from Beira, in

Mozambique. Supplies of refined products are imported from various sources, including South Africa, where associate companies of those operating in Zimbabwe are active.

Coal

Production of coal is centred on the operations of Wankie Colliery Company, which produces two million tonnes of coal per annum. The company is part of the South African based mining group Anglo American Corporation. Output is sold to all sectors of the market in Zimbabwe. The capacity of the Wankie mines is being expanded to provide fuel for new coal-fired capacity at Wankie power station. The first unit of 120 MW is due to come into operation in 1982, with a succession of perhaps three similar units to follow. There are large untapped coal resources in the country, estimated at 28,000 million tonnes.

Electricity

Consumption of electricity exceeds 5,000 GWh per annum. This is supplied principally from the hydro-electric station at Kariba, run by Central African Power Corporation, which is owned jointly by Zimbabwe and Zambia. Power is bought in bulk by the public undertaking, the Electricity Supply Commission, for distribution throughout Zimbabwe. The Commission is also the agent for purchasing imports of power from Zambia and South Africa. The Commission operates a limited amount of thermal plant, but is constructing a large coal-fired station at Wankie, initially of 480 MW, but possibly eventually reaching 1,280 MW, which would give the country a high degree of self-sufficiency in electricity.

ENERGY TRADE

Zimbabwe is a net importer of energy. All of the country's oil requirements are imported, and a significant quantity of electricity is purchased from Zambia, to supplement the availability from hydro-electric plant in Zimbabwe. Currently this import exceeds 2,000 GWh per annum, but is expected to decline as coal-fired capacity at Wankie is developed.

Part Three: Energy Organisations

International

INTER-GOVERNMENTAL AUTHORITES AND AGENCIES

INTERNATIONAL ATOMIC ENERGY AGENCY
Address: Kärntner Ring 11, PO Box 590
A-1011 Vienna, Austria
Telephone (0222) 52 45 11 Telex 112645

The IAEA is a specialised agency of the United Nations concerned with the development of nuclear power for peaceful purposes. Its main objectives are to increase the contribution of atomic energy to peace, health and prosperity throughout the world. A key aspect of its work is the promotion of proper standards of safety and operating procedure for nuclear facilities. This extends to the operation of non-proliferation safeguard arrangements under which research facilities operate in member countries.

INTERNATIONAL ENERGY AGENCY
Address: 2 Rue André Pascal, F-75775 Paris
Cedex 16, France
Telephone (1) 524 82 00

The International Energy Agency (IEA) was established in 1974 as an independent organisation within OECD, specifically to deal with the issues which became apparent as a result of the oil supply crises of 1973-74. Initially these centred on contingency planning in the event of a subsequent sudden shortfall or disruption in crude oil supplies.

Latterly, however, questions of general availability of energy and the role of energy conservation have become more important. The IEA carries out analyses of developments in member countries and co-ordinates programmes of energy conservation and research and development of additional energy resources.

Membership of the IEA is similar to that of OECD, but does not include France, Finland or Iceland.

NUCLEAR ENERGY AGENCY
Address: 2 Rue André Pascal, F-75775 Paris
Cedex 16, France
Telephone (1) 524 82 00

The Nuclear Energy Agency is a semi-autonomous organisation within the OECD. In addition to OECD members the European Commission also participates in NEA activities.

The main purpose of the NEA is to promote co-operation on safety and regulatory aspects of nuclear development and to assess its future role as contributor to economic progress. The NEA co-ordinates international programmes of R&D and monitors technical and economic aspects of the nuclear fuel cycle.

ORGANISATION FOR ECONOMIC CO-OPERATION AND DEVELOPMENT
Address: 2 Rue André Pascal, F-75775 Paris
Cedex 16, France
Telephone (1) 524 82 00

The OECD is the principal inter-governmental organisation of industrial nations in the western world, with a total membership of 24. Membership includes Australia, Japan, New Zealand and Turkey. The two principal objectives of OECD are to help member countries promote economic growth, employment and improved standards of living, and, secondly, to promote sound and harmonious development of the world economy and improve the condition of the poorest economies.

In view of the important role of energy in economic development the OECD monitors closely developments in this sphere. In addition two semi-autonomous agencies exist within the overall structure of OECD: the Nuclear Energy Agency and the International Energy Agency.

ORGANISATION OF ARAB PETROLEUM EXPORTING COUNTRIES
Address: PO Box 20501, Kuwait

OAPEC is an association representing the Arab States within OPEC, the Organisation of Petroleum Exporting Countries, with the addition of Bahrain, Egypt and Syria. It includes a majority of OPEC members and countries which account for a high proportion of total OPEC reserves and production. Member countries are Algeria, Bahrain, Egypt, Iraq, Kuwait, Libya, Qatar, Saudi Arabia, Syria and the United Arab Emirates.

ORGANISATION OF PETROLEUM EXPORTING COUNTRIES
Address: Obere Donaustrasse 93, A-1020 Vienna, Austria
Telephone (0222) 26 55 11 Telex 134474

OPEC was set up in 1960 as an association of states for which the export of petroleum was an important economic activity. Membership is basically open to countries with a substantial net export of crude oil and similar interests to those of existing members. Initial membership comprised Iran, Iraq, Kuwait, Saudi Arabia and Venezuela. Subsequently it has expanded to include Algeria, Ecuador, Gabon, Indonesia, Libya, Nigeria, Qatar and the United Arab Emirates.

The main concern of OPEC is co-ordination of policies on the supply and price of crude oil, but it also operates a fund to channel a proportion of member countries' revenues towards poorer developing countries.

UNITED NATIONS CONFERENCE ON TRADE AND DEVELOPMENT
Address: Palais des Nations, CH-1211, Geneva 10 Switzerland
Telephone (022) 34 60 11 Telex 289696

The United Nations Conference on Trade and Development (UNCTAD) is a standing organisation of the United Nations established as a focus for discussion of economic problems of the developing countries and as a forum for negotiations between developing and developed countries. UNCTAD carries out analyses of issues arising from the energy needs of developing countries, including trading relationships, impact on balance of payments and availability of resources for economic development.

UNITED NATIONS ECONOMIC AND SOCIAL COMMITTEE FOR ASIA AND THE PACIFIC
Address: United Nations Building, Rajadamnern Avenue, Bangkok 2, Thailand
Telephone 282 9161 Telex 7882392

The Economic and Social Committee for Asia and the Pacific (UNESCAP) is the principal specialist organisation within the United Nations dealing with economic problems of Asia, Australasia and the Pacific region. It provides a focus for co-ordinated information and analysis of issues in economic development, including the availability of energy resources and electrification programmes.

UNITED NATIONS ECONOMIC COMMISSION FOR AFRICA
Address: Africa Hall, PO Box 3001, Addis Ababa Ethiopia
Telephone 44 72 00 Telex 21029

The United Nations Economic Commission for Africa (UNECA) is responsible to the Economic and Social Council of the United Nations for preparing and promoting policies and programmes to cope with the economic and energy problems of African countries.

UNECA has formulated a Plan of Action for the Development and Utilisation of New and Renewable Sources of Energy and organised international conferences in Africa to examine the scope of renewable resources for its member countries. Arising from this work the UNECA is to establish a Regional Centre for Solar Energy Research and Development.

UNITED NATIONS ECONOMIC COMMISSION FOR WESTERN ASIA
Address: Nabil Adel Building, Bir Hassan PO Box 4656, Beirut, Lebanon
Telephone 27 82 27

The Economic Commission for Western Asia (UNECWA) is a permanent organ within the United Nations concerned with economic problems within the Western Asian region, which includes most countries of the Middle East.

ENTERPRISES (INTERNATIONAL ENERGY COMPANIES)

ASARCO INCORPORATED
Address: 120 Broadway, New York NY 10271 USA
Telephone (212) 669 1270
Gross Revenue: $1,817.0 million (1980)
Total Assets: $2,044.8 million
Energy Sector: Coal

ASARCO is a major producer of non-ferrous metals, principally silver, copper, lead and zinc, with increasing involvement in energy resources. Associate companies include Southern Peru Copper Corporation, Mexican Desarrollo Industrial Minero and MIM Holdings, the Australian mining enterprise. ASARCO's interest in MIM Holdings is to be reduced from 49 per cent to 44 per cent in order to permit the latter to become more involved in natural resource development in Australia.

MIM Holdings, based originally on metal ore production at Mount Isa, Queensland, is to develop a new coal deposit in Northern Queensland with the object of exporting up to four million tonnes per annum. ASARCO itself already has coal mining operations in the United States.

ATLANTIC RICHFIELD COMPANY
Address: 515 South Flower Street, Los Angeles
California CA 90071, USA
Telephone (213) 486 3511
Gross Revenue: $16,671.7 million (1979)
Total Assets: $13,833.4 million
Energy Sectors: Oil, gas, coal

Atlantic Richfield (ARCO) is a United States based international oil company, with related interests in natural gas production and coal mining. ARCO is involved in all phases of the oil industry and has total product sales of the order of 36 million tonnes per annum. Sales of natural gas are around 1,500 million cubic feet per day. Coal shipments exceed six million tonnes per annum.

ARCO's operations are concentrated in the United States, but the company is operator for an offshore development in Indonesia and is involved in other exploration work, including the China Sea and in Somalia. An ARCO subsidiary has a major holding in a new coal export development in Queensland.

BRITISH PETROLEUM COMPANY LIMITED
Address: Britannic House, Moor Lane
London EC2Y 9BU, England
Telephone (01) 920 8000
Gross Revenue: £25,347.0 million (1980)
Total Assets: £11,455.0 million
Energy Sectors: Oil, gas, coal, uranium, petrochemicals

BP is one of the major international oil companies. From the position of being a key seller of crude oil the group became during the 1970s deficient in crude oil in relation to its product market requirements, despite making significant discoveries of oil in many different parts of the world.

Outside North America and Western Europe BP's largest operation is in Australia, where the company has refining and marketing operations and is building up a position in the expanding coal industry. The company is involved in exploration in several areas, particularly Malaysia and the China Sea, and oil marketing in many countries of Africa, Asia and the Pacific.

Major energy projects in which BP companies are participants include LNG in Abu Dhabi, gas development in New Zealand and a share of an export coal project in South Africa.

BURMAH OIL COMPANY
Address: Burmah House, Pipers Way, Swindon
Wiltshire SN3 1RE, England
Telephone (0793) 30151 Telex 449225
Gross Revenue: £1,407.9 million (1981 excl taxes)
Total Assets: £951.7 million
Energy Sectors: Oil, gas

Burmah Oil Company is a British-based international oil company with interests in exploration, production and transportation. Burmah has also diversified into other related areas such as engineering and automotive products. The company is a leading refiner and supplier of lubricating oils under the Castrol name.

Burmah has long established interests in exploration, production, refining and marketing in Pakistan and India. It holds a 70 per cent interest in Pakistan Petroleum, which owns the Sui gas field, and also has some oil production. Exploration work is being carried out in several countries, notably the Sudan and Gabon.

CALTEX PETROLEUM CORPORATION
Address: 380 Madison Avenue, New York
NY 10017, USA
Energy Sector: Oil

Caltex Petroleum Corporation is jointly owned by Standard Oil Company of California and Texaco Inc, and is a holding company for some 70 individual operating companies engaged in exploration, production, refining and marketing in Africa, Asia and Australasia.

COMPAGNIE FRANÇAISE DES PETROLES
Address: 5 Rue Michel-Ange, F-75781 Paris
Cedex 16, France
Telephone (1) 524 46 46 Telex 611992
Gross Revenue: F40,866 million (1980 excl taxes)
Total Assets: F73,515 million
Energy Sectors: Oil gas, coal, uranium

Compagnie Française des Pétroles (CFP) has developed into an important international company with extensive involvement in exploration, processing and marketing in many parts of the world. The French government has a substantial shareholding with an effective 40 per cent voting right.

Major subsidiary companies are involved in oil refining and marketing in Australia and South Africa. In South Africa the company is also a participant in an export-orientated coal project.

CONOCO INCORPORATED
Address: High Ridge Park, Stamford
Connecticut 06904, USA
Telephone (203) 359 3500 Telex 965905
Gross Revenue: $18,766.3 million (1980)
Total Assets: $11,036.3 million
Energy Sectors: Oil, gas, coal, uranium

Conoco is a leading US oil and coal producing company with additional interests in uranium exploration. The company is actively involved in production in Dubai, Indonesia and Libya and exploration in these and other areas.

CONSOLIDATED GOLD FIELDS LIMITED
Address: 49 Moorgate, London EC2R 6BQ
England
Telephone (01) 606 1020
Gross Revenue: £944.7 million (1981)
Total Assets: £422.8 million
Energy Sectors: Coal, uranium

Consolidated Gold Fields is an international mining company, principally involved in gold production, but also increasing its activities in exploration for, and development of, energy and other natural resources. Main areas of operation are South Africa and Australia.

In South Africa the availability to the company of uranium arises from gold mining. Also in South Africa the group owns Apex Mines which produced 2.7 million tonnes of coal in 1980. Consolidated Gold Fields has an indirect interest of around 25 per cent in the Glendell steam coal project in New South Wales.

ENTE NAZIONALE IDROCARBURI
Address: Piazzale Enrico Mattei 1, 00144 Rome
Italy
Telephone (06) 59001 Telex 610082
Gross Revenue: L26,439.5 billion (1980)
Total Assets: L31,791.7 billion
Energy Sectors: Oil, gas, uranium

ENI is the Italian state hydrocarbons agency and one of the key industrial groupings in the Italian economy. It has become involved in exploration on a wide scale internationally and has refining or marketing operations in many countries. These usually operate under the AGIP name.

ENI subsidiaries are also involved in construction and engineering work associated with energy developments, including refinery construction and pipeline laying. The group is particularly involved in oil industry operations in North Africa, West Africa and East Africa, acting as operator of several national refineries and exploring for oil and gas. In Nigeria AGIP is a participant with Phillips Petroleum in one of the principal oil producing companies.

EXXON CORPORATION
Address: 1251 Avenue of the Americas,
New York, NY 10020, USA
Telephone (212) 398 3000
Gross Revenue: $110,380.6 million (1980)
Total Assets: $56,576.6 million
Energy Sectors: Oil, gas, coal, uranium

Exxon is the largest of the major international oil companies with exploration, production, processing and marketing operations throughout the world. It is a participant in Aramco, the key Saudi Arabian production company.

Availability of oil to Exxon in the Middle East, Africa, Australia and the Far East is around 280,000 barrels per day, with 280 million cubic feet per day of gas.

Affiliates and associates of the Exxon group occupy leading positions in the oil industry in Australia, Japan and Malaysia. Exploration is being undertaken in many areas. In Australia the group partners BHP as producer in the main oil production area of the Bass Strait and is involved in uranium and oil shale projects.

GETTY OIL COMPANY
Address: 3810 Wilshire Boulevard, Los Angeles
California CA 90010, USA
Telephone (213) 381 7151 Telex 910 321 2832
Gross Revenue: $10,436.8 million (1980)
Total Assets: $8,266.7 million
Energy Sector: Oil, gas, coal, uranium

Getty Oil is a United States based oil company with interests in oil, gas and coal operations, and in exploration for uranium, notably in Australia. The company's activities are concentrated in North America, but Getty has important interests in some particular countries.

Getty holds a concession, covering approximately 2,100 square miles in the Partitioned Zone between Saudi Arabia and Kuwait, with proved reserves of 360 million barrels. The company operates a 2.5 million tonne per annum refinery at Mina Saud.

In Japan Getty has a 50 per cent shareholding in Mitsubishi Oil Company, which operates refineries at Kawasaki and Mizushima, and an indirect holding in Tohoku Oil Company's refinery at Sendai. Other exploration and production activities are carried on in Indonesia and Sharjah.

GULF OIL CORPORATION
Address: Gulf Building, Pittsburgh
Pennsylvania 15230, USA
Telephone (412) 263 5000
Gross Revenue: $28,389 million (1980)
Total Assets: $18,638 million
Energy Sectors: Oil, gas, coal, nuclear power, uranium

Gulf Oil is a major energy company, with production of oil, gas, coal and uranium. Production operations are concentrated in North America, but the group has interests in production in Angola, Gabon, Indonesia, Nigeria and Zaire. Exploration is undertaken in a number of other countries in Africa, the Middle East, Asia and Australasia.

Marketing operations in these areas are limited. The company now has no significant interests in oil refining, but marketing and trading companies are active in most countries of Asia and Australasia.

NV KONINKLIJKE NEDERLANDSCHE PETROLEUM MAATSCHAPPIJ
Address: 30 Carel van Bylandtlaan, Den Haag
Netherlands
Telephone (070) 77 66 55
Energy Sectors: Oil, gas coal, nuclear power

Royal Dutch Petroleum Company is a Dutch based company and principal parent company of the Royal Dutch/Shell group. Royal Dutch Petroleum Company holds 60 per cent of the shares in the two holding companies Shell Petroleum NV and Shell Petroleum Company Limited. The two holding companies control the many operating and service companies of the Royal Dutch/Shell group.

LONRHO LIMITED
Address: Cheapside House, 138 Cheapside
London EC2V 6BL, England
Telephone (01) 606 9898
Gross Revenue: £2,100.7 million (1980)
Total Assets: £848.3 million
Energy Sectors: Coal, oil

Lonrho is a diversified company, the operations of which are concentrated in Western Europe and Africa. Mining operations account for only a small proportion of total group turnover, but more than 40 per cent of pre-tax profits.

The company has interests in mining operations in South Africa through its 86 per cent shareholding in Duiker Exploration, which is involved in production of anthracite and steam coal in the Transvaal.

MINATOME SA
Address: 69-73 Rue Dutot, F-75738 Paris
Cedex 15, France
Telephone (1) 539 2260 Telex 250718
Energy Sector: Uranium

Minatome was established in 1975 by the Pechiney-Ugine-Kuhlmann group and Compagnie Française des Pétroles to explore for uranium deposits, particularly with a view to supporting the development of nuclear power in France.

Minatome is involved in uranium exploration and production in West Africa, through SOMAIR, in Niger, and COMUF, in Gabon. The company is also actively involved in examining Australian uranium prospects.

MOBIL CORPORATION
Address: 150 East 42nd Street, New York
NY 10017, USA
Telephone (212) 833 4242 Telex 232561
Gross Revenues: $63,726 million (1980)
Total Assets: $32,705 million
Energy Sectors: Oil, gas

Mobil is one of the major international oil companies with producing, refining and marketing operations throughout the world. Total sales of oil products exceed 2.3 million barrels per day, of which more than 500,000 barrels per day are in the Far East and Australasia. The company is a participant in the Aramco consortium in Saudi Arabia.

Mobil is a partner, with Exxon and Japanese oil companies, in one of the leading refining/marketing groups in Japan. Mobil affiliates are also prominent in Australia, New Zealand and the Philippines. The group operates a major export refinery at Singapore.

Mobil is one of the leading companies involved in exporting LNG from Indonesia to Japan. In Nigeria the group is an important producer of oil. Exploration for oil and gas is continuing in several countries, including Indonesia, Nigeria and Angola.

OCCIDENTAL PETROLEUM CORPORATION
Address: 10889 Wilshire Boulevard, Los Angeles
California CA 90024, USA
Telephone (213) 879 1700 Telex 673389
Gross Revenue: $12,726.3 million (1980)
Total Assets: $6,629.9 million
Energy Sectors: Oil, gas, coal

Occidental Petroleum is a United States based mineral resources group with interests in oil, gas and coal production, and in oil shale development. The company is engaged in transportation and trading of energy products and other natural resources.

The company is one of the leading foreign oil companies involved in Libya, where it has access to some 240,000 barrels per day of crude oil. Exploration is being undertaken in several countries, including Abu Dhabi, Australia, Morocco, Oman, Pakistan and Tunisia.

PETROFINA SA
Address: Rue de la Loi 33, B-1040 Bruxelles
Belgium
Telephone (02) 513 69 00 Telex 21556
Gross Revenue: BF339.3 billion (1980)
Total Assets: BF231.3 billion
Energy Sectors: Oil, gas

Petrofina is one of the largest Belgian companies and a substantial company in the international oil business. It has refining and marketing operations in Europe and integrated oil activities in North America. Outside these areas its attention is concentrated on areas of long standing Belgian interest in central Africa.

Exploration by Petrofina has been successful in Zaire. The company is also a participant in a consortium producing oil in Angola. Other exploration has been carried out in Tunisia.

PHILLIPS PETROLEUM COMPANY
Address: Phillips Building, Bartlesville
Oklahoma 74004, USA
Telephone (718) 661 6600 Telex 492455
Gross Revenue: $13,713 million (1980)
Total Assets: $9,844 million
Energy Sectors: Oil, gas, coal, uranium

Phillips Petroleum is engaged in the exploration for, and production of, oil and gas in countries throughout the world. Exploration is also actively undertaken, principally in North America, to develop coal and uranium resources.

Outside North America and Europe key countries of significance within the group's operations currently are Nigeria, Indonesia and Australia. In Nigeria Phillips is partnered by AGIP in one of the main producing companies. Production in Indonesia is small but exploration activity is continuing. Other areas of exploration include the north-west offshore area of Australia and the Philippines.

RIO TINTO-ZINC CORPORATION LIMITED
Address: 6 St James's Square, London SW1Y 4LD
England
Telephone (01) 930 2399
Gross Revenue: £2,796 million (1980)
Total Assets: £3,990 million
Energy Sectors: Coal, uranium, oil, gas

Rio Tinto-Zinc Corporation (RTZ) is a British-based natural resource development company, with interests in production, processing and manufacturing. Subsidiary and associate companies operate in North America, South America, Europe, Africa, the Far East and Australasia. Interests in Australasia are held through CRA, in which RTZ holds a majority share. This is expected to be reduced in due course to 49 per cent to permit CRA to engage in a wider range of activities. Through CRA the group has coal production in Australia by Kembla Coal and Coke Pty, Blair Athol Coal Pty and an interest in deposits at Tarong, Queensland. CRA has a 51 per cent holding in the Mary Kathleen uranium mine. RTZ has a 46.5 per cent interest in Rössing Uranium Ltd, which operates a large uranium mine in Namibia.

ROYAL DUTCH/SHELL GROUP
Addresses: Shell International Petroleum Co Ltd
Shell Centre, London SE1 7NA, England
Telephone (01) 934 1234 Telex 919651

Shell Internationale Petroleum Mij BV
30 Carel van Bylandtlaan, Den Haag
Netherlands
Telephone (070) 77 66 55
Gross Revenue: £47,673 million (1981)
Total Assets: £34,076 million
Energy Sectors: Oil, gas, coal, nuclear power

The Royal Dutch/Shell companies form part of a group owned by the Royal Dutch Petroleum Company and Shell Transport and Trading Company. Activities are carried out worldwide under the 'Shell' name. Of the international oil companies turnover of the Royal Dutch/Shell group is second only to that of Exxon Corporation.

The Shell group is more extensively involved in Africa, Asia and Australasia than any other individual company. There are few countries in these areas in which Shell affiliates and associates do not play an important part, either in exploration, production, refining or marketing of oil. The group is also prominent in LNG trading, between Brunei and Japan, and is developing its interests in coal, particularly in Australia and South Africa.

In the Middle East Shell is less involved and is not a participant in the Aramco consortium. But it has become closely involved with major projects for utilising oil and gas in Abu Dhabi and Saudi Arabia. Its main production interests in the area are in Oman.

SHELL TRANSPORT AND TRADING COMPANY LIMITED
Address: Shell Centre, London SE1 7NA
England
Telephone (01) 934 1234 Telex 25781
(overseas) 25733 (inland)
Energy Sectors: Oil, gas, coal, nuclear power

Shell Transport and Trading is a British company with a 40 per cent interest in the two holding companies of the Royal Dutch/Shell group, viz Shell Petroleum NV and Shell Petroleum Company Limited, which in turn own the operational companies of the group.

STANDARD OIL COMPANY (INDIANA)
Address: 200 East Randolph Drive, Chicago
Illinois 60601, USA
Telephone (312) 856 3800
Gross Revenue: $27,831.9 million (1980)
Total Assets: $20,167.5 million
Energy Sectors: Oil, gas

Standard Oil Company (Indiana) is one of the leading American oil companies with numerous subsidiaries and associates active in many parts of the world. Subsidiaries invariably operate under the Amoco title.

Amoco companies have explored successfully for oil in Egypt and North Africa, and active areas of exploration include Oman, Tanzania, Zaire, Indonesia, Pakistan and the Philippines. In Australia Amoco has integrated refining/marketing operations, based on the company's wholly owned refinery near Brisbane.

STANDARD OIL COMPANY OF CALIFORNIA
Address: 225 Bush Street, San Francisco
California CA 94104, USA
Telephone (415) 894 7700
Gross Revenue: $42,919 million (1980)
Total Assets: $22,162 million
Energy Sectors: Oil, gas

Standard Oil Company of California is one of the world's major international oil companies with operations in all phases of oil and gas production and supply throughout the world.

The company has important oil producing interests in Saudi Arabia, where it is a participant in the Aramco consortium, and in Indonesia, where it is partnered by Texaco Inc. Standard Oil and Texaco operate together in many countries through Caltex Petroleum Corporation, particularly in refining and marketing of oil products, as well as through PT Caltex Pacific Indonesia.

Principal areas of interest in exploration include Australia, Indonesia, Japan, Korea, Nigeria and the Pacific. The company has also been successful in discovering oil in the Sudan.

TEXACO INC
Address: 2,000 Westchester Avenue, White Plains,
New York NY 10650, USA
Telephone (914) 253 4000
Gross Revenue: $52,484.6 million (1980)
Total Assets: $26,430.4 million
Energy Sectors: Oil, gas

Texaco is a major international oil company with extensive involvement in exploration, refining and marketing in many countries throughout the world.

Texaco has important interests in crude oil availability from Saudi Arabia and Indonesia. Texaco is a participant in the Aramco consortium in Saudi Arabia and is partner with Standard Oil Company of California in PT Caltex Pacific Indonesia, which is responsible for almost 50 per cent of Indonesian output of crude oil.

Through Caltex Petroleum's many subsidiaries and associates Texaco is involved in refining and marketing in Australia, Japan and most other countries of Africa, the Middle East, Asia and Australasia.

UNION OIL COMPANY OF CALIFORNIA
Address: Union Oil Center, 461 South Boylston
Street, Los Angeles, California CA 90017, USA
Telephone (213) 486 7600 Telex 677593
Gross Revenue: $10,436.9 million (1980)
Total Assets: $6,772.1 million
Energy Sectors: Oil, gas, geothermal, uranium

Union Oil Company (UNOCO) is a United States based energy company with interests in oil, gas and uranium in the United States, oil and gas production in Asia and has a major interest in exploiting geothermal energy. Production of oil and gas outside the USA in 1980 amounted to 77,000 barrels per day and 60 million cubic feet respectively.

UNOCO's main interests in oil production are in the Sassan field, off Iran, and in Indonesia. But gas has become of much greater significance with development of three fields discovered by the company in the Gulf of Thailand.

The company has built on its experience of geothermal energy exploitation in California to take a leading role in the Philippine development programme. A UNOCO subsidiary is also examining the scope for geothermal energy production in Japan.

URANGESELLSCHAFT MBH & CO KG
Address: Bleichstrasse 60-62, P O Box 4409
D-6000 Frankfurt am Main, West Germany
Telephone (0611) 2169-1 Telex 0413199
Capital: DM42 million
Energy Sector: Uranium

Urangesellschaft was established in 1967 at the instigation of the Federal government to help procure supplies of natural uranium for nuclear power plants. The company is owned equally by three key West German mining and energy groups, VEBA, STEAG and Metallgesellschaft.

The company has as primary objective the exploration for, and development of, uranium deposits. It may operate mines and mills and trade in uranium ores. Exploration is being carried out in Australia, Indonesia and several African countries as well as in North America and South America.

PROFESSIONAL INSTITUTIONS AND TRADE ASSOCIATIONS

CIGRE
Address: 112 Boulevard Haussmann, F-75008 Paris, France
Telephone (1) 522 65 12
Energy Sector: Electricity

CIGRE, the International Conference on Large High-Voltage Electric Systems, is an organisation bringing together electricity producers, plant manufacturers and consultants involved in the problems of generation and transmission of high voltage electricity. Conferences are held in alternate years at which a wide range of issues are discussed, building on the work of permanent working groups.

CIRED
Address: IEE Conference Department
Savoy Place, London WC2R 0BL, England
Telephone (01) 240 1871 Telex 261176
Energy Sector: Electricity

CIRED is the International Conference on Electricity Distribution. Its main activity is a conference every two years for representatives of all organisations concerned with the distribution of electricity.

INTERNATIONAL GAS UNION
Address: Union Internationale de l'Industrie du Gaz (UIIG), 62 Rue de Courcelles
F-75008 Paris, France
Telephone (1) 766 03 51
Energy Sector: Gas

The IGU is an organisation of the principal gas undertakings in each of 36 member countries. The concern of IGU is to study problems of the industry and its development and to promote communication between technical people in different countries.

Every three years the IGU organises a world congress on gas. At these conferences permanent working groups report on various economic and technical aspects of the gas industry.

INTERNATIONAL INSTITUTE FOR APPLIED SYSTEMS ANALYSIS
Address: Schlossplatz 1, A-2361 Laxenburg
Austria
Telephone (02236) 7115 21 0 Telex 079137
Energy Sector: Energy

The IIASA was set up with support from scientific institutions in 17 nations including eastern bloc countries. Its remit is the evaluation of long-term trends and the inter-relationships of systems. A five-year programme analysing the role of energy and long-term supply and demand prospects was completed in 1980.

UNIPEDE
Address: 39 Avenue de Friedland, F-75008
Paris 8, France
Telephone (1) 256 94 00 Telex 660471
Energy Sector: Electricity

The Union Internationale des Producteurs et Distributeurs d'Energie is an association including electricity undertakings or their representative organisations. It provides communication between members on all technical and technico-economic aspects of electricity generation and distribution.

THE URANIUM INSTITUTE
Address: New Zealand House, Haymarket
London SW1Y 4TE, England
Telephone (01) 930 5726 Telex 917611
Energy Sector: Uranium

The Uranium Institute brings together the various parties concerned with the development of uranium. It includes exploration, development and processing companies, fuel fabricators and electricity generating companies. The Institute provides a co-ordinated view of the uranium industry and an annual assessment of the supply and demand situation.

WORLD ENERGY CONFERENCE
Address: 34 St James's Street, London SW1
England
Telephone (01) 930 3960
Energy Sector: Energy

The World Energy Conference takes place every four years. At these conferences the full range of technical and economic aspects of energy development and use are considered. There is a permanent secretariat and organising committees in many countries work to prepare material for the conferences. Proceedings are published providing a survey of main energy issues.

WORLD PETROLEUM CONGRESS
Address: 61 New Cavendish Street
London W1M 8AR, England
Telephone (01) 636 1004
Energy Sectors: Oil, gas

The World Petroleum Congress takes place every four years. These conferences bring together expert opinion and synthesized information on all aspects of the petroleum sector. Work is carried on under a permanent secretariat based in London and the various national committees which are usually centred on industry professional associations in participating countries.

Middle East

Bahrain

ENTERPRISES (PUBLIC AND PRIVATE SECTOR COMPANIES)

BAHRAIN NATIONAL GAS COMPANY
Address: PO Box 477, Manama
Telephone 24 20 20 Telex 9317
Energy Sectors: Oil, gas

Bahrain National Gas Company (BANAGAS) has been established by the state to extract gas liquids from the associated gas produced from BANOCO's Jebel Al-Dukhan field. The company is a joint venture with Caltex Petroleum, which operates the Bahrain refinery, and Arab Petroleum Investment Corporation (APICORP), which has provided finance for the project. BANAGAS has a 75 per cent interest, the remaining shares divided equally between Caltex and APICORP.

In 1980 BANAGAS took 1.5 billion cubic feet of gas, separating out LPG and naphtha. LPG is exported by Caltex to consumers in Japan.

BAHRAIN NATIONAL OIL COMPANY (BANOCO)
Address: PO Box 504, Manama
Telephone 71 40 81 Telex 8670
Energy Sector: Oil

BANOCO was established in 1976 as the state oil company. It holds the state's 60 per cent interest in the country's refinery, with the right to participate in the full range of company activities including overseas marketing. BANOCO has taken over all of the interests formerly held by BAPCO in exploration and production of oil and natural gas and also handles distribution and marketing of refined products in Bahrain.

BAHRAIN PETROLEUM COMPANY LIMITED
Address: PO Box 609, Awali
Energy Sector: Oil

Bahrain Petroleum Company (BAPCO) owns and operates a large refinery at Sitra, geared to producing a wide range of products for markets in the Middle East, the Indian Ocean area and the Far East. BAPCO is operated by Caltex Petroleum, in which Texaco Inc and Standard Oil Company of California are equal partners. The state holds a 60 per cent majority interest.

A small refinery was built at Sitra in 1936 and has since been expanded to 250,000 barrels per day capacity, equivalent to more than 12 million tonnes per annum. Crude oil is supplied from local fields and from Saudi Arabia where the Caltex companies are participants in Aramco. In 1980 80 per cent of crude oil was imported from Saudi Arabia. BAPCO itself undertakes the marketing of export products, supplying also the needs of the national oil company for the local market.

Iran

ENTERPRISES (PUBLIC AND PRIVATE SECTOR COMPANIES)

LAVAN PETROLEUM COMPANY
Address: PO Box 3243, Tehran
Telephone 01 80 85 Telex 21 24 20
Energy Sector: Oil

Lavan Petroleum Company (LAPCO) is a joint-venture exploration and production company, operating the offshore Sassan field. The company was set up with 50 per cent participation by the state oil company NIOC with 12.5 per cent shares held by each of four United States based oil companies, Atlantic Richfield, Murphy Oil, Sun Oil and Union Oil of California. These companies were notified by the government in 1980 that their assets were being taken over by the state.

LAPCO production in 1979 was 90,000 barrels per day, or some 4.5 million tonnes. This was well below the levels of output achieved in earlier years.

NATIONAL IRANIAN GAS COMPANY
Address: 2 Hamid Street, PO Box 3375,
Roosevelt Avenue, Tehran
Telephone 82 10 31 Telex 2514
Energy Sector: Gas

National Iranian Gas Company (NIGC) is the state-owned undertaking handling all aspects of production, transportation and sales of natural gas. The company's activities are effectively integrated with and controlled by NIOC.

NIGC supplies gas to major industrial and petrochemical plants in Iran, but its principal operation has been the construction of the Iranian Gas Trunkline (IGAT) to the Soviet Union. The IGAT line is fed by a large plant at Bid Boland, where there are five units of 250 million cubic feet per day capacity.

Sales by NIGC to the Soviet Union have been disrupted severely as a result of the reduced amounts of natural gas being made available. This has led to uncertainty over sales via IGAT and cancellation of the project to build a second trunk line. NIGC's projects to export LNG to Japan have also been shelved.

NATIONAL IRANIAN OIL COMPANY
Address: Takhte-Jamshid Avenue, PO Box 1863
Tehran
Telephone 6151 Telex 2514
Energy Sectors: Oil, gas

National Iranian Oil Company (NIOC) has been established for more than 20 years as the state corporation dealing with the oil and gas sectors. It is one of the oldest of the producer country national oil companies and has gradually assumed virtually complete control of all oil and gas operations.

NIOC operates the production, transportation and refining facilities of the former Iranian Consortium areas and has assumed control of other fields in which foreign companies have been involved under joint-venture or service contract arrangements. The precise status of these companies is not as yet fully determined, but Continental Shelf Oil Company has been set up by NIOC to operate the fields.

NIOC has in the past become involved in several overseas ventures, especially in refining in India, Korea, South Africa and, prospectively, Senegal. Following fundamental policy changes in the last two years, however, participations in Korea and South Africa have been relinquished. Its position in the Madras refinery remains uncertain as is its attitude towards the project to build a refinery in Senegal.

PROFESSIONAL INSTITUTIONS AND TRADE ASSOCIATIONS

IRANIAN PETROLEUM INSTITUTE
Address: National Iranian Gas Company,
PO Box 3375, Tehran
Energy Sectors: Oil, gas

The Institute is a representative body for organisations and individuals in the oil and gas industries, providing a focus for information on the industries and for consideration of standards and other technical matters. The Institute is a member organisation of the International Gas Union.

Iraq

GOVERNMENT DEPARTMENTS AND OFFICIAL AGENCIES

MINISTRY OF PETROLEUM
Address: PO Box 6118, Al Mansoor City, Baghdad
Telephone 88 41 61 Telex 2216

The Ministry of Petroleum has wide responsibilities for the oil industry and oversees the activities of a number of state organisations involved in production, transportation and refining of crude oil, distribution of oil products, and oil-related projects. These include the national oil company INOC, the State Establishment for Oil Refining and Gas Processing, the State Establishment for Oil Products and Gas Distribution, and the State Establishment for Oil Projects.

ENTERPRISES (PUBLIC AND PRIVATE SECTOR COMPANIES

IRAQ NATIONAL OIL COMPANY
Address: PO Box 476, Jumhouriah Street, Khullani Square, Baghdad
Telephone 89061 Telex 2204
Energy Sectors: Oil, gas

The Iraq National Oil Company (INOC) was established in 1964 to represent state interests in the oil industry. It has progressively taken over all assets of foreign companies formerly operating in the country, using only service contract arrangements for specific exploration, production, construction or marketing projects.

INOC has built up its own exploration capability and handles all crude oil in Iraq. It operates trunk pipelines to refineries and export terminals and is the state agency for overseas sales of crude oil. It also operates a fleet of oil tankers for ocean transportation.

STATE ESTABLISHMENT FOR OIL REFINING AND GAS PROCESSING
Address: PO Box 3069, Saadoun Street, Baghdad
Telex 2229
Energy Sectors: Oil, gas

The State Establishment for Oil Refining and Gas Processing is directly responsible to the Ministry of Petroleum for oil refining operations in Iraq. This covers the operations of the two principal refineries, at Basrah and Daurah.

Israel

GOVERNMENT DEPARTMENTS AND OFFICIAL AGENCIES

MINISTRY OF ENERGY AND INFRASTRUCTURE
Address: 234 Yafo Road, Jerusalem 94383

The Ministry of Energy and Infrastructure is responsible for formulation and administration of government policy towards all of the energy supply industries, energy conservation and the development of alternative fuels. The Ministry has direct responsibility for the Israel Electric Corporation.

ENTERPRISES (PUBLIC AND PRIVATE SECTOR COMPANIES)

ISRAEL ELECTRIC CORPORATION LIMITED
Address: 16 Hahashmal Street, Tel-Aviv
Telephone 61 43 43 Telex 46507
Gross Revenue: IS 4,813.6 million (1981)
Total Assets: IS 4,363.1 million
Energy Sector: Electricity

Israel Electric Corporation is a public undertaking generating and supplying electricity throughout Israel. Sales of electricity in 1980-81 totalled 12.2 TWh to the corporation's 1.3 million customers.

The corporation operates nearly 2,800MW of generating capacity. This has been entirely oil based, using heavy fuel oil and gas oil but the Hadera power station is being converted to a dual-fired basis so that it can use coal and contracts have been concluded to import coal from Australia and the UK. The corporation holds 26 per cent in the National Coal Supply Company, with the government holding the balance. This company was set up in 1980 as the agency for importing coal for Israel Electric Corporation's power stations.

ISRAEL FUEL CORPORATION LIMITED (DELEK)
Address: 6 Ahuzat Baijit Street, PO Box 1831, Tel-Aviv
Telephone 59421 Telex 33671
Energy Sector: Oil

DELEK is one of the principal supply oranisations for petroleum products in the Israeli market. It is involved in importing crude oil for refining in the Ashdod and Haifa refineries and distribution of products. Affiliated companies include Fuel Oil Trading Company and American Israel Gas Corporation.

ISRAEL NATIONAL OIL COMPANY
Address: PO Box 20115, Maya Building, No 74 Petach Tikva Road, Tel-Aviv
Telephone (03) 33 79 34 Telex 33498
Energy Sectors: Oil, gas

The Israel National Oil Company has been established as a special vehicle for ensuring the state's involvement in the search for, and supply of, oil or gas. Its primary concern is to explore for indigenous oil and gas resources, which are very small in relation to national demand. Current activities include a drilling programme in the Sodom area, where the company has expectations of discovering commercially viable deposits of gas.

Jordan

GOVERNMENT DEPARTMENTS AND OFFICIAL AGENCIES

DEPARTMENT OF ENERGY
Address: Ministry of Industry and Trade,
PO Box 2019, Amman
Telex 21163

The Department of Energy supervises the activities of companies and other organisations in the field of energy, as well as preparing policy advice on the energy sector. The Department also represent the national interest in international relations relating to energy.

NATURAL RESOURCES AUTHORITY
Address: PO Box 2220, Amman
Telephone 44381 Telex 1415

The Natural Resources Authority is a state agency concerned with exploration for, and development of energy resources. It encompasses the functions of geological survey and exploration, which includes direct involvement in exploration as well as supervision of the activities of foreign companies. It currently has under way a programme of exploration for oil shale.

ENTERPRISES (PUBLIC AND PRIVATE SECTOR COMPANIES)

JORDAN ELECTRICITY AUTHORITY
Address: Administrative Department Complex,
PO Box 2310, Jabal Amman
Telex 21259
Energy Sector: Electricity

The Jordan Electricity Authority is a state undertaking responsible for the public electricity supply system. It operates thermal and hydro-electric generating plant and the national transmission system and handles distribution of electricity to final consumers.

JORDAN PETROLEUM REFINERY COMPANY LIMITED
Address: Abu Bakker Al Seddeek Street,
PO Box 1079, Amman
Telephone 30151 Telex 1246
Energy Sector: Oil

Jordan Petroleum Refinery Company operates the national oil refinery, which is located at Zarqa, north-east of the capital, Amman. The state holds a 50 per cent share in the refinery company.

The Zarqa refinery processed 1.8 million tonnes of crude oil in 1980, supplying the country's needs for oil products. Crude oil is supplied to the refinery from Saudi Arabia, through the trunk pipeline to the Mediterranean coast.

Kuwait

GOVERNMENT DEPARTMENTS AND OFFICIAL AGENCIES

MINISTRY OF ELECTRICITY AND WATER
Address: PO Box 12, Safat
Telephone 43 38 21 Telex 300

The Ministry of Electricity and Water is responsible for the provision of electricity within Kuwait. Publicly owned generating plant supplies almost all of Kuwait's needs, including power generated in conjunction with desalination plant.

MINISTRY OF OIL
Address: PO Box 5077, Safat
Telephone 41 52 01 Telex 2363

The Ministry of Oil oversees the activities of several key state corporations involved in exploration, production of oil and gas, oil refining, petrochemicals manufacture and sale of oil and gas. These include the Kuwait National Petroleum Company, Kuwait Oil Company and Kuwait Oil Tanker Company.

ENTERPRISES (PUBLIC AND PRIVATE SECTOR COMPANIES)

ARABIAN OIL COMPANY LIMITED
Address: PO Box 1641, Safat
Telephone 43 92 01 Telex 2095
Energy Sector: Oil

Arabian Oil Company (AOC) is a joint-venture crude oil production company, including the Saudi Arabian and Kuwait governments and a consortium of Japanese consuming companies. Japanese companies involved are Tokyo Electric Power, Nippon Steel, Kansai Electric Power and Nomura Securities.

The company produces Khafji and Hout crude from offshore areas in the Saudi Arabia-Kuwait Neutral Zone. Crude oil produced by AOC is purchased by Japanese companies or run in the company's small 30,000 barrels per day refinery at Ras Al-Khafji for shipment to Japan.

GETTY OIL COMPANY
Address: PO Box 1, Mina Saud
Telephone 43 53 50 Telex 959-3074
Energy Sector: Oil

Getty Oil Company, the United States based international oil company, participates in crude oil production and refining in the Kuwait-Saudi Arabia neutral zone. The company is partnered by the Kuwait state company KOC.

Production by Getty Oil in 1980 exceeded three million tonnes. The crude oil is of high sulphur content and some of it is processed at the company's Mina Saud refinery. Getty's proven crude oil reserves in the area are estimated to be around 360 million barrels.

KUWAIT NATIONAL PETROLEUM COMPANY
Address: PO Box 70, Safat
Telephone 42 01 21 Telex 2000
Energy Sector: Oil

Kuwait National Petroleum Company (KNPC) is a subsidiary of the state holding company Kuwait Petroleum Corporation. KNPC is concerned with refining crude oil and sales of crude oil and finished products both in the domestic market and abroad.

KNPC operates three substantial refineries with an annual throughput capacity in total of approximately 28 million tonnes per annum. These include the Shuaiba complex, which has a capacity to produce 10 million tonnes per annum of petrochemical and conventional petroleum products. Projects are in hand to modernise KNPC's two other refineries, at Mina Al Ahmadi and Mena Abdulla, to provide eventually 25 million tonnes per annum of capacity, with the capability of producing a high proportion of light and middle distillate products.

KUWAIT OIL COMPANY
Address: PO Box 20007, Kuwait Investment Company Building, Central Commercial Area
Telephone 42 52 81 Telex 2046
Energy Sectors: Oil, gas

Kuwait Oil Company (KOC) is the state-owned company responsible for exploration and production of oil and gas. It forms part of the group of companies within Kuwait Petroleum Corporation.

KOC produces a high proportion of the country's crude oil from the fields formerly developed by the company as a joint-venture of British Petroleum Company and Gulf Oil. Crude oil is transferred to Kuwait National Petroleum Company for refining or export sales. KOC also produces large volumes of associated gas which are utilised in petrochemicals and power generation.

Lebanon

ENTERPRISES (PUBLIC AND PRIVATE SECTOR COMPANIES

ELECTRICITE DU LIBAN
Address: Nahr Street, PO Box 131, Beirut
Telephone 22 00 00
Energy Sector: Electricity

Electricité du Liban (EDL) is the public electricity supply undertaking in Lebanon. It operates as an autonomous organisation under the general control of the government.

EDL supplies around 85 per cent of the country, excluding some localities where small franchise holders operate. EDL is expected to assume the responsibility for supply once these franchises have expired.

EDL produced some 2,300 GWh of electricity in 1980 from its thermal and hydro-electric plant. In 1981 an inter-connection with the Syrian grid was completed.

MEDITERRANEAN REFINERY COMPANY (MEDRECO)
Address: Medreco Refinery, Zahrani, Sidon
Energy Sector: Oil

MEDRECO is the operating company for the oil refinery at Zahrani. Mobil Corporation holds a 50 per cent share and Standard Oil Company of California and Texaco the balance through Caltex Petroleum.

The Refinery has a processing capacity of less than one million tonnes per annum. Crude oil is supplied via the Trans Arabia Pipeline (Tapline) from Saudi Arabia to Sidon.

Oman

GOVERNMENT DEPARTMENTS AND OFFICIAL AGENCIES

MINISTRY OF ELECTRICITY AND WATER
Address: PO Box 4491, Ruwi
Telephone 70 22 33

The Ministry of Electricity and Water is concerned with the development of adequate electricity generating capacity and the creation of an efficient inter-connecting grid for distribution of electricity throughout the country.

MINISTRY OF PETROLEUM AND MINERALS
Address: PO Box 551, Muscat
Telephone 70 20 66

The Ministry of Petroleum and Minerals is responsible for control of exploration and development of oil and gas. It has general supervision of the state's interests which are held in Petroleum Development Oman, and of policy towards exploration through the Council for Natural Gas.

ENTERPRISES (PUBLIC AND PRIVATE SECTOR COMPANIES)

PETROLEUM DEVELOPMENT OMAN
Address: PO Box 81, Muscat
Telephone 60 75 65 Telex 3212
Energy Sector: Oil

Petroleum Development Oman (PDO) is the principal oil exploration and production company in Oman. The state holds a 60 per cent interest and the Royal Dutch/Shell group 34 per cent. Other participants are the major French group CFP and Partex.

PDO's production capability continues to be expanded. New fields in the south of the country are producing 70,000 barrels per day, or a quarter of current total production.

Qatar

GOVERNMENT DEPARTMENTS AND OFFICIAL AGENCIES

MINISTRY OF ELECTRICITY AND WATER
Address: PO Box 41, Doha
Telephone 23251 Telex 4478

The Ministry of Electricity and Water is responsible for the provision of electricity in Qatar. Through the Electricity Department it controls generating plant and transmission lines and the distribution of electricity to final consumers.

MINISTRY OF FINANCE AND PETROLEUM
Address: PO Box 83, Doha
Telephone 32 15 25 Telex 4324

The Ministry of Finance and Petroleum is concerned with all aspects of the exploitation of oil and gas in Qatar. It supervises the activities of the Qatar General Petroleum Corporation and its subsidiary agencies and companies.

ENTERPRISES (PUBLIC AND PRIVATE SECTOR COMPANIES)

NATIONAL OIL DISTRIBUTION COMPANY (NODCO)
Address: PO Box 10033, Doha
Telephone 77651 Telex 4324
Energy Sector Oil

NODCO is a wholly owned subsidiary of the state oil company QGPC engaged in oil refining and products distribution in Qatar. The company operates a refinery at Umm Said, the annual throughput capacity of which is around 600,000 tonnes.

QATAR GAS COMPANY
Address: Doha
Telephone 22511 Telex 4201
Energy Sector: Gas

Qatar Gas Company was established as a joint venture by the state and Shell International Gas Company, part of the Royal Dutch/Shell group, to exploit the Khuff gas field in the north-western part of the Qatar offshore area. The state has a 70 per cent shareholding.

QATAR GENERAL PETROLEUM CORPORATION (QGPC)
Address: PO Box 3212, Doha
Telephone 26241 Telex 4343
Energy Sectors: Oil, gas

QGPC is a state corporation which acts as holding company for state interests in all aspects of exploration, production, processing and disposal of oil and natural gas. Wholly owned subsidiaries include Qatar Petroleum Producing Authority and

National Oil Distribution Company and the corporation has a majority holding in Qatar Gas Company.

QGPC also has interests in related activities such as the Arab Shipbuilding and Repair Yard Company (ASRY), the joint-venture shipping company AMPTC, the OAPEC owned Arab Petroleum Services Company and Qatar Fertilizer Company. Other important interests are a shareholding in the SUMED operation and participation in Arab Petroleum Investments Corporation (APICORP).

QATAR PETROLEUM PRODUCING AUTHORITY (QPPA)
Address: PO Box 47, Doha
Telephone 27674 Telex 4253
Energy Sectors: Oil, gas

QPPA is an agency of the state oil corporation QGPC responsible for state interests in exploration, production, treatment, transportation and storage of oil or gas in Qatar. Its organisation is divided into two main divisions, dealing with onshore and offshore operations respectively.

Saudi Arabia

GOVERNMENT DEPARTMENTS AND OFFICIAL AGENCIES

MINISTRY OF INDUSTRY AND ELECTRICITY
Address: PO Box 5729, Omar Bin Al Khatab Road, Riyadh
Telephone 477 2722 Telex 201154

The Ministry of Industry and Electricity is concerned with ensuring the adequate supply of electricity for major industrial projects and general public supply purposes. It oversees the generation and distribution activities of the local electricity companies and of the General Organisation for Electricity, which is concerned with the development of regional grids, the establishment of larger scale modern power plant and rural electrification.

MINISTRY OF PETROLEUM AND MINERAL RESOURCES
Address: PO Box 247, Airport Road, Riyadh
Telephone 476 2552 Telex 200997

The Ministry of Petroleum and Mineral Resources is responsible for controlling all activities in the oil and gas industries. Its main agency for control is the state oil company, Petromin, which occupies a predominant position in production, refining and marketing.

ENTERPRISES (PUBLIC AND PRIVATE SECTOR COMPANIES)

ARABIAN AMERICAN OIL COMPANY (ARAMCO)
Address: PO Box 319, Riyadh
Telex 67020
Energy Sectors: Oil, gas

Aramco is one of the world's leading crude oil exploration and production companies, responsible for more than 95 per cent of Saudi Arabian output. The company is owned by four major American international oil companies, Exxon, Mobil, Standard Oil of California and Texaco. Since 1980, however, ownership of Aramco's assets has been virtually fully taken over by the state, with Aramco fulfilling a management role.

Aramco activities are centred on massive oil and gas reserves located in the north-eastern part of the country and offshore in the Persian Gulf. Production has never reached maximum capability, but in 1977 averaged more than nine million barrels per day and at times during 1980 and 1981 exceeded 10 million barrels per day.

A very high proportion of Aramco's oil is exported, but some is used to supply the company's large refinery at Ras Tanura, where capacity is over 20 million tonnes per annum, as well as smaller refineries supplying the local market. Increasing quantities of crude oil are to be used in new refineries planned or under construction, and other projects are aimed at using large quantities of associated gas for power generation, industrial fuel, and petrochemicals manufacture.

GENERAL ORGANISATION FOR ELECTRICITY
Address: Riyadh
Energy Sector: Electricity

The General Organisation for Electricity is a state undertaking charged with the establishment of effective and adequate electricity generating capacity and transmission networks. These are being formed initially on a regional basis, but a national grid is also being planned.

The Organisation has primarily concentrated on financing major new plant in the eastern region now operated by the Consolidated Electricity Company for the Eastern Province area. Similar companies are being created in four other regions, consolidating numerous small existing companies and establishing large scale gas turbine equipment.

PETROLEUM AND MINERAL ORGANISATION (PETROMIN)
Address: PO Box 757, Riyadh
Telephone 478 1661 Telex 201058
Energy Sector: Oil

Petromin is the key state oil corporation, responsible to the Ministry of Petroleum and Mineral Resources, and holding the state's interests in exploration, production, ocean transportation and marketing.

Petromin handles all state sales of crude oil and controls distribution and marketing of oil products in Saudi Arabia. It owns Riyadh Oil Refining Company and has 75 per cent shareholding in Jeddah Oil Refining Company. It is also 50 per cent participant in the major developments with Shell, at Jubail, and Mobil, at Yanbu, which will produce very large quantitites of petroleum and petrochemical products. Petromin expects to build a further major refinery to process heavy crudes at Juaymah.

South Yemen

ENTERPRISES (PUBLIC AND PRIVATE SECTOR COMPANIES

YEMEN NATIONAL OIL COMPANY
Address: PO Box 5050, Maalia, Aden
Telephone 24151 Telex 215
Energy Sector: Oil

The Yemen National Oil Company is a state corporation controlling all aspects of the oil industry in South Yemen. It has assumed full control of the Aden refinery, which was originally built to supply the bunker trade and export markets, but the operations of which are now primarily geared to meeting local demand, although crude oil is processed on behalf of several governments. These include India, Libya and the Soviet Union.

Syria

GOVERNMENT DEPARTMENTS AND OFFICIAL AGENCIES

MINISTRY OF OIL, ELECTRICITY AND MINERAL RESOURCES
Address: Jumhuriah Street, Damascus
Telex 11256

The Ministry of Oil, Electricity and Mineral Resources is responsible for the provision of all forms of energy in Syria. It controls the operations of Syrian Petroleum Corporation, the country's two refining companies, and the oil marketing company as well as overseeing the activities of the Establishment for Electricity.

ENTERPRISES (PUBLIC AND PRIVATE SECTOR COMPANIES)

BANIAS REFINERY COMPANY
Address: PO Box, Banias
Telex 41050
Energy Sector: Oil

Banias Refinery Company is a state-owned corporation, operating the country's principal refinery at Banias on the Mediterranean coast, which is also the terminal for exporting Syrian and Iraqi crude oil. Capacity of the refinery is around six million tonnes per annum. Production is largely sold to SADCOP for the internal market.

ESTABLISHMENT FOR ELECTRICITY
Address: Mutanabi Street, Damascus
Telex 11056
Energy Sector: Electricity

The Establishment for Electricity is a state agency responsible for ensuring an adquate and economic supply of electricity in Syria. It is under the supervision of the Ministry of Oil, Electricity and Mineral Resources.

HOMS REFINERY COMPANY
Address: PO Box 352, Homs
Telex 41004
Energy Sector: Oil

The Homs Refinery Company operates the state-owned refinery at Homs, in north-western Syria. The refinery has a crude oil thoughput capacity of five million tonnes per annum and is geared to meeting local market demand. Output from the refinery is sold to SADCOP for distribution in Syria.

SYRIAN COMPANY FOR OIL TRANSPORT
Address: PO Box 2849, Damascus
Telex 4102
Energy Sector: Oil

The Syrian Company for Oil Transport is a state corporation. It was formed originally to take over the assets formerly owned by the Iraq Petroleum Company.

The company operates a number of important pipelines linking the oil fields to Syrian refineries and to export terminals on the Mediterranean. Large quantities of crude oil emanating from northern Iraq are also transported to the Banias terminal.

SYRIAN PETROLEUM COMPANY
Address: PO Box 2849, Damascus
Energy Sector: Oil

Syrian Petroleum Company is a state-owned corporation charged with exploration for oil, drilling and production. Crude oil is sold either to the refinery companies to meet local demand or to the Syrian Company for Oil Transport for shipment to export terminals.

The company operates five fields, Suwaidiyah, Karatchik, Jubaisseh, Rumeilan and Al-Hayane. Total output from these fields is around 10 million tonnes per annum.

SYRIAN STORAGE AND DISTRIBUTION COMPANY FOR PETROLEUM PRODUCTS (SADCOP)
Address: PO Box 40, Hidjaz Square, Damascus
Telephone 45 13 48 Telex 11260
Energy Sector: Oil

SADCOP is the national distribution and marketing company for petroleum products. Its activities are supervised by the Ministry of Oil, Electricity and Mineral Resources.

SADCOP's supplies of products are obtained from the two state-owned refineries operated by Banias Refinery Company and Homs Refinery Company.

Turkey

GOVERNMENT DEPARTMENTS AND OFFICIAL AGENCIES

MINISTRY OF ENERGY AND NATURAL RESOURCES
Address: Bakanliklar Milli Müdafa Cad, Ankara

The Ministry of Energy and Natural Resources has wide responsibilities for the development and provision of all main forms of energy. It supervises the activities of several key state economic enterprises, including the Turkish Electricity Authority (TEK), the National Coal Company (TKI), The General Directorate for State Hydraulic Works (DSI) and the national oil corporation TPAO.

ENTERPRISES (PUBLIC AND PRIVATE SECTOR COMPANIES)

ANADOLU TASFIYEHANESI AS (ATAS)
Address: PK 37, Mersin
Telephone 11863
Energy Sector: Oil

ATAS is a joint-venture refinery company operating a refinery at Mersin. Participants in the company include Mobil, with a majority shareholding, Shell and BP. Mobil and Shell have long standing interests in exploration and production in Turkey and are the only private companies to have significant local production. The capacity of the Mersin refinery is around 4.5 million tonnes per annum.

TÜRKIYE ELEKTRIK KURUMU (TEK)
Address: Necatibey C11, Ankara
Telex 42245
Energy Sector: Electricity

The Turkish Electricity Authority, TEK, is one of the foremost state economic enterprises, undertaking generation of electricity for the national supply system, operating the main transmission lines and distributing electricity throughout the country. TEK was established in 1970, absorbing numerous local municipal plants, and is responsible to the Ministry of Energy and Natural Resources.

TEK operates a variety of generating plants using coal, lignite and oil as fuels, as well as a substantial amount of hydro-electric capacity. Annual output is over 20 TWh from capacity of 4.4 GW. The Authority has under way the construction of a large scale lignite fired project at Afsin-Elbistan and is considering a 600 MW nuclear reactor. Hydroelectric plant developed by the General Directorate of State Hydraulic Works (DSI) will also provide a further large tranche of capacity, not dependent on imported fuel.

TÜRKIYE KÖMÜR ISLETMELERI KURUMU (TKI)
Address: Bak Resmi D Sayfa 12, Ankara
Telephone 18 72 10
Energy Sector: Coal

TKI, the Turkish National Coal Company, is a state enterprise responsible for all of the country's output of hard coal and 80 per cent of lignite production. Hard coal production includes important quantities of metallurgical coal for the iron and steel industry as well as steam coal for the power stations of TEK.

TKI is set on an expansion path in order to exploit the country's substantial lignite resources. This includes development of deposits in the Afsin-Elbistan Basin to fuel a 1,400 MW power station now under construction.

TÜRKIYE PETROLLERI ANONIM ORTAKLIGI (TPAO)
Address: Müdafa Caddesi 22, Bakanliklar, Ankara
Telephone 17 91 60 Telex 42426
Energy Sector: Oil

TPAO is a state owned corporation involved in all aspects of the oil industry in Turkey. It produces 35-40 per cent of all indigenous crude oil and operates a large proportion of the country's refining capacity. TPAO has interests in numerous other companies in distribution and marketing, pipelines operation, marine transport, petrochemicals and fertiliser manufacture.

Production of crude oil is largely from the oil fields of the south-east, which supply TPAO's refinery at Batman. Quantities of this crude oil are also shipped to other TPAO refineries at Izmit and Izmir. Batman is a relatively small refinery of around one million tonnes per annum capacity. The Izmit and Izmir refineries have a total capacity of around 13 million tonnes per annum and use a high proportion of imported crude oil. TPAO is also building a new refinery in the Ankara area.

NV TURKSE SHELL
Address: Ataturk Bulvari No 127, Bakanliklar, Ankara
Telephone 47 98 00 Telex 42354
Energy Sector: Oil

Turkse Shell is a wholly owned subsidiary of the Royal Dutch/Shell group involved in oil exploration and production. It is the leading producer of crude oil in Turkey, with output of 20-30,000 barrels per day, equivalent to 40-50 per cent of total indigenous output. Production is from fields in the south-eastern part of the country, where exploration is continuing. Turkse Shell is also exploring in the western part of the country.

United Arab Emirates

GOVERNMENT DEPARTMENTS AND OFFICIAL AGENCIES

MINISTRY OF PETROLEUM AND MINERAL RESOURCES
Address: PO Box 59, Abu Dhabi
Telephone 62810 Telex 22544

The Ministry of Petroleum and Mineral Resources is the principal ministry concerned with the exploitation and use of the country's oil and gas resources. Responsibilities include the activities of the national oil company and its subsidiary and associate companies, and control of foreign involvement in exploration and production.

ENTERPRISES (PUBLIC AND PRIVATE SECTOR COMPANIES)

ABU DHABI COMPANY FOR ONSHORE PETROLEUM OPERATIONS (ADCOPO)
Address: PO Box 303, Abu Dhabi
Telex 22284
Energy Sector: Oil

ADCOPO was established in 1978 as the operating company for the principal onshore oil producing fields. The state oil company ADNOC has a 60 per cent shareholding. Other participants include British Petroleum Company, the Royal Dutch/Shell group and the French international oil company CFP. ADCOPO is responsible for producing approximately half of total national oil output.

ABU DHABI GAS LIQUEFACTION COMPANY
Address: PO Box 3500, Abu Dhabi
Telex 22698
Energy Sector: Gas

Abu Dhabi Gas Liquefaction Company is a joint venture between the Abu Dhabi state oil and gas company ADNOC and a consortium of Japanese and European companies set up to manage a gas liquefaction plant on Das Island. Japanese participants are Mitsui and Bridgestone. British Petroleum Company and the French international oil company CFP are the other partners. ADNOC holds a 51 per cent interest.

The Das Island plant was commissioned in 1977 and has a capacity of 550 million cubic feet per day. The plant is designed to extract more than one million tonnes per annum of LPG and pentane and produce some 2.2 million tonnes of LPG.

ABU DHABI MARINE AREAS OPERATING COMPANY (ADMA-OPCO)
Address: PO Box 303, Abu Dhabi
Telephone 82 66 00 Telex 22284
Energy Sector: Oil

ADMA-OPCO is the operating company for the two key offshore oil fields of Umm Shaif and Zakum. The national oil company holds a 60 per cent interest. Other participants are British Petroleum Company, the French international oil company CFP and the Japan Oil Development Company.

ADMA-OPCO's output of crude oil is of the order of 500,000 barrels per day. This represents around 40 per cent of total crude oil production.

ABU DHABI NATIONAL OIL COMPANY (ADNOC)
Address: PO Box 898, Abu Dhabi
Telephone 45600 Telex 22215
Energy Sectors: Oil, gas

ADNOC is one of the key state organisations concerned with the exploitation of national oil and gas resources. It is involved in exploration, production, refining and marketing operations.

ADNOC holds a 60 per cent interest in the two main oil producing concessions in Abu Dhabi which together yield 90 per cent of national output. The company is correspondingly heavily involved in continuing exploration and development programmes. Overseas interests include a 15 per cent holding in the SUMED pipeline and 40 per cent of the Pak-Arab Refinery Company, which operates a refinery at Karachi.

ADNOC has a number of subsidiary companies specialising in certain aspects of the oil industry. These include ADNATCO, the national tanker company, which is building up a fleet of nine tankers, and a distribution/marketing arm to supply petroleum products to the local market. ADNOC operates the country's two oil refineries, at Ruwais and Umm Al Nar, which have a combined throughput capacity of some 6.5 million tonnes per annum.

ABU DHABI OIL COMPANY LTD
Address: 1-19 Akasaka, 8-chome, Minato-ku, Tokyo, Japan
Telephone 405 7111 Telex 22725
Energy Sector: Oil

Abu Dhabi Oil Company was set up in 1968 as a joint venture between Daikyo Oil, Maruzen and Nikko to develop oil production in Abu Dhabi. The company discovered and developed the Mubarras field, located offshore and producing up to 20,000 barrels per day of crude oil.

DUBAI MARINE AREAS LIMITED
Address: PO Box 569, Dubai
Telex 46453
Energy Sector: Oil

Dubai Marine Areas Limited operates the main oil producing fields in Dubai. Operator for the group is Conoco. Other participants are the French oil company CFP and the Spanish state exploration and production company Hispanoil.

Yemen Arab Republic

ENTERPRISES (PUBLIC AND PRIVATE SECTOR COMPANIES)

YEMEN PETROLEUM COMPANY
Address: PO Box 3360, Hodeida
Telephone 2366
Energy Sector: Oil

The Yemen Petroleum Company is a state-owned organisation responsible for all aspects of oil industry operations in Yemen.

Africa

Algeria

GOVERNMENT DEPARTMENTS AND OFFICIAL AGENCIES

MINISTERE DE L'ENERGIE ET DES INDUSTRIES PETROCHIMIQUES
Address: 80 Av Ahmed Ghermoul, Algiers
Telephone 66 33 00

The Ministry of Energy and Petrochemical Industries is the principal government department concerned with energy development and supply. State undertakings responsible to the Ministry include SONATRACH, the state oil and gas enterprise.

SOCIETE NATIONAL DE RECHERCHE ET D'EXPLOITATION MINIERES (SONAREM)
Address: 127 Boulevard Salah, Bouakouir, Algiers
Telephone 61 05 83

SONAREM is a state agency responsible for control of the exploitation of all natural resources other than oil and gas. Projects which it is assisting include investigations of coal, oil shale and uranium deposits, in conjunction with foreign companies.

ENTERPRISES (PUBLIC AND PRIVATE SECTOR COMPANIES)

COMPAGNIE FRANÇAISE DES PETROLES (ALGERIE)
Address: 8 Av Franklin Roosevelt, Algiers
Telephone 59 49 63
Gross Revenue: FF8,031.2 million (1979)
Energy Sector: Oil

CFP (Algérie) is an affiliate of the French based international oil company Compagnie Française des Pétroles, which holds 85 per cent of the shares. The company holds a one-third interest in joint-production arrangements with the state oil and gas organisation SONATRACH at several locations, including Hassi Messaoud. Availability of crude oil to the French group is around 11 million tonnes per annum.

CFP (Algérie) is exploring with SONATRACH in various parts of Algeria. The French group also has exploration agreements via its subsidiary TOTAL Algérie.

SOCIETE NATIONAL POUR LA RECHERCHE, LA PRODUCTION, LE TRANSPORT, LA TRANSFORMATION ET LA COMMERCIALISATION DES HYDROCARBURES (SONATRACH)
Address: 80 Av Ahmed Ghermoul, Algiers
Telephone 65 67 66 Telex 52790
Energy Sectors: Oil, gas

SONATRACH is one of the key state-owned undertakings of Algeria, with responsibility for the exploitation and commercial development of the country's substantial oil and gas resources. It has full control of exploration, transportation and production, although foreign companies participate in varying degrees in crude oil production arrangements and disposal. The state company handles around 80 per cent of crude oil output, although French groups, particularly Elf-ERAP and CFP, enjoy a special relationship and SONATRACH handles only 51 per cent of production from fields in which these companies are involved.

SONATRACH is engaged in exploration with a number of foreign companies, maintaining a majority interest. It operates 3,000 km of trunk pipelines, capable of handling 65 million tonnes per annum of crude oil. The state company also operates the country's three oil refineries, at Algiers, Arzew and Hassi Messaoud, with a total capacity of around six million tonnes per annum.

SONATRACH is an important company in natural gas production and liquefaction. It operates major liquefaction plants at Arzew and Skikda, supplied from Hassi R'Mel. LNG is exported to several European countries. Natural gas is also sold in bulk to SONELGAZ for distribution to consumers within Algeria.

SOCIETE NATIONALE DE L'ELECTRICITE ET DU GAZ (SONELGAZ)
Address: 2 Boulevard Salah, Bouakouir, Algiers
Telephone 61 29 66
Energy Sectors: Electricity, gas

SONELGAZ is a state undertaking with sole responsibility for production, transmission and distribution of electricity. It handles any cross-border transfers of electricity. It is also responsible for the distribution and sale of natural gas to residential, commercial and other smaller scale consumers. Gas is bought in bulk from the state gas producing company SONATRACH at points along the latter's trunk pipelines from Hassi Messaoud to Arzew and Skikda.

Electricity production by SONELGAZ takes place from a number of large and small power stations. The company has thermal power plant at Oran, Skikda, Annaba and other locations, and also has substantial capacity in the form of gas turbines.

Angola

GOVERNMENT DEPARTMENTS AND OFFICIAL AGENCIES

MINISTERIO DA INDUSTRIA E ENERGIA
Address: Luanda
Telex 3373
Energy Sector: Electricity

The Ministry of Industry and Energy is the ministry responsible for the supply of electricity in Angola and the development of hydro-electric resources, the principal form of power for many industries.

MINSTERIO DOS PETROLEOS
Address: Luanda
Telex 3300

The Ministry of Petroleum is concerned with all aspects of oil and gas development in Angola. The Ministry is responsible for the activities of the state oil company SONANGOL, which is involved in exploration, production, refining and distribution.

ENTERPRISES (PUBLIC AND PRIVATE SECTOR COMPANIES)

CABINDA GULF OIL COMPANY
Address: CP 2950, Luanda
Telephone 30040 Telex 3167
Energy Sector: Oil

Cabinda Gulf Oil Company is an affiliate of the international major oil company Gulf Oil, operating in the Cabinda enclave of Angola. The national oil company SONANGOL now holds a majority 51 per cent share of the company.

Cabinda Gulf is one of the leading oil producing companies of Africa, having struck large reserves of crude oil on the African coast. Output is of the order of 100,000 barrels per day. The field, which has been in production since the mid 1960s is estimated to have reserves of around 300 million tonnes.

COMPANHIA DE PETROLEOS DE ANGOLA (PETRANGOL)
Address: Av Restauradores de Angola, CP 1320 Luanda
Telephone 35802 Telex 3246
Energy Sector: Oil

PETRANGOL is engaged in exploration, production, refining and distribution of oil products in Angola. A majority of the shares (some 74 per cent) are held by the Belgian international oil company Petrofina. Other shares are held by the state oil company SONANGOL.

PETRANGOL has production in the Kwanza and Congo Basins of around 750,000 tonnes per annum. Crude oil is either exported from an ocean terminal or supplied by pipeline to the company's refinery at Luanda. The Luanda refinery has a throughput capacity of around 1.5 million tonnes per annum and is the only one in Angola.

SOCIEDADE NACIONAL DE COMBUSTIVEIS DE ANGOLA
(SONANGOL)
Address: Rua 5 de Outubro, Lobito
Telex 3260
Energy Sectors: Oil, gas

SONANGOL was established in 1976 as a state company to control the exploitation of oil and natural gas in Angola, and to undertake all production and distribution of fuels. It supervises all activities in exploration, development, oil refining and distribution.

SONANGOL also holds the state's interests in existing activities. These include the 51 per cent holding in the country's principal oil producing field operated by Cabinda Gulf Oil Company.

Benin

ENTERPRISES (PUBLIC AND PRIVATE SECTOR COMPANIES)

SOCIETE BENINOISE D'ELECTRICITE ET D'EAU (SBEE)
Address: BP 123, Cotonou
Energy Sector: Electricity

Société Beninoise d'Electricité et d'Eau is a publicly owned undertaking responsible for the production and distribution of electricity throughout Benin. The company's principal source of supply is hydro-electric power from the Akosombo Dam on the Volta River. Benin is entitled to a share of 25 MW from this scheme.

Cameroon

GOVERNMENT DEPARTMENTS AND OFFICIAL AGENCIES

MINISTERE DES MINES ET DE L'ENERGIE
Address: Yaoundé
Telephone 22 48 04 Telex 8504

The Ministry of Mines and Energy is responsible for activities of all energy sectors. These responsibilities include the control of exploration activity, crude oil refining, the distribution of oil products and the provision of electricity through the public supply grid.

ENTERPRISES (PUBLIC AND PRIVATE SECTOR COMPANIES)

SOCIETE CAMEROUNAISE DES DEPOTS PETROLIERS
Address: Immeuble Gulf, Douala
Energy Sector: Oil

The Société Camerounaise des Dépôts Petroliers is a company specially established by the State, in conjunction with oil marketing companies operating in Cameroon, to operate oil supply installations and maintain stocks of petroleum products. This responsibility includes the maintenance of strategic as well as operational stocks.

State agencies hold a majority of the shares in the company. Other participants are the local affiliates of Shell, Mobil, CFP, BP, Texaco and AGIP.

SOCIETE ELF DE RECHERCHES ET D'EXPLOITATION DES PETROLES DU CAMEROUN
(ELF SEREPCA)
Address: 83 Boulevard de la Liberté, BP 2214, Douala
Telephone 42 13 66 Telex 5299
Energy Sector: Oil

ELF SEREPCA is the principal group undertaking oil exploration in Cameroon. The company is partnered by affiliates of the Royal Dutch/Shell group in several concession areas.

The company is 62 per cent owned by the French state oil company Entreprise de Recherches et d'Activités Petrolières (ERAP). The Cameroon government has a 26 per cent interest. The balance of the shares is held by other French exploration investment groups.

SOCIETE NATIONALE DE RAFFINAGE
Address: Cap Limboh, BP 365, Victoria
Telephone 33 22 38/39 Telex 5561
Energy Sector: Oil

Société Nationale de Raffinage (SONARA) owns and operates a modern refinery at Victoria on the coast west of Douala. The state has a 66 per cent shareholding in the company. The balance is held by four international oil companies which are engaged in product marketing in Cameroon, the French companies Total and Elf Aquitaine, Mobil Corporation and the Royal Dutch/Shell group.

The refinery has a capacity of two million tonnes per annum and is designed to process local crude oil and imported grades. The refinery is expected to meet the demand of the local market with some export of fuel oil and LPG.

SOCIETE NATIONALE D'ELECTRICITE DU CAMEROUN (SONEL)
Address: BP 4077, Douala
Telephone 42 54 44 Telex 5551
Gross Revenue: CFA F16,700 million (1979)
Energy Sector: Electricity

SONEL is a state undertaking responsible for the production and distribution of electricity for public supply in Cameroon. The state holds a majority of the shares. A small holding is in the hands of the French industrial company Pechiney, which has industrial energy consuming operations in Cameroon.

SONEL has numerous local power stations, but its availability of power is increasingly coming from a small number of key hydro-electric schemes. These include plants in operation or under construction at Song Loulou and Lagdo. The Song Loulou project will eventually have a capacity of 290 MW, with the object of supplying power to aluminium smelters, the oil refinery and the grid in the central region.

Central African Republic

ENTERPRISES (PUBLIC AND PRIVATE SECTOR COMPANIES)

ENERGIE CENTRAFRICAINE (ENERCA)
Address: BP 880, Bangui
Telephone 61 20 22 Telex 5241
Energy Sector: Electricity

ENERCA is the national electricity supply authority of the Central African Republic, responsible for the production and distribution of electricity. Around 90 per cent of ENERCA's power is derived from the hydro-electric station at Boali. Annual output is of the order of 50 GWh.

Congo

ENTERPRISES (PUBLIC AND PRIVATE SECTOR COMPANIES)

SOCIETE NATIONALE D'ENERGIE
Address: BP 95, Brazzaville
Telephone 81 38 58 Telex 5261
Energy Sector: Electricity

Société Nationale d'Energie is a state undertaking responsible for the production and distribution of electricity in the Congo. Most of the company's supply of electricity is generated by hydro-electric plant at three main locations. Basis of the public supply system has been the dam at Djoué.

A second dam, at Moukoukoulou, was completed in 1979, providing power for an industrial complex. A third hydro-electric dam is at Loubomo, on the Bouenza River, 200 km west of Brazzaville.

SOCIETE NATIONALE DE RECHERCHE ET D'EXPLOITATION PETROLIERES (HYDRO-CONGO)
Address: BP 2008, Brazzaville
Telephone 81 34 59 Telex 5215
Energy Sector: Oil

Hydro-Congo is a state undertaking with responsibility for exploration and production activity in the Congo. Hydro-Congo carries out exploration in conjunction with several foreign oil companies.

The state company operates an oil refinery at Pointe Noire. The refinery has a capacity of one million tonnes per annum and meets the requirements of the inland market. Hydro-Congo holds a monopoly right to the distribution of petroleum products.

Egypt

GOVERNMENT DEPARTMENTS AND OFFICIAL AGENCIES

MINISTRY OF ELECTRICITY AND POWER
Address: Nasr City, Abasseiah, Cairo
Telephone 82 95 65 Telex 92097

The Ministry of Electricity and Power deals with all aspects of electricity development and supply. These include the operations of the Egyptian Electricity Authority, the work of the Qattara Depression Authority in establishing the viability of the scheme to utilise the difference in height between the Depression and the Mediterranean, and the programme of extending rural electrification.

MINISTRY OF PETROLEUM
Address: 5th Floor, Ministry of Industry Building, 2 Latin America Street, Garden City, Cairo
Telephone 29929 Telex 93112

The Ministry of Petroleum is responsible for all activities of the oil and gas sectors. These include control of Egyptian General Petroleum Company and its operating subsidiaries and supervision of the involvement of foreign companies active in Egypt.

ENTERPRISES (PUBLIC AND PRIVATE SECTOR COMPANIES)

ALEXANDRIA PETROLEUM COMPANY
Address: 7 Lazoughli Street, Cairo
Telephone 29860
Energy Sector: Oil

Alexandria Petroleum Company is a wholly owned subsidiary of the state oil company Egyptian General Petroleum Company. It operates an oil refinery at Alexandria with a crude oil throughput capacity of three million tonnes per annum.

ARAB PETROLEUM PIPELINES COMPANY
Address: 9 Amin Yehia Street, Zizinia, PO Box 2056, Alexandria
Telephone 64138 Telex 54295
Energy Sector: Oil

Arab Petroleum Pipelines Company owns and operates the Suez-Mediterranean crude oil pipeline (SUMED). The twin pipelines have a capacity to transport 80 million tonnes per annum between Ain Soukhna, in the Gulf of Suez, and Sidi Krir, west of Alexandria.

The company is owned by the Egyptian state, through EGPC, with a 50 per cent interest and four key Persian Gulf producer countries—Saudi Arabia, Kuwait and Abu Dhabi, each with 15 per cent, and Qatar.

EGYPTIAN ELECTRICITY AUTHORITY
Address: Nasr City, Abasseiah, Cairo
Telephone 82 70 71
Total Sales: E£91.7 million (1977)
Energy Sector: Electricity

The Electricity Authority is a state undertaking engaged in the generation, transmission and bulk supply of electricity. It operates the Unified Power System, which services the main consuming area in the northern part of the country, based on thermal generating stations in that part of the country and the mass transfer of electricity from the Aswan Dam and High Dam on the Nile.

The Authority aims to maximise its availability of hydro-electric power and is working on a project to exploit the Qatarra Depression, which is below the level of the Mediterranean. This could provide a capacity of 8,000 MW in principle. It also has a project for a 600 MW nuclear power station at Sidi Krir, on the coast west of Alexandria.

EGYPTIAN GENERAL PETROLEUM CORPORATION
Address: Osman Abdel Hafiz Street, Nasr City
Telephone 60 32 99 Telex 92049
Energy Sectors: Oil, gas

The Egyptian General Petroleum Corporation (EGPC) is the state oil company of Egypt, extensively involved in all aspects of the oil and gas industry. It has sole rights to handle the importation or exportation of oil and holds a 50 per cent interest in the Suez-Mediterranean crude oil pipeline.

EGPC's subsidiary General Petroleum Company undertakes exploration and production. It holds a 50 per cent interest in Gulf of Suez Petroleum Company and Western Desert Petroleum Company, in which its partners are Standard Oil Company of Indiana and Phillips Petroleum Company respectively.

EGPC subsidiaries operate the country's oil refineries and petroleum products are distributed through other subsidiaries Petroleum Co-operative Company and Misr Petroleum Company. Another subsidiary, Petrogas, has been established to distribute natural gas from the Abu Qir gas field.

EGYPTIAN PETROLEUM DEVELOPMENT COMPANY LIMITED
Address: 9-14 Toranomon 2-chome, Minato-ku, Tokyo, Japan
Telephone 502 0481 Telex 26623
Energy Sector: Oil

Egyptian Petroleum Development Company is an overseas exploration company set up by Japanese groups to search for oil in Egypt. Participants include Teikoku Oil Company and Dai-Ichi Oil Development Company.

The company discovered the West Bakr oil field onshore in the Gulf of Suez area. Production from this field commenced in 1980.

EL NASR PETROLEUM COMPANY
Address: 6 Orabi Street, Cairo
Telephone 74 54 00
Energy Sector: Oil

El Nasr Petroleum Company is a state-owned company operating oil refineries at Suez and Alexandria. It is a wholly owned subsidiary of the national oil company EGPC. The company's total refining capacity is around 5.5 million tonnes per annum.

GULF OF SUEZ PETROLEUM COMPANY
Address: 1097 Cornishe El-Nil Street,
PO Box 2400, Garden City, Cairo
Telephone 31883
Energy Sector: Oil

Gulf of Suez Petroleum Company (GUPCO) is a partnership of the state oil company EGPC and the international oil company Standard Oil of Indiana. Each group has a 50 per cent shareholding.

GUPCO has fields in production in the Gulf of Suez and is actively exploring and proving additional reserves. Production in 1980 was at 144,000 barrels per day, equivalent to more than seven million tonnes through the year.

MISR PETROLEUM COMPANY
Address: 6 Orabi Street, Cairo
Telephone 74 52 81
Energy Sector: Oil

Misr Petroleum Company is a wholly owned subsidiary of the state oil company EGPC. Misr Petroleum distributes petroleum products made available from associate companies operating oil refineries at several locations.

PETROLEUM CO-OPERATIVE COMPANY
Address: 94 El Kasr El Eini Street, Cairo
Telephone 31800
Energy Sector: Oil

The Petroleum Co-operative Company is one of the state-owned organisations distributing petroleum products to the Egyptian market. The company is a wholly owned subsidiary of the state oil company EGPC.

SUEZ CANAL AUTHORITY
Address: Garden City, Cairo
Telephone 20748/49
Energy Sector: Oil

The Suez Canal Authority is a state controlled organisation responsible for operating and developing the Suez Canal. The Canal is a route for substantial quantities of oil moved in both northbound and southbound directions. Northbound oil traffic in 1980 totalled 28 million tonnes, while a further 14 million tonnes transited the Canal in the opposite direction.

SUEZ OIL PROCESSING COMPANY
Address: 15 Nabil El Wakkad Street, Dokki
Telephone 80 21 66
Energy Sector: Oil

Suez Oil Processing Company is a subsidiary of the state oil company EGPC, engaged in refining crude oil at Mostord, Suez and Tanta. The company's principal refinery is at Mostord, which has a throughput capacity of around four million tonnes per annum. Other refineries at Suez and Tanta are each capable of processing only one million tonnes per annum.

WESTERN DESERT PETROLEUM COMPANY
Address: Borg El Thagr Building, PO Box 412,
Alexandria
Telephone 28718
Energy Sector: Oil

Western Desert Petroleum Company (WEPCO) has been producing crude oil from fields near the Libyan border since 1966. The state oil company EGPC holds a controlling interest, and is partnered by Phillips Petroleum Company and the Spanish State oil exploration company Hispanoil.

WEPCO operates the Alamein and Yidma fields, in which Phillips has a 50 per cent interest, and the Umbarka field in the north-west of the country, in which Phillips and Hispanoil are both involved. Output is transported to the terminal at El Hamra for shipment to EGPC's refineries.

Gabon

GOVERNMENT DEPARTMENTS AND OFFICIAL AGENCIES

MINISTERE DE L'ENERGIE ET DES RESSOURCES HYDRAULIQUES
Address: BP 576, Libreville
Telephone 76 16 06

The Ministry of Energy and Hydraulic Resources has principal responsibility for the supply of electric power to the economy and for development of water resources to provide electricity. It supervises the activities of the national electricity supply company Société d'Energie et d'Eau du Gabon.

MINISTERE DES MINES ET DU PETROLE
Address: BP 2110, Libreville
Telephone 72 21 72

The Ministry of Mines and Petroleum is responsible for exploration and development of crude oil or natural gas and for the supply of finished products to the Gabon market. It is involved in the uranium mining and processing activities of COMUF, oil and gas exploration by Elf-Gabon and other companies, and oil refining by Société Gabonaise de Raffinage.

ENTERPRISES (PUBLIC AND PRIVATE SECTOR COMPANIES)

COMPAGNIE DES MINES D'URANIUM DE FRANCEVILLE (COMUF)
Address: BP 260, Libreville
Telephone 72 43 10 Telex 5281
Energy Sector: Uranium

COMUF is engaged in production of uranium from mines in the Franceville area of Gabon. Participants in the company include the state and several French groups concerned with the development and utilisation of uranium. Apart from the Gabon state holding of 25 per cent, the largest shares are held by Compagnie de Mokta (32 per cent), COGEMA (19 per cent) and Minatome (13 per cent).

COMUF's reserves of uranium in the Franceville area are estimated at nearly 30,000 tonnes. Output is currently sold to French agencies.

ELF-GABON
Address: BP 525, Port-Gentil
Telephone 75 23 65 Telex 8210
Energy Sector: Oil

Elf-Gabon is an associate company of the French state oil and gas group Société National Elf Aquitaine (SNEA), which has a 49 per cent shareholding. Gabon state agencies hold a 25 per cent interest.

Elf-Gabon has oil fields in production, with output around nine million tonnes per annum, and is actively exploring. The company has interests in energy intensive operations, through its 32 per cent holding in Gaboren, which produces ammonia, and 26 per cent holding in Sogacel, which produces cellulose and paper. Elf-Gabon is one of the largest shareholders in the industry oil refinery at Port-Gentil. Elf-Gabon also has an interest in the national electricity supply company Société d'Energie et d'Eau du Gabon.

SOCIETE D'ENERGIE ET D'EAU DU GABON (SEEG)
Address: BP 2187, Libreville
Telephone 72 19 11 Telex 5222
Energy Sector: Electricity

SEEG is responsible for the production, transmission and distribution of electricity in Gabon. The state holds a majority 64 per cent interest, but several other major industrial companies also participate, including Elf-Gabon, Société Gabonaise de Raffinage and COMUF.

SEEG has three principal central power generating plants. At Kinguele, near Libreville, it operates a 120 MW hydro-electric station. At Port-Gentil and Libreville there are two thermal power stations. SEEG has under consideration a second major hydro-electric scheme, at Grand Poubara.

SOCIETE GABONAISE DE RAFFINAGE (SOGARA)
Address: BP 564, Port-Gentil
Telephone 75 26 21 Telex 8217
Energy Sector: Oil

SOGARA operates an oil refinery at Port-Gentil supplying the country's basic requirements for finished petroleum products. The refinery has a crude oil processing capacity of one million tonnes per annum.

The SOGARA refinery supplies products to the marketing companies, which hold shares in the refinery company. The state has a 25 per cent shareholding and other participants are local affiliates of Elf, CFP, Mobil, Shell, BP, Petrofina, Texaco and AGIP.

Ghana

GOVERNMENT DEPARTMENTS AND OFFICIAL AGENCIES

GHANA SUPPLY COMMISSION
Address: PO Box 1735, Accra
Telephone 28181

The Ghana Supply Commission is a state agency responsible for all of the government's procurement arrangements. These include the acquisition of crude oil and the supply of petroleum products to the national market.

The Commission purchases crude oil for processing at the state-owned refinery at Tema. Refined products are sold to marketing companies or ex-refinery for export.

ENTERPRISES (PUBLIC AND PRIVATE SECTOR COMPANIES)

ELECTRICITY CORPORATION OF GHANA
Address: PO Box 2394, Accra
Telephone 64907
Energy Sector: Electricity

The Electricity Corporation is a state undertaking responsible for the transmission, distribution and retail sale of electricity in Ghana. The Corporation's network has been extended to include around one-quarter of the total population.

The Corporation's supply of electricity is provided by the Volta River Authority, which operates the Akosombo Dam and is building a second hydro-electric plant on the Volta River at Kpong.

GHANA OIL COMPANY
Address: Boundary Road/Adjabeng Road, PO Box 3183, Accra
Telephone 28822
Energy Sector: Oil

Ghana Oil Company came into being in 1976 on completion of arrangements between the government of Ghana and the Italian state-owned company AGIP, whereby the latter's interests in refining and marketing were transferred to the state. The former AGIP Ghana formed the basis for the development of a major state presence in distribution and marketing and Ghana Oil Company is now the largest oil marketing company.

GHANAIAN ITALIAN PETROLEUM COMPANY LIMITED (GHAIP)
Address: PO Box 599, Industrial Area, Tema
Telephone 2881
Energy Sector: Oil

GHAIP is the state-owned company operating the country's only oil refinery, at Tema, near Accra. The company was acquired completely by the state as part of a general arrangement with the Italian State oil company AGIP.

The GHAIP refinery has a throughput capacity of around 1.3 million tonnes per annum and meets most of the needs of the domestic market. Petroleum products are delivered to local marketing companies for distribution.

VOLTA RIVER AUTHORITY
Address: PO Box M77, Accra
Telephone 64941
Energy Sector: Electricity

The Volta River Authority was set up by the Government of Ghana to develop, manage and operate hydro-electric plant on the Volta River. The Authority manages the Akosombo Dam, which has a capacity over 900 MW and is building another 150 MW plant at Kpong. Output is around 4.6 TWh per annum.

Output from the Volta River Authority supplies virtually all of Ghana's electricity, with a major part transferred to Volta Aluminium Company's smelting plant. Power is also exported to Togo and Benin under long-term arrangements.

Ivory Coast

GOVERNMENT DEPARTMENTS AND OFFICIAL AGENCIES

MINISTERE DES MINES
Address: BP V50, Abidjan
Telephone 32 50 03 Telex 2262

The Ministry of Mines is responsible for the activities of extractive industries in Ivory Coast. These interests include notably the licencing of oil and gas exploration and control of the development of known deposits. It supervises the state oil agency PETROCI.

ENTERPRISES (PUBLIC AND PRIVATE SECTOR COMPANIES)

SOCIETE ENERGIE ELECTRIQUE DE LA COTE D'IVOIRE (EECI)
Address: BP 1345, Abidjan
Telephone 32 02 33 Telex 3738
Energy Sector: Electricity

EECI is the national electricity supply company for Ivory Coast. The state finance company SONAFI holds a majority 76 per cent share in the company, but there is also participation by private shareholders and by the French state electricity company Electricité de France.

EECI's supply of electricity is derived from hydro-electric and thermal power stations. Principal hydro-electric plant is at Kossou, on the Bandama

River, where there is capacity of 175 MW. There are other plants on the Bandama and Sassandra rivers. Thermal generating capacity has also been built in the Abidjan area, with a capacity of 200 MW.

SOCIETE IVOIRIENNE DE RAFFINAGE
Address: BP 1269, Abidjan
Telephone 36 91 99 Telex 746
Energy Sector: Oil

Société Ivoirienne de Raffinage owns and operates the country's only oil refinery at Abidjan. Capacity of the refinery is two million tonnes per annum. The state oil company PETROCI has a 42.5 per cent interest. Other shares are held by affiliates of foreign oil companies which are involved in distribution and marketing in Ivory Coast. These include BP, CFP, Exxon, SNEA, Mobil, Shell and Texaco.

The Upper Volta government holds a small interest as the refinery is an important source of petroleum products for that country. Some products are also supplied to the Mali market. Crude oil is imported from a number of African and Middle East sources, as well as from Venezuela, including heavier crude to provide feedstock for the bitumen refinery of Société Multinationale de Bitumes.

SOCIETE NATIONALE D'OPERATIONS PETROLIERES DE LA COTE D'IVOIRE
Address: BP V194, Abidjan
Telephone 32 40 58 Telex 2135
Energy Sectors: Oil, gas

PETROCI was established as the state oil company to participate in all aspects of oil and gas exploration and development and in the transportation and supply of crude oil and finished products. The company is largest shareholder in the country's oil refinery at Abidjan.

PETROCI holds a 10 per cent interest in each of the groups developing oil resources. This share will increase as output rises. In the Bélier field it is partnered by Shell and Esso, in the Espoir field its partners are Phillips, Sedco and AGIP. The Espoir field also has significant quantities of associated gas.

Kenya

GOVERNMENT DEPARTMENTS AND OFFICIAL AGENCIES

MINISTRY OF ENERGY
Address: Box 30582, Electricity House, Harambee Avenue, Nairobi
Telephone 27553

The Ministry of Energy is responsible for overseeing developments in all sectors of energy supply, and for the formulation of future energy policy. Apart from control of the activities of Kenya Power Company and East African Power and Lighting Company the Ministry is involved in assessing the potential for exploitation of geothermal energy and new hydro-electric schemes.

ENTERPRISES (PUBLIC AND PRIVATE SECTOR COMPANIES)

EAST AFRICAN OIL REFINERIES LIMITED
Address: Box 90401, Changamwe, Mombasa
Telephone 43 35 11
Energy Sector: Oil

East African Oil Refineries (EAOR) owns and operates the country's only oil refinery, at Mombasa. Participants in the refinery company include those involved in distribution and marketing of finished products or associate companies.

The Mombasa refinery has a crude oil throughput capacity of four million tonnes per annum. The refinery has been operating at well below capacity necessitating some importation of lighter products. There is a project under consideration to build a cracking plant at the refinery, so as to match output more closely with market requirements.

EAST AFRICAN POWER AND LIGHTING COMPANY LIMITED
Address: Box 30177, Electricity House, Harambee Avenue, Nairobi
Telephone 21276
Energy Sector: Electricity

East African Power and Lighting Company is the sole distributor of electricity for public consumption in Kenya. Power is supplied to the company by Kenya Power Company from its hydro-electric stations on the Tana River and on import from the Owen Falls plant in Uganda.

ESSO STANDARD KENYA LIMITED
Address: Esso House, Box 30200, Mama Ngina Street, Nairobi
Telephone 33 13 11
Energy Sector: Oil

Esso Standard Kenya is a subsidiary of Exxon Corporation involved in distribution and marketing of oil products in Kenya. Products are obtained from the Mombasa refinery of East African Oil Refineries Ltd.

KENYA POWER COMPANY LIMITED
Address: POB 7936, Electricity House,
Harambee Avenue, Nairobi
Telephone 21251
Energy Sector: Electricity

Kenya Power Company is responsible for the production of electricity for the public supply system. Electricity is transmitted in bulk to the East African Power and Lighting Company for sale to industrial and other consumers. It also handles the importation of power from the Owen Falls plant in Uganda.

The company uses hydro-electric capacity for most of its production. Around 85 per cent is produced at the two main plants on the Tana River.

KENYA SHELL LIMITED
Address: Shell & BP House, Harambee Avenue,
Box 43561, Nairobi
Telephone 29222
Energy Sector: Oil

Kenya Shell is a marketing company jointly owned by the Royal Dutch/Shell group and British Petroleum. It distributes products throughout Kenya and is supplied mainly from the Mombasa refinery of East African Oil Refineries Ltd, in which BP and Shell are both participants.

Liberia

GOVERNMENT DEPARTMENTS AND OFFICIAL AGENCIES

MINISTRY OF LANDS, MINES AND ENERGY
Address: Capital Hill, PO Box 9024, Monrovia
Telephone 22 24 78

The Ministry of Lands, Mines and Energy is concerned with questions of oil supply, exploration for oil, gas and minerals and the electricity supply industry, as well as general matters of energy policy.

ENTERPRISES (PUBLIC AND PRIVATE SECTOR COMPANIES)

LIBERIA ELECTRICITY CORPORATION
Address: PO Box 165, Maxwell Building,
Ashmun Street, Monrovia
Telephone 22 18 90 Telex 4288
Energy Sector: Electricity

The Liberia Electricity Corporation is the principal organisation responsible for public electricity supply. Its role and the extent of its unified grid is being continually expanded, and rural electrification programmes are under way.

The position of the Corporation has been built on the development of the Mount Coffee hydro-electric plant, which provides three-quarters of all electricity in the public supply system. A second hydro-electric scheme is under consideration as part of a joint-venture with Ivory Coast at Nyaabé on the Cavally River.

LIBERIA PETROLEUM REFINING
Address: PO Box 90, Bushrod Island, Monrovia
Telephone 22 27 30 Telex 4254
Energy Sector: Oil

Liberia Petroleum Refining operates an oil refinery near Monrovia. The refinery, which is the only one in Liberia has a crude oil processing capacity of around 750,000 tonnes per annum and meets most of the country's product requirements.

Libya

ENTERPRISES (PUBLIC AND PRIVATE SECTOR COMPANIES)

LIBYAN NATIONAL OIL CORPORATION
Address: Benghazi
Energy Sectors: Oil, gas

The Libyan National Oil Corporation (LNOC) was established as the vehicle for state control of the country's oil and gas resources. It has a majority shareholding in all oil and gas production, in the transportation and liquefaction of natural gas and owns the country's two oil refineries.

LNOC has entitlement to a large proportion of production from most fields, although only 51 per cent under some arrangements with Occidental Petroleum Corporation. LNOC sells some of its entitlement to overseas buyers, but also resells substantial volumes to its partners for disposal.

In 1979 LNOC moved to acquire 51 per cent of the interests of the Exxon group in natural gas liquefaction facilities at Marsa El Brega. Since that time Exxon has withdrawn from all of its oil and gas operations in Libya, so that LNOC now is responsible for LNG operations and sales. LNG is currently being sold to Spain and Italy.

OASIS OIL COMPANY OF LIBYA
Address: PO Box 395, Tripoli
Telephone 31116 Telex 20158
Energy Sector: Oil

Oasis Oil Company is one of the key crude oil producing groups in Libya, with output in 1980 at a level of some 660,000 barrels per day. The Libyan National Oil Company has a majority shareholding. The balance is held by three United States based oil companies, Marathon Oil Company, Conoco Incorporated and Amerada Hess Corporation.

OCCIDENTAL OF LIBYA INCORPORATED
Address: Occidental International Oil Inc, 16 Palace Street, London SW1, England
Telephone 01-828 5600 Telex 918818
Energy Sector: Oil

Occidental of Libya is a wholly owned subsidiary of the United States based international oil company Occidental Petroleum Corporation. The company has extensive oil production interests in Libya in association with the National Oil Corporation.

Occidental retains a 49 per cent interest in its oldest concession areas, but under more recent production sharing arrangements this proportion is only 19 per cent, and in prospective areas now under evaluation this level will be only 10-15 per cent. However, Occidental also buys back substantial volumes of oil from the National Oil Company. Total offtake by the company in 1980 was approximately 200,000 barrels per day.

Malagasy Republic

ENTERPRISES (PUBLIC AND PRIVATE SECTOR COMPANIES)

JIRO SY RANO MALAGASY (JIRAMA)
Address: 149 Rue Rainandriamampandry, Antananarivo
Telephone 200-31 Telex 22235-200
Energy Sector: Electricity

JIRAMA is a publicly owned company responsible for the supply of electricity and water throughout Madagascar. JIRAMA operates some 65 MW of capacity, including hydro-electric and thermal plant.

SOLITANA MALAGASY (SOLIMA)
Address: 2 Avenue Grandidier, BP 140, Antananarivo
Telex 22222
Energy Sector: Oil

SOLIMA is a state corporation with the monopoly of crude oil importation, oil refining and disposal of products on the home and export markets. The company operates a refinery at Tamatave.

Malawi

GOVERNMENT DEPARTMENTS AND OFFICIAL AGENCIES

DEPARTMENT OF ENERGY AND CONTINGENCY PLANNING
Address: Gemini House, Box 30301, Capital City, Lilongwe 3
Telephone 73 13 77

The Department of Energy and Contingency Planning is concerned with the main aspects of energy policy and the general activities of the Electricity Supply Commission. The Department is directly responsible to the Office of the President.

ENTERPRISES (PUBLIC AND PRIVATE SECTOR COMPANIES)

ELECTRICITY SUPPLY COMMISSION OF MALAWI
Address: Escom House, PO Box 2047, Blantyre 3
Telephone 63 34 55 Telex 4312
Energy Sector: Electricity

The Electricity Supply Commission is responsible for the production of electricity for the public supply system and the development of an interconnected grid throughout the country. The Commission uses a combination of thermal and hydro-electric capacity totalling some 80 MW. Total sales in 1979 were 356 GWh.

The Commission's production of electricity is concentrated in the central part of the country, where there are several hydro-electric plants on the Shire River. Further hydro-electric projects are planned for this area.

Mali

ENTERPRISES (PUBLIC AND PRIVATE SECTOR COMPANIES)

SOCIETE ENERGIE DU MALI
Address: BP69 Av Lyautey, Bamako
Telephone 22 30 20 Telex 587
Energy Sector: Electricity

Energie du Mali is a publicly controlled company responsible for the production and distribution of electricity in Mali. The state holds 55 per cent of the shares. The French state electricity undertaking Electricité de France has a small interest.

The company's main power station is at Bamako. A 45 MW hydro-electric plant is being built at Selingué and another is being built at Manantali, jointly with Senegal.

Mauritania

ENTERPRISES (PUBLIC AND PRIVATE SECTOR COMPANIES)

SOCIETE NATIONALE D'EAU ET D'ELECTRICITE (SONELEC)
Address: BP 355, Nouakchott
Telephone 52308 Telex 587
Energy Sector: Electricity

SONELEC is a publicly owned undertaking responsible for the public electricity supply system in Mauritania. Technical and operating advice is provided by the French state electricity authority Electricité de France.

Morocco

GOVERNMENT DEPARTMENTS AND OFFICIAL AGENCIES

MINISTERE DE L'ENERGIE
Address: Quartier des Ministères, Rabat-Chellah
Telephone 65951 Telex 31910

The Ministry of Energy has two principal departments concerned with the development of energy, the Bureau de Recherches et de Participations Minières (BRPM) and the recently created Office National de Recherche et d'Exploitation Petrolière (ONAREP).

BRPM has been responsible for activities concerning all minerals except phosphate. ONAREP consolidates the state's interests in exploration and development of hydrocarbons, including oil shale.

ENTERPRISES (PUBLIC AND PRIVATE SECTOR COMPANIES)

OFFICE NATIONAL DE L'ELECTRICITE
Address: BP 498, 65 Rue Aspirante Lafuente, Casablanca
Telephone 22 41 65 Telex 22780
Energy Sector: Electricity

The Office National de l'Electricité (ONEL) is a state agency responsible for production and distribution of electricity through the public supply system. ONEL's production in 1980 totalled 4,760 GWh.

ONEL's generating capacity consists of thermal power stations based on coal and fuel oil and hydro-electric plant. Hydro-electric stations supplied more than 30 per cent of electricity supplies during 1980.

SOCIETE ANONYME MAROCAINE ITALIENNE DE L'INDUSTRIE DU RAFFINAGE (SAMIR)
Address: Route Côtière de Casablanca, BP 89, Mohammedia
Telephone 25 01 Telex 21882
Energy Sector: Oil

SAMIR is the operating company for the oil refinery at Mohammedia. A controlling interest is held by the state. The SAMIR refinery is the principal refinery in Morocco, with a crude oil processing capacity of 2.5 million tonnes per annum.

SOCIETE NATIONALE DES PRODUITS PETROLIERS
Address: 42 Av de l'Armée Royale, Casablanca
Energy Sector: Oil

The Société Nationale des Produits Petroliers is a state-owned company involved in the distribution and marketing of petroleum products.

Mozambique

GOVERNMENT DEPARTMENTS AND OFFICIAL AGENCIES

NATIONAL DIRECTORATE OF ENERGY
Address: Av Agostinho Neto 70, CP 2447, Maputo
Telephone 74 20 11 Telex 6-407

The National Directorate of Energy forms part of the Ministry of Industry and Energy and is responsible for activities of the oil industry and electricity supply organisations. These include the national oil company PETROMOC, the electricity distribution company Electricidade de Moçambique and the state's interests in Hidroelectrica de Cabora Bassa.

NATIONAL DIRECTORATE OF MINES
Address: Av Filipe Magaia 528-2º, Maputo
Telephone 22447 Telex 6-413

The National Directorate of Mines is part of the Ministry of Industry and Energy. It has responsibility for coal and other mining activities in Mozambique. Currently it is co-ordinating the development of new coal mines in the south-west of the country.

ENTERPRISES (PUBLIC AND PRIVATE SECTOR COMPANIES)

ELECTRICIDADE DE MOÇAMBIQUE
Address: Av Agostinho Neto No 70, PO Box 2447, Maputo
Telephone 74 20 11 Telex 6-407
Energy Sector: Electricity

Electricidade de Moçambique is a state controlled undertaking responsible for the public supply of electricity. The company operates its own generating capacity, but also relies to a large extent on production from the Cabora Bassa dam, which is operated by Hidroelectrica de Cabora Bassa.

EMPRESA NACIONAL DE CARVÃO DE MOÇAMBIQUE (CARBOMOC)
Address: PO Box 1152, Maputo
Telex 6-413
Energy Sector: Coal

CARBOMOC is the state corporation responsible for coal production. Activities are centred on mines in the Moatize area of Tete Province. Production is to be increased from 0.5 million tonnes per annum, to fuel development projects and supply export markets.

EMPRESA NACIONAL DE PETROLEOS DE MOÇAMBIQUE (PETROMOC)
Address: Praça dos Trabalhadores 9, Box 417, Maputo
Telephone 27191/4 Telex 6-382
Energy Sector: Oil

PETROMOC has been established as the state company responsible for operational aspects of the oil industry in Mozambique. This includes control of the country's refinery, the importation of crude oil and disposal of refined products.

HIDROELECTRICA DE CABORA BASSA
Address: CP 4120, Maputo
Telex 6-467
Energy Sector: Electricity

Hidroelectrica de Cabora Bassa operates and manages the important hydro-electric plant at the Cabora Bassa dam. The company is a mixed public and private operation, but ownership is being gradually transferred to the Mozambique government.

The Cabora Bassa dam has a generating capacity of 2,000 MW and a large proportion of production is transmitted to South Africa. The company also supplies the national electricity distribution undertaking Electricidade de Moçambique.

Niger

GOVERNMENT DEPARTMENTS AND OFFICIAL AGENCIES

OFFICE NATIONAL DES RESSOURCES MINIERES (ONAREM)
Address: Niamey
Telex 5300

ONAREM is a state agency concerned with the exploration for, and exploitation of, minerals in Niger. Its principal interests lie in supervision of the uranium producing operations of SOMAIR and COMINAK. ONAREM has a 33 per cent shareholding in SOMAIR.

ENTERPRISES (PUBLIC AND PRIVATE SECTOR COMPANIES)

COMPAGNIE MINIERE D'AKOUTA (COMINAK)
Address: BP 839, Niamey
Telephone 73 34 25 Telex 5269
Energy Sector: Uranium

COMINAK is the operating company at Niger's second uranium mine at Akouta. It represents a combination of state and overseas interests from France, Japan and Spain. The state agency Uraniger has a 31 per cent shareholding. Principal overseas holdings are in the hands of the Commissariat à l'Energie Atomique of France, with 34 per cent, and the Japan Overseas Uranium Resources Development Corporation, with 25 per cent. The Spanish state uranium procurement agency ENUSA holds the remaining 10 per cent.

COMINAK's mining and processing plant commenced operations in 1978. Production is expected to rise eventually to a rate of 2,000 tonnes per annum.

SOCIETE DES MINES DE L'AIR (SOMAIR)
Address: BP 892, Niamey
Telex 5240
Energy Sector: Uranium

SOMAIR is the operating company for uranium mines at Arlit in central Niger. The company extracts uranium ore from an open pit and operates a processing mill at the site.

SOMAIR is owned by a consortium of overseas companies with the national agency ONAREM holding a 33 per cent interest. Largest shareholder after ONAREM is Cogéma of France with 27 per cent and other French groups hold a further 27 per cent. Agip Nucleare and Urangesellschaft also have interests.

SOCIETE NIGERIENNE DE CHARBON D'ANOU ARAREN (SONICHAR)
Address: BP 724, Niamey
Telex 5296
Energy Sector: Coal

SONICHAR has been established by a group of companies and state interests to exploit coal deposits at Anou Araren for electricity generation. The state mining agency ONAREM has a 26 per cent shareholding. Other principal shareholders are the two uranium mining companies SOMAIR and COMINAK, which will consume most of the power generated. The national electricity supply company NIGELEC also has an interest.

SOCIETE NIGERIENNE D'ELECTRICITE (NIGELEC)
Address: BP 202, Niamey
Telephone 72 26 92 Telex 5224
Energy Sector: Electricity

NIGELEC is a state controlled undertaking producing, transmitting and distributing electricity through the public supply system. The state holds 95 per cent of the shares of NIGELEC. The company obtains the major part of its electricity supplies from the Kaindji dam in Nigeria.

Nigeria

GOVERNMENT DEPARTMENTS AND OFFICIAL AGENCIES

FEDERAL MINISTRY OF MINES AND POWER
Address: 6 Storey Building, Broad Street, Lagos
Telephone 63 26 64

The Ministry of Mines and Power has wide responsibilities for the co-ordination of the activities of key state undertakings in the oil, gas, mining and electricity supply industries. It is also closely involved in the activities of private companies in exploration for, and development and disposal of, energy products. Organisations dealing directly with the Ministry include the Nigerian National Petroleum Corporation, the National Electric Power Authority, Nigerian Mining Corporation and Nigerian Coal Corporation.

ENTERPRISES (PUBLIC AND PRIVATE SECTOR COMPANIES)

AFRICAN PETROLEUM LIMITED
Address: AP House, 54/56 Broad Street,
PO Box 512, Lagos
Telephone 63 20 03 Telex 21242
Gross Revenue: N168.1 million (1980)
Total Assets: N22.6 million
Energy Sector: Oil

African Petroleum Limited is a state owned company engaged in distribution and marketing of petroleum products throughout Nigeria. The company was established following the take-over of the assets and operations of the Nigerian affiliate of British Petroleum Company.

African Petroleum is developing as one of the principal marketing companies in Nigeria, supplying automotive, aviation and industrial markets. Products are obtained from state oil refineries and the company also manufactures lubricating oils itself.

GULF OIL COMPANY (NIGERIA) LIMITED
Address: Gocon House, 19 Tinuba Square, PMB 2469, Lagos Island, Lagos
Telephone 65 08 20 Telex 21314
Energy Sector: Oil

Gulf Oil Company's Nigerian affiliate operates the second largest crude oil producing operation in Nigeria. It has a 40 per cent interest in the venture, the balance being held by the state oil company.

Gulf Oil's share of production in 1980 was over six million tonnes, including important offshore fields. Further exploration and development work is being carried out and the company expects a 50 per cent increase in its crude oil availability by 1985.

MOBIL OIL NIGERIA LIMITED
Address: 50 Broad Street, PMB 12054, Lagos
Telephone 63 51 71 Telex 21228
Energy Sector: Oil

Mobil Oil Nigeria is a wholly owned subsidiary of the major international oil company Mobil Corporation. It has a 40 per cent interest in important oil producing operations, in which it is partnered by the state oil company. Production by the group in 1980 exceeded 10 million tonnes.

NATIONAL ELECTRIC POWER AUTHORITY (NEPA)
Address: 24-25 Marina, PMB 1203, Lagos
Telephone 51370
Energy Sector: Electricity

The NEPA is a state-owned undertaking responsible for the public supply of electricity in Nigeria. It produces the major part of its electricity requirements from its own generating plant, which comprises both thermal and hydro-electric capacity. Approximately half of NEPA's electricity comes from hydro-electric plant, including the important Kaindji Dam scheme on the Niger River. A further 1,100 MW of hydro-electric capacity is planned.

NATIONAL OIL AND CHEMICAL MARKETING COMPANY LIMITED
Address: PO Box 2052, 38-39 Marina, Lagos
Telephone 58520-9
Energy Sector: Oil

The National Oil and Chemical Marketing Company came into being in 1975, following the acquisition of a majority shareholding in the former Shell Nigeria by the state oil company. Shell retains a 40 per cent interest.

The company is a distributor and marketer of petroleum products throughout Nigeria, acquiring products from the several state-owned oil refineries in the country.

NIGERIAN COAL CORPORATION
Address: PMB 1053, 29 Okpara Avenue, Enugu, Anambra State
Telephone 25 56 91 Telex 51115
Energy Sector: Coal

The Nigerian Coal Corporation is a state-owned undertaking with the responsibility for revitalising existing coal-producing operations and exploiting any other commercial possibilities for mining. Currently the Corporation operates mines in the Enugu area, the production capacity of which is estimated to be around two million tonnes per annum.

NIGERIAN MINING CORPORATION
Address: 24 Naraguta Avenue, PMB 2154, Jos, Plateau State
Telephone (073) 2149 Telex 81139
Energy Sector: Uranium

Nigerian Mining Corporation is a state-owned company established to promote the exploration for, and development of, minerals other than coal. It has set up a subsidiary, Nigerian Uranium Mining Company (NUMCO), to explore for uranium, on its own account or in association with other companies. NUMCO is currently undertaking detailed examination of uranium prospects in the eastern part of the country in partnership with the French company Minatome.

NIGERIAN NATIONAL PETROLEUM CORPORATION
Address: PMB 12701, Broad Street, Ikoyi, Lagos
Telex 21126
Energy Sector: Oil

The National Petroleum Corporation is the principal state organisation in the oil industry. It holds the state's shareholdings in wholly and partly owned companies involved in production of crude oil, oil refining and the marketing of refined products.

The Corporation has acquired the ownership of the Port Harcourt refinery, which is managed by its subsidiary Nigerian Petroleum Refinery Company, and has constructed new refineries at Warri and Kaduna. The Corporation handles crude oil supplied to the refineries and disposes of large volumes on the international market.

NIGERIAN PETROLEUM REFINERY COMPANY LIMITED
Address: PO Box 585, Alesa-Eleme, Port Harcourt, Rivers State
Telephone 22 22 10
Energy Sector: Oil

Nigerian Petroleum Refinery Company (NPRC) is a wholly owned subsidiary of the state oil company Nigerian National Petroleum Corporation. NPRC manages the country's oldest oil refinery at Port Harcourt, which has a crude oil processing capacity of around three million tonnes per annum.

PHILLIPS OIL COMPANY (NIGERIA) LIMITED
Address: 3rd Floor, Western House, 8-10 Broad Street, PMB 12612, Lagos
Telephone 63 75 73 Telex 21390
Energy Sector: Oil

Phillips Oil Company (Nigeria) is a subsidiary of the United States based Phillips Petroleum Company. The company has crude oil producing operations in which it is partnered by AGIP and the Nigerian National Petroleum Corporation. Phillips' interest is 20 per cent.

Production of crude oil by the group in 1980 was around 10 million tonnes, of which Phillips' entitlement was some two million tonnes. Production is concentrated in onshore fields in the south-east, between Warri and Port Harcourt.

SHELL PETROLEUM DEVELOPMENT COMPANY OF NIGERIA LIMITED
Address: Freeman House, 21-22 Marina, PMB 2418, Lagos Island, Lagos
Telephone 26861 Telex 21235
Energy Sectors: Oil, gas

Shell Petroleum Development Company is a subsidiary of the Royal Dutch/Shell group. Operating in association with the Nigerian National Petroleum Corporation, it produces around 50 per cent of the country's crude oil, which in 1980 approached 60 million tonnes. Shell's interest in the joint operation is 20 per cent.

The company is actively involved in further exploration for oil and gas, both onshore and offshore. The fields operated by Shell are capable of producing large volumes of natural gas, for which export-orientated LNG schemes have been considered.

Senegal

ENTERPRISES (PUBLIC AND PRIVATE SECTOR COMPANIES)

SOCIETE AFRICAINE DE RAFFINAGE (SAR)
Address: 15 Boulevard de la Republique, PO Box 203, Dakar
Telephone 22 46 84 Telex 527
Gross Revenue: CFA F22,700 million (1978)
Energy Sector: Oil

SAR is the operating company for the country's oil refinery at M'Bao. The refinery has a crude oil throughput capacity of 1.2 million tonnes per annum and serves the national market and the export bunker trade.

Participants in the refinery company include the foreign based companies involved in marketing in Senegal, with a 10 per cent Senegalese shareholding. Largest participant is the French State group SNEA, with 30 per cent.

SOCIETE SENEGALAISE DE DISTRIBUTION D'ENERGIE ELECTRIQUE (SENELEC)
Address: BP 93, 28 Rue Vincens, Dakar
Telephone 22315 Telex 661
Energy Sector: Electricity

SENELEC is a state controlled undertaking which provides power through the public supply system. Its operations include generation, transmission and distribution. Other public authorities are also shareholders in the company.

SENELEC's generating capacity is concentrated at Dakar, where it has 155 MW plant. Based on Dakar the company has established a supply system extending to Rufisque, Thies, St Louis, Louga and Kaolack.

Sierra Leone

ENTERPRISES (PUBLIC AND PRIVATE SECTOR COMPANIES)

SIERRA LEONE PETROLEUM REFINING COMPANY LIMITED
Address: BP Building, Siaka Stevens Street, Freetown
Telex 3246
Energy Sector: Oil

The Sierra Leone Petroleum Refining Company operates an oil refinery at Freetown, which supplies most of the country's requirements of petroleum products. The refinery's capacity is approximately half a million tonnes per annum. The state holds a 50 per cent share in the refinery and controls the refining company's operations.

South Africa

GOVERNMENT DEPARTMENTS AND OFFICIAL AGENCIES

ATOMIC ENERGY BOARD
Address: Pelindaba, Private Bag X256, Pretoria 0001, Transvaal
Telephone (012) 79 44 41 Telex 3-0253

The Atomic Energy Board is a statutory authority responsible for all aspects of nuclear power development in South Africa. The Board undertakes fundamental research, research into technologies, materials and nuclear fuel processing and is responsible for supervising the safety of uranium production and consumption activities.

The AEB participates in international discussions on the nuclear fuel cycle and is closely concerned in the bringing into operation of a nuclear power station outside Cape Town. The AEB itself undertakes the evaluation of indigenous uranium resources.

DEPARTMENT OF MINERAL AND ENERGY AFFAIRS
Address: North Vaal Building, Vermeulen Street, Pretoria, Transvaal
Telephone 48 55 56

The Department has wide responsibilities for the activities of energy supply industries and the administration and formulation of policies relating to the sector. The Department is responsible in particular for key enterprises, including ESCOM, the national electricity supply undertaking, and SASOL, which utilises large quantities of coal for conversion to oil products and petrochemical feedstocks.

Also under the control of the Department are the Atomic Energy Commission, the Nuclear Fuels Corporation and SOEKOR, the oil and gas exploration company. Responsibilities extend to consumer prices, the general control of mining and the Geological Survey of South Africa.

FUEL RESEARCH INSTITUTE OF SOUTH AFRICA
Address: 21 Lynnwood Road, PO Box 217, Pretoria 0001, Transvaal
Telephone 74 31 26 Telex 3-0430

The Fuel Research Institute was set up under statute with the responsibility of examining the fuel resources of the country and undertaking research on all matters relating to fuels. Individual divisions of the Institute are concerned with surveying and grading of coal, coal preparation, carbonisation and briquetting, combustion and chemistry.

ENTERPRISES (PUBLIC AND PRIVATE SECTOR COMPANIES)

ANGLO AMERICAN COAL CORPORATION (AMCOAL)
Address: 44 Main Street, Johannesburg 2001, Transvaal
Telephone 838 8111
Energy Sector: Coal

AMCOAL is one of the principal operating companies of the leading mining finance company Anglo American Corporation, which holds 51 per cent of the issued shares. AMCOAL is the largest coal producing company in South Africa, with production of 34 million tonnes in 1980. Output from all mines administered by the group totalled 37 million tonnes.

AMCOAL operates 14 collieries producing steam coal, metallurgical coal and anthracite in Transvaal, Orange Free State and Natal. A high proportion of output is sold to ESCOM for power generation and to the steel producer ISCOR. Sales to ESCOM in 1980 totalled 22.6 million tonnes and large scale expansion of production for ESCOM's Tutuka and Lethabo power stations is planned at the New Denmark and New Vaal collieries at Standerton (Transvaal) and Sasolburg (Orange Free State) respectively.

ANGLO AMERICAN CORPORATION OF SOUTH AFRICA LIMITED
Address: 44 Main Street, Johannesburg 2001, Transvaal
Telephone 838 8111 Telex 7167
Gross Revenue: R5,253.2 million (1981)
Total Assets: R760.6 million
Energy Sectors: Coal, uranium

Anglo American Corporation (AAC) is one of the leading mining finance houses of South Africa, holding investments in mining, financial, industrial and commercial companies both in Southern Africa and abroad. It is extensively involved in the production of diamonds, gold, uranium, coal and other metals and minerals.

AAC holds the majority interest in Anglo American Coal Corporation (AMCOAL) which is the largest coal producer in South Africa. Output from mines managed by AMCOAL was nearly 37 million tonnes in 1980 and increasing as demand rose from domestic and export markets.

Uranium is produced at several of the gold mines operated by AAC.

BARLOW RAND LIMITED
Address: Barlow Park, Katherine Street,
Sandton 2196, Transvaal
Telephone (011) 786 3470
Gross Revenue: R4,571.9 million (1981)
Total Assets: R3,507.0 million
Energy Sectors: Coal, uranium

Barlow Rand is a large South African company with interests in mining, exploration, metal manufacturing and engineering, production of building materials and other industrial and consumer goods. Mining and exploration activities account for eight per cent of turnover and 16 per cent of assets.

Mining operations are undertaken by the subsidiary company Rand Mines Limited. Rand Mines manages the collieries of Transvaal Consolidated Land and Exploration Company, in which Barlow Rand holds a 63 per cent interest, and two gold mining companies, which produce uranium, Blyvooruitzicht Gold Mining Company and Harmony Gold Mining Company.

BP SOUTHERN AFRICA (PTY) LIMITED
Address: BP Centre, 214 West Street, Box 1806,
Durban, Natal
Telephone 32 92 11
Energy Sectors: Oil, coal

BP Southern Africa is a wholly owned subsidiary of the major international oil company British Petroleum. It is involved in oil refining and marketing and has interests in export orientated coal mining capacity in the Transvaal.

The BP affiliate is leading marketer of petroleum products, which are obtained from the state-owned oil-from-coal producer SASOL and from the Durban refinery in which it is a partner.

BP Southern Africa is the principal participant in the development of a new mine at Middelburg, Transvaal. The mine will be operated and managed by Rand Mines.

ELECTRICITY SUPPLY COMMISSION
Address: Megawatt Park, Maxwell Drive, Sandton,
Transvaal
Telephone 80 08 11
Gross Revenue: R1,772.0 million (1980)
Total Assets: R8,972.8 million (1980)
Energy Sectors: Electricity, coal

The Electricity Supply Commission (ESCOM) is a state undertaking established in 1922 to ensure the supply of electricity to the South African economy. As such it is one of the cornerstones of the economy which relies heavily on electrical energy. ESCOM is responsible for the production of around 60 per cent of all electricity produced in Africa, and 93 per cent of electricity supplied in South Africa. This includes a share of the electricity derived from the Cabora Bassa dam in Mozambique.

Total quantity of electricity supplied by ESCOM in 1980 was 87.5 TWh, compared with only 35 TWh in 1970. This rapid rate of growth is continuing with a high level of investment by ESCOM in generating capacity and transmission lines. ESCOM operates 20 coal fired power stations with several of 1,000-3,000 MW. There are also four gas turbine and hydro-electric stations. Total generating capacity at end 1980 was 18.4 GW.

ESCOM's national transmission network includes 14,600 km of very high tension lines of 220-400 kV and a further 1,030 km of 533 kV DC lines, integrating the activities of the regional undertakings.

GENERAL MINING UNION CORPORATION (GENCOR)
Address: 6 Hollard Street, Johannesburg 2001,
Transvaal
Telephone 836 1121
Gross Revenue: R3,201.2 million (1981)
Total Assets: R2,922.2 million
Energy Sectors: Coal, uranium

GENCOR was formed in 1980 on the merging of two of South Africa's largest mining finance houses, General Mining and Finance Corporation and Union Corporation Limited. A controlling interest in GENCOR is held by Federale Mynbou, the investment and holding company.

GENCOR is extensively involved in the mining of gold and uranium, coal, platinum and other metals and minerals. The group produces around one-quarter of South African uranium output and is the second largest coal producer, with sales of 25 million tonnes per annum by its affiliate Trans-Natal Coal Corporation. Of this total in 1980 almost 18 million tonnes was supplied to ESCOM power stations or to the steel producer ISCOR. Total output from mines managed by GENCOR is now around 35 million tonnes per annum.

Uranium is produced as a by-product by West Rand Consolidated Mines Limited and Chemwes, a joint operation of Stilfontein Gold Mining Company and Buffelsfontein Gold Mining Company, both of which are managed by GENCOR and produce uranium from slimes. In addition the company established in 1981 a new mine, the Beisa division of St Helena Gold Mining Company, which is primarily a uranium mining operation.

GOLD FIELDS OF SOUTH AFRICA LIMITED
Address: Gold Fields Building, 75 Fox Street, Johannesburg 2001, Transvaal
Gross Revenue: R210.9 million (1981)
Total Assets: R434.1 million
Energy Sectors: Coal, uranium

Gold Fields of South Africa, in which the UK based mining company Consolidated Gold Fields has a 48 per cent interest, is a producer of uranium, as a by-product of its gold mining operations, and of coal, through its ownership of Apex Mines Limited.

Uranium is produced by the Blyvooruitzicht Gold Mining Company and Driefontein Consolidated Limited. The Greenside Colliery of Apex Mines produces some 2.7 million tonnes per annum for various domestic and overseas markets.

JOHANNESBURG CONSOLIDATED INVESTMENT COMPANY LIMITED
Address: Consolidated Building, Fox Street, Johannesburg, Transvaal
Telephone 836 2571
Pre-Tax Profit: R128.8 million (1981)
Total Assets: R878.5 million
Energy Sectors: Coal, uranium

Johannesburg Consolidated Investment Company is a leading mining finance house with interests in a broad range of mining, industrial, property and financial enterprises. The company manages Randfontein Estates Gold Mining Company and Western Areas Gold Mining Company, both of which are producers of uranium. The company has also acquired the full ownership of Tavistock Collieries Limited which produced over four million tonnes of coal in 1980.

Tavistock Collieries is wholly engaged in coal mining and is the vehicle for Johannesburg Consolidated to extend its activity in this area. The company has entered into a partnership with TOTAL Exploration South Africa to develop additional capacity for the export market. Other coal rights are held in eastern Transvaal, Natal and KwaZulu.

KANGRA HOLDINGS (PROPRIETARY) LIMITED
Address: PO Box 2465, Johannesburg 2000, Transvaal
Telephone 838 5375
Energy Sector: Coal

Kangra Holdings owns several coal mines in Natal and Transvaal, with total sales in 1979 of 3.5 million tonnes. Output includes anthracite at three collieries in Natal. This is handled by the Anthracite Producers Association for disposal in the inland general trade and for export. The company owns the Spitzkop Colliery at Ermelo, Transvaal, which produces over a million tonnes per annum of steam coal for export by the Shell group.

NATIONAL PETROLEUM REFINERS OF SOUTH AFRICA (PTY) LIMITED (NATREF)
Address: Jan Haak Road, Sasolburg, Orange Free State 9570
Telex 84377
Energy Sector: Oil

NATREF operates an oil refinery at Sasolburg with an annual processing capacity of 3.5 million tonnes. A majority shareholding is held by the state-owned oil-from-coal producer Sasol Limited. A 30 per cent interest is held by the French oil company CFP. Technically the Iranian state oil company NIOC retains a minority interest.

RÖSSING URANIUM LIMITED
Address: Unicorn House, 70 Marshall Street Johannesburg, Transvaal
Telephone 836 1641
Energy Sector: Uranium

Rössing Uranium Limited was set up by a group of companies to operate the uranium development at Rössing in Namibia. The Rössing deposit is located 70km from Swakopmund in the Namibian desert and necessitated the construction of all related infrastructure. Commercial production began in 1976 and in 1980 reached 5,000 short tonnes of uranium oxide.

The RTZ group is a major shareholder in Rössing Uranium, with a 46.5 per cent interest. Other partners include the South African Gencor group and Industrial Development Corporation. The French oil group CFP also has an interest through the 10 per cent holding of Total Compagnie Minière et Nucléaire.

SASOL LIMITED
Address: 55 Commissioner Street Johannesburg 2001, Transvaal
Telephone 836 7414 Telex 8-9931
Gross Revenue: R1,331.0 million (1980)
Total Assets: R1,080.7 million
Energy Sectors: Coal, oil, gas

SASOL is the holding company for separate operating companies managing the three oil-from-coal plants which are in operation or under construction at Sasolburg and Sekunda. Sasol One Pty owns the established oil conversion plant at Sasolburg. Sasol Two Pty and Sasol Three Pty have been set up as corporate vehicles for the two major new works at Sekunda.

SASOL is itself owned by state agencies, but Sasol Two and Sasol Three are to have a broader ownership in order to draw on outside sources of finance. Coal for the three plants is produced by SASOL, which owns the Sigma and Bosjesspruit collieries. Sigma Colliery produced 5.3 million tonnes in 1980-81, and Bosjesspruit Colliery 8 million tonnes. The latter is expected to produce 27 million tonnes per annum when Sasol Two and Sasol Three plants are fully operational.

Output from SASOL's operations includes feedstocks, gas and finished products. The company has a subsidiary, South African Gas Distribution Corporation (GASCOR) which supplies works gas to a number of industrial consumers in the Rand, and has a majority shareholding in the only South African controlled refinery at Sasolburg.

SHELL SOUTH AFRICA (PTY) LIMITED
Address: Box 740, Silverton, Transvaal
Telephone 83 41 71
Energy Sectors: Oil, coal

Shell South Africa is a subsidiary of the Royal Dutch/Shell group, engaged in oil refining and product marketing in South Africa. The company is also involved in exporting coal from Spitzkop Colliery and its joint-venture with Barlow Rand at Rietspruit.

Shell is one of the principal marketers of petroleum products in South Africa. Products are obtained from the state oil-producing organisation and from the Durban refinery which Shell manages. The refinery is the largest in South Africa, with a crude oil throughput capacity of nearly eight million tonnes per annum.

SOUTH AFRICAN IRON AND STEEL CORPORATION LIMITED (ISCOR)
Address: Waggonwheel Circle, PO Box 450
Pretoria 0001, Transvaal
Telephone 41-4111 Telex 3-672
Gross Revenue: R2,184.8 million (1981)
Total Assets: R3,041.3 million
Energy Sector: Coal

ISCOR is the state-owned steel producer, which is one of the largest individual consumers of coal after ESCOM and SASOL. ISCOR also operates its own mines producing coking coal. Of total consumption of 6.3 million tonnes in 1980-81 two million tonnes were produced by ISCOR's subsidiary Durban Navigation Collieries Pty at Dannhauser, Natal, and the Grootegeluk Colliery at Waterberg, Transvaal.

SOUTHERN OIL EXPLORATION CORPORATION (PTY) LIMITED (SOEKOR)
Address: PO Box 3087, Johannesburg, Transvaal
Telephone 724 7307 Telex 430186
Energy Sectors: Oil, gas

SOEKOR is a state-owned corporation responsible for exploration for oil and gas in South Africa and its offshore areas. To date it has identified few deposits of hydrocarbons and none of any commercial significance. SOEKOR shoulders virtually the whole burden of the exploration programme as the international oil companies are not involved.

TAVISTOCK COLLIERIES LIMITED
Address: PO Box 590, Johannesburg
Transvaal 2000
Telephone 836 2571
Energy Sector: Coal

Tavistock Collieries is a wholly owned subsidiary of Johannesburg Consolidated Investment Company, engaged in coal mining in the Transvaal. The company operates mines at Witbank and Bethal producing over four million tonnes per annum of bituminous coal.

Output is sold mainly to the internal market, but some coal is exported through the company's membership of Transvaal Coal Owners Association in which it has a 13.4 per cent interest. The Tavistock Collieries group of mines is expanding its investment in production and transportation facilities to export around 100,000 tonnes per month from the Richards Bay Terminal.

TRANS-NATAL COAL CORPORATION LIMITED
Address: PO Box 61824, Marshalltown 2107
Transvaal
Telephone 836 1121
Gross Revenue: R379.5 million (1981)
Total Assets: R328.6 million
Energy Sector: Coal

Trans-Natal Coal Corporation is a major coal producing company, managed by GENCOR, which holds a 42 per cent interest. The Corporation is involved in prospecting, the acquisition of coal rights, production and marketing.

The Corporation operates eight collieries in Transvaal and three in Natal. Sales in 1981 totalled 28 million tonnes, of which 17 million tonnes were fed to ESCOM's power stations. More than four million tonnes were exported compared with 2.6 million tonnes the previous year. Collieries linked to ESCOM are at Usutu, Kilbarchan, Blinkpan, Optimum and Matla, which is being substantially expanded. Kilbarchan, Hlobane and Northfield collieries supply coal to the steel making company ISCOR. The general trade and export market is supplied from Transvaal Navigation, Delmas and Ermelo collieries.

TRANSVAAL CONSOLIDATED LAND AND EXPLORATION COMPANY LIMITED
Address: PO Box 62370, Marshalltown 2107
Transvaal
Telephone 836 1166
Gross Revenue: R380.0 million (1981)
Energy Sector: Coal

Transvaal Consolidated Land and Exploration Company is an affiliate of the Barlow Rand group, which holds 63 per cent of the shares. The company holds major interests in four important coal producing companies, Witbank Colliery Limited, Welgedacht Exploration Company, Rietspruit Opencast Services and Duvha Opencast Services.

Currently Witbank Colliery is the principal coal producer, with output of more than 11 million tonnes in 1981, but extraction from the Duvha mine is building up rapidly towards an eventual level of more than 10 million tonnes per annum as ESCOM's Duvha power station achieves full load. Rietspruit Opencast Services is a joint venture with Shell to produce five million tonnes per annum of coal for export. The group's total coal reserves are estimated to be 13,000 million tonnes.

TREK BELEGGINGS LIMITED
Address: 46th Floor, Carlton Centre
Johannesburg 2001, Transvaal
Telephone 21 56 41 Telex 8-6696
Gross Revenue: R309.4 million (1981)
Total Assets: R78.5 million
Energy Sector: Oil

Trek Beleggings was established in 1970 as a South African based marketer of petroleum products and has developed a network of 320 service stations for automotive fuels and distribution channels for other products. The company is now a wholly owned subsidiary of the leading mining company GENCOR.

WITBANK COLLIERY LIMITED
Address: PO Box 62370, Marshalltown 2107
Transvaal
Telephone 836 1166
Gross Revenue: R158.6 million (1981)
Total Assets: R83.2 million
Energy Sector: Coal

Witbank Colliery is the principal coal producing subsidiary of the Barlow Rand group, operating several mines in northern Transvaal. Total sales in 1981 amounted to 11.4 million tonnes, of which 5.3 million tonnes were exported.

Sales to ESCOM for electricity generation increased from 1.4 to 2.9 million tonnes and are expected to increase to 9.5 million tonnes by 1985, with the continuing development of the Duvha opencast mine.

PROFESSIONAL INSTITUTIONS AND TRADE ASSOCIATIONS

ANTHRACITE PRODUCERS ASSOCIATION
Address: 13th Floor, Bank of Lisbon Building
37 Sauer Street, Johannesburg, Transvaal
Telephone 836 9861 Telex 87112
Energy Sector: Coal

The Anthracite Producers Association is an organisation of colliery operating companies in Natal, acting as a co-ordinated channel for marketing production to both internal and export markets. The Association handles output of Natal Anthracite Colliery Limited (Amcoal group), Alpha Anthracite Company (Lonrho group), and anthracite mines belonging to Kangra Holdings (Pty) Limited.

CHAMBER OF MINES OF SOUTH AFRICA
Address: 5 Hollard Street, PO Box 809
Johannesburg, Transvaal
Telephone 838 8211 Telex 87057
Energy Sectors: Coal, uranium

The Chamber of Mines is a representative organisation of mining companies, but with an extensive role in certain operational aspects of the industry. The Chamber also has some powers delegated to it by the Department of Mineral and Energy Affairs, with regard to the operation of mines and other facilities. The Chamber provides facilities for recruitment, training and health care on behalf of the industry and is the forum for the development of certain mining issues on a national basis. The Nuclear Fuels Corporation (NUFCO) has been formed under the aegis of the Chamber of Mines.

NATAL ASSOCIATED COLLIERIES (PTY) LIMITED
Address: 12th Floor, Bank of Lisbon Building
37 Sauer Street, Johannesburg, Transvaal
Telephone 833 6200 Telex 80005
Energy Sector: Coal

Natal Associated Collieries is an organisation of coal producing companies set up to provide a marketing channel for coal, principally in the domestic market. The group handles steam coal, coking coal and anthracite from colliery companies managed by GENCOR, Rand Mines and Kangra Holdings.

SOUTH AFRICAN INSTITUTE OF MINING AND METALLURGY
Address: PO Box 61019, Marshalltown 2107
Transvaal
Energy Sectors: Coal, uranium

The Institute of Mining and Metallurgy is a professional organisation representing a forum for research and information on technical and economic issues. It produces a regular journal discussing new developments in mineral and metal exploration, extraction and processing.

TRANSVAAL COAL OWNERS ASSOCIATION PROPRIETARY LIMITED (TCOA)
Address: 12th Floor, Bank of Lisbon Building
Corner Sauer and Market Streets
Johannesburg, Transvaal
Telephone 834 5151/7 Telex 80005
Energy Sector: Coal

TCOA is an organisation of the main coal producing companies in Transvaal and Orange Free State, established to co-ordinate distribution and marketing of output. There are 24 member collieries of TCOA.

Since 1970 TCOA has been actively building up its export trade and commissioned the Richards Bay export terminal in 1976.

Sudan

GOVERNMENT DEPARTMENTS AND OFFICIAL AGENCIES

MINISTRY OF ENERGY AND MINING
Address: PO Box 410, Khartoum
Telex 256

The Ministry of Energy and Mining has general responsibility for state-owned undertakings and supervision of private sector companies involved in the oil industry, mineral exploration and development and electricity supply. This includes control of the Public Petroleum Corporation and the Public Corporation for Oil Products and Pipelines.

ENTERPRISES (PUBLIC AND PRIVATE SECTOR COMPANIES)

PORT SUDAN REFINERY LIMITED
Address: PO Box 354, Port Sudan
Telephone 3991 Telex 504
Energy Sector: Oil

Port Sudan Refinery Limited operates the country's only oil refinery, at Port Sudan. The state holds a controlling interest and is partnered by affiliates of the international oil groups Royal Dutch/Shell and British Petroleum.

The Port Sudan refinery has an annual throughput capacity of over one million tonnes per annum and is principal source of supply for oil marketing companies in the country.

PUBLIC ELECTRICITY AND WATER CORPORATION
Address: PO Box 1380, Khartoum
Telex 642
Energy Sector: Electricity

The Public Electricity and Water Corporation is a state undertaking producing and distributing electricity through the public supply system. This system excludes most of the larger industrial power consumers, who operate their own generating capacity.

The Corporation has an important hydro-electric power station on the Blue Nile at Roseires, the capacity of which is being taken up to 250MW. Other stations use fuel oil and diesel. The Corporation's system links the principal towns of the Sudan.

Tanzania

GOVERNMENT DEPARTMENTS AND OFFICIAL AGENCIES

MINISTRY OF WATER AND ENERGY
Address: Tancot House, PO Box 9153
City Drive, Dar-es-Salaam
Telephone 27811

The Ministry of Water and Energy is concerned with the utilisation of water resources and development of hydro-electric potential as a principal source of electricity in Tanzania. The

Ministry is responsible for the activities of the national public electricity supply company TANESCO.

ENTERPRISES (PUBLIC AND PRIVATE SECTOR COMPANIES)

STATE MINING CORPORATION
Address: 417-8 United Nations Road
PO Box 4958, Dar-es-Salaam
Telephone 28781 Telex 41354
Energy Sector: Coal

The State Mining Corporation has been established to plan and co-ordinate the exploration for, and development of, indigenous mineral or metal deposits. Its principal concern currently is the exploitation of coal deposits in the south-west of the country, greatly expanding the existing level of output.

TANZANIA ELECTRIC SUPPLY COMPANY LIMITED (TANESCO)
Address: Box 9024, Independence Avenue
Dar-es-Salaam
Telephone 27281 Telex 41318
Energy Sector: Electricity

TANESCO is a state undertaking producing electricity and distributing it through the public supply system. The company is under the general control of the Ministry of Water and Energy.

TANESCO uses thermal and hydro-electric plant to produce electricity, but is planning to rely on hydro-electric plant and additional coal-fired stations for future supplies.

TANZANIAN ITALIAN PETROLEUM REFINERY (TIPER)
Address: Kiganibou, Dar-es-Salaam
Energy Sector: Oil

TIPER is the operating company for the country's only refinery at Dar-es-Salaam. The refinery has an annual crude oil treatment capacity of around 700,000 tonnes.

The state maintains control of the operations of TIPER. The Italian state oil company AGIP also has an interest and provides technical assistance to the company.

Togo

ENTERPRISES (PUBLIC AND PRIVATE SECTOR COMPANIES)

COMPAGNIE ENERGIE ELECTRIQUE DU TOGO (CEET)
Address: BP 42, Lomé
Energy Sector: Electricity

CEET is a state-owned undertaking producing electricity and operating the public supply system in Togo. The major part of the company's generating capacity is the 120 MW thermal generating station located near the country's oil refinery on the coast, and supplied with fuel oil from the refinery.

CEET also has available 25 MW of capacity from the Akosombo dam in Ghana and is considering with Benin the construction of a hydro-electric scheme on the Mono River.

SOCIETE TOGOLAISE DES HYDROCARBURES
Address: BP 3283, Lomé
Telex 5210
Energy Sector: Oil

Société Togolaise des Hydrocarbures was set up in 1976 to represent the state's interests in the oil and gas sectors. The state has an 80 per cent shareholding.

The company has responsibility for control of exploration and exploitation of oil and gas and for oil refining and is involved in marketing. It owns a recently completed refinery at Lomé, which has a throughput capacity of one million tonnes per annum.

Tunisia

GOVERNMENT DEPARTMENTS AND OFFICIAL AGENCIES

DIRECTION DE L'ENERGIE
Address: Rue de la Kasbah, Tunis
Telephone 26 11 21

The Direction de l'Energie is the principal government department responsible for energy policy and energy developments. It forms part of the Ministère de l'Economie Nationale.

ENTERPRISES (PUBLIC AND PRIVATE SECTOR COMPANIES)

SOCIETE DE RECHERCHE ET D'EXPLOITATION DES PETROLES EN TUNISIE (SEREPT)
Address: 6 Rue de Venezuela, BP 409, Tunis
Telephone 28 32 88 Telex 20393
Energy Sector: Oil

SEREPT is an oil exploration and development company in which the State is partnered by several French oil groups. Major shareholdings are held by the French state company Société National Elf-Aquitaine, with 50 per cent and Compagnie Française des Pétroles.

SOCIETE TUNISIENNE DE L'ELECTRICITE ET DU GAZ (STEG)
Address: 38 Rue K Ataturk, Tunis
Telephone 24 35 22 Telex 120206
Energy Sectors: Electricity, gas

STEG is a publicly owned undertaking responsible for the production and distribution of electricity and gas to all categories of consumers in Tunisia.

SOCIETE TUNISIENNE DE RAFFINAGE (SOTURAF)
Address: 11 Av Kherreddine Pacha, Tunis
Telephone 28 71 88 Telex 12128
Energy Sector: Oil

SOTURAF is the operating company of the state-owned oil refinery at Bizerta. The refinery has a crude oil processing capacity of around 1.5 million tonnes per annum, with output supplied to local marketing companies.

Upper Volta

ENTERPRISES (PUBLIC AND PRIVATE SECTOR COMPANIES)

SOCIETE VOLTAIQUE D'ELECTRICITE (VOLTELEC)
Address: BP 54, Ouagadougou
Energy Sector: Electricity

VOLTELEC is a state-owned undertaking providing electricity for the public supply system. The company is responsible to the Ministry of Public Works.

VOLTELEC has several small power stations providing a supply network for some of the main centres and has been pursuing a programme of electrification and capacity construction. Two additional hydro-electric stations with a total annual production capacity of 75 GWh are planned for the Komtenga and White Volta rivers.

Zaire

ENTERPRISES (PUBLIC AND PRIVATE SECTOR COMPANIES)

SOCIETE NATIONAL D'ELECTRICITE (SNEL)
Address: 49 Bd du 30-Juin, BP 500, Kinshasa
Telephone 24127 Telex 21570
Energy Sector: Electricity

SNEL is a state controlled organisation responsible for the public electricity supply. The state holds a 70 per cent shareholding.

SNEL generates almost all of its electricity from hydro-electric power stations. With a capacity of some 1,000 MW annual production is of the order of 4,000 GWh. Main areas of electricity production are Shaba and Bas Zaire. SNEL is involved in the project for a substantial further development of hydro-electric power through the Inga project.

SOCIETE ZAIRO-ITALIENNE DE RAFFINAGE (SOZIR)
Address: Av Tombalbaye, BP 1478, Kinshasa
Telex 21119
Energy Sector: Oil

SOZIR owns and operates the country's only refinery, at Moanda. The refinery is owned jointly by the state of Zaire and the Italian national oil company AGIP, which also provides operating expertise.

The Moanda refinery has a crude oil processing capacity of around three-quarters of a million tonnes per annum and operates on locally produced crude oil.

ZAIRE GULF OIL COMPANY
Address: Centre UZB, BP 7189, Kinshasa
Telex 21528
Energy Sector: Oil

Zaire Gulf Oil Company is an affiliate of the United States based major oil company Gulf Oil Corporation. The company participates in crude oil production and is actively exploring in the country.

ZAIRE PETROLEUM COMPANY LIMITED
Address: Avenue du Port 14-16, Kinshasa
Telex 21454
Energy Sector: Oil

Zaire Petroleum Company is a consortium of foreign based companies involved in exploration and production of crude oil. The company has Japanese interests through the participation of Teikoku Oil and the international trading company Mitsui. Gulf Oil Corporation also has an interest.

Production from Zaire Petroleum's fields commenced in 1975. A substantial proportion of the output is shipped to Japanese oil refineries.

Zambia

GOVERNMENT DEPARTMENTS AND OFFICIAL AGENCIES

MINISTRY OF MINES
Address: Box 31969, Chilufya Mulenga Road
Lusaka
Telephone 21 12 20

The Ministry of Mines is responsible for general policy towards mineral exploration and production and supervision of the activities of the state industrial and mining corporation ZIMCO.

MINISTRY OF POWER, TRANSPORT AND COMMUNICATIONS
Address: Box RW 50065, Fairley Road, Lusaka
Telephone 21 32 11

The Ministry of Power, Transport and Communications is responsible for the general policies and development of the state electricity supply undertaking ZESCO, including investment in power generation plant and programmes for rural electrification.

ENTERPRISES (PUBLIC AND PRIVATE SECTOR COMPANIES)

CENTRAL AFRICAN POWER CORPORATION
Address: Woodgate House, Box 30233, Cairo Road
Lusaka
Telephone 21 42 82
Energy Sector: Electricity

Central African Power Corporation is a company jointly owned by the states of Zambia and Zimbabwe, set up to own and operate the original Kariba hydro-electric station on the Zambezi River. Capacity of the station is 660 MW.

The power station is located on the Zimbabwe side of the Zambezi and, following the development of other major hydro-electric schemes on the north bank at Kariba and at Kafue Gorge, mainly supplies power to the Zimbabwe supply network.

INDENI PETROLEUM REFINERY COMPANY LIMITED
Address: Bwana Mkubwa, PO Box 1869, Ndola
Telephone 3277 Telex 34221
Energy Sector: Oil

Indeni Petroleum Refinery Company operates the national refinery at Ndola. The state has a controlling interest, with the participation of the Italian state oil company AGIP, which provides technical expertise.

The Ndola refinery has a crude oil processing capacity of around 1.2 million tonnes per annum. Crude oil and other feedstock is supplied by pipeline from Tanzania.

TAZAMA PIPELINES LIMITED
Address: Tazama House, Buteko Avenue
PO Box 1651, Ndola
Telephone 4695 Telex 34160
Gross Revenue: K13.1 million (1981)
Total Assets: K52.7 million
Energy Sector: Oil

Tazama Pipelines owns and operates the pipeline linking the Zambian refinery at Ndola to the port of Dar-es-Salaam in Tanzania. The company is a wholly owned subsidiary of the state holding company Zambia Industrial and Mining Corporation. Pipeline throughput in 1980-81 was 729,000 tonnes of crude oil and feedstock.

ZAMBIA ELECTRICITY SUPPLY CORPORATION LIMITED (ZESCO)
Address: Stand No 6949, Great East Road
PO Box 33304, Lusaka
Telephone 82091 Telex 40150

Gross Revenue: K49.6 million (1979)
Total Assets: K386.8 million
Energy Sector: Electricity

ZESCO came into operation in 1970 as licensee under Act of Parliament to generate, transmit and distribute electricity in Zambia. It rationalised the previously existing local distribution systems and extended the process of electrification throughout the country. ZESCO's shares are held by the state holding company ZIMCO.

ZESCO operates power stations at Kafue Gorge and Victoria Falls, with total capacity of 1,560 MW. The Kafue Gorge plant has a capacity of 900 MW. Additional supplies to the Corporation's grid are obtained from the Central African Power Corporation and Kariba North Bank Company which operate the hydro-electric plants at Kariba.

Most sales are made directly to consumers by ZESCO, but a large amount of power is channeled through Copperbelt Power Company to the energy-intensive mining establishments in the Copperbelt.

ZAMBIA INDUSTRIAL AND MINING CORPORATION (ZIMCO)
Address: Zimco House, PO Box 30090
Cairo Road, Lusaka
Telephone 72981 Telex 4180
Energy Sectors: Coal, electricity

ZIMCO is a state corporation controlling important mineral and metal production and processing operations, including some of the leading energy consuming establishments such as Nchanga Consolidated Copper Mines, which produces large amounts of power for its own use. ZIMCO is the holding company for the national electricity supply corporation ZESCO.

Maamba Collieries, a wholly owned subsidiary company of ZIMCO, produces coal at Maamba near the south-eastern border of the country. Output is of the order of one million tonnes per annum.

Zimbabwe

GOVERNMENT DEPARTMENTS AND OFFICIAL AGENCIES

MINISTRY OF MINES
Address: Earl Grey Building, Corner Livingstone Avenue/4th Street, P/Bag 7709, Salisbury
Telephone 79 01 31

The Ministry of Mines has general responsibility for the development of the country's mineral and metal resources, the evaluation of resources, control of coal production and prices, refining transportation of energy and market aspects.

The Ministry includes departments of metallurgy and mining engineering and the Zimbabwe Geological Survey. The Ministry also provides support for the Mining Promotion Corporation.

ENTERPRISES (PUBLIC AND PRIVATE SECTOR COMPANIES)

ELECTRICITY SUPPLY COMMISSION
Address: PO Box 377, Electricity Centre
Samora Machel Avenue, Salisbury
Telephone 70 38 41 Telex 43231
Gross Revenue: Z$ 72.5 million (1981)
Total Assets: Z$ 220.1 million
Energy Sector: Electricity

The Electricity Supply Commission is a state undertaking with broad responsibilities for the generation, transmission and distribution of electricity in Zimbabwe. Its licensed area covers the whole of the country except the city areas of Bulawayo, Gwelo, Salisbury and Umtali.

The Commission supplies around 70 per cent of all electricity used in Zimbabwe. Sales in 1980-81 totalled 4.9 TWh, of which 4.0 TWh was consumed by mining and other industrial operations. The Commission has three thermal power stations, but the bulk of its supply is derived from the Kariba hydro-electric station of Central African Power Corporation.

WANKIE COLLIERY COMPANY LIMITED
Address: PO Box 1108, 70 Samora Machel Avenue
Central, Salisbury C4
Telex 4394
Gross Revenue: Z$36.8 million (1981)
Total Assets: Z$59.9 million
Energy Sector: Coal

Wankie Colliery Company operates open-cast and deep-mining coal mines at Wankie, in north-west Zimbabwe. The company is controlled by the South African mining company Anglo American Corporation.

Production by the company in 1980-81 totalled three million tonnes, for use in electricity generation, in industry and as input to coking plants.

PROFESSIONAL INSTITUTIONS AND TRADE ASSOCIATIONS

CHAMBER OF MINES
Address: Chamber of Mines Building
Gordon Avenue, PO Box 712, Salisbury
Telephone 70 28 43
Energy Sector: Coal

The Chamber of Mines is a representative organisation for mining companies in Zimbabwe providing a co-ordinated approach to government on general mining issues. It also acts as a means of information for its members and other interested parties about technical, commercial and political developments.

Asia/Pacific

Australia

GOVERNMENT DEPARTMENTS AND OFFICIAL AGENCIES

AUSTRALIAN ATOMIC ENERGY COMMISSION
Address: Cliffbrook, 45 Beach Street, Coogee, New South Wales 2034
Telephone (02) 665 1221 Telex 20273

The Atomic Energy Commission is a statutory organisation under the general responsibility of the Department of National Development and Energy concerned with all aspects of the development, operation and use of nuclear power in Australia.

DEPARTMENT OF NATIONAL DEVELOPMENT AND ENERGY
Address: Tasman House, Hobart Place, Canberra City, ACT 2601
Telephone (062) 458211 Telex 62101

The Department of National Development and Energy is responsible for overseeing all aspects of energy development and supply, including the establishment of basic information on the country's resources, formulating policy in such areas as financing and exports, monitoring oil and gas industry operations and activity in research and development.

The Department includes the Bureau of Mineral Research, Geology and Geophysics. Other leading bodies linked to the Department include the National Energy Advisory Committee and the National Energy Research and Development Committee.

STATE ELECTRICITY COMMISSION OF QUEENSLAND
Address: Corner Warry Street and Gregory Terrace, Brisbane, Queensland 4000
Telephone (07) 253 9811 Telex 41772

The State Electricity Commission is responsible for planning and ensuring the proper development and co-ordination of the electricity supply industry in Queensland. Particular functions concern safety regulations, the establishing of tariffs, raising finance and administering all relevant legislation.

ENTERPRISES (PUBLIC AND PRIVATE SECTOR COMPANIES)

AAR LIMITED
Address: AMP Place, 10 Eagle Street, Brisbane, Queensland 4000
Telephone (07) 221 2366 Telex 42395
Energy Sectors: Oil, gas

Wholly owned subsidiary of CSR engaged in petroleum production and exploration for oil, gas and other minerals. AAR group companies are involved in drilling and exploration. Associated Pipelines Ltd, in which AAR has an 85 per cent interest, is responsible for operating the Roma-Brisbane natural gas trunk pipeline.

AAR itself holds a share of the Roma gasfield which amounted to 149 million cubic metres in the year 1980-81. AAR also handles around 300,000 barrels per day of Seram crude oil.

AMOCO AUSTRALIA LIMITED
Address: 201 Pacific Highway, North Sydney, New South Wales 2060
Telephone (02) 923 7509
Energy Sector: Oil

Amoco Australia is an affiliate of the United States based international oil company Standard Oil Company of Indiana. The State Government Insurance Office of Queensland has a 20 per cent interest in the company as a result of its investment in the expansion of Amoco's Australian refinery.

Amoco Australia operates a refinery on Bulwer Island near Brisbane, of capacity around two million tonnes per annum. The refinery forms the base for Amoco's marketing activities throughout the country.

AMPOL LIMITED
Address: 84 Pacific Highway, North Sydney, New South Wales 2060
Telephone (02) 929 6222
Gross Revenue: A$672.1 million (1980)
Total Assets: A$576.2 million
Energy Sector: Oil

Ampol is one of the leading Australian based oil companies with interests in exploration, production, refining and marketing. Substantial shares in Ampol are held by three companies, not otherwise primarily involved in the oil industry: Ansett Transport Industries (20 per cent), Pioneer Concrete Services (12 per cent) and Brambles Industries (12 per cent).

Ampol's marketing operations are widespread and backed up by its three million tonnes per annum refinery at Lytton, north-east of Brisbane. Petroleum products are distributed throughout the country. The company also has interests in other transport related activities.

The company has holdings in some important Australian energy development groups. It is a participant in WAPET which is producing crude oil from the Barrow Island field, off Western Australia, and has a nine per cent holding in Kathleen Investments which has a share of Mary Kathleen Uranium Ltd.

AUSTRALIAN GAS LIGHT COMPANY
Address: AGL Centre, Corner Pacific Highway and Walker Street, North Sydney, New South Wales 2060
Telephone (02) 922 0101
Gross Revenue: A$133.6 million (1980)
Total Assets: A$433.0 million
Energy Sector: Gas

Australian Gas Light Company (AGL) is the largest private enterprise public utility in Australia, distributing natural gas and LPG in New South Wales. AGL's sales of natural gas increased sharply in 1980 to 55,600 terajoules (equivalent to around 1,500 million cubic metres of gas). The company's LPG division supplies some 65,000 customers in New South Wales and Queensland and several towns gas systems based on LPG.

AGL's natural gas is supplied under long term contract from producing companies in the Cooper Basin area. The company itself holds a 12 per cent interest in Santos, the major member and operator of the Cooper Basin consortium. AGL is also involved with Bridge Oil and Cluff Oil (Australia) in exploration in the Surat/Eromanga basins of northern New South Wales.

BLAIR ATHOL COAL PROPRIETARY LIMITED
Address: 17th Floor, AMP Place, 10 Eagle Street, Brisbane, Queensland 4000
Telephone (07) 229 3900 Telex 40813
Energy Sector: Coal

Blair Athol Coal is owned by a consortium of companies, with a 62 per cent shareholding by CRA, the mining group in which RTZ has a majority interest. Also participating are Arco (Queensland) Coal Pty Ltd and the Electric Power Development Company of Japan.

The principal market for Blair Athol Coal is expected to be Japan, but other overseas exports are also likely. Shipments could begin in 1984 rising to a rate of five million tonnes per annum.

BOND CORPORATION HOLDINGS LIMITED
Address: International House, 26 St George's Terrace, Perth, Western Australia 6000
Telephone (09) 325 4555
Pre-Tax Profits: A$13.2 million (1981)
Total Assets: A$293.7 million
Energy Sectors: Oil, gas, coal

Bond Corporation is a diversified company with interests in minerals and energy developments, either directly or through its shareholdings in Santos, Endeavour Resources, Reef Oil, Basin Oil and Pacific Copper. A controlling interest is held by Mr Alan Bond.

A key element in the company's portfolio of energy interests is its effective participation in Cooper Basin production of oil and gas. It benefits particularly from the decision to commercialise the gas liquids available in the area. Santos, Reef Oil and Basin Oil are all involved, and as the project develops Bond Corporation will build up an eight per cent interest in gas liquids production.

BORAL LIMITED
Address: Boral House, 221 Miller Street, North Sydney, New South Wales 2060
Telephone (02) 920951 Telex 20702
Gross Revenue: A$608.1 million (1981)
Total Assets: A$585.5 million
Energy Sector: Gas

Boral is a diversified company with interests in quarrying, building materials, related engineering and other activities. The Energy Division of the company accounts for around ten per cent of total annual turnover.

Boral's principal energy interests lie in the supply and distribution of gas. It is a leading supplier of liquefied petroleum gas, for direct consumption, towns gas-making and as an automotive fuel. It now operates the first Australian flag LPG carrier for coastal operations and to back up its distribution network in the Pacific Islands.

Gas utility operations service some 100,000 domestic and industrial consumers in Brisbane, through its subsidiary, Brisbane Gas Company. Boral supplies numerous other centres in Queensland, using towns gas, natural gas and LPG. LPG deliveries are also made in New South Wales, ACT, Western Australia and Northern Territory.

BRIDGE OIL LIMITED
Address: CBA Centre, 60 Margaret Street, Sydney, New South Wales 2000
Telephone (02) 20 549 Telex 26518
Gross Profit: A$4.8 million (1981)
Total Assets: A$56.4 million (1981)
Energy Sectors: Oil, gas, coal

Bridge Oil is engaged in exploration for minerals and oil, with interests in fields producing oil and gas. Although concentrating on its extensive interests in mineral and petroleum tenements in Australia Bridge Oil is also participating in exploration work in the USA and Indonesia. Australian Metal Holdings holds 44 per cent of the shares of Bridge Oil and the balance is held by numerous investment funds.

The company's principal producing interests are in the Cooper Basin and the Surat Basin. Revenue is earned from gas production in the Cooper Basin, but Bridge Oil will benefit particularly from the project now under way to extract and market gas liquids from the area. The production of gas from the Surat Basin for supply to Brisbane Gas Co is being increased, involving Bridge Oil in association with Hartogen Energy, Offshore Oil and Moonie Oil.

Bridge Oil has a prospective coal development project at Alpha, in central Queensland, where proven reserves of steam coal are 1.25 billion tonnes. Bridge Oil is partnered by French state energy companies in this venture. The company is also validating coal deposits in New South Wales.

BRITISH PETROLEUM COMPANY OF AUSTRALIA
Address: 1 Albert Road, Melbourne, Victoria 3004
Telephone (03) 268 4111
Gross Revenue: A$1,942.6 million (1980)
Total Assets: A$1,546.8 million
Energy Sectors: Oil, coal, gas

British Petroleum Company of Australia is a wholly owned subsidiary of British Petroleum and the holding company for the group's energy interests in the country. These activities include oil refining and marketing, coal exploitation and participation in natural gas development off the north-west coast.

In the oil sector the BP group operates a refinery at Westernport near Fremantle, Western Australia. The refinery has a capacity of over five million tonnes per annum and underpins the company's oil product marketing operations throughout the country. Principal coal interest lies in Clutha Development Pty, which is wholly owned. Clutha operates seven underground mines in the Burragorang Valley and several other mines in New South Wales. The group also has a 49 per cent interest in the Clarence Mine, being developed in partnership with Oakbridge. The company also has a 16.7 per cent interest in the major offshore gas field which is to supply both markets in Western Australia and an export LNG plant.

BROKEN HILL PROPRIETARY COMPANY LIMITED
Address: BHP House, 140 William Street, Melbourne, Victoria 3000
Telephone (03) 60 0701
Gross Revenue: A$4,577.7 million (1981)
Total Assets: A$6,069.9 million
Energy Sectors: Oil, gas, coal

BHP is one of Australia's leading industrial companies, with broad interests in the production and treatment of minerals and energy and the manufacture of steel and steel products. Production of coal by the Steel Division was some 7.4 million tonnes in 1981, with a further 4.2 million tonnes extracted by the Minerals Division. The Oil and Gas Division handled around 185,000 barrels per day of crude oil and delivered 2,300 million cubic metres of natural gas. Important quantities of LPG (800,000 tonnes) and ethane were also produced.

BHP's prominence in oil and gas production stems from its 50/50 partnership with Esso in the offshore discoveries in the Gippsland Basin. Exploitation of these fields has been basic to the establishment of Australia's high level of independence from the international oil market. But BHP also has interests in Cooper Basin gas fields and the large field discovered off the north-west coast, which could lead to the export of liquefied natural gas.

The group's coal mining activities are centred on Queensland and New South Wales. Group mines at Gregory, Queensland are the main source of coking coal. Through its shareholding in Thiess Dampier Mitsui BHP has interests in the Moura and Kianga mines and in the Riverside development. Additionally the company is developing coal at Saxonvale (New South Wales) and holds a 50 per cent share in the Boggabri coal deposit.

CALTEX OIL (AUSTRALIA) PROPRIETARY LIMITED
Address: Caltex House, 167 Kent Street, Sydney, New South Wales 2000
Telephone (02) 20555
Energy Sector: Oil

Caltex Oil (Australia) is part of the Caltex Petroleum group of companies, and affiliated to the two major international oil companies Standard Oil Company of California and Texaco. Caltex Oil is principally involved in refining and marketing activities.

The company has an oil refinery at Kurnell, New South Wales, operated by its subsidiary company Australian Oil Refining Pty. This is currently the largest refinery in Australia, with a capacity of nearly seven million tonnes per annum. The refinery is a complex one, producing a full range of products for the company's marketing operations.

COAL AND ALLIED INDUSTRIES LIMITED
Address: Gold Fields House, 1 Alfred Street, Sydney, New South Wales 2000
Telephone (02) 27 86 41 Telex 21226
Gross Revenue: A$212.5 million (1981)
Total Assets: A$226.2 million
Energy Sector: Coal

Coal and Allied Industries (C&A) is engaged primarily in the extraction of coal from mines in New South Wales, but with other related interests in engineering and transportation. Major shareholders in C&A are the Howard Smith group, with 50 per cent, and CRA, the Australian arm of the international Rio Tinto-Zinc group, which holds just under 14 per cent.

Output from C&A's mines in the year 1980-81 totalled 5.6 million tonnes and total tonnage handled was 6.1 million tonnes. The company's Hunter Valley mine has a capacity of 2.5 million tonnes per annum and this capacity is to be doubled. A second mine is planned in the Hunter Valley, with reserves of some 350 million tonnes, most of which can be obtained by open-cut operation. The company has several other mines, including deep-mining operations and has part holdings in other joint-venture mines, principally a 50 per cent interest in Hebden Mining Company, which supplies 1.5-2.0 million tonnes per annum to the Electricity Commission of New South Wales.

The larger part of C&A's sales, 4.4 million tonnes in 1980-81, is exported. Coking coal is sold to steel mills in Japan. Steam coal is sold to both Japan and Western European countries.

CRA LIMITED
Address: 55 Collins Street, Melbourne, Victoria 3001
Telephone (03) 658 3333
Gross Revenue: A$1,828.1 million (1981)
Total Assets: A$3,951.8 million
Energy Sectors: Coal, uranium

CRA is a major Australian minerals exploration and production company. Although currently owned 61 per cent by the Rio Tinto-Zinc Corporation the parent company has agreed to move towards a position of 51 per cent public (Australian) ownership. Group interests include processing, smelting and fabrication, and although concentrated in Australia, also extend to other parts of the Far East/Pacific area as well as to the USA and Europe.

CRA has varied interests in metals and minerals including base and precious metals. Energy interests lie in the Mary Kathleen uranium deposit and in several important coal developments now under way in New South Wales and Queensland. These include the operations of Kembla Coal & Coke Pty, a 62 per cent share in Blair Athol Coal Pty, deposits at Tarong (Queensland) and a 25 per cent interest in the Hail Creek development, also in Queensland.

CSR LIMITED
Address: 1-7 O'Connell Street, Sydney, New South Wales 2000
Telephone (02) 237 5231 Telex 20285
Gross Revenue: A$1,827.8 million (1981)
Total Assets: A$2,298.1 million
Energy Sectors: Oil, gas, coal, uranium

CSR is one of the largest Australian industrial companies with extensive involvement in natural resources. Apart from its traditional base in sugar it has developed interests in iron ore, alumina, chemicals and other minerals. Group companies are active in coal mining, oil and gas production and energy transportation. The Energy Division contributed 14 per cent of group profit in 1981, but accounted for half of CSR's investment in that year.

CSR has interests via AAR in the production and transportation of oil (Seram crude oil) and natural gas (from the Roma field). But of increasing importance is the company's involvement in coal production for both Australian and export markets. Group output of coal rose from less than one million tonnes in 1976 to 7.4 million tonnes in 1981. Numerous new projects are planned and under discussion, to develop coal deposits at several locations in Queensland and also in New South Wales and Western Australia.

Subject to final government approval it is planned to construct a pilot plant for uranium production at Honeymoon. CSR holds a 26 per cent interest and is exploring for uranium in most of the Australian states.

ELECTRICITY TRUST OF SOUTH AUSTRALIA
Address: 220 Greenhill Road, Eastwood,
South Australia 5063
Telephone (08) 223 0383 Telex 88655
Gross Revenue: A$214.1 million (1980)
Energy Sectors: Electricity, coal

The Electricity Trust of South Australia operates under several statutes and is responsible for supplying almost all of the State's electricity requirements. Its output in 1980 amounted to 95 per cent of all electricity supplies, and 99 per cent of electricity in the public supply system. The Trust supplies directly or indirectly to virtually all households in the state.

The Trust has a total generating capacity of some 1,890 MW, operating on a wide range of fuels, including oil, natural gas and coal. Natural gas from the Cooper Basin is used in two power stations at Adelaide. At Port Augusta coal is burned, which is produced at the Trust's own mine at Leigh Creek, 270 km away. Other stations use fuel oil or diesel.

ENDEAVOUR RESOURCES LIMITED
Address: 136 Exhibition Street, Melbourne,
Victoria 3000
Telephone (03) 654 3377
Pre-tax Operating Profit: A$0.6 million (1981)
Total Assets: A$103.9 million
Energy Sectors: Oil, gas, coal

Endeavour Resources has varied interests in mineral and energy prospects and is actively engaged in exploration in Australia and overseas, including Egypt, Indonesia, the Philippines and USA. The Bond group has a substantial shareholding in the company (39 per cent) and Endeavour has direct or indirect interests in Santos, Reef Oil, Basin Oil and Pacific Copper.

The company's principal energy producing operation is Rhondda Collieries near Ipswich, Queensland. In 1981 some 1.3 million tonnes of coal was mined and supplied to the Queensland Electricity Generating Board. Prospects for sharing in oil or gas production currently centre on developments off the north-west coast and in the Surat and Cooper Basins.

ENERGY AUTHORITY OF NEW SOUTH WALES
Address: 1 Castlereagh Street, Sydney,
New South Wales 2000
Telephone (02) 239 0311
Gross Revenue: A$795.8 million (1981)
Total Assets: A$2,083.7 million
Energy Sector: Electricity

The Energy Authority of New South Wales combines the activities of the former Electricity Commission with the additional responsibilities for regulation and use of energy and energy resources arising from the Energy Authority Act of 1976. The Commission is the principal generating authority in New South Wales, supplying in bulk to a number of local authority distribution undertakings and some industrial consumers.

The Commission operates most of the state's 7,800 MW of generating capacity, mainly from thermal power stations fuelled by black coal produced in the state. It is also entitled to two-thirds of the output from the Snowy Mountains scheme after Commonwealth Territory requirements have been met.

ESSO AUSTRALIA
Address: Esso House, 127 Kent Street, Sydney,
New South Wales
Telephone (02) 236 2911
Energy Sectors: Oil, gas

Esso Australia is a subsidiary of the major international oil company Exxon Corporation. The company has important interests in oil and gas production and is a leading refining and marketing company.

Production interests arise primarily from the prolific discoveries in the Bass Strait/Gippsland Basin area, made in partnership with Broken Hill Proprietary. The company has a 35 per cent holding in Petroleum Refineries (Australia) which operates refineries at Altona, Victoria, and Port Stanvac, South Australia. Participation at these refineries provides Esso with some 2.5-3.0 million tonnes per annum of oil products for its distribution and marketing operations.

GAS AND FUEL CORPORATION OF VICTORIA
Address: 171 Flinders Street, Melbourne,
Victoria 3000
Telephone (03) 63 0391
Gross Revenue: A$319.3 million (1981)
Energy Sector: Gas

The Gas and Fuel Corporation of Victoria operates under statute of the state parliament. The state holds 96 per cent of the shares, with the balance held by the public. The Corporation's principal responsibility is the general distribution of gas throughout the state.

Natural gas is received from the Esso/BHP group at its Longford terminal. In 1980-81 deliveries totalled 186 million gigajoules, a 25 per cent increase on the previous year, backing up the Corporation's programme of extending the natural gas network.

GRIFFIN COAL MINING COMPANY
Address: 24 Kings Park Road, West Perth, Western Australia 6005
Telephone (09) 322 5022 Telex 93751
Gross Revenue: A$34.9 million (1981)
Total Assets: A$41.2 million
Energy Sector: Coal

Griffin Coal Mining Company owns open cut mining operations in the Collie Basin of Western Australia and has a programme of further development in that area, with interests in exploring for coal and other energy resources in several other parts of the country. Seeko Industries holds a 30 per cent interest in the company.

Open cut mining operations are now producing over two million tonnes per annum, with output destined for the Muja power station of the Western Australian State Energy Commission. The power station is being further expanded and additional workings of coal in the area are being undertaken.

Griffin Coal Mining has some interests in oil and gas exploration through its 49 per cent holding in North West Mining NL, one of the participant companies in Strata Oil.

HARTOGEN ENERGY LIMITED
Address: Hartogen House, 15 Young Street, Sydney, New South Wales 2000
Telephone (02) 27 21 21 Telex 22481
Operating Profit (Pre-Tax): A$2.9 million (1980)
Total Assets: A$26.0 million
Energy Sectors: Oil, gas, coal

Hartogen Energy is involved in exploration for oil, gas and coal in several of the main areas of interest in Australia, including the Bonaparte Basin off the north-west coast. The company has production of oil, gas and coal. Almost 50 per cent of the shares of Hartogen Energy are held by the Genoa Oil group, with which it is associated in several exploration areas. The company has undertaken a rapid increase in exploration since 1978.

A major part of Hartogen's revenue derives from its 50 per cent interest in the Wambo Colliery in the Hunter Valley, New South Wales. Output of coking/steam coal is around one million tonnes per annum. Wambo Mining Corporation itself holds a 20 per cent interest in other deposits in the Hunter Valley, which may be developed in conjunction with Miners Federation and Agip Australia.

Production of gas, and to a limited extent of oil, is obtained from the Surat Basin, Queensland. This availability is the basis for Hartogen's contract for supplying five billion cubic feet per annum to Consolidated Fertilizers Ltd in Brisbane.

HYDRO-ELECTRIC COMMISSION
Address: 4-16 Elizabeth Street, Hobart, Tasmania 7000
Telephone 30 1101
Gross Revenue: A$113.8 million (1980)
Energy Sector: Electricity

The Hydro-Electric Commission is organised under statute and is solely responsible for generation, transmission and retail distribution in the state. In 1980 total production amounted to 7.8 TWh.

The Commission's system is based almost entirely on hydro-electric generating plant. The Commission's 22 plants range in capacity from 10 MW to 288 MW. A 240 MW oil-fired station is maintained at Bell Bay to supply peak load requirements and compensate for any disruptions to the hydro-electric system.

KEMBLA COAL AND COKE PROPRIETARY LIMITED
Address: Level 4, Crown Central, Corner Crown and Keira Streets, Wollongong, New South Wales 2500
Telephone (042) 2874 55 Telex 29172
Gross Revenue: A$140.6 million (1980)
Energy Sector: Coal

Kembla Coal & Coke Pty (KCC) is a wholly owned subsidiary of CRA, the mining exploration and production company. KCC production in 1980 totalled 2.75 million tonnes of coal and nearly three million tonnes of coking coal.

Coking coal is sold to Australian Iron & Steel Pty at Port Kembla and also exported. Overseas markets are being further expanded with initial deliveries to South Korea and Pakistan, while the long term contract to supply China Steel (Taiwan) has been increased.

MARY KATHLEEN URANIUM LIMITED
Address: 95 Collins Street, Melbourne, Victoria 3000
Telephone (03) 63 0491
Gross Revenue: A$70.3 million (1980)
Total Assets: A$67.0 million
Energy Sector: Uranium

Mary Kathleen Uranium (MKU) is the operating company exploiting uranium deposits at Mary Kathleen, Queensland. Mining operations recommenced only in 1976 after being shut down for some 13 years. The mine was open from 1958 to 1963 for the supply of uranium oxide to the UK.

MKU is owned 51 per cent by CRA, in which the RTZ group has a majority interest. But the Australian Atomic Energy Commission holds 41.6 per cent indicating the depth of state involvement in uranium development. The balance of the shares is held by Kathleen Investments (Australia) 2.6 per cent and the Australian public 4.8 per cent.

Output of uranium oxide amounted to some 835 tonnes in 1980. Material is supplied under long term contract for processing to nuclear fuel.

MEEKATHARRA MINERALS LIMITED
Address: 175 Pitt Street, Sydney,
New South Wales 2000
Telephone (02) 232 3188 Telex 70913
Total Assets: A$2.4 million (1981)
Energy Sector: Coal

Meekatharra Minerals has been built up as an energy resource exploration and development company, holding permits in New South Wales, Tasmania and South Australia. A 44 per cent shareholding is held by Mr Donald O'Callaghan.

In 1980 the company discovered a major deposit of sub-bituminous black coal in the Arckaringa Basin of South Australia. The area over which Meekatharra Minerals has a 50 or 100 per cent interest extends to some 14,000 square kilometers. The deposit is being evaluated as steaming coal and for its suitability for liquefaction.

MIM HOLDINGS LIMITED
Address: MIM Building, 160 Ann Street,
Brisbane, Queensland 4000
Telephone (07) 228 1122 Telex 40160
Gross Revenue: A$721.4 million (1981)
Total Assets: A$1,600.1 million
Energy Sector: Coal

MIM Holdings is a major Australian based mining company, built upon the extensive operations in lead-zinc-silver ore extraction at Mount Isa, Queensland. A major shareholding is held by the United States minerals company ASARCO. ASARCO's holding is being reduced from 49 per cent to 44 per cent in order to permit MIM Holdings to become more involved in resource development in Australia.

The main new area of development of MIM Holdings is in coal mining. Production of coal at Collinsville amounted to 1.2 million tonnes in 1980, but major developments are in hand to expand the capacity of Collinsville and exploit steam coal at Newlands and coking coal at Oaky Creek. The Newlands project is intended to provide exports of up to four million tonnes per annum by 1985. The Oaky Creek Project, in which MIM Holdings has a 78 per cent interest, is designed to export three million tonnes per annum.

MOBIL OIL AUSTRALIA LIMITED
Address: 2 City Road, South Melbourne,
Victoria 3001
Telephone (03) 62 02 31
Gross Revenue: A$1,629.9 million (1980)
Total Assets: A$822.9 million
Energy Sectors: Oil

Mobil Oil Australia, a subsidiary company of the major international oil company, is a leading distributor of oil products throughout Australia. The company holds a 65 per cent interest in Petroleum Refineries (Australia) Pty (PRA), which operates refineries at Altona, Victoria, and Port Stanvac, South Australia. Combined capacity of the refineries is more than eight million tonnes per annum.

Mobil does not produce oil in Australia. It is now actively exploring for oil or gas, particularly in Western Australia. Exploration is also directed towards coal and other minerals, including uranium.

MOONIE OIL COMPANY LIMITED
Address: 229 Robinson Road East, Geebung,
Queensland 4034
Telephone (07) 265 1999 Telex 40265
Gross Revenue: A$22.7 million (1981)
Total Assets: A$57.1 million
Energy Sectors: Oil, gas, coal, uranium

Moonie Oil Company operates the Moonie oilfield in Queensland, and is increasingly engaged in exploration and development work for oil, gas and coal in several parts of the country. Major shareholdings are held by Lowell Pty, with 27 per cent, and Paribas Investments Australia, with 22 per cent. Moonie Oil also has interests in gas, coal and uranium through its 23 per cent shareholdings in the Oilmin group and Flinders Petroleum.

The major source of group income is the Moonie oilfield, where output declined to only 350,000 barrels in 1981, but the company has other interests in oil and gas fields in the Surat Basin, including the Alton oil field, the Boxleigh and Kincora gas/condensate fields, and Thomby Creek oil field. A minor interest is held indirectly in the Palm Valley gas field, which is to supply an average 65 million cubic feet per annum to the Northern Territory Electricity Commission. Uranium interests are held by the Oilmin group which has a 50 per cent holding in the Beverley uranium deposits in South Australia.

NORTHERN TERRITORY ELECTRICITY COMMISSION
Address: 2nd Floor, CML Building,
59 Smith Street, Darwin, NT 5794
Telephone 80 8511 Telex 85395
Gross Revenue: A$25.1 million (1980)
Energy Sector: Electricity

The Northern Territory Electricity Commission is generally responsible for the public supply of electricity throughout the Northern Territory. It was established as a statutory authority only in 1978, with responsibilities concerning safety, operations, transmission and distribution and advice on tariffs to the state government.

As the population of the Territory is relatively small and widely distributed, there is no inter-connected grid. Instead, each centre has its own local generation facilities and distribution system. Largest plant is 140 MW at Darwin. Diesel and gas turbine units are widely used.

OAKBRIDGE LIMITED
Address: Oakbridge Building, 52 Phillip Street, Sydney, New South Wales 2000
Telephone (02) 241 1231 Telex 21549
Gross Revenue: A$107.7 million (1981)
Total Sales: A$165.5 million
Energy Sector: Coal

Oakbridge is a diversified company with particular interests in coal and mineral development. The company owns or has interests in several mines in New South Wales, including the Clarence mine which is being developed in partnership with British Petroleum's Australian subsidiary.

Production of saleable coal from all mines was just under two million tonnes in 1980-81. Sales by the group were 1.7 million tonnes, accounting for half of total revenue. A major programme is being undertaken to raise mine capacity to 5.5 million tonnes per annum, exporting steam coal to numerous countries in the Far East and Europe.

QUEENSLAND ELECTRICITY GENERATING BOARD
Address: 255 Adelaide Street, Brisbane, Queensland 4000
Telephone (07) 228 2111
Energy Sector: Electricity

The Electricity Generating Board is a public authority with primary responsibility for electricity generation and bulk transmission in Queensland. It operates power stations and main transmission lines in an inter-connected grid from Cooktown in the north to the New South Wales border and centres in the west of the state, such as Julia Creek, Clermont and Dulacca. Electricity is supplied in bulk to the seven boards which distribute at retail level.

The Board's generating capacity of over 3,000 MW consists almost entirely of thermal power stations fuelled with hard coal produced within the state. A small amount of hydro-electric capacity is operated in the north of Queensland.

SANTOS LIMITED
Address: 183 Melbourne Street, North Adelaide, South Australia 5006
Telephone (08) 267 5000 Telex 82716
Gross Revenue: A$50.2 million (1981)
Total Assets: A$161.8 million
Energy Sector: Gas

Santos has emerged as one of the leading Australian based energy companies with important interests in the production of natural gas and a major exploration programme.

Santos is a 45 per cent participant in the Moomba gas field in the Cooper Basin, which supplies large volumes of gas to Adelaide and Sydney. Output from Moomba reached 355 million cubic feet per day in 1980. From 1983 gas liquids extraction is expected to amount to some 25,000 barrels per day. The Santos share of recoverable reserves of gas and gas liquids at end 1980 was estimated at 40 billion cubic metres and 121 million barrels respectively.

Exploration for petroleum is continuing throughout the Cooper Basin and to this is added the company's interests in coal and mineral exploration at various locations.

SHELL AUSTRALIA LIMITED
Address: 155 William Street, PO Box 872K, Melbourne, Victoria 3001
Telephone (03) 609 1711 Telex 30560
Gross Revenue: A$2,489.7 million (1980)
Total Assets: A$2,026.2 million
Energy Sectors: Oil, coal, gas

Shell Australia is the principal affiliate company in Australia of the Royal Dutch/Shell group, with subsidiary and associate companies involved in exploration for oil, coal, gas and minerals, oil refining and marketing and coal production.

The company's petroleum operations are based on refineries at Clyde, New South Wales, and Geelong, Victoria. The Clyde refinery is the oldest refinery in the country, and has a throughput capacity of three million tonnes per annum, but the Geelong refinery is substantially larger, with an annual capacity of five million tonnes. Shell companies market the full range of finished products. The group has a 19 per cent shareholding in the North West Shelf gas project and is a participant in WAPET which is involved in exploration and production of oil and gas in Western Australia. Through WAPET Shell has an interest in the Barrow Island oilfield.

Shell has interests in several coal producing areas. It has major shareholdings in Austen & Butta and Bellambi Coal Company, and is participating in potential development projects in New South Wales, Queensland, Victoria, Western Australia and Tasmania.

H C SLEIGH LIMITED
Address: 160 Queen Street, Melbourne, Victoria 3001
Telephone (03) 609 6222 Telex 30358
Gross Revenue: A$727.3 million (1981)
Total Assets: A$274.0 million
Energy Sector: Coal

H C Sleigh is engaged in a diverse range of activities including mining, forestry and construction. It also has interests in transportation and finance with a long established role in many trading activities.

The firm was for many years involved in petroleum marketing, but in 1981 disposed of its subsidiary Golden Fleece to Caltex. Sleigh's primary involvement in the energy sector now lies in its 40 per cent share in the Warkworth mine (New South Wales), which commenced operations in 1981. The mine will supply steam and coking coal for export to Japan, as well as around 250,000 tonnes per annum to the Electricity Commission of New South Wales.

SNOWY MOUNTAINS HYDRO-ELECTRIC AUTHORITY
Address: Monaro Highway, Cooma North, New South Wales 2630
Telephone (0648) 21777 Telex 61025
Energy Sector: Electricity

The Snowy Mountains Authority is a statutory corporation of the Australian government formed to develop the water resources of the Snowy Mountain area for power generation and for irrigation in the Murray and Murrumbidgee Valleys. Total capacity of the scheme is 3,740 MW, which is used to supply the needs of the Capital Territory, with the balance utilised by the states of New South Wales (two-thirds) and Victoria (one-third).

The Authority operates under the supervision of the Snowy Mountains Council, which consists of representatives of the Commonwealth government, the Authority and the states of New South Wales and Victoria.

SOUTH AUSTRALIAN GAS COMPANY
Address: 35 Waymouth Street, Adelaide, South Australia 5000
Telephone (08) 510141 Telex 28917
Gross Revenue: A$61.3 million (1980)
Energy Sector: Gas

The South Australian Gas Company is incorporated by act of parliament of South Australia, with the exclusive right to supply mains gas throughout the state. Its principal source of natural gas is the Cooper Basin in the northern part of South Australia, but it also distributes liquefied petroleum gas widely throughout the state.

STATE ELECTRICITY COMMISSION OF VICTORIA
Address: 15 William Street, Melbourne, Victoria 3000
Telephone (03) 615 0433
Gross Revenue: A$739.2 million (1980)
Energy Sectors: Electricity, coal

The State Electricity Commission (SEC) is organised under statute with the authority to build and operate power stations and supply electricity throughout the state. It is also permitted to develop and exploit coal deposits to provide fuel for the power stations. Electricity is sold to regional and municipal undertakings for retail distribution.

The SEC operates several large coal fired power stations, including Hazelwood (1,600 MW) and two stations at Yallourn of total capacity 1,250 MW. In total the Commission has generating capacity of over 3,000 MW, almost all of which is based on coal. The SEC is thus also a major coal producer and consumer, using up to 30 million tonnes per annum. In addition it is entitled to one-third of the production of the Snowy Mountains Hydro-Electric Authority, after meeting the requirements of the Capital Territory.

STATE ENERGY COMMISSION OF WESTERN AUSTRALIA
Address: 365 Wellington Street, Perth, Western Australia 6000
Telephone (09) 326 4194 Telex 92674
Gross Revenue: A$307.7 million (1981)
Total Assets: A$856.5 million
Energy Sectors: Electricity, gas

The State Energy Commission was formed in 1975 combining the activities of the former State Electricity Commission and the Fuel and Power Commission, with a view to better co-ordination and utilisation of energy resources in Western Australia. The SEC accordingly has a wide range of responsibilities, supplying electricity throughout the state, distributing gas from the Dongara field and absorbing the major part of coal production.

The SEC has power stations at Muja, Kwinana, South Fremantle and Bunbury feeding power into the grid covering the metropolitan area and the south-west of the state. The Commission also operates 36 isolated power stations servicing individual localities.

The SEC supplies natural gas to more than 100,000 customers in the metropolitan area and to the Fremantle Gas and Coke Company. This system will expand when North West Shelf gas becomes available. The SEC will be responsible for transporting, distributing and marketing the natural gas. In total the SEC will purchase around 3.8 billion cubic metres per annum.

STRATA OIL NL
Address: 24 Kings Park Road, West Perth, Western Australia 6005
Telephone (09) 322 5022 Telex 94306
Total Assets: A$11.9 million (1981)
Energy Sectors: Oil, gas

Strata Oil is an exploration company holding permits for onshore and offshore exploration, particularly concentrating on oil and gas. The company now includes the petroleum exploration interests formerly held by Haoma Gold Mines NL and North West Mining NL. Substantial shareholdings in Strata Oil are held by these companies, and also by investment funds.

Exploration is being undertaken in Western Australia and Queensland and off the north coast. The company has a 27 per cent interest in the Woodada natural gas field to the north of Perth. Partial development of the field has taken place, which may be used to supply Perth prior to the arrival of North West Shelf gas in the mid 1980s.

THIESS DAMPIER MITSUI COAL PROPRIETARY LIMITED
Address: T & G Building, Queen and Albert Streets, Brisbane, Queensland 4000
Telephone (07) 221 5222 Telex 40706
Energy Sector: Coal

Thiess Dampier Mitsui (TDM) is a consortium company organised to develop coal deposits particularly with a view to supplying export markets. The major shareholding (68 per cent) is held by BHP, with a further 22 per cent in the hands of CSR. The balance of 20 per cent is held by the Japanese trading company Mitsui.

The Moura and Kiangi mines in Queensland are currently in operation with a total output in 1981 of over 2.4 million tonnes. Total production by TDM will more than double when a proposed development of coking coal at Riverside, in the Nebo area west of Mackay, Queensland, is brought on stream. Output from Riverside, which is destined mainly to supply steel mills in Japan, is expected to average well over three million tonnes per annum.

THIESS HOLDINGS LIMITED
Address: 146 Kerry Road, Archerfield, Queensland 4108
Telephone (07) 277 3011 Telex 40169
Energy Sector: Coal

One of the principal holding companies within the CSR group, providing civil engineering, oil industry services, transportation and related activities. Thiess Holdings is the operating arm of CSR responsible for coal production at Callide and South Blackwater, Queensland. In 1981 output at Callide was 2.4 million tonnes and will be greatly increased to supply the requirements of Queensland Alumina Ltd and the Queensland electricity generating companies. The South Blackwater mine produced 1.4 million tonnes in 1981, mostly coking coal for export.

Through Thiess Holdings the CSR group holds a 22 per cent interest in Thiess Dampier Mitsui Coal Pty (TDM). TDM is to develop a coking coal extraction scheme at Riverside, Queensland, with output destined for Japanese steel mills.

UTAH DEVELOPMENT COMPANY
Address: 56 Pitt Street, Sydney, New South Wales 2000
Gross Revenue: A$663.0 million (1980)
Energy Sector: Coal

Utah Development, in which General Electric Company of the United States holds an 89 per cent interest, is operator for Central Queensland Coal Associates. Utah development has a 76 per cent participation in this group. The balance is held by Mitsubishi Development Pty and AMP Society, with a small share held directly by UMAL Consolidated. UMAL also has a 10 per cent interest in Utah Development itself.

Shipments of coal from CQCA mines in 1980 were 16.4 million tonnes of coking grades from mines at Goonyella, Peak Downs, Norwich Park, Saraji and Blackwater.

WEST AUSTRALIAN PETROLEUM PTY LIMITED
Address: 12 St George's Terrace, Perth, Western Australia
Telephone (09) 325 0181 Telex 92008
Energy Sectors: Oil, gas

WAPET is a joint-venture company engaged in exploration for, and production of, oil and gas in Western Australia. Participants include three major international oil companies, Shell, Standard Oil Company of California and Texaco each with a holding of just under 29 per cent. Ampol Petroleum, the only Australian based group with refining capacity in the country, holds the balance of 14 per cent.

The WAPET group has been responsible for developing the oilfield off Barrow Island, one of the earliest producing fields in the country, and the Dongara gasfield north of Perth. Production at Barrow Island amounted to 28,000 barrels per day in 1980, an increase on the previous year. Gas sales from the Dongara field were also higher, at 900 million cubic metres.

WESTERN MINING CORPORATION HOLDINGS LIMITED
Address: 360 Collins Street, Melbourne, Victoria 3000
Telephone (03) 602 0300
Gross Revenue: A$303.6 million (1981)
Total Assets: A$1,024.2 million
Energy Sectors: Coal, uranium

The Western Mining group is one of Australia's largest home-based companies, with broad involvement in metals and minerals extraction and processing. Through BH South, in which WMC has an 80 per cent interest, the group has holdings in Alcoa of Australia and CRA.

The WMC group has been building up holdings in energy projects and is involved with Shell, BP, Esso and Urangesellschaft in the exploitation of coal and uranium. WMC holds majority interests in the Olympic Dam project at Roxby Downs, South Australia, where uranium is to be produced in association with gold and copper, and at Yeelirrie, Western Australia, where uranium production is expected to start in 1986. WMC also has a wholly owned brown coal prospect near Kingston, South Australia, which could be developed to supply fuel for electricity generation.

WOODSIDE PETROLEUM LIMITED
Address: 459 Collins Street, Melbourne, Victoria 3000
Telephone (03) 62 59 51
Energy Sector: Gas

Woodside Petroleum is the operating company for the development of major offshore gas reserves in the North Rankin field, off Western Australia, which are to be used both in Western Australia itself and as the basis for a liquefaction plant to export LNG to Japan.

Shareholdings in Woodside Petroleum include most of the leading oil companies active in Australia. A 43 per cent interest is held by North West Shelf Development Pty, which is a partnership of Shell and BHP. British Petroleum, Standard Oil Company of California and Texaco, the last two through a Caltex company, hold most of the remaining shares.

PROFESSIONAL INSTITUTIONS AND TRADE ASSOCIATIONS

AUSTRALIAN COAL ASSOCIATION
Address: 2nd Floor, National Mutual Centre, 44 Market Street, Sydney, New South Wales 2000
Telephone (02) 29 7202 Telex 27470
Energy Sector: Coal

The Australian Coal Association is an organisation formed to represent the interests of the coal mining industry acting as a forum for some 70 member companies in dealings with government and associated industries.

AUSTRALIAN GAS ASSOCIATION
Address: Gas Industry House, 7 Moore Street, Canberra City, ACT 2601
Telephone (062) 47 3955 Telex 62137
Energy Sector: Gas

The Australian Gas Association was established to bring together companies and undertakings of the gas industry in order to promote its general development. It also provides a co-ordinating role for some technical and operational aspects of the industry.

AUSTRALIAN INSTITUTE OF PETROLEUM LIMITED
Address: Wales Corner, 227 Collins Street, Melbourne, Victoria 3000
Telephone (03) 632756 Telex 33421
Energy Sectors: Oil, gas

The Australian Institute of Petroleum is a representative organisation for the oil industry and its personnel. It acts as a source of information for the industry and about the industry, publishing statistical and other information on oil and gas production, refining, marketing and related data. The Institute produces regular bulletins and information pamphlets, notably the Petroleum Gazette and Oil and Australia.

AUSTRALIAN PETROLEUM EXPLORATION ASSOCIATION
Address: Suite 807, London Assurance House, 20 Bridge Street, Sydney, New South Wales 2000
Telephone (7) 27 6718 Telex 23033
Energy Sectors: Oil, gas

The Australian Petroleum Exploration Association (APEA) was formed to bring together companies involved in all phases of exploration and development of oil and gas, including exploration, drilling, transportation and refining. The APEA also fulfils a training role in providing advanced technical courses and holding conferences and seminars.

ELECTRICITY SUPPLY ASSOCIATION OF AUSTRALIA
Address: Swann House, 22 William Street, Melbourne, Victoria 3000
Telephone (03) 62 4641
Energy Sector: Electricity

The Electricity Supply Association is a voluntary body comprising the various state regulatory authorities, the major generating, transmitting and distribution authorities, together with several Commonwealth authorities. The Association's function is to provide an interchange of information and the agency for joint action on matters affecting the electricity supply industry generally.

The Association has a number of standing committees, dealing with: generation; transmission and distribution; financial management; marketing, service and load development; human resources. It also prepares statistical and other information on a country-wide basis.

Bangladesh

GOVERNMENT DEPARTMENTS AND OFFICIAL AGENCIES

MINISTRY OF PETROLEUM AND MINERAL RESOURCES
Address: Bangladesh Secretariat, Dacca

The Ministry of Petroleum and Mineral Resources has wide responsibilities for the development and supply of oil and gas, and the formulation of government policy on energy.

The Ministry is closely involved in the oil and gas industries through key state enterprises which are responsible for exploration, production, oil refining, importing and marketing. Reporting to the Ministry are the Bangladesh Petroleum Corporation and the Bangladesh Oil and Gas Corporation.

ENTERPRISES (PUBLIC AND PRIVATE SECTOR COMPANIES)

BANGLADESH OIL AND GAS CORPORATION
Address: Chambers Building, 122-124 Motijheel Commercial Area, Dacca 2
Telephone 24 57 17 Telex 725
Energy Sectors: Oil, gas

The Bangladesh Oil and Gas Corporation (Petrobangla) is the principal state undertaking for the gas sector, responsible for exploiting the known deposits of gas and exploring for further reserves. Already 10 fields have been discovered. The Titas field is producing 80 million cubic feet per day and a large field at Bakhrabad has not yet been developed.

Operational activities in the production, transmission and distribution of gas are mainly undertaken by two subsidiaries of Petrobangla, the Titas Gas Transmission and Distribution Company and Jalalabad Gas Distribution System.

BANGLADESH PETROLEUM CORPORATION
Address: 28 Bandabanghu Avenue, GPO 2003, Dacca 2
Telephone 25 00 72/3 Telex 65607
Energy Sector: Oil

Bangladesh Petroleum Corporation is a state owned company concerned with the importation of oil, oil refining and marketing of finished products. The Corporation holds the state's interest in the Chittagong refinery of Eastern Refinery Ltd and is responsible for organising or controlling imports.

The Corporation has been responsible for arranging for processing at Singapore of additional crude oil to supplement the output of ERL.

BANGLADESH POWER DEVELOPMENT BOARD
Address: WAPDA Building, Motijheel Commercial Area, Dacca,
Telex 5617
Energy Sector: Electricity

The Power Development Board is a state undertaking responsible for ensuring an adequate supply of electricity in Bangladesh in co-operation with the Rural Electrification Board.

Total electricity generated by the BPDB in 1979-80 was over 2,100 GWh mainly from thermal power stations. These are increasingly using natural gas as fuel. A part of the Board's electricity is derived from the Karnafuli hydroelectric power station.

BURMAH EASTERN LIMITED
Address: P O Box 4, Strand Road, Chittagong
Telephone 81851/3 Telex 65756
Energy Sector: Oil

Burmah Eastern is one of the leading marketers of petroleum products in Bangladesh. The company is an affiliate of Burmah Oil Company of the UK, which has a 49 per cent shareholding, the balance is held by the State. Burmah Oil also holds a 30 per cent interest in Eastern Refinery Ltd, which operates the refinery at Chittagong.

Total sales of all products in 1979-80 amounted to 570,000 tonnes or nearly 39 per cent of the market. Burmah Eastern is sole supplier of certain special products including aviation fuel and naphtha and is also strong in the market for motor gasoline.

EASTERN REFINERY LIMITED
Address: PO Box 35, North Patenga, Chittagong
Telephone 25 15 21
Energy Sector: Oil

Eastern Refinery Ltd (ERL) is the operating company for the oil refinery at Chittagong. The State holds a majority (70 per cent) interest in ERL. The remaining 30 per cent is held by Burmah Oil Company.

The Chittagong refinery is the only refinery in Bangladesh. Capacity is around 1.5 million tonnes per annum. The refinery is operated virtually at full capacity supplying products to the main marketing companies.

JALALABAD GAS DISTRIBUTION SYSTEM
Address: Chambers Building, 122-124 Motijheel Commercial Area, Dacca 2
Telephone 24 57 17
Energy Sector: Gas

Jalalabad Gas Distribution System is an operating division of Bangladesh Oil and Gas Corporation responsible for supplying gas from the Sylhet gas field to Sylhet and adjoining areas. Gas is also piped to the Habiganj area.

Output of gas from the Sylhet field in 1979-80 was around 7,000 million cubic feet. A high proportion of this gas was sold to the fertiliser factory at Fenchuganj.

JAMUNA OIL COMPANY LIMITED
Address: Amin Court, 62/63 Motijheel Commercial Area, Dacca 2
Telephone 24 66 58 Telex 765
Gross Revenue: Taka 1,455.2 million (1980)
Total Assets: Taka 424.6 million
Energy Sector: Oil

Jamuna Oil Company is one of the leading distributors and marketers of oil products in Bangladesh, accounting for one-third of total product sales. The Company has above average sales to power stations and caters for a large part of the international bunker oil trade.

TITAS GAS TRANSMISSION AND DISTRIBUTION COMPANY
Address: Chambers Building, 122-124 Motijheel Commercial Area, Dacca 2
Telephone 24 57 17 Telex 725
Energy Sector: Gas

Titas Gas Transmission and Distribution Company is a wholly owned subsidiary of Bangladesh Oil and Gas Corporation, responsible for operating the gas supply system of the Titas gas field.

Output from the Titas field in 1979-80 was some 29,000 million cubic feet, which was transported to consumers in the Dacca area. The major proportion of gas was consumed in power stations, a fertiliser plant and by other industrial users. The company's 115,000 commercial and residential customers account for only a small proportion of total sales.

Brunei

ENTERPRISES (PUBLIC AND PRIVATE SECTOR COMPANIES)

BRUNEI SHELL PETROLEUM COMPANY LIMITED
Address: Seria, State of Brunei
Telex 3313
Energy Sectors: Oil, gas

Brunei Shell Petroleum Company is an associate company of the Royal Dutch/Shell group engaged in exploration and production of oil and gas in Brunei. The State of Brunei holds a 50 per cent interest.

Brunei Shell Petroleum produces crude oil from the Seria oil field, which is transported to the export terminals of Lutong and Miri in Sarawak. The company also produces large volumes of natural gas which are sold to the Lumut liquefaction plant of Brunei LNG Ltd for shipment to Japan.

Burma

GOVERNMENT DEPARTMENTS AND OFFICIAL AGENCIES

MINISTRY OF NUMBER 2 INDUSTRY
Address: Ministers Office, Rangoon

The Ministry of Number 2 Industry is responsible for the activities of the key state organisations of the energy sector, including the Electric Power Corporation, Myanma Oil Corporation and Petrochemical Industries Corporation, which operates the country's oil refineries.

ENTERPRISES (PUBLIC AND PRIVATE SECTOR COMPANIES)

ELECTRIC POWER CORPORATION
Address: 197-9 Lower Kemmendine Road, Rangoon
Telephone 15366
Energy Sector: Electricity

The Electric Power Corporation is a state undertaking responsible for generation and transmission of electricity throughout Burma. Output of the public supply system is around 1,000 GWh per annum.

MYANMA OIL CORPORATION
Address: PO Box 1049, 604 Merchant Street, Rangoon
Telephone 15266 Telex 21307
Energy Sectors: Oil, gas

Myanma Oil Corporation is the state owned company responsible for exploration and production of oil and gas and the supply of crude oil and products to the Burmese market.

PETROCHEMICAL INDUSTRIES CORPORATION
Address: 33 York Road, Rangoon
Telephone 14467
Energy Sector: Oil

Petrochemical Industries Corporation is a state-owned enterprise responsible for operating oil refineries at Syrian and Chauk. The major refinery is at Syrian with a capacity of around one million tonnes per annum. The small refinery at Chauk has a capacity of only 300,000 tonnes per annum.

China (People's Republic of)

GOVERNMENT DEPARTMENTS AND OFFICIAL AGENCIES

MINISTRY OF COAL INDUSTRY
Address: 16 Heping Li, Beijing
Telephone 46 12 23

The Ministry of Coal Industry is responsible for all aspects of coal industry activities in China. These responsibilities include: education, safety and training; scientific and technological development; procurement of plant and equipment; supervision of operations; geological research, information and exploration.

Also falling under the aegis of the Ministry are numerous institutes concerned with technical and scientific aspects of mining, and the China National Coal Development Corporation.

MINISTRY OF ELECTRIC POWER
Address: Beijing
Telephone 33 21 83

The Ministry of Electric Power is involved in operational and planning aspects of electricity supply throughout the country. The Ministry controls the grids in the North, North-East, East, North-West and Central regions. It is also responsible for the planning and construction of hydro-electric schemes and nuclear power station projects.

MINISTRY OF PETROLEUM INDUSTRY
Address: Liu Pu Kuang, Beijing
Telephone 44 26 31 Telex 22314

The Ministry of Petroleum Industry is concerned with all aspects of oil and gas exploration and development and the supply of finished products. Several individual corporations operate under the control of the Ministry, including: National Oil and Gas Exploration and Development Corporation; National Petroleum Refining Corporation; Petroleum Corporation of the PRC.

STATE ENERGY COMMISSION
Address: Liu Pu Kuang, Beijing

The State Energy Commission was established in 1980 to provide greater overall planning and control of energy development and use. It is responsible for co-ordinating the activities of the Ministries of Coal, Electric Power and Petroleum.

Specific responsibilities of the Commission include formulation of longer term plans, development of new sources of energy, energy conservation, the dissemination of information on new technologies and scientific research.

ENTERPRISES (PUBLIC AND PRIVATE SECTOR COMPANIES)

NATIONAL COAL DEVELOPMENT CORPORATION
Address: PO Box 1401, 16 Heping Bei Lu, Andingmen Wai, Beijing
Telephone 46 12 23

The National Coal Development Corporation operates under the general control of the Ministry of Coal Industry. The Corporation is responsible for a number of key areas of trading relationships with overseas companies and organisations, including: compensation trade; joint ventures, technical co-operation and contracts for engineering equipment or services.

NATIONAL OIL AND GAS EXPLORATION AND DEVELOPMENT CORPORATION
Address: PO Box 766, Beijing
Telephone 44 46 31 Telex 22312

The Oil and Gas Exploration and Development Corporation operates under the supervision of the PCPRC and is responsible for operational activities in oil and gas development, including most fields where foreign interests are likely to be involved.

Specific areas of concern to the Corporation are in drilling, supply of equipment, geological exploration, oil field construction works, and offshore activity, as well as oil and gas production.

PETROLEUM CORPORATION OF THE PRC
Address: PO Box 766, Beijing
Telephone 44 43 13

The Petroleum Corporation (PCPRC) was established in 1977 in order to promote the use of advanced technology from abroad in developing the domestic oil industry. The Corporation is also concerned with the country's crude oil export business. PCPRC operates under the general control of the Ministry of Petroleum Industry.

Hong Kong

ENTERPRISES (PUBLIC AND PRIVATE SECTOR COMPANIES)

CHINA LIGHT AND POWER COMPANY
Address: 147 Argyle Street, Kowloon, Hong Kong
Telephone (3) 711 5111 Telex 74488
Gross Revenue: HK$ 4,309.1 million (1981)
Total Assets: HK$ 4,131.5 million
Energy Sector: Electricity

China Light and Power is the sole supplier of electricity in the Kowloon Peninsula and the New Territories of Hong Kong. Total sales in 1980-81 exceeded 8,500 GWh to the company's 950,000 customers.

Electricity for China Light and Power is generated at Tsing Yi and Hok Un by power stations operated by Peninsula Electric Power Company Ltd (PEPCO) and Kowloon Electric Supply Company Ltd (KESCO). Both companies are owned 60 per cent by Eastern Energy Ltd, a subsidiary of Exxon Corporation, and 40 per cent by China Light and Power. Capacity of the PEPCO stations is 1,800 MW and of KESCO's stations 500 MW, with a 1,400 MW plant under construction, for firing on oil or coal.

Since 1979 the company's system has been connected to that of Guangdong Province in China, with a net flow from the company. There is also increasing co-ordination between China Light and Power and Hong Kong Electric Company, with completion of a submarine cable in 1981.

THE HONG KONG AND CHINA GAS COMPANY LIMITED
Address: Leighton Centre, 77 Leighton Road, Causeway Bay, Hong Kong
Telephone (5) 79 14 33 Telex 86086
Gross Revenue: HK$ 332.6 million (1980)
Total Assets: HK$ 398.4 million
Energy Sector: Gas

Hong Kong and China Gas Company supplies towns gas on Hong Kong island, Kowloon and the New Territories. Sales in 1980 amounted to 33 million therms to the company's 170,000 customers.

The company produces gas from naphtha at its plant at Mau Tau Kok. Hong Kong island is connected by a submarine pipeline. New reforming plant is being added to take total production capacity up to 80 million cubic feet per day.

HONG KONG ELECTRIC HOLDINGS LIMITED
Address: Electric House, 44 Kennedy Road, Hong Kong
Telex 73071
Gross Revenue: HK$ 1,376.6 million (1980)
Total Assets: HK$ 5,711.1 million
Energy Sector: Electricity

The principal company of the Hong Kong Electric Holdings group is Hong Kong Electric Company Ltd which is sole supplier of electricity on Hong Kong island, Ap Lei Chau and Lamma. The company is a private company operating under franchise from the government, which exercises some control on tariffs.

Hong Kong Electric operates generating plant at Ap Lei Chau power station of capacity 1,056 MW. Maximum demand in 1980 reached nearly 800 MW from a total of 311,000 customers. The company is planning additional capacity of 250 MW based on coal as fuel, with conversion of a further 250 MW by 1983. The Hong Kong system is connected to that of China Light and Power on the mainland, providing an increased level of security and efficiency.

India

GOVERNMENT DEPARTMENTS AND OFFICIAL AGENCIES

DEPARTMENT OF ATOMIC ENERGY
Address: Chhatrapati Shivaji Maharaj Marg,
Bombay 400039
Telephone 37 17 37

The Department of Atomic Energy is responsible for all aspects of the nuclear industry, including fuel supply and processing, the nuclear power programme, power station operations, engineering equipment, and education and training. The Department provides the secretariat for the Atomic Energy Commission, which is responsible for formulating policy on nuclear power.

MINISTRY OF ENERGY AND IRRIGATION
Address: Shram Shakti Bhavan -1,
New Delhi 110001
Telephone 38 53 78

The Ministry of Energy and Irrigation is responsible for much of the country's indigenous energy production, excluding oil and gas. Included in the Ministry are departments concerned with coal and electricity.

The responsibilities of the Ministry recognise the need for co-ordination of the use of water power for electricity generation and for irrigation and flood control. Coal is the base-load fuel in electricity production and the central government is becoming increasingly involved in ensuring electricity supplies through its 'super' thermal power stations and regional grids.

MINISTRY OF PETROLEUM, CHEMICALS AND FERTILISERS
Address: Shastri Bhavan -1, Dr Rajendra Prasad Road, New Delhi 110001
Telephone 38 35 01

The Ministry of Petroleum, Chemicals and Fertilisers is responsible for policy development with regard to oil and gas and for the operations of key state undertakings, including the Oil and Natural Gas Commission, Indian Oil Corporation and Oil India Ltd.

ENTERPRISES (PUBLIC AND PRIVATE SECTOR COMPANIES)

BHARAT COKING COAL LIMITED
Address: Sijua, Vihar Building, Jharia PO,
Dhanbad 826001, Bihar
Gross Revenue: Rs 1,837.9 million (1980)
Total Assets: Rs 5,439.4 million
Energy Sector: Coal

Bharat Coking Coal is a subsidiary of the state corporation Coal India, and operates mines in the Bihar region. Output, which is largely of coking coal has been maintained at around 20 million tonnes per annum, although the company's mines have the potential for greater production.

BHARAT PETROLEUM CORPORATION LIMITED
Address: Bharat Bhavan 4/6, Currimbhoy Road,
Ballard Estate, Bombay 400038
Telephone 26 22 81 Telex 12242
Gross Revenue: Rs 9,361.8 million (1980)
Total Assets: Rs 1,016.1 million (1980)
Energy Sector: Oil

Bharat Petroleum Corporation was established as a state company to take over the Burmah-Shell Refineries, since which time it has developed as an integrated refining and marketing company.

The company operates one of the country's largest refineries, of capacity 5.3 million tonnes per annum, at Bombay. The refinery is being specially equipped to process offshore oil produced from the Bombay High field.

BONGAIGAON REFINERY AND PETROCHEMICALS LIMITED
Address: Suryakiran Building, 19/114 Kasturba
Gandhi Marg, New Delhi 110001
Telephone 35 31 81 Telex 314315
Total Assets: Rs 9.1 million (1980)
Energy Sector: Oil

Bongaigaon Refinery and Petrochemicals is a state-owned company set up to own and operate the new refinery which has been constructed at Bongaigaon in Assam. The refinery is relatively small, with a crude oil throughput capacity of only one million tonnes.

CENTRAL COALFIELDS LIMITED
Address: Dharbhanga House, Ranchi, Bihar
Telex 625201
Gross Revenue: Rs 2,540.5 million (1980)
Total Assets: Rs 6,504.7 million
Energy Sector: Coal

Central Coalfields is a subsidiary of the state coal corporation Coal India, responsible for operating mines in the central part of the country. The company operates more than 60 mines and is developing new ones. Output in 1979-80 reached 24 million tonnes, following a steady upward trend from 1976.

COAL INDIA LTD
Address: Coal Bhavan, 10 Netaji Subhas Road, Calcutta 700001
Gross Revenue: Rs 603.0 million (1980)
Total Assets: Rs 16,840.2 million
Energy Sector: Coal

Coal India is one of the country's key state undertakings, responsible for coal production throughout India. Production is organised into four operating subsidiaries, largely on a geographical basis, Bharat Coking Coal Ltd, Central Coalfields Ltd, Eastern Coalfields Ltd and Western Coalfields Ltd. Coal India was established in 1975 as the holding company following nationalisation of non-coking coal mines in 1973, and incorporating Bharat Coking Coal Ltd, which was set up in 1972.

Total production of the Coal India group in 1979-80 amounted to over 91 million tonnes, a slight increase on the previous year's performance. Lower production from deep-mining operations was more than offset by a 10 per cent increase in open-cast extraction, which now accounts for more than one-third of the group's output. Output is mainly of steam coal, but more than 21 million tonnes of coking coal was produced, almost entirely by Bharat Coking Coal.

COCHIN REFINERIES LIMITED
Address: Postbag 2, Ambalamugal 682302, Ernakulam District, Kerala
Gross Revenue: Rs 3,854.9 million (1980)
Total Assets: Rs 543.3 million
Energy Sector: Oil

Cochin Refineries is a joint venture company between the government of India, Phillips Petroleum Co of the United States and the Calcutta based company of Duncan Brothers. The company operates a refinery of 3.5 million tonnes per annum capacity, where a cracking unit is being installed in order to provide a high yield of light and middle distillate products.

DAMODAR VALLEY CORPORATION
Address: Bhabani Bhavan, Alipore, Calcutta 700027
Gross Revenue: Rs 949.3 million (1980)
Total Assets: Rs 4,526.3 million
Energy Sector: Electricity

The Damodar Valley Corporation (DVC) was constituted by the state in 1948, and is responsible for the unified development of the Damodar Valley in the states of Bihar and West Bengal. Principal responsibilities of the Corporation include flood control, irrigation, power generation, transmission and distribution and other economic and environmental aspects.

The DVC has completed the construction of four multi-purpose dams, a barrage at Durgapur, hydro-electric plants and three large thermal power stations at Chandrapura, Bokaro and Durgapur. DVC's generating capacity consists of 1,200 MW of thermal capacity and 100MW of hydro-electric. Total power generation in 1979-80 was 4,620 GWh.

EASTERN COALFIELDS LIMITED
Address: Sanctoria, PO Dishergarh 718333, Burdwan District, West Bengal
Gross Revenue: Rs 2,203.5 million (1980)
Total Assets: Rs 4,513.3 million
Energy Sector: Coal

Eastern Coalfields is a subsidiary of the state coal company Coal India, and operates numerous mines in West Bengal. Output in 1979-80 was only just over 20 million tonnes and has been on a downward trend, partly as a result of power shortages and industrial troubles in the region. In the period 1975-78 Eastern Coalfields was the largest individual producing company in the Coal India group.

HINDUSTAN PETROLEUM CORPORATION LIMITED
Address: Petroleum House, 17 Jamshedji Tata Road, Bombay 400020
Telephone 299151 Telex 2414
Gross Revenue: Rs 11,717.8 million (1980)
Total Assets: Rs 2,083.0 million
Energy Sector: Oil

Hindustan Petroleum Corporation is a wholly state owned company formed to take over the refining interests of Caltex Oil Refining. Hindustan Petroleum operates a refinery of 3.5 million tonnes per annum capacity at Bombay and a smaller refinery, of 1.5 million tonnes per annum capacity, at Vizagapatnam.

HYDRO-CARBONS INDIA LIMITED
Address: Tel Bhavan, Dehra Dun 248 001, Uttar Pradesh

Total Assets: Rs 531.7 million (1980)
Energy Sector: Oil

Hydro-Carbons India is a state-owned company set up to participate in a joint-venture arrangement for oil exploration and production in Iran. Partners of Hydro-Carbons India are Agip and Phillips Petroleum.

Due to disruption of economic and oil industry activity in Iran during 1979 and 1980 the group was unable to lift any crude oil from Iran and no sales of crude oil were made.

INDIAN OIL CORPORATION LIMITED
Address: 254C Dr Annie Besant Road, Prabhadevi, Bombay 400025
Telephone 45 33 11
Gross Revenue: Rs 44,854.1 million (1980)
Total Assets: Rs 10,755.1 million
Energy Sector: Oil

Indian Oil Corporation (IOC) was formed to co-ordinate the refining and marketing activities of the public sector. The Refineries and Pipelines Division of IOC operates refineries at Gauhati (Assam), Barauni (Bihar), Jawahar Nagar (Gujarat), Haldia (West Bengal) with a total crude oil throughput capacity of more than 13 million tonnes per annum, but the Gauhati refinery has a capacity of less than one million tonnes per annum. A fifth refinery is under construction at Mathura (Uttar Pradesh).

IOC's Marketing Division sold 18 million tonnes of products in 1980, making it by far the largest marketer, with a market share of around 60 per cent.

INDO-BURMA PETROLEUM COMPANY LIMITED
Address: Gillander House, Netaji Subhas Road, Calcutta 700001
Gross Revenue: Rs 2,056.3 million (1980)
Total Assets: Rs 135.6 million
Energy Sector: Oil

Indo-Burma Petroleum Company is a state-owned company engaged in marketing of oil and chemicals and in engineering. Sales of oil products are of the order of one million tonnes per annum.

MADRAS REFINERIES LIMITED
Address: Refinery House, Manali, Madras 600068
Telephone 55 31 51 Telex 658
Gross Revenue: Rs 4,345.5 million (1980)
Total Assets: Rs 689.4 million
Energy Sector: Oil

Madras Refineries was set up as a joint venture to build and operate a refinery in Madras. Participants with the government are the National Iranian Oil Company and Amoco. Production at the refinery was 2.8 million tonnes in 1980, close to the maximum capacity of the refinery.

MINERAL EXPLORATION CORPORATION LIMITED
Address: Seminary Hill, Nagpur 440006
Energy Sector: Coal

The Mineral Exploration Corporation is a state company set up to plan, promote and organise exploration programmes for minerals both in India and abroad. Activities include surveys, contract drilling and mining operations etc., either independently or on behalf of other organisations. The Corporation has been responsible for verifying some 1,500 million tonnes of coal reserves.

MINERALS AND METALS TRADING CORPORATION OF INDIA LIMITED
Address: Express Building, 9-10 Bahadur Shah Zaffar Marg, New Delhi 110002
Telephone 27 88 51 Telex 2285
Gross Revenue: Rs 14,099.2 million (1980)
Total Assets: Rs 1,384.1 million
Energy Sector: Coal

Minerals and Metals Trading Corporation is a state owned trading agency handling imports and exports of iron ore, barytes, chrome ore and other metals, minerals, fertilisers and industrial raw materials. Exports of coal form a relatively small proportion of the corporation's activities, with only 75,000 tonnes exported in the year 1979-80, although shipments were substantially higher in the previous year.

NATIONAL HYDRO-ELECTRIC POWER CORPORATION LIMITED
Address: 57 Nehru Place, New Delhi 110024
Telephone 68 29 20 Telex 213477
Total Assets: Rs 1,730.6 million (1980)
Energy Sector: Electricity

The National Hydro-Electric Power Corporation is a state undertaking organised to plan, promote and organise the integrated and efficient development of hydro-electric power. It has a remit to investigate, construct and operate the necessary power stations and transmission lines and engage in distribution and sale of electricity.

The Corporation now has responsibility for the Loktak and Baira Sul hydro-electric projects, and is also involved in the Salal and Devighat (Nepal) schemes.

NATIONAL THERMAL POWER CORPORATION LIMITED
Address: Kailash Building, 26 K Ghandi Marg -1, New Delhi
Telephone 68 31 04 Telex 312266

Total Assets: Rs 2,597.8 million (1980)
Energy Sector: Electricity

The National Thermal Power Corporation was set up by the Indian government in 1975 to complement the activities of state electricity boards, through the construction of so-called 'super' power stations which would have a major impact on regional electricity supply and strengthen the inter-state network.

The Corporation operates some 5,000 MW of capacity, including the Badarpur power station which is being expanded to a total of 700 MW and is planning and constructing four large scale thermal power stations at Singrauli, Korba, Ramagundam and Farakka. The construction programme also involves 3,660 km of 400 kV high tension transmission lines.

NEYVELI LIGNITE CORPORATION LIMITED
Address: Neyveli—607801, South Arcot District, Tamil Nadu
Telex 417350
Total Assets: Rs 4,511.0 million (1980)
Energy Sector: Coal

Neyveli Lignite Corporation is a publicly owned company organised to exploit lignite deposits around Neyveli, near Madras. The lignite reserves are exploited as part of an overall scheme to supply a thermal power station and a fertiliser works.

Output in 1979-80 was just under three million tonnes, representing a 10 per cent fall from the previous year, and less than half the rate designed to meet the full requirements of power generation and fertiliser manufacture. Since 1975 the power station has run partially on fuel oil.

A second mine is being constructed with a production capacity of 4.7 million tonnes per annum, associated with a 630 MW power station, with the prospect of further increases in output of lignite and electricity generating capacity. The Neyveli project supplied 1.8 TWh to the Tamil Nadu Electricity Board in 1979-80.

NORTH EASTERN ELECTRIC POWER CORPORATION LIMITED
Address: Shillong-Jowai Road, Laitumkhrah, Shillong 793003, Meghalaya
Total Assets: Rs 549.8 million (1980)
Energy Sector: Electricity

The North-Eastern Electric Power Corporation was incorporated as a publicly owned agency in 1976 to undertake the Kopil hydro-electric project and related electricity transmission system. The Corporation is also investigating a number of other sites in the States of Arunachal Pradesh and Mizoram for hydro-electric potential.

OIL AND NATURAL GAS COMMISSION
Address: PO Box No 20, 43 Rajpur Road, Dehra Dun 248003, Uttar Pradesh
Telex 595207
Gross Revenue: Rs 4,332.6 million (1980)
Total Assets: Rs 15,911.9 million
Energy Sectors: Oil, gas

The Oil and Natural Gas Commission is a statutory body with the sole responsibility for oil and gas exploration and development. The ONGC has production both onshore and offshore totalling 9.5 million tonnes in 1979-80, representing more than 80 per cent of total Indian production.

Onshore output was 5.1 million tonnes, lower than in previous years, but exploitation of offshore resources in the Bombay High field increased to 4.4 million tonnes and is expected to rise to much higher levels. The Bombay High field also produces substantial volumes of gas, the 1979-80 level being 363 million cubic metres.

The ONGC has extensive drilling programmes onshore and offshore, including activity off the Krishna Godawari coast, Kerala coast and in the Andaman Islands. ONGC also carries out exploration assignments in other countries.

OIL INDIA LIMITED
Address: Allahabad Bank Building, 17 Parliament Street, PO Box 203, New Delhi 1
Telephone 31 08 41
Gross Revenue: Rs 1,102.0 million (1980)
Total Assets: Rs 2,768.0 million
Energy Sector: Oil

Oil India was originally formed as a joint-venture company between the government of India with a one-third share and Burmah Oil Co holding two-thirds to explore for oil in the eastern part of India. Since 1961 ownership has been on a 50/50 basis.

Oil India is involved in exploration onshore and offshore along the Orissa coast. Several fields have been discovered by the company. In 1979-80 production was 2.2 million tonnes, but has been on a declining trend for a number of years. Oil India also transports crude oil supplied by the Oil and Natural Gas Commission to the Gauhati and Barauni refineries via its pipeline system.

RURAL ELECTRIFICATION CORPORATION LIMITED
Address: DDA Building, Nehru Place, New Delhi 110019
Telephone 68 26 01 Telex 31/4405
Gross Revenue: Rs 447.9 million (1980)
Total Assets: Rs 8,128.9 million
Energy Sector: Electricity

The Rural Electrification Corporation was set up by the federal government to provide financial assistance to state electricity boards, rural electricity co-operatives and state governments for rural electrification projects.

In 1979-80 the Corporation approved 776 projects for rural electrification, involving loan assistance amounting to Rs 2,090 million. During that year more than 13,000 additional villages were covered.

SINGARENI COLLIERIES COMPANY LIMITED
Address: Mehar Manjil, Red Hills, Khairtabad, Hyderabad 500004
Telex 155321
Gross Revenue: Rs 890.9 million (1980)
Total Assets: Rs 1,801.5 million
Energy Sector: Coal

Singareni Collieries is owned jointly by the state of Andhra Pradesh and the Indian government, with the state holding a majority interest. Singareni Collieries produced 9.4 million tonnes of coal in 1979-80, of which two million tonnes was obtained by open-cast method. Output from both deep-mined and open-cast operations is on a rising trend.

TATA HYDRO-ELECTRIC POWER SUPPLY COMPANY LIMITED
Address: Bombay House, 24 Homi Mody Street, Fort, Bombay 410023
Telephone 25 91 31 Telex 0113863
Gross Revenue: Rs 366.1 million (1980)
Total Assets: Rs 282.4 million
Energy Sector: Electricity

Tata Hydro-Electric Power Supply Company is part of the Tata group, which includes numerous substantial manufacturing and engineering enterprises. The company generates electricity for group operations and for the public supply system in Western Maharashtra.

TATA POWER COMPANY LIMITED
Address: Bombay House, 24 Homi Mody Street, Fort, Bombay 410023
Telephone 25 91 31 Telex 0113863
Gross Revenue: Rs 905.0 million (1980)
Total Assets: Rs 687.3 million
Energy Sector: Electricity

Tata Power Company is part of the major Indian industrial manufacturing and engineering group of Tata. The Tata group includes some large energy using establishments, but also produces electricity for the public supply system in the Bombay area of Western Maharashtra.

URANIUM CORPORATION OF INDIA LIMITED
Address: Jaduguda Mines, Singhbhum 832102, Bihar
Telex 626285
Gross Revenue: Rs 65.9 million (1980)
Total Assets: Rs 185.3 million
Energy Sector: Uranium

The Uranium Corporation of India is a state company set up in 1967 to provide for the country's uranium requirements. The company produces uranium concentrates, from its plant at Jaduguda. Uranium is also recovered from tailings at the state-owned Hindustan Copper Ltd, also in the Jaduguda area.

All uranium concentrates produced by the company are acquired compulsorily by the Department of Atomic Energy, which is responsible for the use of nuclear materials and electricity generating plant.

WESTERN COALFIELDS LIMITED
Address: Bisesa House, Temple Road, Nagpur 440001
Telex 36352
Gross Revenue: Rs 1,893.3 million (1980)
Total Assets: Rs 4,451.9 million
Energy Sector: Coal

Western Coalfields is a subsidiary of the state coal company Coal India, responsible for operating mines in the western part of the country. Production in 1979-80 exceeded 26 million tonnes, making the company the largest producer within the Coal India group.

Indonesia

GOVERNMENT DEPARTMENTS AND OFFICIAL AGENCIES

MINISTRY OF MINING (DEPARTAMEN PERTAMBANGA)
Address: Jalan Merdeka Selatan 18, Jakarta
Telephone 40686 Telex 44363

The Ministry of Mining is responsible for many aspects of the energy sector. Its responsibilities include supervision of the state oil and gas undertaking Pertamina and exploration and development in general. It is also concerned with the increasing exploitation of Indonesia's coal deposits through the state company PT Batubara.

MINISTRY OF PUBLIC WORKS AND POWER (DEPARTAMEN PEKERJAAN UMUM DAN TENAGA)
Address: Jalan Patimura 20, Jakarta
Telephone 71538 Telex 47131

The Ministry of Public Works and Power is one of the two ministries principally concerned with the energy sector. Its responsibilities include the operations of the state-owned electricity supply undertaking PLN and the development of hydroelectric and geothermal energy resources.

ENTERPRISES (PUBLIC AND PRIVATE SECTOR COMPANIES)

PT CALTEX PACIFIC INDONESIA
Address: Jalan Kebon Sirih 52, PO Box 158, Jakarta Pusat
Telephone 37 69 08 Telex 44386
Energy Sector: Oil

PT Caltex Pacific Indonesia is a joint affiliate of the two major United States international oil companies Standard Oil Company of California and Texaco. The company, operating under contract to the state oil company Pertamina, is a major producer of crude oil in Indonesia. Output handled by the company is almost as great as that of Pertamina, of the order of 750,000 barrels per day.

PERUSAHAAN GAS NEGARA
Address: PO Box 119, Jalan Ridwan Rais 2, Jakarta Pusat
Telephone 37 48 09
Energy Sector: Gas

Perusahaan Gas Negara (PGN) is a state-owned corporation responsible for public gas supply in Indonesia. The company supplies gas to some 21,000 customers in Jakarta and seven other major towns. Total sales of gas during 1980 amouted to some 3.3 million cubic metres.

The largest gas using centre is Cirebon which is supplied with natural gas, as is also Jakarta. Other towns are supplied with manufactured gas with oil-based feedstock.

PERUSAHAAN PERTAMBANGAN MINYAK DAN GAS BUMI NEGARA (PERTAMINA)
Address: PO Box 12, Jalan Merdeka Timur No 1, Jakarta Pusat
Telephone 30 32 300 Telex 44152
Energy Sector: Oil

Pertamina is the principal state-owned undertaking in the oil and gas sectors, with responsibility for exploration and production, and the supply of oil products for the national market. It produces more than half of total crude oil output itself, approximately 40 million tonnes per annum, and authorises other production through contract arrangements.

Almost all of the country's refinery capacity is owned by Pertamina. This includes five principal refineries, at Plaju, Cilacap, Dumai, Sungei Gerong and Balikpapan. Together with other smaller refineries Pertamina's total refining capacity is around 25 million tonnes per annum.

PERUSAHAAN UMUM LISTRIK NEGARA (PLN)
Address: Blok MI/135, Jalan Trunojojo, Kebayoran Baru, Jakarta Selatan
Telephone 71 57 08
Energy Sector: Electricity

Perusahaan Umum Listrik Negara (PLN) is a state undertaking responsible for generating electricity and distributing it throughout Indonesia. It produces almost all of the electricity used in the public supply system. Total sales of electricity are of the order of 5,000 GWh per annum.

PLN uses mainly oil or gas fired power stations, with a substantial number of smaller diesel engine or gas turbine units. Approximately 25 per cent is met from hydro-electric plant. In the longer term PLN is looking towards indigenous coal resources as its main source of fuel.

Japan

GOVERNMENT DEPARTMENTS AND OFFICIAL AGENCIES

AGENCY FOR SCIENCE AND TECHNOLOGY
Address: 2-2-1 Kasumigaseki, Chiyoda-ku, Tokyo 100
Telephone (03) 581 5271

The Agency for Science and Technology, which is responsible direct to the Prime Minister's office, is concerned with many aspects of nuclear power development in Japan. Subordinate organisations include the Atomic Energy Bureau and the Nuclear Safety Bureau as well as research institutes for radiology and metals.

The Atomic Energy Bureau provides the secretariat for the Atomic Energy Commission which is the body advising the government on nuclear issues. The AEB deals with policy development, reactors and technology, and nuclear fuel.

AGENCY OF NATURAL RESOURCES AND ENERGY
Address: 3-1 Kasumigaseki 1-chome, Chiyoda-ku, Tokyo 100
Telephone (03) 501 1511

The Agency of Natural Resources and Energy is the principal government organisation dealing with energy supply industries. It forms part of the Ministry of International Trade and Industry.

POWER REACTOR AND NUCLEAR FUEL DEVELOPMENT CORPORATION
Address: Sankaido Building, 1-9-13 Akasaka, Minato-ku, Tokyo
Telephone (03) 582 1241

The Power Reactor and Nuclear Fuel Development Corporation is a specialist research and development corporation under the direct control of the Prime Minister's Office and the Science and Technology Agency.

ENTERPRISES (PUBLIC AND PRIVATE SECTOR COMPANIES)

CHUBU ELECTRIC POWER COMPANY
Address: 1 Higashishincho, Higashi-ku, Nagoya 461-91
Telephone (052) 951 8211 Telex 04444405
Gross Revenue: Yen 1,495.0 billion (1981)
Total Assets: Yen 2,401.8 billion
Energy Sectors: Electricity, nuclear power

Chubu Electric Power Company is one of the largest electricity producing and distributing companies in Japan, supplying areas of central Japan between Tokyo and Nagoya.

The company relies on thermal power stations for its electricity, the principal ones being at Shinnagoya, Chita, Taketoyo and Nishinagoya. A nuclear power station is under construction at Hamaoka. It is also intended to step up the use of liquefied natural gas as a pollution-free fuel. The company's programme includes three LNG-fired units.

CHUGOKU ELECTRIC POWER COMPANY
Address: 4-33 Komachi, Nakaku, Hiroshima City, Japan 732
Telephone (0822) 41 0211 Telex 06653945
Gross Revenue: Yen 785.8 billion (1981)
Total Assets: Yen 1,255.9 billion
Energy Sectors: Electricity, nuclear power

The Chugoku Electric Power Company supplies electricity in the westernmost part of Honshu.

The company is reliant on fuel oil for well over half of its electricity but is moving to diversify its pattern of fuels, and has two coal-fired plants and one LNG-fired plant under construction. It also has a nuclear power station in operation at Shimane.

DAIKYO OIL COMPANY LIMITED
Address: Jyowa-Yaesu Building, 4-1 Yaesu 2-chome, Chuo-ku, Tokyo
Telephone (03) 274 5211 Telex 2226517
Gross Revenue: Yen 1,003.5 billion (1981—nine months)
Total Assets: Yen 563.4 billion
Energy Sector: Oil

Daikyo Oil is one of the leading wholly Japanese owned oil companies involved in all phases of oil supply. There are more than 50 companies affiliated to Daikyo Oil, including exploration and development in the Middle East, Africa, Asia and North America.

The company owns a large refinery at Yokkaichi, of throughput capacity around 10 million tonnes per annum. In addition it has interests in Fuji Oil Company and Kashima Oil Company, which operate refineries at Sodegaura and Kashima respectively. Daikyo's interests in these companies provide access to a further 3-3.5 million tonnes per annum throughput.

Daikyo has production interests in the Middle East. It holds 25.6 per cent in Abu Dhabi Oil Company and 17.6 per cent in Mubarras Oil Company, which are both exploiting offshore resources in Abu Dhabi.

GENERAL OIL COMPANY LIMITED
Address: 2-8-6 Nishi-Shimbashi, Minato-ku, Tokyo 105
Telephone (03) 595 8300 Telex 02522910
Gross Revenue: Yen 784.0 billion (1981)
Total Assets: Yen 388.0 billion
Energy Sector: Oil

General Oil Company is a distributor and marketer of refined products, in which the Exxon group holds a 49 per cent interest, and is one of the sources of product for Exxon's local affiliate. General Oil's principal sources of products are the two refineries operated by its subsidiary company, General Oil Refining. These are located at Sakai (six million tonnes per annum) and Kawasaki (over 2.5 million tonnes per annum).

General Oil Company also has a majority holding in Nansei Oil Company, which owns a four million tonnes per annum refinery at Nishihara.

IDEMITSU KOSAN
Address: 1-1 Marunouchi 3-chome,
Chiyoda-ku, Tokyo
Telephone (03) 213 3111 Telex 22219
Energy Sector: Oil

Idemitsu Kosan is the leading wholly Japanese owned oil company, supplying approximately 13 per cent of the Japanese oil market. Its operations are backed up by several major refineries, including one of the country's largest refineries, at Chiba, which has a throughput capacity of over 15 million tonnes per annum.

Other refineries are located at Tokuyama, Hyogo, Hokkaido and Aichi. Total capacity of the group's refineries is around 38 million tonnes per annum, and will be increased by a further five million tonnes following an expansion of the Hyogo refinery.

C ITOH AND COMPANY LIMITED
Address: 5-1 Kita-Aoyama 2-chome,
Minato-ku, Tokyo 107
Telephone (03) 639 3574 Telex 2295
Gross Revenue: Yen 11,190.9 billion (1981)
Total Assets: Yen 3,462.4 billion
Energy Sectors: Oil, uranium

C Itoh and Company is one of Japan's major trading companies, with indirect and direct interests in energy developments both in Japan and abroad. The Energy and Chemicals Division accounts for 30 per cent of all transactions.

Through C Itoh Fuel Company and various affiliate companies the group operates some 1,800 service stations and has the largest network of LPG outlets. Large quantities of oil are sold wholesale, particularly as a result of C Itoh's international trading and importing activities. The group is involved in exploration in a number of countries and played a central role in organising the participation of Japanese electric power companies in developing the Ranger uranium deposit in Australia.

JAPAN NATIONAL OIL CORPORATION
Address: 2-2-2 Uchisaiwaicho, Chiyoda-ku,
Tokyo 100
Telephone (03) 580 5411 Telex 22166

Japan National Oil Corporation (JNOC) is a state agency operating under statute with the objectives of supplying funds and facilitating the financing of petroleum exploration and development in the national interest and in assisting oil stockpiling. It assumed the role of the former Japan Petroleum Development Corporation together with an increased role in stockpiling.

JNOC is currently providing finance or guarantees for 53 companies, most of which are involved in oil exploration and production overseas, but also including nine exploring the Japanese continental shelf and two in LNG projects. Assisted companies are exploring in North and South America, West Africa, the Middle East and the Far East.

In 1978 JNOC began a programme to stockpile 10 million cubic metres of oil on its own account. This is necessary in order to take total stockpiles above the level of 90 days supply—the extent of obligations imposed on the private sector.

JAPAN PETROLEUM EXPLORATION COMPANY LIMITED
Address: 6-1 Ohtemachi 1-chome,
Chiyoda-ku, Tokyo
Telephone (03) 201 7571 Telex 25408
Energy Sector: Oil

Japan Petroleum Exploration Company (JAPEX) was organised to carry out exploration and development of oil and gas, particularly in Japan. But it has also been the vehicle for Japanese participation in a number of successful ventures overseas. Japex operates several onshore fields in Japan.

KANSAI ELECTRIC POWER COMPANY
Address: 3-22 Nakanoshima 3-chome,
Kita-ku, Osaka 530
Telephone (06) 441 8821 Telex 5248320
Gross Revenue: Yen 1,791.4 billion (1981)
Total Assets: Yen 3,552.6 billion
Energy Sector: Electricity, nuclear power

Kansai Electric is one of the largest electricity utilities in Japan with sales in 1980-81 of over 81 TWh to 9.4 million customers. The region supplied by the company covers the central part of mainland Japan including the cities of Osaka, Kyoto and Kobe.

Total plant capacity of Kansai Electric exceeds 23,000 MW including 5,700 MW of nuclear plant and 5,800 MW of hydro-electric capacity. But almost 12,000 MW consists of thermal plant using oil, natural gas or coal. The company also has three pumped storage schemes. New capacity of all types is under construction, including an extension to the nuclear power station at Takahama.

KYODO OIL COMPANY LIMITED
Address: 2-11-2 Nagata-Choo, Chiyoda-ku,
Tokyo
Telephone (03) 580 3711 Telex 25790
Energy Sector: Oil

Kyodo Oil Company is a leading supplier of oil products. It is one of the principal wholly Japanese owned companies supplying approximately 13 per cent of the market. It is the largest shareholder in Kashima Oil Company, which owns and operates a nine million tonnes per annum refinery at Kashima.

KYUSHU ELECTRIC POWER COMPANY
Address: 2-1-82 Watanabedori, Chuo-ku, Fukuoka 810
Telephone (092) 761 3031
Gross Revenue: Yen 911.8 billion (1981)
Total Assets: Yen 1,745.4 billion
Energy Sectors: Electricity, nuclear power, geothermal

Kyushu Electric Power Company is responsible for the public electricity supply system on Kyushu. It uses several forms of energy for its electricity production.

The company's main thermal generating stations are at Karatsu, Karita and Oita, using oil, coal and LNG. It also operates a nuclear power station at Genkai. Kyushu Electric Power Company has also taken a leading role in the exploitation of geothermal energy.

MARUZEN OIL COMPANY LIMITED
Address: 1-3 Nagahoribashi-Suji, Minami-ku, Osaka 542-91
Telephone (06) 271 1251 Telex 2222881
Gross Revenue: Yen 1,721.1 billion (1981)
Total Assets: Yen 1,112 billion
Energy Sector: Oil

Maruzen Oil Company is a wholly Japanese owned oil refining and marketing company. The company also has access to overseas crude oil through its participation in Arabian Oil Company, which has crude oil production in the Kuwait/Saudi Arabian Neutral Zone.

Principal facilities of the group are the refineries at Chiba, Sakai, Matsuyama and Shimotsu with a total crude oil throughput capacity of nearly 20 million tonnes per annum, of which the Chiba refinery accounts for almost half.

MITSUBISHI OIL COMPANY LIMITED
Address: 2-4, 1-chome, Toranomon, Minato-ku, Tokyo
Telephone (03) 595 7663 Telex 0222-4104
Gross Revenue: Yen 1,319.4 billion (1981)
Total Assets: Yen 816.1 billion
Energy Sector: Oil

Mitsubishi Oil Company is involved in oil exploration, refining and marketing in Japan. The company is owned equally by the Mitsubishi group and Getty Oil Company of the United States. Sales of oil products by the company exceed 18 million cubic metres per annum. Products are obtained from the company's large refineries at Kawasaki and Mizushima, with combined capacity of 375,000 barrels per day, equivalent to nearly 19 million tonnes per annum. The company also has a 57 per cent interest in Tohoku Oil Company, which operates a five million tonnes per annum refinery at Sendai.

MITSUI AND COMPANY LIMITED
Address: 2-1 Ohtemachi 1-chome, Chiyoda-ku, Tokyo
Telephone (03) 285 1111 Telex 22253
Gross Revenue: Yen 14,929.7 billion (1981)
Total Assets: Yen 4,938.0 billion
Energy Sectors: Oil, gas, coal

Mitsui and Company is the primary organisation within the Mitsui group, which is one of Japan's leading industrial groups. It is directly involved in energy developments, but is also an important international trading company, handling crude oil, petroleum products and liquefied gas for Japanese and overseas markets.

In Japan Mitsui holds a 50 per cent interest in Kyokuto Petroleum Industries, which operates a 7.5 million tonnes per annum refinery at Chiba. Mitsui markets part of the products produced at the refinery. The group also owns the Hokkaido Colliery and Steamship Company, which operates coal mines on Hokkaido.

Mitsui's most significant direct involvement in energy is its 22 per cent holding in Abu Dhabi Gas Liquefaction Company. This company operates a major plant for production of LNG, which is shipped to Japan, and of LPG. Mitsui holds a dominant role in the importation of coking coal into Japan and is engaged in exploration overseas for uranium and other natural resources.

NIPPON OIL COMPANY LIMITED
Address: 3-12 Nishi-Shimbashi 1-chome, Minato-ku, Tokyo 105
Telephone (03) 502 1111 Telex 27237
Gross Revenue: Yen 3,946.0 billion (1980—net of excise taxes)
Total Assets: Yen 1,547.4 billion
Energy Sectors: Oil, coal

Nippon Oil Company is the largest oil company in Japan with subsidiary and associate companies involved in exploration, importation, refining, trading and transportation. The company is closely linked to Caltex Petroleum, the partnership of Standard Oil Company of California and Texaco. Caltex supplies Nippon Oil with 80 per cent of the group's crude oil requirements, and the two groups are building up a coal import business based on Australian export availability.

Total sales of oil products in 1980-81 amounted to some 52 million cubic metres. These were supplied principally from the refineries of Nippon Petroleum Refining Company, Nihonkai Oil Company and Koa Oil Company. Total capacity of these refineries, together with Nippon Oil's own small refinery at Niigata, is close to 900,000 barrels per day (45 million tonnes per annum).

NIPPON PETROLEUM REFINING COMPANY LIMITED
Address: 3-12 Nishi-Shimbashi 1-chome, Minato-ku, Tokyo 105
Telephone (03) 502 1111 Telex 27237
Gross Revenue: Yen 897.9 billion (1981)
Energy Sector: Oil

Nippon Petroleum Refining Company is one of the largest oil refining companies in Japan. It is owned by Nippon Oil Company and Caltex. Caltex is the main supplier of crude oil for the Nippon Oil group, which markets the output of the refining company.

Nippon Petroleum Refining operates five refineries. Largest is at Negishi with a throughput capacity of 330,000 barrels per day, more than 16 million tonnes per annum. Other refineries are located at Muroran, Yokohama, Kudamatsu and Okinawa.

OSAKA GAS COMPANY LIMITED
Address: 1 Hirano-machi 5-chome, Higashi-ku, Osaka 541
Telephone (06) 202 2221 Telex 0522 5275
Gross Revenue: Yen 572.6 billion (1981)
Total Assets: Yen 609.8 billion
Energy Sector: Gas

Osaka Gas Company is the second largest gas supply company in Japan, serving Osaka and surrounding areas. The company operates five gas plants and has a terminal at Himeji for receipt and regasification of imported LNG.

SHELL OIL COMPANY
Address: 5-2 Kasumigaseki 3-chome, Chiyoda-ku, Tokyo 100
Energy Sector: Oil

Shell Oil Company (Shell Sekiyu KK) is a subsidiary of the Royal Dutch/Shell group engaged in distribution and marketing of oil products in Japan. The company is associated particularly with Showa Oil Company. Shell Oil has a 25 per cent participation in one of the country's largest refineries, belonging to Showa Yokkaichi Oil Company.

SHOWA OIL COMPANY LIMITED
Address: 7-3 Marunouchi 2-chome, Chiyoda-ku, Tokyo 100
Telephone (03) 231 0311 Telex 28765
Gross Revenue: Yen 1,050.0 billion (1981)
Total Assets: Yen 570.9 billion
Energy Sector: Oil

Showa Oil Company is involved in the importation and refining of crude oil and the distribution of oil products. It is closely linked to companies of the Royal Dutch/Shell group, which is the company's main supplier of crude oil.

The company owns refineries at Kawasaki and Niigata, with a total throughput capacity of close to 10 million tonnes per annum. In addition it has a 50 per cent interest in Showa Yokkaichi Oil Company which operates a 15 million tonnes per annum refinery, at Yokkaichi, west of Nagoya.

TOA NENRYO KOGYO
Address: Palace Side Building, 1-1 Hitotsubashi 1-chome, Chiyoda-ku, Tokyo 100
Telephone (03) 213 2211 Telex 2227055
Gross Revenue: Yen 1,250.0 billion (1981)
Total Assets: Yen 627.0 billion
Energy Sector: Oil

Toa Nenryo Kogyo is the central company in a group affiliated to the international Exxon and Mobil companies. Both Exxon and Mobil have 25 per cent holdings in the group, which supplies products to the Esso and Mobil marketing companies.

Toa Nenryo owns three refineries with a total crude oil capacity of more than 20 million tonnes per annum. The refineries at Kawasaki and Wakayama each have capacities of 9-10 million tonnes per annum.

TOHO GAS COMPANY LIMITED
Address: 19-18 Sakurada-cho, Atsuka-ku, Nagoya 456
Telephone (052) 871 3511 Telex 04477 651
Gross Revenue: Yen 126.2 billion (1981)
Total Assets: Yen 150.4 billion
Energy Sector: Gas

Toho Gas Company is one of the largest gas utility companies in Japan, serving the whole of Aichi Prefecture and part of Gifu in the Nagoya region of Honshu. The company operates four gas-making plants and has a terminal for receipt and regasification of liquefied natural gas (LNG). From 1983 the company is to import some 500,000 tonnes per annum of LNG from Indonesia.

TOHOKU ELECTRIC POWER COMPANY
Address: 3-7-1 Ichibancho, Sendai City 980
Telephone (0222) 25 2111 Telex 0852 655
Gross Revenue: Yen 943.6 billion (1981)
Total Sales: Yen 1,627.6 billion
Energy Sector: Electricity

Tohoku Electric Power Company generates and distributes electricity in the Tohoku region of northern Honshu. Its main thermal power stations are at Hachinoke, Akita, Sendai, Shim-Sendai and Niigata. The company uses oil for more than 60 per cent of its electricity generation.

Tohoku Electric has a nuclear power station at Onagawa and has joint-venture projects with Tokyo Electric Power Company for a new coal-fired power station and for the importation of LNG.

TOKYO ELECTRIC POWER COMPANY
Address: 1-3 Uchisaiwai-cho 1-chome, Chiyoda-ku, Tokyo 100
Telephone (03) 501 8111 Telex 0222 4045
Gross Revenue: Yen 3,135.2 billion (1981)
Total Assets: Yen 6,589.6 billion
Energy Sectors: Electricity, nuclear power

Tokyo Electric Power Company is the largest producer and distributor of electricity in Japan, and is thought to be the largest privately owned electricity utility in the world. Annual output of electricity is of the order of 170 TWh.

The company has its principal power stations at Yokosuka, Arugasaki, Kashima and Oita and a major nuclear power plant at Fukushima. Tokyo Electric has been in the forefront of the construction of nuclear power stations and the use of LNG as a pollution-free fuel in other power stations.

TOKYO GAS COMPANY LIMITED
Address: 2-16 Yaesu 1-chome, Chuo-ku, Tokyo 103
Telephone (03) 273 0111
Gross Revenue: Yen 638.4 billion (1981)
Total Assets: Yen 634.2 billion
Energy Sector: Gas

Tokyo Gas Company is the largest gas supply utility in Japan, supplying the Tokyo metropolitan area and surrounding districts. The company has six gas plants but is switching extensively to the use of liquefied natural gas and reducing its use of oil and coal. Tokyo Gas Company participates in some of the principal overseas LNG development projects along with other gas and electricity generation companies.

PROFESSIONAL INSTITUTIONS AND TRADE ASSOCIATIONS

ATOMIC ENERGY RELATIONS ORGANISATION
Address: 1-13-1 Shimbashi, Minato-ku, Tokyo
Telephone (03) 504 1381
Energy Sector: Nuclear power

The Atomic Energy Relations Organisation has been established by the nuclear power industry to disseminate information to the mass media, local authorities and municipalities, communities and teachers and to the public at large.

ATOMIC INDUSTRY FORUM
Address: 1-13-1 chome, Shimbashi, Minato-ku, Tokyo
Telephone (03) 591 6121
Energy Sector: Nuclear power

The Atomic Industry Forum is a non-profit organisation including a wide range of interests as a forum for interchange of ideas and information about nuclear power. The approximately 700 member groups include electricity utilities, manufacturing companies, construction firms, banks, economic institutions, research and development agencies, local authorities and information organisations.

JAPAN GAS ASSOCIATION
Address: 12-15 Toranomon 1-chome, Minato-ku, Tokyo 105
Energy Sector: Gas

The Japan Gas Association is a representative organisation for the gas industry in Japan. It is the Japanese member of the International Gas Union.

JAPAN PETROLEUM INSTITUTE
Address: 17F Shin-Aoyama East Building, 1-1 Minami-Aoyama 1-chome, Minato-ku, Tokyo
Telephone: (03) 475 1235
Energy Sectors: Oil, gas

The Japan Petroleum Institute is the principal organisation of personnel and companies in the oil and gas industries, providing a representative forum for discussion of technical and other issues and formulation of proposals to official bodies.

PETROLEUM PRODUCERS ASSOCIATION OF JAPAN
Address: 1-9-4 Ohtemachi, Chiyoda-ku, Tokyo
Telephone (03) 279 5841
Energy Sector: Oil

The PPAJ is a private, non-profit organisation of Japanese companies involved in petroleum exploration and production. The main purposes of the Association are to recommend appropriate policies to government to promote oil development and inform the general public on oil exploration questions.

Member companies produced 8,000 barrels per day of oil in Japan in 1980 and some 400,000 barrels per day from overseas fields. Companies are involved in many countries in the Middle East, Africa and Asia.

Korea (Republic of)

GOVERNMENT DEPARTMENTS AND OFFICIAL AGENCIES

KOREA ENERGY MANAGEMENT CORPORATION
Address: 98-5 Wooni-Dong, Chongro-ku, Seoul
Telephone 764 2517

The Korea Energy Management Corporation (KEMCO) was set up under the Energy Rationalisation Law of 1980 to foster conservation and more efficient use of energy across all sectors of the economy. Functions of KEMCO include the introduction of energy audits, promotion of the use of new technologies, dissemination of information and training.

KOREA MINING PROMOTION CORPORATION
Address: 679-1 Shindaebang-Dong, Dongjak-ku 151, Seoul
Telephone 833 0040/9

The Korea Mining Promotion Corporation was established in 1967 and has assumed increasing responsibility in subsequent years for operational standards, safety, improved mining technology and technical research to aid the domestic coal mining industry and promote involvement of Korean companies in overseas resource development ventures.

The Corporation undertakes surveys of resource deposits, finances mining developments and acts as the source of information and advice on overseas developments and project possibilities.

MINISTRY OF ENERGY AND RESOURCES
Address: 35-34 Tongui-dong, Chongro-ku, Seoul
Telephone 72 52 41 Telex 23472

The Ministry of Energy and Resources has a wide range of responsibilities for the supply of energy and other resources. These include control of resource development, production and importation. Separate divisions deal with the coal and electricity industries. The Ministry has offices dealing with planning and resources policy, as well as questions of mining operations and safety.

ENTERPRISES (PUBLIC AND PRIVATE SECTOR COMPANIES)

DAI-HAN COAL CORPORATION
Address: 1-888 Youido-dong, Yongdungpo-gu, Seoul
Telephone 782 4211 Telex 23621
Energy Sector: Coal

Dai-Han Coal Corporation is a state-owned coal company, operating seven mines and producing a signficiant proportion of state sector output. The company forms one of the bases for the future expansion of the Korean coal industry under the country's long term development plan and the sponsorship of the Korea Mining Promotion Corporation.

HONAM OIL REFINERY COMPANY LIMITED
Address: Kuk Dong Building, 60-1, 3-ka Choongmu-ro, Choong-ku, Seoul
Telephone 77155 Telex 23261
Energy Sector: Oil

Honam Oil Refinery Company is a joint-venture between Korean interests and an affiliate of Caltex Petroleum Company. It operates a refinery at Yosu of capacity around 19 million tonnes per annum, making it one of the major oil refineries of the Far East.

The company is one of Korea's principal importers of crude oil. Products are sold to all sectors of the home market directly or via associate companies.

KOREA ELECTRIC COMPANY
Address: 18 Chungjin-dong, Chongro-ku, Seoul
Telephone 75 54 15/9 Telex 26334
Energy Sectors: Electricity, nuclear power

Korea Electric Company (KECO) is a state undertaking responsible for the generation, transmission and distribution of electricity throughout the country. Total sales of electricity by KECO in 1980 were 32.7 TWh.

The company has for a long time been highly dependent on oil-fired generating capacity, but is in the process of making a major transition to other fuels. The first nuclear power station was commissioned in 1978 and eight more are under construction, with an eventual total capacity of 7-8,000 MW.

Uranium is currently purchased by KECO under long-term contracts from Australia and Canada, but it is actively searching for new sources, notably in Gabon. KECO is negotiating for LNG from Indonesia to provide an alternative to oil. Yet other power stations will be built to run on indigenous anthracite or imported coal.

KOREA OIL CORPORATION
Address: 10-1, 2-Ka Namdaemun-Ro, Choong-ku, Seoul
Telephone 776 0051 Telex 23654
Energy Sector: Oil

Korea Oil Corporation owns and operates one of the country's two major oil refineries, located at Ulsan. The Ulsan refinery has a crude oil throughput capacity of some 14 million tonnes per annum. The refinery provides a wide range of products for the national market and feedstock for petrochemical operations.

The Ulsan refinery was originally built and operated as a joint-venture between the United States based international company Gulf Oil and the state-owned Korea Stock Holding Company. In 1980 its 50 per cent interest was sold by Gulf Oil to its partner, but subsequently transferred to another Korean company Sunkyong Ltd.

KYUNG-IN ENERGY COMPANY LIMITED
Address: Pacific Construction Building, CPO Box 7373, 35 Seosomun-dong, Choong-ku, Seoul
Telephone 77151 Telex 27375
Energy Sectors: Oil, electricity

Kyung-In Energy Company is a joint-venture between Union Oil Company of California and the Korean Kyung-In group, operating an oil refinery at Inchon. The refinery has a capacity of 60,000 barrels per day (three million tonnes per annum) and in 1980 operated at 70 per cent of this capacity.

Associated with the company's refinery is a 325 MW power station. This supplies not only the electricity requirements of the refinery, but also contributes electricity to Korea Electric Company's public supply system.

SSANYONG CORPORATION
Address: CPO Box 409, 24-1 Zuh-dong 2-ka, Choong-ku, Seoul
Telephone 260 4081/9 Telex 24270
Energy Sector: Oil

Ssanyong Corporation is a diversified company with interests in the cement, paper, construction and shipping industries, heavy engineering and energy. Through its subsidiary, Ssanyong Oil Refining Company, it owns an oil refinery at Onsan.

Capacity of the Onsan refinery, which was completed in 1980, is 60,000 barrels per day (approximately three million tonnes per annum). The refinery was originally started as a joint-venture with the National Iranian Oil Company, which withdrew from the venture in 1980.

Malaysia

GOVERNMENT DEPARTMENTS AND OFFICIAL AGENCIES

MINISTRY OF ENERGY, TELECOMMUNICATIONS AND POSTS
Address: 1st Floor, Wisma Damansara, Jalan Semantan, Kuala Lumpur

The Ministry of Energy, Telecommunications and Posts is concerned with general development of the energy industries throughout Peninsular Malaysia, Sarawak and Sabah.

MINISTRY OF WORKS AND UTILITIES
Address: Jalan Tun Ismail, Kuala Lumpur
Telephone 20 01 99

The responsibilities of Ministry of Works and Utilities include supervision of the operations and plans of the public electricity supply undertakings in Peninsular Malaysia, Sarawak and Sabah.

ENTERPRISES (PUBLIC AND PRIVATE SECTOR COMPANIES)

ESSO MALAYSIA BERHAD
Address: 3rd Floor, Kompleks Antarabangsa, Jalan Sultan Ismail, Kuala Lumpur 04-03
Telephone 422322
Energy Sectors: Oil, gas

Esso Malaysia is a subsidiary of the major international oil company Exxon Corporation. It is engaged in all aspects of the oil industry in Malaysia and participates in gas discoveries off the east coast of Peninsular Malaysia.

The company's oil supply operations are based on a refinery at Port Dickson on the south western coast of the Malay Peninsula. The refinery has a throughput capacity of around two million tonnes per annum.

LEMBAGA LETRIK NEGARA
(NATIONAL ELECTRICITY BOARD)
Address: PO Box 1003, 126 Jalan Bangsar,
Kuala Lumpur
Telephone 27791
Energy Sector: Electricity

The National Electricity Board is responsible for the public supply system of Peninsular Malaysia. It accounts for 95 per cent of all electricity supplies.

The Board has an important hydro-electric plant at Temengor, of 340 MW capacity, but two-thirds of its capacity is oil-fired, including the 480 MW power station at Port Dickson, close to two of the country's oil refineries. In the medium term the NEB is likely to construct a gas-fired station, based on the large offshore reserves discovered to the east of the Malay Peninsula.

MALAYSIA LNG BERHAD
Address: Petronas, 136 Jalan Pudu,
Kuala Lumpur 05-3
Telephone 48 90 66 Telex 31123
Energy Sector: Gas

Malaysia LNG was formed in 1978 to own and operate the liquefaction plant being constructed at Bintulu, Sarawak. A majority interest is held by Petronas, with 75 per cent. Shell Gas, an affiliate of the Royal Dutch/Shell group, and Mitsubishi Corporation of Japan hold the remaining shares.

Malaysia LNG is to acquire gas from the producing group of Petronas and Shell and export LNG to Japan from 1983. At full capacity the plant will process some six million tonnes per annum of LNG.

PETROLIAM NASIONAL BERHAD
Address: PO Box 2444, 136 Jalan Pudu,
Kuala Lumpur 05-3
Telephone 20 36 44 Telex 31123
Energy Sectors: Oil, gas

Petroliam Nasional Berhad (Petronas) is a state corporation, set up to implement government policies in the oil and gas sectors. Its principal concerns currently are in exploration and development of oil and gas resources. Private companies operate under contract to Petronas.

Petronas holds a majority share of Malaysia LNG, the company which operates the Bintulu LNG terminal now under construction and is the partner of Shell Sarawak in the production of gas for the LNG plant.

The state corporation has so far limited its involvement in downstream oil operations, but it is expected to initiate construction of new refining capacity to keep pace with the needs of the national market. Plans exist for a 165,000 barrels per day (just over eight million tonnes per annum) refinery at Melaka, and a smaller refinery at Kerteh.

SARAWAK ELECTRICITY SUPPLY
CORPORATION
Address: Peti Surat No 149, Kuching
Telephone 53211 Telex 70100
Gross Revenue: M$ 70.8 million (1980)
Total Assets: M$ 197.5 million
Energy Sector: Electricity

Sarawak Electricity Supply Corporation (SESCO) is a publicly owned undertaking responsible for generation, transmission and distribution of electricity in Sarawak. The larger part of SESCO's supplies are consumed in the domestic and commercial sectors, as the oil and gas industry meets its own power needs.

The Corporation has a generating capacity of 139 MW of which 77 MW is at the Kuching power station. The Corporation has around 65,000 customers with a total consumption of 275 GWh.

SARAWAK SHELL BERHAD
Address: Lutong, Sarawak
Telephone Miri 36144 Telex 74235
Energy Sectors: Oil, gas

Sarawak Shell is a subsidiary of the Royal Dutch/Shell group engaged in exploration and production in Sarawak. Oil has been produced by the company in Sarawak since 1910.

Currently the company's production is from offshore fields and exported via a terminal at Lutong. Production in 1980 exceeded 90,000 barrels per day. Sarawak Shell is also producing natural gas for liquefaction at the Bintulu LNG plant being built by Malaysia LNG Berhad.

Sarawak Shell operates a small refinery at Lutong. The refinery meets primarily the needs of Sarawak and Sabah, exporting a proportion to Peninsular Malaysia and Singapore.

SHELL MALAYSIA TRADING SENDIRIAN
BERHAD
Address: Bangunan Shell Malaysia, Off Jalan
Semantan, Damansara Heights, Kuala Lumpur
23-03
Telephone 94 91 44 Telex 30414
Energy Sector: Oil

Shell Malaysia Trading is the Royal Dutch/Shell group's subsidiary responsible for distribution and marketing of oil products in Peninsular Malaysia. The company supplies more than 40 per cent of the total market, drawing mainly on output from the group's refinery at Port Dickson.

SHELL REFINING COMPANY BERHAD
Address: Batu 1, Jalan Pantai, Port Dickson
Telephone (06) 791 311 Telex 63852
Gross Revenue: M$ 1,517.9 million (1980)
Total Assets: M$ 464.5 million
Energy Sector: Oil

Shell Refining Company is a subsidiary of the Royal Dutch/Shell group, operating a refinery at Port Dickson. The refinery has a throughput capacity of 4.5 million tonnes per annum and provides products for Shell's marketing operations in Peninsular Malaysia. Crude oil is also processed on behalf of Petronas, BP and Mobil.

New Zealand

GOVERNMENT DEPARTMENTS AND OFFICIAL AGENCIES

LIQUID FUELS TRUST BOARD
Address: Greenock House, Lambton Quay, Box 17, Wellington
Telephone 726 108 Telex 31488

The Liquid Fuels Trust Board was established in 1978 as a state agency charged with promoting, encouraging, financing, undertaking and co-operating in any activity which is aimed at reducing the use of imported fuels for transportation. The Board's operations are financed by a levy of 0.1 cent per litre, levied on motor gasoline and automotive diesel oil.

The Board's initial work concentrated on the potential of the newly developed natural gas resources for transportation needs and a programme of activities is being supported to develop the use of compressed natural gas (CNG). Of perhaps equal importance is the possible application of LPG availability for the same end.

Attention is now moving towards the longer term potential of indigenous resources for production of automotive fuels. This includes evaluation of Southland lignite, biomass and alcohol fuels.

MINISTRY OF ENERGY
Address: Lambton House, 152-172 Lambton Quay, Wellington
Telephone 727 044 Telex 31488

The Ministry of Energy came into being in 1978 on the merging of the former Ministry of Energy Resources with the Electricity Department and the Mines Department. It therefore has a central role in the supply of energy as well as the responsibility for preparing overall energy policies and programmes. It is also responsible for the activities of a number of key agencies and corporations, including the Liquid Fuels Trust Board, Petroleum Corporation of New Zealand and the NZ Energy Research and Development Committee.

Arising from the establishment of the Ministry there was prepared in 1980 the country's first long term energy plan, which examined the possible development of energy supply and consumption through to the 1990s.

NEW ZEALAND SYNTHETIC FUELS CORPORATION
Address: Liquid Fuels Trust Board, Greenock House, Lambton Quay, Wellington
Telephone 726 108

The Synthetic Fuels Corporation was set up in 1980 to carry out design work and a full appraisal of the project to construct a synthetic gasoline plant, with the objective of producing up to 570,000 tonnes of synthetic fuel annually. Production route would be through initial conversion of Maui gas to methanol and thence to gasoline, consuming in principle up to 16 per cent of the Maui field's output.

Mobil Oil, which is one of the participants in the Marsden point oil refinery, and a leading marketer of petroleum products, is involved in the project in partnership with government agencies.

ENTERPRISES (PUBLIC AND PRIVATE SECTOR COMPANIES)

AUCKLAND GAS COMPANY LIMITED
Address: PO Box 34, 26 Wyndham Street, Auckland
Telephone 32 269
Gross Revenue: NZ$ 18.5 million (1981)
Total Assets: NZ$ 19.1 million
Energy Sector: Gas

Auckland Gas Company is a private undertaking franchised to sell gas in the Greater Auckland region. Total sales of gas in 1980-81 were 5,086 terajoules, equivalent to 120 million cubic metres of natural gas.

The company's operations and sales are expanding rapidly on the base of new natural gas supplies becoming available from the offshore Maui field. Many new customers are being taken on and the mains system is being upgraded to handle the additional natural gas which will be supplied under long-term contract by the Natural Gas Corporation.

BP NEW ZEALAND LIMITED
Address: 20 Customhouse Quay, Wellington
Telephone 729 729
Energy Sectors: Oil, gas

BP New Zealand is a subsidiary of the British Petroleum Company, responsible primarily for marketing oil products. These are obtained mainly from the industry refinery at Marsden Point, Whangarei, in which BP has an interest.

BP is a participant in Shell BP and Todd Oil Services, which is engaged in exploration for oil and gas and is partnering the state oil corporation in development of the Maui field. LPG from the Maui field is to be marketed through a new company Liquigas, in which BP has a shareholding.

FLETCHER CHALLENGE LIMITED
Address: 105-109 The Terrace, PO Box 1696, Wellington
Telex 3418
Gross Revenue: NZ$ 2082.1 million (1981)
Total Assets: NZ$ 980.1 million
Energy Sectors: Oil, gas

Fletcher Challenge Limited represents the merger of three leading New Zealand companies, Fletcher Holdings, Challenge Corporation and Tasman Pulp and Paper Company to form the largest listed New Zealand company. The new grouping has interests in rural trading, forest industries, construction, property and manufacturing as well as the energy sector.

The group includes Rockgas, which is a major distributor of LPG in New Zealand. It will have a 16.5 per cent participation in Liquigas, the consortium company set up to distribute and market the increased quantities of LPG produced by the Maui development. Fletcher Challenge is also associated with Hematite Petroleum NZ, a subsidiary of Broken Hill Proprietary, and the state oil company Petrocorp in oil and gas exploration off Taranaki province.

LIQUIGAS LIMITED
Address: Petrocorp House, 86 Lambton Quay, Wellington
Telephone 739 812 Telex 31146
Energy Sector: Gas

Liquigas Limited was formed in 1981 by a group of energy companies, including the state-owned Petroleum Corporation of New Zealand, with the objective of developing the market for liquefied petroleum gas as an alternative motor fuel and establishing the necessary distribution facilities. Petrocorp, through its subsidiary Offshore Mining Company, holds a 25 per cent interest. Other participants are the leading oil marketers Shell, BP and Todd, together with NZ Industrial Gases and the diversified New Zealand based company Fletcher Challenge.

NATURAL GAS CORPORATION OF NEW ZEALAND LIMITED
Address: Petrocorp House, 86 Lambton Quay, PO Box 10148, Wellington
Telephone 723 258 Telex 31146
Gross Revenue: NZ$ 29.6 million (1981)
Energy Sector: Gas

Natural Gas Corporation, now a wholly-owned subsidiary of Petroleum Corporation of New Zealand, was set up in 1967 to represent the state's interests in the development of the Kapuni gas/condensate field. Since the discovery of the Maui field NGC's activities have expanded considerably.

NGC supplies nine North Island gas utility companies and has an increasing number of direct sales contracts with bulk consumers. NGC operates the main onshore pipeline network which has expanded from 536 km in 1978 to around 1,300 km, with further extensions planned. The corporation operates the Kapuni processing plant, which annually treats some 640 million cubic metres of gas and will be involved in further downstream petrochemical operations of the Petrocorp group.

NEW ZEALAND ELECTRICITY
Address: Rutherford House, Lambton Quay, Wellington
Telephone 723 550
Gross Revenue: NZ$ 546.8 million (1981)
Total Assets: NZ$ 2,779.2 million
Energy Sectors: Electricity, geothermal

NZ Electricity is the operating arm of the Ministry of Energy's Electricity Division, responsible for bulk generation and transmission of electricity for all types of user. NZ Electricity's operations include generation of heat and electricity from geothermal sources. The Division is also involved with the work of the Rural Electrical Reticulation Council, which is concerned with supplying facilities to bring electricity to the remotest areas still unconnected to the grid.

NZ Electricity has over 30 generating stations throughout New Zealand. Manapouri hydroelectric station produces approximately 20 per cent of total supply. Benmore hydro station accounts for more than 10 per cent. Only three other stations, New Plymouth, Wairakei and Roxburgh, produced more than 1,000 GWh in 1980-81, with a combined total output equivalent to nearly 20 per cent of NZ Electricity's availability in that year.

NEW ZEALAND INDUSTRIAL GASES LIMITED
Address: 181-195 Wakefield Street, PO Box 3337, Wellington
Telephone 859 839 Telex 31234
Gross Revenue: NZ$ 57.0 million (1981)
Total Assets: NZ$ 59.2 million
Energy Sectors: Oil, gas

NZ Industrial Gases (NZIG) distributes a range of gases throughout New Zealand to industrial users. BOC holds a combined total of 62.5 per cent of the company's shares.

NZIG is a participant in the recently formed company, Liquigas, which will distribute liquefied petroleum gas to bulk installations in both North Island and South Island, for onward sale to automotive fuel outlets.

NEW ZEALAND REFINING COMPANY LIMITED
Address: Marsden Point, PO Box 44, Whangarei
Telephone Ruakaka 27011 Telex 2275
Gross Revenue: NZ$ 41.5 million (1980)
Total Assets: NZ$ 107.2 million
Energy Sector: Oil

New Zealand Refining Company (NZRC) operates the country's only oil refinery, at Whangarei, North Island. Participants include affiliates of major international oil companies BP, Shell and Mobil.

NZRC meets the major part of New Zealand's requirements for petroleum products having a crude oil throughput capacity of around three million tonnes per annum. The refinery is linked by pipeline to Marsden power station and a pipeline for clean products is being constructed to Auckland. Investment in the refinery will expand its capacity and enable it to meet the prospective pattern of product demand, thereby eliminating the need for imports of any of the main oil products.

OFFSHORE MINING COMPANY LIMITED
Address: Petrocorp House, 86 Lambton Quay, Wellington
Telephone 739 812
Gross Revenue: NZ$ 39.5 million (1981)
Energy Sectors: Oil, gas

Offshore Mining Company is a wholly owned subsidiary of Petroleum Corporation of New Zealand.

OMC was formed in 1973 but was purchased by the state when arrangements by Maui participants were completed and now represents the State's 50 per cent shareholding in the field. The company also holds Petrocorp's 25 per cent interest in Liquigas Ltd, the consortium company set up in association with Shell, BP, Todd, Fletcher Challenge and New Zealand Industrial Gases, to market liquefied petroleum gas as a transport fuel.

PETROLEUM CORPORATION OF NEW ZEALAND LIMITED
Address: Petrocorp House, 86 Lambton Quay, PO Box 5082, Wellington
Telephone 739 812 Telex 31146
Gross Revenue: NZ$ 69.3 million (1981)
Total Assets: NZ$ 790.0 million
Energy Sectors: Oil, gas

Petroleum Corporation of New Zealand (Petrocorp) was established in 1978 as the state undertaking responsible for ensuring that onshore and offshore petroleum resources are discovered and exploited in line with government energy plans. Petrocorp has four principal subsidiary companies, Natural Gas Corporation of New Zealand, Offshore Mining Company, Petroleum Corporation of New Zealand (Exploration) and Petrochemical Corporation of New Zealand.

Petrocorp's activities currently centre on the development of the Maui field and the utilisation of the gas as fuel and feedstock, as well as in the expanding natural gas network serving residential and commercial sectors. The corporation will hold a substantial interest in petrochemical plants utilising the gas, e.g. for methanol and fertilisers, and it has a 25 per cent interest in Liquigas Ltd, which has been formed to distribute and market gas liquids as a transport fuel.

SHELL OIL NEW ZEALAND LIMITED
Address: PO Box 2091, 96-102 The Terrace, Wellington
Telephone 720 080 Telex 3331
Energy Sectors: Oil, gas

The principal activity of Shell Oil New Zealand is the marketing of petroleum products, most of which are obtained from the New Zealand Refining Company, in which Shell has an interest.

Shell companies in New Zealand are also involved in exploration and production. Shell holds a 37.5 per cent interest in Shell BP and Todd Oil Services Ltd, and thus an 18.7 per cent interest in the Maui field development.

STATE COAL MINES
Address: Mines Division, Ministry of Energy, Anvil House, Wakefield Street, PO Box 6342, Wellington
Telephone 735 755 Telex 31341
Gross Revenue: NZ$ 37.7 million (1980)
Total Assets: NZ$ 130.9 million
Energy Sector: Coal

State Coal Mines is the operating arm of the Ministry of Energy's Mines Division responsible for operating state-owned coal mines, which produce around 1.3 million tonnes per annum and account for more than half of total national production. The Division oversees the activities of the private coal mining companies and is responsible for licensing mines and ensuring standards of safety and operations. By virtue of the Geothermal Energy Act and the Atomic Energy Act the Mines Division is also involved in some aspects of geothermal and nuclear development.

Of the Division's total production more than half is obtained by open-cast extraction. The long established coal-producing area of Waikato is being revitalised with major investment in modern mechanical mining equipment to provide the large quantities of coal required by the Huntly power station. An extensive programme of exploration and evaluation of reserves throughout the country is being undertaken.

PROFESSIONAL INSTITUTIONS AND TRADE ASSOCIATIONS

COAL MERCHANTS FEDERATION
Address: 25 Waterloo Road, L Hutt, Wellington
Telephone 664 762
Energy Sector: Coal

The Coal Merchants Federation represents the interests of distributors of coal, and forms an interface with the state coal mines organisation and private sector mining companies.

ELECTRICAL SUPPLY AUTHORITIES ASSOCIATION OF NEW ZEALAND
Address: 60 Kent Terrace, Wellington
Telephone 859 632
Energy Sector: Electricity

The Electrical Supply Authorities Association is an organisation of undertakings involved in the supply, distribution and retailing of electricity.

GAS ASSOCIATION OF NEW ZEALAND
Address: PO Box 10340, 155-161 The Terrace, Wellington
Telephone 725 850
Energy Sector: Gas

The Gas Association brings together undertakings involved in gas supply and marketing and provides a forum for discussion of technical and economic issues of general interest.

Pakistan

GOVERNMENT DEPARTMENTS AND OFFICIAL AGENCIES

MINISTRY OF PETROLEUM AND NATURAL RESOURCES
Address: Pakistan Secretariat, Islamabad
Telephone 21 50 1

The Ministry has wide responsibilities for the development of oil and gas resources and supervision of public and private sector companies involved in production and distribution of oil, gas and coal.

MINISTRY OF WORKS AND IRRIGATION
Address: Pakistan Secretariat, Islamabad
Telephone 29 25 9

The Ministry of Works and Irrigation includes within its responsibilities the development of water resources for power production as well as irrigation and it has a general responsibility for the activities of the electricity generation and transmission companies.

PAKISTAN MINERAL DEVELOPMENT CORPORATION
Address: PIDC House, Dr Ziauddin Ahmed Road, Karachi 4
Telex 24385

The Mineral Development Corporation is a state agency responsible for promoting exploration for minerals, including coal, and the development of related production and treatment facilities.

The Corporation is investing in the expansion of output at Sharigh Collieries and a washing plant which will provide coal suitable for Pakistan Steel Mills. Investment is also being put into expansion of Makerwal Collieries and development of the Jhimpir-Meting coal field.

ENTERPRISES (PUBLIC AND PRIVATE SECTOR COMPANIES)

ATTOCK OIL COMPANY LIMITED
Address: PO Rawalpindi
Energy Sector: Oil

Attock Oil Company is a private company operating a refinery at Morgah, between Meyal and Rawalpindi. The refinery has a throughput capacity of 36,000 barrels per day, approximately 1.8 million tonnes per annum, and runs entirely on locally produced crude oil.

KARACHI ELECTRIC SUPPLY CORPORATION LIMITED
Address: Aimai House, Abdullah Haroon Road, Karachi
Telephone 51 67 11 Telex 25601
Gross Revenue: R 1,130.4 million (1980)
Total Assets: R 4,291.6 million
Energy Sector: Electricity

The Karachi Electric Supply Corporation (KESCO) generates and distributes electricity to retail consumers, with total sales of over 2,000 GWh. Installed capacity is 670 MW, based largely on natural gas from the Sui field. The balance of fuel input is oil.

KESCO also incorporates electricity produced from the KANUPP nuclear reactor near Karachi. This reactor has a capacity of 137 MW and has at times provided up to 30 per cent of KESCO's electricity requirements.

NATIONAL REFINERY LIMITED
Address: 7-B Korangi Industrial Zone, Karachi-31
Telephone 31 10 71 Telex 741
Energy Sector: Oil

National Refinery Limited operates a refinery at Karachi with a crude oil throughput capacity of two million tonnes per annum and providing a full range of finished products. The refinery is under state control.

OIL AND GAS DEVELOPMENT CORPORATION
Address: Shafi Chambers, Club Road, Karachi
Telephone 51 061 Telex 23651
Energy Sectors: Oil, gas

The Oil and Gas Development Corporation is a state undertaking responsible for exploration and development of oil and gas. The Corporation carries out major programmes on its own account and has a number of international companies operating under contract.

The Corporation has discovered important oil and gas deposits including a large gas/condensate field at Dhodak. The Corporation is also building up oil production from the Toot field.

PAKISTAN BURMAH-SHELL LIMITED
Address: PBS House, 6 Ch Khaliquzzaman Road, PO Box 3901, Karachi 04-01
Telephone 51 13 76-79 Telex 2612
Gross Revenue: R 5,040.0 million (1980)
Total Assets: R 455.9 million
Energy Sector: Oil

Pakistan Burmah Shell (PBS) is one of the leading marketers of petroleum products in Pakistan. Burmah Oil Company and Shell Petroleum Company each holds a 24.5 per cent interest and National Investment Trust a further 22 per cent. The remaining 29 per cent is widely held.

PBS has depots at Karachi and other main centres in the Punjab and North West Frontier provinces, and also at Quetta (Baluchistan).

PAKISTAN OILFIELDS LIMITED
Address: State Life Building, The Mall, Rawalpindi
Telex 5684
Energy Sector: Oil

Pakistan Oilfields operates oil and gas producing fields in the Meyal area of the northern Punjab. Production of oil from this area amounts to around 250,000 tonnes per annum. Gas production is of the order of 400 million cubic metres per annum. The state has a 40 per cent holding in the company.

PAKISTAN PETROLEUM LIMITED
Address: PO Box 3942, PIDC House, Dr Ziauddin Ahmed Road, Karachi
Telephone 51 13 30 Telex 869
Gross Revenue: R 160.6 million (1980)
Total Assets: R 387.0 million
Energy Sector: Gas

Pakistan Petroleum is the operator for the Sui gas field in Baluchistan. Burmah Oil Company holds 70 per cent of the company. Almost all of the balance is in the hands of the state.

The Sui field accounts for around 80 per cent of the country's natural gas production.

PAKISTAN REFINERY LIMITED
Address: PO Box 4612, Korangi Creek Road, Karachi 2
Telephone 31 01 83 Telex 25453
Energy Sector: Oil

Pakistan Refinery Limited operates an industry refinery at Karachi. Refinery capacity is around two million tonnes per annum. Participants include the international oil companies Burmah Oil, Shell and Caltex and the state holds the share formerly held by Esso Eastern. Offtake of finished products is by marketing affiliates of these companies.

PAKISTAN STATE OIL COMPANY
Address: Dawood Centre, Moulvi Tamizuddin Khan Road, Karachi
Telephone 824
Energy Sector: Oil

Pakistan State Oil Company is a state owned company handling a major proportion, around 57 per cent, of total finished product sales in Pakistan. The company represents the consolidation of distribution and marketing operations formerly owned by Esso Eastern Inc and two locally based marketing companies. The company obtains product supplies from National Refinery Ltd at Karachi and by import.

WATER AND POWER DEVELOPMENT AUTHORITY
Address: WAPDA House, Lahore
Telex 4869
Energy Sector: Electricity

The Water and Power Development Authority (WAPDA) is a public undertaking responsible for the major part of electricity generation and transmission in Pakistan. Its total generating capacity is 3,000 MW, including a substantial element of hydro-electric power.

The Tarbela scheme, which is still being developed represents the country's largest individual centre of electricity production. Other important hydro-electric plant is located at Mangla and Warsak. WAPDA has plans for a large coal-fired station in the Lakhra area and will also use natural gas as a fuel.

WAPDA is constructing a network of high voltage transmission lines including a 500 kV line to Faisalabad, and eventually to Karachi.

PROFESSIONAL INSTITUTIONS AND TRADE ASSOCIATIONS

PETROLEUM INSTITUTE OF PAKISTAN
Address: 4th Floor, PIDC House, Dr Ziauddin Ahmed Road, Karachi
Telephone 510 391
Energy Sectors: Oil, gas

The Petroleum Institute is the representative organisation for companies and personnel engaged in the oil and gas industries. The Institute is a member of the International Gas Union.

Papua New Guinea/ Pacific Islands

GOVERNMENT DEPARTMENTS AND OFFICIAL AGENCIES

DEPARTMENT OF MINERALS AND ENERGY
Address: PO Box 2352, Konedobu, Papua New Guinea
Telephone 21 40 11

The Department of Minerals and Energy has a wide range of responsibilities concerning mineral exploration and development and the formulation of policy on energy. The Department's Energy Planning Unit has carried out many studies into the scope for production and conservation of energy in Papua New Guinea.

NATIONAL ENERGY DEVELOPMENT COMPANY
Address: Boroko, Papua New Guinea

The National Energy Development Company has been established by the government to stimulate the development of the renewable energy resources of Papua New Guinea. The company is able to provide finance for viable projects.

ENTERPRISES (PUBLIC AND PRIVATE SECTOR COMPANIES)

ELECTRIC POWER CORPORATION
Address: PO Box 879, Apia, Western Samoa
Telephone 22261 Telex 63
Energy Sector: Electricity

The Electric Power Corporation is a publicly owned utility responsible for generation, distribution and sale of electricity in Western Samoa. The Corporation's annual system output is around 30 GWh. Generation of electricity is mainly by diesel generators, but the Corporation is also actively developing small hydro-electric plants.

PAPUA NEW GUINEA ELECTRICITY COMMISSION
Address: PO Box 1105, Boroko, Papua New Guinea
Telephone 25 58 33
Energy Sector: Electricity

The Electricity Commission is a public authority with responsibilities for providing a public supply of electricity. The Commission is involved in all aspects of electricity supply, from power station construction to electricity retailing. Generating plant is operated for the larger centres of population from hydro-electric, oil-fired plant or diesel generators.

PUBLIC UTILITIES BOARD
Address: Betio, Tarawa, Kiribati
Energy Sector: Electricity

The Public Utilities Board is responsible for the generation and distribution of electricity in Kiribati (formerly the Gilbert Islands). The Board is a statutory and non-profit making organisation. Supply of electricity is by small localised diesel generating plant.

Philippines

GOVERNMENT DEPARTMENTS AND OFFICIAL AGENCIES

MINISTRY OF ENERGY
Address: 7901 Makati Avenue, Makati, Metro Manila

The Ministry of Energy was formed in 1979, consolidating the activities of several pre-existing agencies, and with a broad responsibility for energy policy and development activities. The Ministry has two principal divisions, the Bureau of Energy Development and the Bureau of Energy Utilization.

The Bureau of Energy Development is responsible for exploration, development and production aspects of all forms of energy. The Bureau of Energy Utilization is concerned particularly with downstream activities of the oil industry in the refining and marketing sectors. Also reporting to the Ministry of Energy are the Philippine Atomic Energy Commission and the Board of Energy, which is a quasi-judicial body controlling prices of petroleum products, piped gas and electricity tariffs.

Principal programmes which the Ministry is currently developing include the 'alcogas' programme for broadening the use of alcohol as a fuel and the project for a major petrochemical plant.

NATIONAL ELECTRIFICATION ADMINISTRATION
Address: CDFC Building, 1050 Quezon Avenue, Quezon City, Metro Manila
Telephone 99 87 81

The National Electrification Administration is a state agency established to organise and promote the provision of electricity to the rural population of the Philippines. Its main strategy is to set up or expand local co-operatives as the owners and operators of small scale generating plant. There are now 114 of these throughout the Philippines.

The NEA is involved in promoting two major developments for use in local power supply. These are mini-hydro plants and dendro-thermal plants utilising renewable forest resources. The NEA anticipates having 500 of these plants in operation by 1987.

ENTERPRISES (PUBLIC AND PRIVATE SECTOR COMPANIES)

BATAAN REFINING CORPORATION
Address: Limay, Bataan
Net Revenue: P9,800 million (1980)
Energy Sector: Oil

Bataan Refining Corporation operates the country's largest refinery, at Limay, where capacity was raised in 1980 to 155,000 barrels per day. The refinery is under the joint ownership of Philippine National Oil Corporation and an affiliate of Mobil Corporation, with PNOC holding a majority 60 per cent share.

Completion of an offshore buoy mooring system has enabled the refinery to receive Very Large Crude Carriers of up to 300,000 dwt. Currently the refinery supplies a large part of national requirements for petroleum products with output of around 100,000 barrels per day, equivalent to five million tonnes per annum.

CALTEX (PHILIPPINES) INC
Address: PO Box 783, 540 Padre Faura, Metro Manila
Telephone 59 70 31
Energy Sector: Oil

Caltex (Philippines) is a wholly owned subsidiary of Caltex Petroleum Corporation, the joint affiliate of Standard Oil Company of California and Texaco. The company operates a refinery at San Pascual, on Luzon, with a throughput capacity of over three million tonnes per annum. The refinery forms the base for Caltex's marketing operations in the Philippines.

MALANGAS COAL CORPORATION
Address: Merritt Road, Fort Bonifacio,
Metro Manila
Telephone 85 89 61 Telex 22259
Energy Sector: Coal

Malangas Coal Corporation is a subsidiary of PNOC, the state oil and energy company, which is engaged in exploration and mining of coal and ores, metals and minerals. Its main project is in the coal areas of Malangas, in Zamboanga del Sur. Two collieries are being developed in this 9,000 hectare area, at Diplahan and Little Baguio. Other mines are being developed at Uling (Cebu), Bislig (Surigao del Sur) and Calatrava (Negros Occidental).

MANILA ELECTRIC COMPANY
Address: Meralco Center, Pasig Rizal, Manila
Telephone 79 92 11
Energy Sector: Electricity

Manila Electric Company (Meralco) is the largest distributor of electricity in the Philippines serving customers in the Metro Manila area. Until the mid 1970s it was also the operator for several large power stations and produced a large part of Philippine electricity. The generation function has since been relinquished to the National Power Corporation which supplies Meralco in bulk from its grid.

NATIONAL POWER CORPORATION
Address: PO Box 2123, Bonifacio Drive,
Port Area, Manila
Telephone 47 21 40/41 Telex 7420120
Net Income: P401 million (1980)
Energy Sector: Electricity

The National Power Corporation (NPC) was set up in 1972 on nationalisation of the Philippine electricity generation sector. The Corporation now operates most of the generation capacity in the Philippines and is undertaking major programmes of development of generating capacity and the transmission network. NPC is directly responsible to the Ministry of Energy and acts in close cooperation with the Ministry's other main agency involved in power development, the Philippine Atomic Energy Commission.

Electricity production by the NPC in 1980 totalled 15.6 TWh, almost all of the public supply. Its generating capacity exceeded 4,000 MW. Approximately 70 per cent of this capacity is oil-fired, with balance hydro-electric or geothermal.

PHILIPPINE GEOTHERMAL INC
Address: Davies Far East Building, Makati,
Metro Manila
Telephone 88 03 86
Energy Sector: Geothermal

Philippine Geothermal Inc (GPI) is a subsidiary company of Union Oil Company of California, which is an important producer of geothermal energy in the United States. PGI is developing two major geothermal fields on Luzon Island, under an agreement with National Power Corporation. At Tiwi, there are four units with total capacity of 220 MW with further units under construction and planned. There is also capacity of 220 MW at Bulalo, with more units planned.

PHILIPPINE NATIONAL OIL CORPORATION
Address: PNOC Building, 7901 Makati Avenue,
Makati, Metro Manila
Telephone 85 90 61 Telex 22259
Gross Revenue: P18,576 million (1980)
Total Assets: P12,700 million
Energy Sectors: Oil, coal, geothermal.

Philippine National Oil Corporation (PNOC) is the state oil company, acting as one of the key agencies for policy and programme implementation directly responsible to the Ministry of Energy. PNOC was set up in 1973, initially with the objectives of ensuring adequate and stable supplies of crude oil and petroleum products to the Philippine market. Since that time, however, its responsibilities have been broadened to include the development of indigenous energy resources.

PNOC has subsidiary companies active in oil, coal, uranium and geothermal resources. PNOC Exploration Corporation is concerned with oil and gas, PNOC Energy Development Corporation is involved in exploration and development of coal, uranium and geothermal resources, and Malangas Coal Corporation is developing coal deposits in Zamboanga. There are also operating subsidiaries in oil refining and petroleum marketing.

PNOC ALCOHOL CORPORATION
Address: Merritt Road, Fort Bonifacio,
Metro Manila
Telephone 85 89 61 Telex 22259
Energy Sector: Renewables

PNOC Alcohol Corporation is a subsidiary company of PNOC, the state oil company. It has been set up to take the leading role in extending the use of alcohol as a fuel, with a remit covering manufacture of alcohol and the production of 'alcogas' fuel for motor vehicles.

PNOC COAL CORPORATION
Address: Merritt Road, Fort Bonifacio,
Metro Manila
Telephone 85 89 61 Telex 22259
Energy Sector: Coal

PNOC Coal Corporation is a subsidiary of the state-owned Philippine National Oil Corporation responsible for exploration and mining of coal and managing the national coal logistics programme. This programme is designed to establish facilities to ensure the handling, transport and distribution capacity necessary for a large scale expansion of the use of coal.

The programme, which is expected to cost P1.35 billion over the period 1982-84, includes expenditure on terminals at Batangas, Naga, Malangas, Poro, Iligan, Surigao and Davao, loading ports at Bislig, Polillo and East and West Batan, and investment in rail depots and coal barges.

Singapore

ENTERPRISES (PUBLIC AND PRIVATE SECTOR COMPANIES)

BP REFINERY SINGAPORE PRIVATE LIMITED
Address: 1 Pasir Panjang Road, Singapore 5
Telephone 63 28 11
Energy Sector: Oil

BP Refinery Singapore Ltd is a wholly owned company of the international oil company British Petroleum. It operates a small refinery at Pasir Panjang, of capacity under 1.5 million tonnes per annum supplying the local marketing affiliate BP Singapore.

ESSO SINGAPORE PRIVATE LIMITED
Address: San Centre, Chin Swee Road, Singapore 0316
Telephone 919100 Telex 21252
Energy Sector: Oil

Esso Singapore is a wholly owned subsidiary of Exxon Corporation engaged in crude oil refining storage, transportation and distribution. It operates a large refinery at Pulau Ayer Chawan, the capacity of which is over 12 million tonnes per annum. Products from the refinery are supplied to the local market and the bunker trade, but a major part of output is shipped to affiliates and other companies for markets in the Far East and Pacific areas.

MOBIL OIL SINGAPORE PRIVATE LIMITED
Address: 18 Pioneer Road, Jurong Town, Singapore 2262
Telephone 65 00 00 Telex 21327
Energy Sector: Oil

Mobil Oil Singapore is a wholly owned subsidiary of Mobil Corporation of New York. It is involved in refining, distribution and marketing in Singapore. The company's refinery at Jurong, has a throughput capacity of nearly 10 million tonnes per annum, and serves other Mobil affiliates in many other countries of the Far East.

PUBLIC UTILITIES BOARD
Address: PUB Building, Somerset Road, Singapore 9
Telephone 235 8888
Gross Revenue: S$1,165.8 million (1980)
Total Assets: S$3,132.6 million
Energy Sectors: Electricity, gas

The Public Utilities Board (PUB) is a publicly owned undertaking responsible for the supply of gas, electricity and water throughout Singapore.

The PUB operates three main power stations, at Pasir Panjang, Jurong and Senoko. Total capacity is just under 2,000 MW, of which 1,100 MW is at Senoko. Further units are being added at Senoko and the PUB has invited tenders to build a coal-fired station. Electricity generation in 1980 amounted to 6,200 GWh.

The PUB's gas system is supplied from a gas works at Kallang, using naphtha as feedstock for reforming. Capacity of the plant is 150,000 cubic metres per day. Production in 1980 was 552 million cubic metres.

SHELL EASTERN PETROLEUM (PRIVATE) LIMITED
Address: 1 Bonham Street, Raffles Place, Singapore 0104
Telephone 43 21 61 Telex 21251
Energy Sector: Oil

Shell Eastern Petroleum is a member company of the Royal Dutch/Shell group. It operates one of the world's key international export refineries at Pulau Bukom. The refinery, which has a total crude oil throughput capacity of around 23 million tonnes per annum is the largest refinery in the Asia/Pacific area and is a source of finished and semi-finished products for Shell affiliates in Singapore, Malaysia and many other countries as well as a centre for international bulk trading.

SINGAPORE REFINING COMPANY
Address: PO Box 225, Jurong Town, Singapore 9161
Telephone 337 4181 Telex 21833
Energy Sector: Oil

Singapore Refining Company operates a consortium-owned refinery at Pulau Merlimau of capacity eight million tonnes per annum. Participants in the refinery are Singapore Petroleum

Company Pte, British Petroleum and Caltex Petroleum. The refinery is geared to producing higher value products for affiliates of the participants and bulk trading. Recent expansion work at the refinery included installation of a 100,000 barrels per day visbreaking unit.

Sri Lanka

GOVERNMENT DEPARTMENTS AND OFFICIAL AGENCIES

MINISTRY OF POWER AND ENERGY
Address: Ceylon Electricity Board Building,
50 Sir Chittampalam A Gardiner Mawatha,
Colombo 2
Telephone 22 051/3

The Ministry of Power and Energy is responsible for all aspects of energy supply and distribution, including the activities of the Ceylon Electricity Board and Ceylon Petroleum Corporation.

ENTERPRISES (PUBLIC AND PRIVATE SECTOR COMPANIES)

CEYLON ELECTRICITY BOARD
Address: 50 Sir Chittampalam A Gardiner Mawatha,
Colombo 2
Telephone 22051/3 Telex 21368
Energy Sector: Electricity

Ceylon Electricity Board is a publicly owned undertaking responsible for generation and distribution of electricity throughout Sri Lanka. The Board uses hydro-electric plant to produce most of its electricity. Hydro-electric capacity is some 330 MW. The Board also maintains 90 MW of thermal plant for peak and standby purposes.

Further major hydro-electric developments are planned by the Board, utilising the resources of the Mahaveli River and its tributaries. By 1984 two projects should have added a further 450 MW of capacity.

CEYLON PETROLEUM CORPORATION
Address: 113 Galle Road, Colombo 3
Telephone 25231 Telex 21167
Energy Sector: Oil

Ceylon Petroleum Corporation is a state-owned company responsible for all aspects of the oil industry. It operates the country's oil refinery, at Sapugaskanda, north of Colombo and distributes oil products throughout the country. The refinery has a capacity of around 2.5 million tonnes per annum, exporting some products to countries in the Indian Ocean area and the Far East.

The Corporation also has responsibility for exploration activity in Sri Lanka. It is planning a programme of exploration activity.

Taiwan

ENTERPRISES (PUBLIC AND PRIVATE SECTOR COMPANIES)

CHINESE PETROLEUM CORPORATION
Address: PO Box 135, 83 Chung Hwa Road,
Taipei
Telephone 361 0221 Telex 11215
Energy Sectors: Oil, gas

Chinese Petroleum Corporation (CPC) is a state-owned company with wide responsibilities in the oil and gas sectors. Established in 1946 it is involved in exploration, production, oil refining, distribution and marketing of oil and gas. It has also assumed responsibility for development of the petrochemicals industry.

Production operations of CPC yield currently some 200,000 tonnes per annum of oil and 1,700 million cubic metres per annum of natural gas. CPC is carrying out intensive exploration operations.

The Corporation imports large quantities of oil which it refines at Kaohsiung and Taoyuan. The Kaohsiung refinery is one of the largest in Asia and has a capacity of around 23 million tonnes per annum.

TAIWAN POWER COMPANY
Address: 39 Ho Ping East Road, Section 1,
Taipei
Telephone 351 8011 Telex 11520
Gross Revenue: NT$ 104.8 billion (1981)
Total Assets: NT$ 333.4 billion
Energy Sectors: Electricity, nuclear power

Taiwan Power Company (Taipower) is a state undertaking with sole responsibility for the development of power resources, construction and generation, distribution and sale of electricity. Sales of electricity exceed 35 TWh per annum. Taipower has achieved almost complete electrification of Taiwan.

Taipower operates some 30 hydro-electric power stations and 16 thermal stations with a capacity of 8,200 MW. Hydro-electric potential is being further exploited, but the main base for future capacity is a programme of nuclear power plant construction. Taipower's first nuclear unit was completed in 1978, a second is being completed in 1982 and a third plant should be completed by the mid 1980s.

Taipower is also planning a substantial increase in coal-fired plant. From a base of only 900 MW using coal at present several major new coal-fired stations are to be built and the Hsinta station is to be converted from oil to coal.

Thailand

GOVERNMENT DEPARTMENTS AND OFFICIAL AGENCIES

NATIONAL ENERGY ADMINISTRATION
Address: Pibultham Villa, Rama I Road, Bangkok 2
Telephone 223 0021

The National Energy Administration is part of the Ministry of Science, Technology and Energy. It is responsible for co-ordinating the activities of the state undertakings involved in energy supply and formulating energy policy.

ENTERPRISES (PUBLIC AND PRIVATE SECTOR COMPANIES)

ELECTRICITY GENERATING AUTHORITY OF THAILAND
Address: 53 Charan Sanitwongse Road, Bangkruai, Nonthaburi, Bangkok
Telephone 424 32 13 Telex 271
Gross Revenue: Baht 10,286.6 million (1980)
Total Assets: Baht 35,934.0 million
Energy Sectors: Electricity, coal

The Electricity Generating Authority of Thailand (EGAT), is a state undertaking responsible for generation and transmission of electricity throughout Thailand. Apart from a small number of local turbine generators operated by the Provincial Electricity Authority, EGAT operates all capacity for the public supply system. This totalled 3,240 MW at the end of 1980, of which 55 per cent was thermal (oil or lignite based) and 39 per cent hydro-electric. EGAT operates lignite mines dedicated to supplying power stations.

In 1980 EGAT power stations generated 14.8 TWh, of which more than 80 per cent was derived from thermal plant. This proportion was 10 per cent higher than in 1979 as a result of poor availability of hydro potential. EGAT's output is mainly sold to the Metropolitan Electricity Authority, for supply in the wider Bangkok area, and to the Provincial Electricity Authority, for other parts of the country. Some three per cent is sold direct to bulk industrial consumers.

METROPOLITAN ELECTRICITY AUTHORITY
Address: 121 Chakphet Road, Bangkok 2
Telephone 212111-20 Telex 2746
Gross Revenue: Baht 13,540.9 million (1981)
Total Assets: Baht 10,065.7 million
Energy Sector: Electricity

The Metropolitan Electricity Authority (MEA) was established as a state enterprise in 1958 with the responsibility for distributing electricity to the capital city area and the provinces of Samutprakarn, Nonthaburi and parts of Pathumthani, a total area of 3,200 square kilometers. Electricity is obtained from the Electricity Generating Authority of Thailand.

The MEA serves a population of more than six million and now has well over 700,000 customers. However, its area includes a high proportion of electricity consumers and energy consuming industries and in 1980 sales of electricity rose to nearly 7,900 GWh, approximately 50 per cent of national consumption. A programme is being completed to bring electricity to all parts of the MEA area.

PETROLEUM AUTHORITY OF THAILAND
Address: 1 Soi Yasoob 1, Viphavadi Rangsit Road, Bangkok 9
Telephone 279 3742 Telex 87940
Energy Sectors: Oil, gas

The Petroleum Authority of Thailand is a state agency established to safeguard national interests in the supply of oil to the Thailand market. The Authority is involved in contracting for crude oil and product supplies and is also exploring for additional oil or gas reserves.

The Authority is co-ordinating the development of the offshore gasfields discovered by Union Oil Company of California. It is constructing the pipeline linking the field to the Bangkok area and principal bulk consumers.

THAI OIL REFINERY COMPANY LIMITED
Address: Sarasin Building, 14 Surasak Road,
Silom, Bangrak, Bangkok 5
Telephone 233 9781 Telex 2695
Energy Sector: Oil

Thai Oil Refinery Company owns and operates an oil refinery at Sriracha. The Royal Dutch/Shell group has a 30 per cent shareholding, with the balance held by Thai shareholders.

Capacity of the Sriracha refinery is just over three million tonnes per annum. Products from the refinery support Shell's marketing operations in Thailand.

Index to Organisations

D

E

F

G

H

I

J

K

L

M

N

O

P

Q

R

S

T

U

V

W

Y

Z

Part Four: Energy Publications

International

001 AFRICA ECONOMIC DIGEST
Middle East Economic Digest, London

English weekly £115.00 p.a.

Commentary and news of current economic developments throughout Africa. Includes information on the many energy-related projects planned or in progress.

002 AFRICA GUIDE
World of Information, Saffron Walden

English annual 300pp

A compendium of economic and social data on a country by country basis with a review of developments in energy and other sectors of the economy.

003 AFRICA—MIDDLE EAST PETROLEUM DIRECTORY
Penn Well Publishing Co, Tulsa

English annual 134pp $30.00

Lists petroleum companies active in Africa and the Middle East, including firms engaged in exploration, drilling production, transportation, petrochemicals, pipeline engineering and construction, equipment and services.

004 AFRICAN BUSINESS
IC Magazines, London

English monthly c80pp £20.00 p.a.

Includes energy news from all African countries, reporting on policies, trade and investment projects.

005 AFRICAN STATISTICAL YEARBOOK
United Nations Economic Commission for Africa, Addis Ababa

English annual 4 vols

Statistics on all sections of the economy, including sections on industry and trade. Individual volumes cover North Africa, East Africa, West Africa, Central Africa and other countries.

006 L'AFRIQUE NOIRE POLITIQUE ET ECONOMIQUE
Ediafric-La Documentation Africaine, Paris

French annual 300pp

Review of developments in the countries of French-speaking West and Central Africa. Contains information on economic projects and government policies and programmes.

007 AGIP—ANNUAL REPORT
AGIP SpA, Rome

English annual 90pp

Contains a review of the international energy situation and information about operations in markets of concern to the AGIP group.

008 ANNUAIRE DE L'AFRIQUE ET DU MOYEN ORIENT
Jeune Afrique, Paris

French annual 300pp

Contains many basic statistical data and other information relating to each country, including French speaking Africa, dealing with energy and other economic developments.

009 ANNUAL REVIEW OF ENERGY
Annual Reviews Inc, Palo Alto

English annual 550pp

Presents a comprehensive review of energy topics, including energy resources, supply technology, end-use, conservation and energy policies.

010 THE ARAB ECONOMIST
Center for Economic, Financial and Social Research and Documentation SAL, Beirut

English monthly 40pp $150.00 p.a.

Journal of economic and business developments. Includes a section relating to oil developments.

011 ARAB OIL AND GAS
The Arab Petroleum Research Center, Paris

English/French/Arabic fortnightly 34pp
$390.00 p.a.

Commentary on developments in oil and gas industries, particularly in the Middle East, news and OPEC and OAPEC statistics.

012 ARAB OIL AND GAS DIRECTORY
The Arab Petroleum Research Center, Paris

English 1978 400pp $135.00

Survey of oil and gas activities in individual Arab countries, including statistics on production and exports, revenues and developments in the energy sector.

013 ASIA AND THE PACIFIC
World of Information, Saffron Walden

English annual 320pp

Compendium of basic data for each country of the Asia/Pacific area, with a review of recent developments in key sectors of the economy.

014 ASIA—PACIFIC PETROLEUM DIRECTORY
Penn Well Publishing Co, Tulsa

English annual 215pp $35.00

Provides a list of petroleum companies active in Asia and the Pacific. Includes firms engaged in drilling, exploration, production, refining, marketing, transportation, petrochemical manufacturing, pipeline engineering and construction, equipment and services.

015 ASIA YEARBOOK
Far East Economic Review, Hong Kong

English annual 280pp

Contains a country by country review of developments in Asia, noting basic data and activities in the energy sector.

016 ASIAN DEVELOPMENT BANK—ANNUAL REPORT
Asian Development Bank, Manila

English annual

Includes details of many energy projects which the Bank is providing financial assistance for.

017 AVAILABILITY OF WORLD ENERGY RESOURCES
Graham & Trotman Ltd, London

English 1980 355pp £15.00

Survey by D C Ion of resources of hydrocarbons, solid fuels, coal, uranium and thorium. Estimates of total energy content, proved reserves. Reserves, production and consumption figures in key countries. Also covers production methods, energy conversion and transportation.

018 THE BALANCE OF SUPPLY AND DEMAND 1978-1990
The Uranium Institute/Mining Journal Books Ltd

English 1979 60pp £8.50

Report analysing demand on the basis of installed nuclear capacity and enrichment capacity and review of supply capabilities and constraints. Includes a number of key tables on capacity of nuclear plant, enrichment capacity and uranium production.

019 BIOMASS
Applied Science Publishers, Barking

English quarterly c80pp £43.00 p.a.

International journal essentially concerned with the use of biomass as an alternative energy source. Topics covered include its significance in developed and less developed countries, global recycling, sources, production, conversion to heat, production of new chemicals and energy balances and economics.

020 BIOMASS AS FUEL
Academic Press Ltd, London

English 1982 222pp $29.50

Examines the resources currently and potentially available in developing and developed countries and the technologies necessary to exploit them.

021 BIOMASS ENERGY PROJECTS
Pergamon Press, Oxford

English 1980 300pp £12.50

Case studies of biomass research studies in various parts of the world. Compares the approaches to the management of biomass energy projects and identifies key policy issues.

022 BRITISH PETROLEUM COMPANY—ANNUAL REPORT
British Petroleum Co Ltd, London

English annual c40pp

Reviews group activities in oil, gas, coal, minerals and petrochemicals throughout the world, with data on crude oil reserves and availability by main areas/ countries and on refining and product sales in Africa/ Middle East and Australasia/Far East.

023 BULLETIN DE L'INDUSTRIE PETROLIERE
SOCIDOC, Paris

French daily 10-20pp

Bulletin of news on developments in all phases of the oil and gas industries, including statistical and survey supplements.

024 CAPITAL INVESTMENT IN THE WORLD OIL INDUSTRY
Chase Manhattan Bank, New York

English annual c40pp

Regular review and analysis of capital investment by oil companies in each phase of the oil industry and on a regional basis.

025 CHEMICAL COMPANY PROFILES—AFRICA, ASIA AUSTRALASIA
Chemical Data Services, London

English 1978 £40.00

Basic information on principal companies involved in the production and sale of chemicals and petrochemicals.

026 CLEAN FUEL SUPPLY
OECD, Paris

English 1978 104pp $6.25

Analyses the technological and economic factors affecting the availability of low sulphur fuels and the introduction of desulphurisation technologies in OECD countries, including Australia, Japan, New Zealand and Turkey, up to the mid 1980s. Provides forecasts of emission levels and costs of desulphurisation.

027 COAL AND ENERGY QUARTERLY
National Coal Board, London

English quarterly 35pp free

Contains articles on technological and economic developments in key coal producing and consuming countries.

028 COAL: BRIDGE TO THE FUTURE
Ballinger Publishing Co, Cambridge (Mass.)

English 1980 247pp £7.00

Report of the World Coal Study involving 16 major coal-using and coal-producing countries. Examines the role that coal might play in meeting world energy needs during the next 20 years.

029 COAL—ENERGY FOR THE FUTURE
Shell International Petroleum Co Ltd, London

English 1980 10pp free

Summarises the key data and findings of the study Coal: Bridge to the Future, which assessed the prospects for coal in the period up to 2000. Published in the series 'Shell Briefing Service'.

030 COAL INTERNATIONAL
Zinder-Neris Inc, Washington

English monthly $170.00 p.a.

Compilation of economic and statistical information on coal, derived from US and other sources, including government publications.

031 COAL: ITS ROLE IN TOMORROW'S TECHNOLOGY
Pergamon Press, Oxford

English 1978 313pp £46.00

A sourcebook on global coal resources, this book establishes the need for coal and the role it will play in the years ahead.

032 COAL STATISTICS INTERNATIONAL
McGraw-Hill Publications, Washington

English monthly

Monthly data on coal production, imports/exports and markets in countries throughout the world. Subscription $217.00 p.a. in North America, $247.00 elsewhere.

033 COAL—THE GROWTH TRADE
Shell International Petroleum Co Ltd, London

English 1981 10pp free

Summary data on coal trade and major developments in terminals, transportation and other aspects of the coal supply chain. Issued in the series 'Shell Briefing Service'.

034 CRUDE OIL IMPORT PRICES 1973-78
OECD, Paris

English/French 1979 22pp

Price data on imports into OECD countries, including Australia, Japan, New Zealand and Turkey, collated from quarterly information gathered through the OECD's Oil Marketing Information System.

035 DEMAND FOR WORLD COAL THROUGH 1995
US Department of Energy, Washington

English 1979 36pp

Collates the results of several models of coal demand in the long term.

036 DEVELOPMENT THROUGH COOPERATION
Ente Nazionale Idrocarburi, Rome

English 1981 2 vols

Study by ENI of the economic interdependence between Arab oil exporting countries and the industrialised countries. Vol 1—the Interdependence Model, including energy and other economic data. Vol 2—proceedings of a seminar between OAPEC and South European countries.

037 DIRECTORY OF SHIPOWNERS, SHIPBUILDERS AND MARINE ENGINEERS
IPC Industrial Press Ltd, London

English annual 1500pp £15.00

Shipping company data with sections on each main type of vessel; shipbuilders/repairers and their dock facilities; engine builders and their licensees; classification societies and their offices worldwide; consultants, government departments, associations and experimentation tanks.

038 ECONOMIC AND SOCIAL SURVEY OF ASIA AND THE PACIFIC
United Nations Economic and Social Commission for Asia and the Pacific, Bangkok

English annual 160pp

General economic review of Asia and the Pacific; including analysis of the impact of energy on economic development and energy demands of the region's economies.

039 ECONOMIC DEVELOPMENT OF THE MIDDLE EAST OIL EXPORTING STATES
Economist Intelligence Unit, London

English 1978 84pp

Special Report No. 58 of the EIU, including a review of trends in the petroleum sector and other economic developments, with comment on the future prospects for each country.

040 ELECTRIC POWER IN ASIA AND THE PACIFIC 1977 AND 1978
United Nations Economic and Social Commission for Asia and the Pacific, Bangkok

English 1981 90pp $10.00

Information on recent and planned power developments in many individual countries of the UNESCAP region, with statistical data on types and capacities of power plant, electricity generated, transmission systems, tariffs and fuel consumption. Issued biennially.

041 THE ELECTRICITY SUPPLY INDUSTRY
OECD, Paris

English/French 1978 82pp $7.50

Results and analysis of the 24th, 25th and 26th annual enquiries into electricity supply industries in OECD countries, including Australia, Japan, New Zealand and Turkey. Statistical data on plant, operations, investments, production and use.

042 THE ELECTRICITY SUPPLY INDUSTRY IN OECD COUNTRIES 1974-76 AND PROSPECTS TO 1980/1985/1990
OECD, Paris

English/French 1978 190pp $9.00

Review of electricity supply in individual OECD member countries, including Australia, Japan, New Zealand and Turkey, covering: capacity, production, consumption, investment, fuel used in generation, and analysis of power consumption in transport and industry.

043 ENCYCLOPEDIA OF ENERGY
McGraw-Hill Publications, Washington

English 1976 785pp

Contains some 300 articles written by specialists on all aspects of energy, including economic, political, technological and environmental.

044 ENERGY
Pergamon Press, Oxford

English monthly 100-200pp

Journal carrying substantial articles on issues and developments in the energy sector of an economic, technical and political nature.

045 ENERGY: A GLOBAL OUTLOOK
Pergamon Press, Oxford

English 1981 300pp £17.00

Written by the Governor of the Saudi Arabian state oil company Petromin. Contains detailed supply and demand analyses for each main region with relevant statistical background.

046 ENERGY AND HYDROCARBONS
Ente Nazionale Idrocarburi, Rome

English/Italian annual 238pp

Basic statistics on: energy consumption and production; reserves, production, trade and consumption of oil; refineries; nuclear power; uranium reserves and production.

047 ENERGY AND THE ECONOMY
OECD, Paris

English 1981 114pp $7.50

Report on the OECD/IEA Symposium in June 1981, Discusses the economic situation and energy policies, the consequences of changes in the international petroleum market, the alternatives to oil, the problems related to energy as a critical political element.

048 ENERGY AND THE INVESTMENT CHALLENGE
Shell International Petroleum Co Ltd, London

English 1979 8pp free

Review of the changing energy situation and the financial implications of developing new and higher cost resources of energy. Includes indicative figures for investment costs of different forms of energy and total investments by the world petroleum industry. Published in the series 'Shell Briefing Service.'

049 ENERGY AND THE OCEANS
IPC Publications Ltd, London

English 1981 $32.50

Updated version of the French edition. Provides comprehensive and objective surveys of the contributions which the oceans can make to satisfy energy demands.

050 ENERGY ASPECTS OF THE FOREST INDUSTRIES
Pergamon Press, Oxford

English 1979 418pp £27.00

Proceedings of a seminar organized by the Timber Committee of the United Nations Economic Commission for Europe. Discusses the worldwide role of wood and wood residues as a source of energy, and the possibilities for improved efficiency in energy use in the different branches of the forest industry.

051 ENERGY BALANCES OF OECD COUNTRIES 1960-74
OECD, Paris

English/French 1976 512pp $25.00

Long-run statistics for primary and secondary energy sources on a common basis. Details of consumption of fuels by main sectors of the economy. Figures for individual countries, including Australia, Japan, New Zealand and Turkey, and regional groupings. Tables of economic growth rates and energy:GNP coefficients.

052 ENERGY BALANCES OF OECD COUNTRIES 1975/1979
OECD, Paris

English/French 1981 168pp $15.00

Statistics for the four-year period 1975-79 for individual OECD countries, including Australia, Japan, New Zealand and Turkey, and regional groupings, covering: primary and secondary energy sources; consumption of fuels by sectors; pattern of electricity production.

053 ENERGY CONSERVATION—THE ROLE OF DEMAND MANAGEMENT IN THE 1980s
OECD, Paris

English 1981 70pp $6.00

Analysis of past development and prospects for Energy Demand Management policies. Shows that effective management of energy demand is essential to maintain balance in the world oil market as it is expected to develop through the 1980s.

054 ENERGY CONSERVATION IN INDUSTRY
OECD, Paris

English 1979 76pp

Analysis of policies of member countries, including Australia, Japan, New Zealand and Turkey, towards energy conservation in industry under the following headings: financial and fiscal incentives; target setting, reporting and auditing; information and advice; combined heat and power production; utilisation of waste heat and waste fuels. Also reviews price developments and the relationship of energy consumption to industrial production.

055 ENERGY CONSERVATION IN THE IEA
OECD, Paris

English 1979

Review of progress in OECD member countries, including Australia, Japan, New Zealand and Turkey, towards achieving energy conservation objectives. Analyses of measures in effect and proposed.

056 ENERGY DEVELOPMENTS
Organization of Arab Petroleum Exporting Countries, Kuwait

English 1978 90pp

Surveys world developments in energy consumption in the period 1973-1977, energy policies in the main industrialised countries and developments of energy resources. Statistical appendix includes figures for exploration and production activity, reserves, consumption and international trade.

057 ENERGY EFFICIENCY
Shell International Petroleum Co Ltd, London

English 1979 82pp

Reviews the potential for improved energy efficiency and the available technologies for reducing energy consumption. Considers the economics of capital investment for energy conservation and other social and psychological factors which need to be taken into account when formulating policy towards energy conservation.

058 ENERGY EXPLORATION AND EXPLOITATION
Graham and Trotman Ltd, London

English quarterly 84pp

Periodical covering developments in economics, technology and trade in energy, with articles and reports.

059 ENERGY FOR INDUSTRY
Pergamon Press, Oxford

English 1979 420pp £33.00

Concerned with the utilization of energy at maximum efficiency in industry, with several case studies.

060 ENERGY FOR INDUSTRY AND COMMERCE
Energy Publications, Cambridge (England)

English annual c100pp £17.50

Review of energy supply and consumption, including developments in international markets and the outlook for supplies and prices, with particular reference to purchasers in industry and commerce.

061 ENERGY FOR INDUSTRY AND COMMERCE QUARTERLY BULLETIN
Energy Publications, Cambridge (England)

English quarterly 16-24pp £36.00 p.a.

Quarterly briefing service for industrial and commercial managers, carrying a review of the general situation in the international oil market and an assessment of trends in prices of fuel supplies.

062 ENERGY FOR RURAL AND ISLAND COMMUNITIES
Pergamon Press, Oxford

English 1981 250pp £18.50

Identifies practical solutions to the urgent energy supply problems, both technical and economic, experienced by small, isolated communities.

063 ENERGY FROM BIOLOGICAL PROCESSES: TECHNICAL AND ENVIRONMENTAL ANALYSES
Ballinger Publishing Co, Cambridge (Mass.)

English 1981 240pp £11.95

Detailed investigation of nearly all biomass sources, including unconventional sources such as oil-bearing and aquatic plants. Each source is evaluated for its potential energy yield and the feasibility and the environmental impacts of growing and harvesting, or collecting it.

064 ENERGY FROM BIOMASS
Applied Science Publishers, Barking

English 1981 994pp £39.00

Proceedings of the 1st European Communities International Conference on Biomass, 1980. Covers subjects such as prospects for energy from biomass, sources and conversion methods, forestry for energy, fermentation to ethanol and biogas, thermochemical routes to gaseous and liquid fuels, new concepts in fuels by biological routes and implementation in both the developed and undeveloped countries.

065 ENERGY FROM BIOMASS
Shell International Petroleum Co Ltd, London

English 1980 8pp free

Summary of the energy content of biomass in the world and possibilities for using it as an energy source. Sections on: sugar cane and ethanol; wood; aquatic resources; bio-gas from wastes. Published in the series 'Shell Briefing Service'.

066 ENERGY FROM THE BIOMASS
The Watt Committee on Energy Ltd, London

English 1979 76pp £10.50

Examination of the potential of biomass as a source of energy.

067 ENERGY FROM THE WAVES
Pergamon Press, Oxford

English 1979 120pp £10.00

Largely non-technical account of the historical background, and current research and development in the field of wave energy and its planned use.

068 ENERGY—GLOBAL PROSPECTS 1985-2000
McGraw-Hill Publications, Washington

English 1977 292pp $14.95

General analysis of the energy situation under the project Workshops on Alternative Energy Strategies at the Massachusetts Institute of Technology. Examines each fuel, energy demand, conservation, geothermal and solar energy.

069 **ENERGY IN A FINITE WORLD**
Ballinger Publishing Co, Cambridge (Mass.)

English 1981 2 vols £12.50/£29.00

Report by the Energy Systems Program Group of the International Institute for Applied Systems Analysis on its five-year energy study. Vol 1: Paths to a Sustainable Future, examines supply and demand for seven major regions. Vol 2: Global Systems Analysis.

070 **ENERGY IN THE ARAB WORLD**
Arab Fund for Economic and Social Development/ OAPEC

English 1979 $56.00

Proceedings of the First Arab Energy Conference, 1979. Detailed analysis of energy-related data in the Arab countries as a basis for estimating present and future energy requirements and supplies and for suggesting national and regional energy policies to optimize the use of energy sources within the framework of Arab development goals.

071 **ENERGY IN THE WORLD ECONOMY**
Resources for the Future Inc, Washington

English 1971 876pp £10.70

Statistical review of trends in output, trade and consumption. Figures for individual countries/areas and types of fuels. Most series from 1925.

072 **ENERGY INTERNATIONAL**
Miller Freeman Publications Inc, San Francisco

English monthly 32pp

Publications containing general energy news, and reports on technical and technological developments. Distributed free within the industry. Available on subscription at $25 p.a. outside the industry.

073 **ENERGY PERSPECTIVES AFTER OIL AND NATURAL GAS**
The Institution of Gas Engineers, London

English 1979 24pp £1.50

Paper analysing the pattern of demands for energy in an industrial community and the development of new integrated energy systems.

074 **ENERGY POLICIES AND PROGRAMMES OF IEA COUNTRIES**
OECD, Paris

English annual 336pp $18.00

Highlights progress of IEA countries, including Australia, Japan, New Zealand and Turkey, in achieving structural change to move towards economies with minimum reliance on oil. Contains an overview to 1990, statistics and a country-by-country assessment of national energy programmes, with recommendations for further action.

075 **ENERGY POLICY**
IPC Science and Technology Press Ltd, Guildford

English quarterly 80pp £60.00 p.a.

Journal with analytical articles and comment on questions of energy policy.

076 **ENERGY PRODUCTION AND ENVIRONMENT**
OECD, Paris

English 1977 108pp $5.50

Examines the questions of: siting major energy facilities; the environmental impact arising from development of offshore oil and gas; problems associated with coal exploitation; interdependence of energy policies and those concerned with sulphur pollution.

077 **ENERGY PROSPECTS TO 1985**
OECD, Paris

English/French 1974 2 vols $11.25

Detailed analysis of energy resources and prospective demand. Studies the issues of energy conservation in electricity generation, industry and the transport, residential and commercial sectors. Vol 1 main report; Vol 2 annexes and statistics.

078 **ENERGY R & D**
OECD, Paris

English 1975 244pp $8.75

Analyses issues in energy research and development and the policies and programmes of OECD member countries, including Australia, Japan, New Zealand and Turkey.

079 **ENERGY REPORT**
Microinfo Ltd, Alton

English monthly 16pp £35.00 p.a.

Regular news service on a wide range of energy matters of a general, political, economic and semi-technical nature.

080 **ENERGY STATISTICS 1973-75**
OECD, Paris

English/French 1977 244pp $10.00

Statistics for all OECD member countries, including Australia, Japan, New Zealand and Turkey, of primary and secondary energy: analyses the sources of energy and pattern of consumption in individual sectors. Covers only the three-year period 1973-75.

081 **ENERGY STATISTICS 1975-79**
OECD, Paris

English/French 1981 296pp $20.00

Detailed statistics for the four years 1975-79 of primary and secondary energy sources. With analysis of consumption by sector. Figures for individual OECD member countries, including Australia, Japan, New Zealand and Turkey, and regions. Some price tables.

082 **ENERGY: THE NEXT TWENTY YEARS**
Ballinger Publishing Co, Cambridge (Mass)

English 1979 656pp $27.00

Report sponsored by the Ford Foundation and administered by Resources for the Future on longer term energy prospects.

083 **ENERGY WORLD**
Institute of Energy, London

English monthly 30pp

Bulletin on energy news and developments, containing articles on individual topics in supply and end-use of energy.

084 **ENI REPORT**
Ente Nazionale Idrocarburi, Rome

English annual 80pp

Annual report on the activities of the ENI group, reviewing energy developments internationally.

085 **ENVIRONMENTAL BENEFITS AND COSTS OF SOLAR ENERGY**
Lexington Books/Gower Publishing Co

English 1981 160pp $26.00

Compares individual solar-energy systems with conventional technologies according to environmental impact per unit of delivered energy.

086 **EXXON CORPORATION—ANNUAL REPORT**
Exxon Corporation, New York

English annual c50pp

Includes a general review of developments in the oil industry and notes on Exxon activity throughout the world.

087 **FAR EAST ECONOMIC REVIEW**
Far East Economic Review, Hong Kong

English monthly c80pp

Regular report and analysis of energy and other economic and political developments throughout the Far East including the People's Republic of China.

088 **FINANCIAL AND OPERATIONAL INFORMATION**
Royal Dutch/Shell Group, London/The Hague

English annual 24pp free

Supplement to the annual report of the Royal Dutch Petroleum Co and Shell Transport and Trading Co Ltd, containing long-run data on the group's income, assets, liabilities, capital expenditure, oil production, tanker and dry cargo fleet and geographical extent of group exploration and production, natural gas processing, coal, nuclear energy, chemicals, metals and research activity.

089 **FINANCIAL TIMES MINING INTERNATIONAL YEARBOOK**
Longman Group Ltd, Harlow

English annual c650pp

Contains details of many international companies involved in mining and associated activities.

090 **FINANCIAL TIMES OIL & GAS INTERNATIONAL YEARBOOK**
Longman Group Ltd, Harlow

English annual 637pp

Provides information on the major oil and gas companies including directors and registered offices, exploration news, subsidiaries and affiliates, production figures and financial data.

091 **FINANCIAL TIMES WHO'S WHO IN WORLD OIL & GAS**
Longman Group Ltd, Harlow

English annual 620pp £24.00

A biographical listing of over 4,000 senior personnel in over 2,000 organisations throughout the world, including senior executives, government officials, academics, engineers and consultants.

092 **FIRST WORLD-WIDE STUDY OF THE PETROCHEMICAL INDUSTRY 1975-2000**
United Nations Industrial Development Organisation, Vienna

English 1978 255pp

Studies the increasing role of the developing countries in the petrochemical sector. Appendices contain statistics on production capacities, consumption, international trade and trends in final demand. Also includes information on R&D organisations in developing countries, sources of finance, policies and environmental problems.

093 **FOREIGN TRADE STATISTICS FOR AFRICA**
United Nations Economic Commission for Africa, Addis Ababa

English annual

Regular analyses of figures for the region and for individual countries distinguishing imports and exports of minerals and fuels.

094 **FOREIGN TRADE STATISTICS OF ASIA AND THE PACIFIC**
United Nations Economic and Social Commission for Asia and the Pacific, Bangkok

English irregular

The United Nations produces detailed trade statistics for the region and individual countries of Asia and the Pacific, with figures for value and volume of main energy imports/exports.

095 **FUEL AND ENERGY ABSTRACTS**
IPC Science and Technology Press Ltd, Guildford

English 6 issues p.a. £100 p.a.

Published on behalf of the Institute of Energy, London, it contains brief abstracts from over 800 world-wide publications on the scientific, technical, economic, social and policy aspects of fuel and energy.

096 **FUTURE COAL PROSPECTS**
Ballinger Publishing Co, Cambridge (Mass.)

English 1980 577pp £25.00

Contains a detailed description of the prospects for coal prepared by teams from each of the 16 countries participating in the World Coal Study, as well as studies for other regions of the world.

097 **FUTURE COAL PROSPECTS COUNTRY AND REGIONAL ASSESSMENTS**
Ballinger Publishing Co, Cambridge (Mass.)

English 1980 608pp $37.50

Comprehensive study of the international coal trade, based on assessments of supply and demand in individual countries and regions.

098 **FUTURE COAL SUPPLY FOR THE WORLD ENERGY BALANCE**
Pergamon Press, Oxford

English 1979 720pp £54.00

Discusses the most recent technical developments in the supply of coal and the long-term future world coal supply.

099 **FUTURE ENERGY CONSUMPTION OF THE THIRD WORLD**
Pergamon Press, Oxford

English 1981 300pp £20.00

Analysis of energy supply and demand in the Third World and individual evaluation of future energy consumption in 156 countries. Special reference to nuclear power.

100 **THE FUTURE ROLE OF GASIFICATION PROCESSES**
The Institution of Gas Engineers, London

English 1979 32pp £1.50

Paper outlining the scope for applying gasification processes: to upgrade heavy oil; for advanced power cycles; for the ironmaking industry; as substitute natural gas.

101 **THE GASOHOL HANDBOOK**
Holt-Saunders Ltd, Eastbourne

English 1982 580pp £18.00

Provides the first contemporary comprehensive collection and synthesis of this renewable energy resource. Also reviews issues which have to be faced when investing in the gasohol industry and the market potential for gasohol.

102 **GASTECH 78**
Gastech Ltd, Rickmansworth

English 1979 346pp

Report of proceedings at the 1978 international LNG/LPG conference. Covers: resources, country reviews, trade, transportation, markets, pricing and safety aspects.

103 **GAS WORLD**
Benn Publications Ltd, London

English monthly 40pp £26.00 p.a.

Journal containing news and articles on the gas industry. Subscription £32.00 outside UK.

104 **GENERAL REVIEW OF THE WORLD COAL INDUSTRY**
Muir Coal Industry Information Service, Hazlemere

English half-yearly 86pp

Commentary on economic and technological developments in the coal industry and international coal trade.

105 **GEOTHERMAL ENERGY PROSPECTS**
Pergamon Press, Oxford

English 1980 300pp £15.00

An overview of the development of the geothermal industry, backed up by case studies. Published in co-operation with the East-West Centre, Hawaii.

106 GEOTHERMAL RESOURCES
Applied Science Publishers, Barking

English 1979 246pp £20.30

Covers the geology and the source of heat, geothermal systems, geysers, geothermal resources and water, geothermal energy and the environment, research and recent development and the future of geothermal energy.

107 GLÜCKAUF
Verlag Glückauf GmbH, Essen

German/English DM162.00 p.a.

Journal carrying economic and technical articles on mining, current statistics and news. An additional English translation series is available at a total cost of DM396.00 p.a.

108 THE GULF STATES
Metra Consulting Group Ltd, London

English 1981 246pp

Produced as part of the series 'Business Opportunities in the 1980s'. Covers Bahrain, Kuwait, Oman, Qatar and the United Arab Emirates. Includes sections on oil, gas, minerals and electric power developments.

109 HYDRIDES FOR ENERGY SOURCE
Pergamon Press, Oxford

English 1978 610pp £39.00

Published on behalf of the International Association for Hydrogen Energy, it is the proceedings of the International Symposium on Hydrides for Energy Storage held in Norway in 1977. Includes prospects and applications of hydrogen as a carrier for the future.

110 THE HYDROGEN ENERGY ECONOMY
Holt-Saunders Ltd, Eastbourne

English 1977 332pp £16.25

Discusses the economic feasibility and costs of a hydrogen-based economy, dealing with technical obstacles, safety issues, public attitudes and the interests of oil and gas industries.

111 HYDROGEN ENERGY SYSTEM
Pergamon Press, Oxford

English 1979 5 vols £150.00

Proceedings of the 2nd World Hydrogen Energy Conference, covering: methods of hydrogen production; transmission, distribution and storage; uses of hydrogen in technical processes and the energy sector; special applications of hydrogen; overall systems economics and environmental aspects.

112 HYDRO-POWER
Pergamon Press, Oxford

English 1980 560pp £39.00

Account of the use of water as an alternative source of energy and its future prospects.

113 IEA-ANNUAL REPORT
International Energy Agency, Paris

English annual 20pp

Report of the IEA on its activities and progress with regard to the objectives of its establishment including energy conservation, oil substitution and emergency preparedness in member countries, including Australia, Japan, New Zealand and Turkey.

114 IMPROVED ENERGY EFFICIENCY
Shell International Petroleum Co Ltd, London

English 1979 10pp free

Identifies the principal areas in which energy conservation would be significant in industry, domestic and transport sectors. Published in the series 'Shell Briefing Service.'

115 INFORMATION HANDBOOK 1979-80
Shell International Petroleum Co Ltd, London

English 1979 128pp

Pocket handbook of information about oil and natural gas, coal, nuclear as well as petrochemicals and metals industries, with particular reference to activities of the Royal Dutch/Shell group of companies. Chapters deal with: exploration and production, marine transportation, refining, demand and supply, natural gas, coal, and nuclear energy. Contains statistics on many key aspects of energy supply.

116 INTERNATIONAL ATOMIC ENERGY AGENCY—ANNUAL REPORT
International Atomic Energy Agency, Vienna

English annual c60pp

Includes a survey of developments in nuclear engineering and nuclear power generation, reports on international nuclear programmes and statistics of nuclear installations in individual countries.

117 INTERNATIONAL COAL TRADE
Bureau of Mines, Washington

English monthly 40pp

Contains statistics and analysis of the coal trade on a country-to-country basis.

118 INTERNATIONAL ENERGY POLICY
Lexington Books/Gower Publishing Co

English 1980 234pp $29.50

Discussion of energy policy from a global perspective including profiles of Persian Gulf countries, Mexico, and China.

119 INTERNATIONAL ENERGY STATISTICAL REVIEW
Central Intelligence Agency, Washington

English 1981 29pp

A review of production, consumption and international trade, stocks and market prices. Covers oil, natural gas and nuclear power.

120 INTERNATIONAL GAS TECHNOLOGY HIGHLIGHTS
International Gas Technology, Chicago

English fortnightly 4pp $35.00 p.a.

Current information service covering gas plant, contracts, supply situation and general energy developments.

121 INTERNATIONAL PETROLEUM ANNUAL
Energy Information Administration, Washington

English annual 34pp

Contains data on production, consumption, oil refining and reserves throughout the world, with tables of indicative prices for principal oil products.

122 INTERNATIONAL PETROLEUM ENCYCLOPEDIA
Penn Well Publishing Co, Tulsa

English 1981 455pp

Covers all stages of oil, gas and petrochemical processing industries on a country by country basis with detailed information on plant capacities, field production and other geological and technical information.

123 INTERNATIONAL PETROLEUM REVIEW
McGraw-Hill Publications, Washington

English annual 400pp $97.00

Compilation of official US government intelligence reports on the petroleum industry. Analyses developments in selected developing countries. Contains an industry outlook and report on petroleum related equipment in 32 countries.

124 INTERNATIONAL PETROLEUM TIMES
IPC Industrial Press Ltd, London

English fortnightly c45pp $46.00 p.a.

Journal of news and articles of an economic and semi-technical nature dealing with oil and gas developments.

125 INTERNATIONAL POWER GENERATION
Fuel & Metallurgical Journals Ltd, Redhill

English monthly c70pp $46.00 p.a.

Periodical with articles on technological developments in electricity production. Includes reports on electricity and the general energy situation in individual countries and news of new projects.

126 KEY INDICATORS OF DEVELOPING MEMBER COUNTRIES OF THE ASIAN DEVELOPMENT BANK
Asian Development Bank, Manila

English half-yearly 220pp

Statistical series for the economies of many countries, including production of oil, coal, electricity and gas. Issued in April and October.

127 LLOYD'S REGISTER OF SHIPPING—ANNUAL REPORT
Lloyd's, London

English annual c50pp

Includes a brief review of developments in the shipping sector, including bulk carriers and tankers.

128 LLOYD'S REGISTER OF SHIPPING—STATISTICAL TABLES
Lloyd's, London

English annual c80pp

Contains details of the world's merchant fleets, new-buildings, ships lost, broken-up and laid up. Fleet analysed by type, size, age and tonnage.

129 LNG REVIEW
Energy Economics Research Ltd, Wokingham

English 1977 136pp

Review of the technical, economic, market and financial aspects of liquefied natural gas. Analysis of LNG projects, plant, trade and transportation.

130 MARINE SOURCES OF ENERGY
Pergamon Press, Oxford

English 1979 169pp £12.50

Analyses the technical feasibility and cost-benefits of various marine energy sources, as well as their capacity to integrate into a variety of industrial developments.

131 MARITIME TRANSPORT
OECD, Paris

English annual c170pp £4.30

Annual report of the Maritime Transport Committee of OECD. Review of developments and longer-term trends in shipping; policies towards shipping; demand for shipping services; shipping availability; the freight markets. Annex carries statistics on: trade; flag, size and structure of fleets; indices of freight market rates.

132 METHANE PRODUCTION FROM AGRICULTURAL AND DOMESTIC WASTES
Applied Science Publishers, Barking

English 1981 280pp £20.00

Gives an introduction to the subject and covers such topics as production of fuels from biomass, biogas production and energy production by practical scale digesters.

133 MIDDLE EAST ECONOMIC DIGEST
Middle East Economic Digest, London

English weekly 76pp £112.00 p.a.

Specialist periodical dealing with economic developments in the Middle East, containing detailed analysis of individual economies.

134 MIDDLE EAST ECONOMIC SURVEY
Middle East Petroleum & Economic Publications, Nicosia

English weekly c24pp $600.00 p.a.

Regular news and commentary on events in the Middle East and North Africa, including oil and gas projects, supplies and production/export statistics.

135 MIDDLE EAST REVIEW
World of Information, Saffron Walden

English annual 350pp

Review of economic and political developments in individual countries of the Middle East. Includes many basic data on the economies of these countries.

136 MIDDLE EAST YEARBOOK
IC Magazines, London

English annual 304pp

Contains a country by country review of political and economic events with background information and statistical data on major sectors of the economy.

137 MINERAL TRADE NOTES
Bureau of Mines, Washington

English monthly free

Contains information derived from US government foreign services and other sources not readily available elsewhere.

138 MINING ANNUAL REVIEW
Mining Journal Ltd, Edenbridge

English annual c630pp $46.00

Survey of activities and developments in the mining industry during the past year. General articles, reviews of individual countries. Information on companies and mines.

139 MINING HEAVY CRUDE AND TAR SAND
Bowker Publishing Co, Epping

English 1980 500pp £47.00

Details developments, future prospects, problems and technical advances needed in order to exploit heavy oils and tar sands.

140 MINING JOURNAL
Mining Journal Ltd, Edenbridge

English weekly 26pp

Contains news of mining companies, economic and technological developments in mining, and information about contracts and markets. Available only on a combined subscription including Mining Magazine and Mining Annual Review at a cost of £62.00 p.a. (non-UK $145.00).

141 MINING MAGAZINE
Mining Journal Ltd, Edenbridge

English monthly c100pp $23.00

Carries articles on mining and specific mining development or projects. A large part is devoted to coal industry matters.

142 MOBIL CORPORATION—ANNUAL REPORT
Mobil Corporation, New York

English annual c50pp

Includes notes on Mobil activities in Africa, Asia and the Pacific and general comment on developments in the international oil industry.

143 NATIONAL OIL COMPANIES
John Wiley & Sons Ltd, Chichester

English 1981 269pp

Contains information on comparative energy policies of consuming countries and case studies of most of the important national oil companies in consumer countries.

144 NATURAL GAS
Scientific Press Ltd, Beaconsfield

English 1979 368pp

Third edition of the general survey of natural gas by E N Tiratsoo. Deals with: geology; technical background; pipelines and transmission systems for liquefied natural gas; potential sources of gas, both conventional and non-conventional.

145 NEW AFRICAN YEARBOOK
IC Magazines, London

English annual 300pp

Basic data on individual countries throughout Africa, with tables and maps. Review of developments affecting energy supply and demand and other sectors of the economy.

146 THE NEXT TWENTY YEARS: BACKGROUND PAPERS FOR ENERGY
Ballinger Publishing Co, Cambridge (Mass.)

English 1980 464pp $35.00

Report sponsored by the Ford Foundation and administered by Resources for the Future. Also contains policy recommendations.

147 NORTH AFRICA
Metra Consulting Group Ltd, London

English 1979 360pp

Produced in the series 'Business Opportunities in the 1980s' the report summarises and analyses developments and plans in key sectors of the economies of North African countries.

148 NUCLEAR ENERGY AGENCY—ACTIVITY REPORT
Nuclear Energy Agency, Paris

English/French annual 116pp

Annual review of the activities of the Nuclear Energy Agency, with particular regard to member countries, including Australia, Japan, New Zealand and Turkey.

149 NUCLEAR ENGINEERING INTERNATIONAL
International Publishing Corporation Ltd, London

English monthly 70pp £40.00 p.a.

Journal containing articles and news of developments in nuclear industries; profiles of individual power plants.

150 NUCLEAR FUEL
McGraw Hill Publications, Washington

English fortnightly

Regular newsletter dealing with news and developments worldwide affecting the nuclear industry, including project plans, construction and exploration contracts, policy developments, mining economics. Subscription is $635 p.a. in North America and $735 p.a. elsewhere.

151 THE NUCLEAR FUEL BANK ISSUE AS SEEN BY URANIUM PRODUCERS AND CONSUMERS
Uranium Institute, London

English 1979 12pp free

Considers the practicability of an internationally regulated bank for nuclear fuel and outlines several alternative lines of approach designed to minimise disruption to the international trade in uranium and nuclear fuel.

152 NUCLEAR FUEL CYCLE REQUIREMENTS AND SUPPLY CONSIDERATIONS THROUGH THE LONG TERM
Nuclear Energy Agency, Paris

English 1978 84pp $8.75

Special study by the Nuclear Energy Agency of the OECD and the International Atomic Energy Agency, complementing the regular NEA/IAEA review of uranium resources, examining all fuel cycle services. In the first part fuel cycle requirements and forecasts for 1977-2000 are examined. Second part deals with estimates of nuclear power growth between 2000 and 2025 and long term reactor strategies.

153 NUCLEAR POWER
Graham & Trotman Ltd, London

English 1979 220pp £16.00

Analyses trends in world energy demand and supply. Discusses environmental and safety aspects of nuclear power generation.

154 NUCLEAR POWER AND THE ENERGY CRISIS: POLITICS AND THE ATOMIC INDUSTRY
Trade Policy Research Centre, London

English 1978 367pp £12.00

Analysis of decision-making in Western Europe and the United States in the field of nuclear power policy.

155 NUCLEAR POWER IN PERSPECTIVE
Kogan Page Ltd, London

English 1982 240pp £9.95

Contains chapters on The Nature of Nuclear Power, Why We Need Nuclear Power, possible long-term nuclear strategy and social and environmental considerations. Also covers the whole spectrum of conflicting opinions about nuclear power.

156 **OAPEC BULLETIN**
Organisation of Arab Petroleum Exporting Countries, Kuwait

English monthly 32pp

News and analysis of energy issues affecting OAPEC member countries, information on projects, statistics. Annual subscription $48.00 to institutions, $24.00 to individuals.

157 **OECD ECONOMIC OUTLOOK**
OECD, Paris

English half-yearly c150pp $21.50 p.a.

Review of the economic situation in OECD member countries including Australia, Japan, New Zealand and Turkey, and the international situation facing their economies, an important aspect of which is the cost and uncertain availability of imported energy.

158 **OECD OBSERVER**
OECD, Paris

English/French bi-monthly c35pp $9.00

General periodical carrying news of OECD activities and publications. Includes articles on energy in OECD and summaries of major reports on energy.

159 **OFFSHORE CONTRACTORS AND EQUIPMENT WORLDWIDE DIRECTORY**
Penn Well Publishing Co, Tulsa

English annual 455pp $50.00

Listing of contractors involved in the off-shore petroleum industry together with brief histories or descriptions of their activities.

160 **OFFSHORE DEVELOPMENT**
The Financial Times Ltd, London

English 1979 3 vols £80.00

Comprises papers from five conferences on offshore development organised by the Financial Times. Vol 1: Engineering and supply—progress in technology and construction. Vol 2: Engineering and supply—development of exploration and production activity in the Celtic Sea. Vol 3: Finance, taxation and government—financial and political issues relevant to oil and gas companies.

161 **OFFSHORE ENGINEER**
Thomas Telford Ltd, London

English monthly 80pp £25.00 p.a.

Contains news and information about offshore activities throughout the world.

162 **OFFSHORE INSPECTION AND MAINTENANCE**
The Financial Times Ltd, London

English 1978 130pp £50.00

Review of inspection and maintenance aspects of offshore development on the basis of experience in the Gulf of Mexico, Alaska, Venezuela and the North Sea. Examines the requirements and buying practices of operating companies.

163 **THE OIL AND GAS DIRECTORY 1981-82**
Oil & Gas Directory, Houston

English 1981 464pp $30.00

Guide to companies and individuals connected with petroleum exploration, drilling and production worldwide.

164 **OIL AND GAS JOURNAL**
Penn Well Publishing Co, Tulsa

English weekly c250pp $65.00 p.a.

Periodical carrying worldwide news of the oil and gas industries, covering offshore activity, petrochemicals, technological developments, general political and economic news. Annual review supplements, with statistics on main aspects of the oil and gas industries. Subscription within the industries is $24.00 p.a.

165 **OIL AND GAS STATISTICS**
OECD, Paris

English/French annual c560pp $32.50

Provides detailed statistics for OECD countries, including Australia, Japan, New Zealand and Turkey, on the supply and disposal of crude oil, natural gas liquids, refinery feedstocks, and all major finished oil products as well as natural gas.

166 **OIL CRISIS . . . . AGAIN?**
British Petroleum Co Ltd, London

English 1979 20pp

Study of the oil supply situation in the non-communist world, dealing with: oil production and reserves; the influence of the United States and OPEC; the impact of the 1973 crisis; the potential of oil, nuclear power and coal in the period 1980-85.

167 **OIL ECONOMISTS' HANDBOOK**
Applied Science Publishers, Barking

English 1977 192pp

Contains a large statistical section with some long-run data, covering main aspects of oil industry operations. Some price information.

168 **OIL, ENERGY AND GROWTH: THE NEXT TWENTY YEARS**
EuroEconomics, Paris

English/French 1978 61pp FF1,200.00

Study of the long-term energy supply and demand balance. Sections on: petroleum potential, conventional and non-conventional; other energy sources; prospective growth in demand; analysis of price trends.

169 OIL FIELDS OF THE WORLD
Scientific Press Ltd, Beaconsfield

English 1976 400pp $32.00

Survey of the background to development of oilfields in individual countries of the world. Statistics of reserves and production.

170 OIL GAS
Urban-Verlag, Hamburg/Vienna

English monthly 55pp

International edition of 'Erdoel Erdgas Zeitschrift', covering general news and developments in geology, production, engineering, processing, refining and marketing.

171 OIL IN THE MIDDLE EAST
Economist Intelligence Unit, London

English quarterly 36pp £30.00 p.a.

Review of exploration, production, refining, consumption and export trade of Middle East countries. Annual supplement provides a summary of statistics and developments.

172 OIL STATISTICS
OECD, Paris

English/French annual c280pp $20.00

Contains data for all OECD countries, including Australia, Japan, New Zealand and Turkey, on supply and disposal of crude oil, feedstocks, natural gas liquids, natural gas and 17 different petroleum products; exports and imports identified by 58 different origins and destinations; shows refinery input and output. Consumption of main petroleum products is broken down into 28 different end-use sectors.

173 OIL TANKERS AND COMBINED CARRIERS
E A Gibson Ltd, London

English monthly

Regular report and statistics of vessels, owners, tonnages owned by major and other oil companies and by state organisations. Also details of deliveries of new tankers, analysis by size and distribution by flag.

174 ON FOSSIL FUEL RESERVES AND RESOURCES
International Institute for Applied Systems Analysis, Laxenburg

English 1978 35pp

Summarises the results of the studies of three independent groups of experts to assess world fossil fuel resources and reserves, dealing with coal, oil and gas. Estimates of maximum production to the year 2020 are made, based on economic and technical considerations.

175 OPEC
Deutsche Bank AG, Frankfurt/Main

English/German 1978 68pp

Analyses the financial and economic impact of higher oil prices since 1973. Includes figures for oil reserves of exporting countries and balance of payments flows attributable to oil exports.

176 OPEC—ANNUAL REPORT
Organisation of Petroleum Exporting Countries, Vienna

English annual 186pp

Includes a review of the general economic situation in the industrialised and developing economies; developments in the petroleum industry; worldwide exploration, production and reserves; the world tanker fleet; the oil industry in OPEC member countries. Statistical section gives figures for production, reserves, refinery operations, consumption and exports of OPEC countries.

177 OPEC BULLETIN
Organisation of Petroleum Exporting Countries, Vienna

English monthly

News digest of matters relevant to OPEC member countries, commentary and analysis. Most issues include a supplement on operations of OPEC or topical issues.

178 OPEC, OFFICIAL RESOLUTIONS AND PRESS RELEASES 1960-1980
Pergamon Press, Oxford

English 1981 224pp £18.00

Contains resolutions adopted by the OPEC conferences and a selection of Press Releases which announce important developments in OPEC's evolution. Documents are arranged chronologically and there is a name and subject index.

179 OPEC OIL REPORT
Petroleum Press Bureau Ltd, London

English 1979 302pp £42.00

Compendium of data on OPEC member countries with sections on: revenues, prices; world energy consumption; international trade; natural gas and LNG availability and trade; refining and petrochemicals developments.

180 OPEC, THE GULF AND THE WORLD PETROLEUM MARKET
Holt-Saunders Ltd, Eastbourne

English 1982 256pp £13.50

Traces in detail all the main aspects of OPEC and its impact on the world petroleum market, the determining of prices, the special deals arranged by trading partners, OPEC's self-image and likely future, the possibilities of cartelization of gas and the dilemmas of recycling of petro-dollar surpluses and the need for new initiatives.

181 OPEC: TWENTY YEARS AND BEYOND
Holt-Saunders Ltd, Eastbourne

English 1982 240pp £14.95

Covers the variety of major issues arising from the new power ascribed to OPEC and the international, economic and political ramifications of OPEC.

182 OPERATING EXPERIENCE WITH NUCLEAR POWER STATIONS IN MEMBER STATES
International Atomic Energy Agency, Vienna

English annual 324pp $24.00

Contains operating information on over 160 units in countries subject to IAEA supervision.

183 OPERATING EXPERIENCE WITH NUCLEAR POWER STATIONS IN MEMBER STATES—PERFORMANCE ANALYSIS REPORT
International Atomic Energy Agency, Vienna

English annual 27pp $4.00

Analytical report based on operating experience in IAEA countries. Covers: construction time span; performance factors, unavailability of plant; comparisons with conventional thermal power plant.

184 ORGANISATION OF ARAB PETROLEUM EXPORTING COUNTRIES—ANNUAL REPORT
Organisation of Arab Petroleum Exporting Countries, Kuwait

English annual c104pp

Reviews: trends in energy consumption in the principal economies; export destinations of Arab crude oil; exploration, production and reserves situation. Reports on developments in the Arab world affecting the international oil situation, with statistics on most aspects.

185 OUR INDUSTRY—PETROLEUM
British Petroleum Co Ltd, London

English 1977 600pp

Handbook on oil industry operations, with special reference to BP. Includes a statistical section with basic figures for production, reserves and consumption in the main countries and regions of the world.

186 OUTLOOK FOR WORLD OIL INTO THE TWENTY FIRST CENTURY
Electric Power Research Institute, Palo Alto

English 1978 180pp

Presents forecasts of oil supply and demand in the non-communist world for the periods 1976-90 and 1990-2005, treating oil as the balancing source of energy. Special emphasis on oil from OPEC countries.

187 THE PETROCHEMICAL INDUSTRY: TRENDS IN PRODUCTION AND INVESTMENT TO 1985
OECD, Paris

English 1979 76pp $8.00

Covers trends in demand, availability of feedstock and investment in the petrochemical industry in OECD countries, including Australia, Japan, New Zealand and Turkey, to 1985. Brief analysis of likely developments in non-member countries over the same period.

188 PETROLEUM DIRECTORY—EASTERN HEMISPHERE
Petroleum Publishing Co, Tulsa

English 550pp $50.00

Directory of companies engaged in exploration, production, drilling, pipelines, processing, engineering, construction and services. Indexed by country and type of activity.

189 PETROLEUM ECONOMIST
Petroleum Press Bureau Ltd, London

English/French/Japanese monthly c40pp $94.00 p.a.

Periodical with news of the oil and gas industries and articles on specific energy topics. Includes reports on the international petroleum products and freight markets.

190 PETROLEUM INTELLIGENCE WEEKLY
Petroleum Intelligence Weekly Inc, New York

English weekly c12pp $950.00 p.a.

Worldwide news service for oil and gas developments. Regularly includes information on crude oil and product prices at export terminals and in the principal bulk markets.

191 PETROLEUM NEWS SOUTH EAST ASIA
Petroleum News, Hong Kong

English monthly c80pp

Regular periodical specialising in developments in the oil and gas industries in South East Asia.

192 PIPELINE AND GAS JOURNAL
Harcourt Brace Jovanovich, New York

English 10 issues p.a. c80pp $25.00 p.a.

Journal containing news and articles on the gas supply industry, pipelines operations, engineering and construction. Reduced subscription rates apply within the petroleum industry.

193 PLATTS OILGRAM NEWS SERVICE
McGraw-Hill Publications, Washington

English daily 4-8pp $707.00 p.a.

Daily news intelligence service covering political and economic developments worldwide in or affecting the oil industry.

194 PLATTS OILGRAM PRICE SERVICE
McGraw-Hill Publications, Washington

English daily 4-8pp $707.00 p.a.

Daily news intelligence service carrying information on prices of crude oil and products in international and domestic markets throughout the world.

195 PORTS OF THE WORLD
Benn Publications Ltd, Tunbridge Wells

English annual 1,092pp £25.00

Details of over 2,000 ports giving: latitude and longitude; authority; documents; approach; accommodation (depths), wharves, storage, equipment, provisions, water; container and ro/ro facilities; ore and bulk cargo facilities; tanker and liquefied gas terminals; bunkers; offshore facilities development; ship repairs; charges; towage; pilotage; traffic figures; medical facilities; airport; working hours; local holidays; cargo handled in a working day; officials (Harbour Master, Lloyd's Agents etc).

196 POWER REACTORS IN MEMBER STATES
International Atomic Energy Agency, Vienna

English 1978 125pp $9.00

Fourth issue of the IAEA's computer-based listing of civilian nuclear reactors as at May 1978, including plants in operation, under construction, planned or shut down.

197 THE PRICING OF CRUDE OIL
Praeger Publishers Inc, New York

English 1975 401pp

Subtitled 'Economic and Strategic Guidelines for an International Energy Policy', examines crude oil prices from the producer countries' point of view, pricing patterns inside and outside the Gulf area, sulphur differentials, relative values of crude oil, gravity differentials.

198 PROCEEDINGS OF THE 10TH WORLD PETROLEUM CONGRESS
Heyden & Son Ltd, London

English 1980 6 vols £420.00

Proceedings of the congress in the following volumes: Vol 1 General; Vol 2 Exploration Supply and Demand; Vol 3 Production; Vol 4 Storage, Transport and Processing; Vol 5 Conservation, Environment, Safety and Training; Vol 6 Index. Volumes are available individually.

199 PROCEEDINGS OF THE 10TH WORLD PETROLEUM CONGRESS VOL 2 EXPLORATION, SUPPLY AND DEMAND
Heyden & Son Ltd, London

English 1980 464pp £98.00

Includes papers on: exploration in new areas; drilling in hostile environments; world reserves of oil and gas; world supply and demand for oil and gas.

200 PROCEEDINGS OF THE 10TH WORLD PETROLEUM CONGRESS VOL 3 PRODUCTION
Heyden & Son Ltd, London

English 1980 424pp £120.00

Papers deal with: technical and economic evaluation of enhanced recovery of crude oil; economics of synthetic hydrocarbons; production from oil sands, oil shale and coal.

201 PROCEEDINGS OF THE 10TH WORLD PETROLEUM CONGRESS VOL 4 STORAGE, TRANSPORT AND PROCESSING
Heyden & Son Ltd, London

English 1980 486pp £120.00

Includes papers on the following subjects: technical and economic aspects of the storage and transport of oil and gas; developments in natural gas liquefaction, handling and transport; technology of processing heavy crude oils and residuals.

202 PROCEEDINGS OF THE 10TH WORLD PETROLEUM CONGRESS VOL 5 CONSERVATION, ENVIRONMENT, SAFETY AND TRAINING
Heyden & Son Ltd, London

English 1980 372pp £98.00

Includes papers dealing with: environmental standards of the petroleum and petrochemical plants in the 1980s and their cost implications; energy conservation in refining and its cost benefits.

203 PRODUCTION AND UTILISATION OF SYNTHETIC FUELS: AN ENERGY ECONOMICS STUDY
Applied Science Publishers, Barking

English 1981 265pp £20.00

Covers synthetic fuels production from other sources of energy, energy requirements evaluation; energy economics and cost estimates of conversion processes.

204 QUARTERLY BULLETIN OF STATISTICS FOR ASIA AND THE PACIFIC
United Nations Economic and Social Commission for Asia and the Pacific, Bangkok

English quarterly 80pp

Includes production data for oil, natural gas, petroleum products, electricity and manufactured gas. Figures for general imports/exports of fuels.

205 QUARTERLY OIL STATISTICS
OECD/International Energy Agency, Paris

English/French quarterly c300pp $50.00

Information for member countries, including Australia, Japan, New Zealand and Turkey, on supply and disposal of crude oil products and natural gas, including production, refinery output, trade, bunkers, refinery fuel and losses and stock changes. Detailed import and export data for 48 origins and 31 destinations for crude oil and products.

206 RENEWABLE ENERGY PROSPECTS
Pergamon Press, Oxford

English 1980 340pp £25.00

Proceedings of a conference on Non-fossil Fuel and Non-Nuclear Fuel Energy Strategies held in the USA in 1979. Discusses the broad energy perspectives, renewable energy resources and energy policy and strategy.

207 RESEARCH ON THE PRODUCTION OF HYDROGEN
International Gas Union, Paris

French 1979 22pp

Reviews existing processes for hydrogen production, concentrating on the hydrolysis of water. Gives indicative figures for investment and operating costs; discusses storage and transport questions. Paper given at the IGU's 1979 international conference.

208 REVIEW OF MARITIME TRANSPORT
United Nations, New York

English annual 95pp $7.00

Annual survey of maritime fleets and principal freight markets, including bulk carriers and tankers.

209 A REVIEW OF THE POTENTIAL FOR THE USE OF NUCLEAR REACTORS FOR HYDROGEN PRODUCTION
Royal Society of Arts, London

English 1977 22pp

Paper reviewing the technology for producing hydrogen, with indications of costs of fuel requirements of the processes.

210 REVIEW OF THE WORLD COAL INDUSTRY TO 1990
Miller Freeman Publications Inc, San Francisco

English 1975 134pp

General survey of the coal industry. Part I survey of the main consuming countries. Part II the major export sources.

211 REVUE DE L'ENERGIE
Les Editions Techniques et Economiques, Paris

French monthly 55pp FF317.00 p.a.

Journal containing articles on energy subjects of an economic or technical nature.

212 THE SEABORNE TRANSPORTATION OF LIQUEFIED NATURAL GAS 1977-85
H P Drewry Ltd, London

English 1977 116pp $140.00

Analysis of the likely costs and revenues of vessels engaged in moving liquefied natural gas, and review of trading patterns in the period to 1985.

213 SHELL BRIEFING SERVICE
Shell International Petroleum Co Ltd, London

English irregular

Series of publications on key aspects of energy supply and markets. Produced for internal information purposes, Shell Briefing Service publications can be obtained on request.

214 SHIPPING STATISTICS AND ECONOMICS
H P Drewry Ltd, London

English monthly 60pp £240.00 p.a.

Continuously up-dated statistics on tanker tonnage, new buildings, orders, scrappage, sales, tankers laid-up, fixings and market freight rates. Figures analysed by type or size of vessel.

215 SHIPPING STATISTICS YEARBOOK
Institute of Shipping Economics, Bremen

English 1981 440pp

Statistics of the world merchant fleet, seaborne trade, ship-building etc. Analysed by country, by port and by sea route.

216 SITING OF MAJOR ENERGY FACILITIES
OECD, Paris

English 1979 126pp $7.50

Describes the technical environmental and socio-economic effects of major energy facilities. Identifies the causes of public opposition, describes the development of regulatory procedures and policies on siting.

217 SMALL-SCALE MINING
Bowker Publishing Co, Epping

English 1980 129pp £6.50

The planning, research and new developments in the mining of small scale deposits of coal, metals and minerals, presented by some 79 engineering and scientific authorities at the United Nation's Symposium on Mining and Small Deposits.

218 SNG FROM RESIDUAL OIL
International Gas Union, Paris

English 1979 16pp

Reviews the state of processes for producing synthetic natural gas from heavy distillates. Paper given at the IGU's 1979 international conference.

219 SOLAR ENERGY
Shell International Petroleum Co Ltd, London

English 1982 10pp free

Identifies the distribution of solar energy throughout the main regions of the world and the means of producing heat and electricity from solar radiation. Published in the series 'Shell Briefing Service'.

220 SOLAR ENERGY IN DEVELOPING COUNTRIES
Pergamon Press, Oxford

English 1979 206pp £26.00

Information on solar energy research and development activities in the developing countries together with details of international organisations engaged in solar activities, literature and information sources and the main suppliers of solar hardware and equipment.

221 SOLAR ENERGY: INTERNATIONAL PROGRESS
Pergamon Press, Oxford

English 1979 2400pp £150.00

Set of four volumes covering the proceedings of the International Symposium-Workshop on Solar Energy, Egypt, 1978. Special emphasis on the near term high priority applications, particularly in the Middle East.

222 SOLAR-HYDROGEN ENERGY SYSTEMS
Pergamon Press, Oxford

English 1979 264pp £21.00

Review of water-splitting systems by solar beam and solar heat, hydrogen production, storage and utilisation.

223 SOLAR VERSUS NUCLEAR: CHOOSING ENERGY FUTURES
Pergamon Press, Oxford

English 1979 250pp £17.50

In-depth examination, initiated by the Swedish Secretariat for Future Studies, of the prospects for replacing the diminishing supplies of fossil fuel with either solar or nuclear energy as a primary energy source.

224 SOME PERSPECTIVES ON OIL AVAILABILITY FOR THE NON-OPEC LDCs
National Foreign Assessment Centre, Washington

English 1980 48pp

Presents data and analysis of the petroleum supply problems of the non-OPEC less developed countries (LDCs), centred on balances of production, consumption and trade of 58 countries.

225 STANDARD OIL COMPANY OF CALIFORNIA—ANNUAL REPORT
Standard Oil Co of California, San Francisco

English annual c50pp

Contains a review of Standard Oil Co activities and general oil and energy developments.

226 STATISTICAL INDICATORS FOR ASIA AND THE PACIFIC
United Nations Economic and Social Commission for Asia and the Pacific, Bangkok

English monthly 80pp

Data for recent months and years on production of individual fuels, electricity and gas in countries of the Asia and Pacific areas.

227 STATISTICAL REVIEW OF THE WORLD OIL INDUSTRY
British Petroleum Co Ltd, London

English annual 32pp free

Annual statistics on world oil production, reserves, refining, transportation, trade and consumption. Also gives data on total primary energy, natural gas, coal and hydro/nuclear power consumption.

228 STATISTICAL YEARBOOK FOR ASIA AND THE PACIFIC
United Nations Economic and Social Commission for Asia and the Pacific, Bangkok

English/French annual 500pp

Statistics on an annual basis for main sectors of the economy, including production of oil, coal, electricity (in public systems and by industrial plants). General fuel trade figures and energy balances.

229 STATISTICS OF FOREIGN TRADE—MONTHLY BULLETIN
OECD, Paris

English/French monthly c116pp S45.00 p.a.

Quarterly figures for value of foreign trade by country for each OECD member country, including Australia, Japan, New Zealand and Turkey. Trade indices. Total trade of EEC, OECD and rest of world. Analysis of trade by broad tariff heading.

230 STATISTICS OF FOREIGN TRADE—TABLES BY COMMODITIES
OECD, Paris

English/French annual 2 vols $30.00

Details for each member country of foreign trade by country of origin/destination under main SITC tariff headings. Vol 1: Imports. Vol 2: Exports.

231 STATISTICS OF FOREIGN TRADE—TABLES BY REPORTING COUNTRY
OECD, Paris

English/French annual 4 vols $30.00

Foreign trade statistics of member countries by main SITC tariff heading and by country of origin/destination.

232 STEAM COAL AND EUROPEAN ENERGY NEEDS TO 1985
Economist Intelligence Unit, London

English 1978 £25.00

An EIU special report (number 52) setting out the likely demand for steam coal to 1985 in 11 West European countries. Reviews electricity production prospects and the competition likely from nuclear power and fuel oil.

233 STEAM COAL PROSPECTS TO 2000
OECD/International Energy Agency, Paris

English 1979 160pp $12.00

Projections of steam coal demand and trade to the year 2000 in an overall energy context. Analyses constraints on the expansion of coal use and trade and identifies policies to encourage the substitution of steam coal for oil.

234 STRATEGIC ISSUES IN THE MIDDLE EAST
Croom Helm Ltd, London

English 1982 192pp £11.95

Considers the major issues in strategic thinking in the Middle East and contains an economic analysis of the Middle East oil situation.

235 SUN: MANKIND'S FUTURE SOURCE OF ENERGY
Pergamon Press, Oxford

English 1978 3 vols £140.00

Proceedings of the International Solar Energy Congress of 1978, covering: international and national plans and programmes; economic and policy aspects; technology of harnessing solar energy; ocean thermal gradients; wind power; agricultural and industrial applications.

236 SUN POWER
Pergamon Press, Oxford

English 1982 220pp £10.00

An introduction to the applications of solar energy. Includes historical background, solar radiation, water and air heating applications, space heating applications, thermal power, economic analysis, biological conversion systems and wind power.

237 SURVEY OF ENERGY RESOURCES
World Energy Conference, London

English 1981

Compendium of detailed information and statistics of resources by country for each main energy type. Compiled for the World Energy Conference of 1980. Part A Text, Part B Appendices and Maps.

238 TANKERS FOR THE 1980s
H P Drewry Ltd, London

English 1980 100pp $295.00

Analysis of the supply and demand situation in the tanker market with an assessment of the market for its investment potential.

239 TEXACO INC—ANNUAL REPORT
Texaco Inc, White Plains

English annual c60pp

Includes notes on Texaco activities and data on refinery runs in principal areas of operation throughout the world.

240 TOWARDS AN ENERGY POLICY FOR TRANSPORT
The Watt Committee on Energy Ltd, London

English 1980 92pp £22.50

Includes discussions on fuel types and future availability, road vehicles of the future, prospects for energy conservation in the railways, energy saving in ships, air transport requirements for 2025, the telecommunications dimension and transport and the consumer.

241 TRANSPORTATION POLICIES AND ENERGY CONSERVATION
Bowker Publishing Co, Epping

English 1976 71pp £3.00

Takes a look at uses of energy in transportation, compares the efficiency of various transportation systems and studies public policies and subsidies received by each.

242 ULCC TRADING OPPORTUNITIES
H P Drewry Ltd, London

English 1977 70pp $150.00

An assessment of the market for tankers of over 300,000 deadweight tonnes.

243 URANIUM AND NUCLEAR ENERGY 1981
Butterworth & Co Ltd, Sevenoaks

English 1982 £27.50

Proceedings of the 1981 annual symposium of the Uranium Institute. Sections deal with: World Energy Perspectives; Uranium Supply and Demand; Nuclear Acceptance; Uranium Production and Exploration; International and Commercial Aspects of Uranium Policy.

244 URANIUM AND NUCLEAR POWER
Union Bank of Switzerland, Zurich

English 1976 39pp

Discusses principal aspects of nuclear energy. Includes an historical section on uranium deposits, the nuclear fuel cycle, uranium as a source of energy, nuclear power plants and the environment, economics of nuclear power plants, current and future uranium supply.

245 THE URANIUM EQUATION—THE BALANCE OF SUPPLY AND DEMAND 1980-95
Mining Journal Books Ltd, Edenbridge

English 1981 58pp £12.00

Published for the Uranium Institute. Contains an analysis of uranium demand, stockpiling policies, reactor programmes, supply capabilities and other operating parameters.

246 URANIUM ORE PROCESSING
International Atomic Energy Agency, Vienna

English 1976 238pp $19.00

Proceedings of an IAEA advisory group examining the capacity to meet an expanding demand for uranium and processes available to new milling plants. Also deals with in-situ treatment and uranium from sea water.

247 URANIUM—RESOURCES, PRODUCTION AND DEMAND
Nuclear Energy Agency/International Atomic Energy Agency, Paris/Vienna

English or French 1979 196pp

Regular biennial review by the two specialist intergovernmental organisations dealing with the uranium situation. Reviews of the availability and cost of resources, growth of demand, constraints on availability and government policies. Includes sections on each individual country involved in uranium exploration and production, with notes on thorium.

248 URBAN DISTRICT HEATING USING NUCLEAR HEAT
International Atomic Energy Agency, Vienna

English/French/Russian 1977 207pp $15.00

Proceedings of an IAEA advisory group covering technological, economic and environmental aspects of the utilisation of waste heat from nuclear power stations for district heating purposes.

249 WATER AND BIOMASS AS ENERGY RESOURCES
International Gas Union, Paris

English 1979 14pp

Comparative economic analysis of costs of producing energy from different types of biomass. Paper given at the IGU's 1979 conference.

250 WESTERN ENERGY POLICY
Macmillan Press Ltd, London

English 1978 198pp £10.00

Part I looks at past policies in a comparative assessment and energy trends in the European Community. Part II examines specifically policies in the United Kingdom, West Germany and the United States; prospects for global energy 1985-2000.

251 WINDOW ON OIL
The Financial Times Ltd, London

English 1978 188pp £50.00

Study by B F Grossling of the prospects for oil production outside the Middle East.

252 WINNING MORE OIL: INCREASED IMPORTANCE OF ENHANCED OIL RECOVERY
Financial Times Business Information Ltd, London

English 1981 238pp $195.00

Review of technologies for enhanced oil recovery and the potential for their use worldwide.

253 **WORKSHOPS ON ENERGY SUPPLY AND DEMAND**
OECD, Paris

English 1979 502pp $30.00

Set of 20 papers presented in workshops held in 1976 and 1977, dealing with aspects of present and future world energy supply and demand.

254 **WORLD COAL**
Miller Freeman Publications Inc, San Francisco

English monthly 100pp

Journal containing news and reports on coal developments. Distributed free within the industry. Subscription outside the industry $30.00 p.a.

255 **WORLD COAL INDUSTRY REPORT AND DIRECTORY 1979-80**
Miller Freeman Publications Inc, San Francisco

English 1979 328pp

Survey of the coal industry in principal countries throughout the world, dealing with: reserves, development, operations and trade. Describes individual mines. Section on the international coal trade, with buyers guide.

256 **WORLD DEVELOPMENT REPORT**
Oxford University Press, Oxford

English annual 200pp

Annual assessment by the World Bank of main development issues, including problems arising from rising oil prices, world energy demand, the potential of new energy sources and the programmes and investment needed to support economic growth in developing countries.

257 **WORLD ENERGY BOOK**
Kogan Page Ltd, London

English 1978 264pp £12.00

Reference source on energy terms and technologies, with a substantial statistical section and maps covering resources and production of energy.

258 **WORLD ENERGY OUTLOOK**
Exxon Corporation, New York

English annual 32pp free

Assesses the general situation of the world economy and energy demand, prospective energy supply, the outlook for electricity, nuclear power, coal, gas, oil and synthetic fuels. Reviews the energy situation in the United States, Europe and Japan.

259 **WORLD ENERGY OUTLOOK**
OECD, Paris

English/French 1977 106pp $12.00

Review of energy supply and demand prospects for OECD countries up to 1985 and in the period beyond 1990. Examines the potential for energy conservation in industry, transport, commerce and the residential sectors. Annexes include energy balance sheets.

260 **WORLD ENERGY RESOURCES 1985-2000**
IPC Science & Technology Press Ltd, Guildford

English 1978 260pp

Executive summaries of papers presented at the World Energy Conference, covering reserves, demand and conservation.

261 **WORLD ENERGY SUPPLIES 1950-74**
United Nations, New York

English 1976 826pp $38.00

Long-run data on production, consumption and trade in solid fuels, crude petroleum, petroleum products, gaseous fuels, electrical energy and uranium, for individual countries throughout the world. Includes figures for electricity generation plant by type in public and industrial sectors.

262 **WORLD ENERGY SUPPLIES 1973-78**
United Nations, New York

English 1979 234pp $20.00

Up-dated statistics for the period 1973-78 covering: production, consumption and trade of solid fuels, crude petroleum, petroleum products, gaseous fuels, electrical energy and uranium. Includes data on capacity in oil refining and electricity generation.

263 **WORLD GAS REPORT**
Noroil Publishing House, Stavanger

English weekly 10pp £15.00 p.a.

Journal containing news items and comment on discoveries, project developments, supply and pricing in international gas trade.

264 **WORLD INDUSTRY IN 1980**
United Nations Industrial Development Organisation, Vienna

English 1981 $16.00

Report on the UNIDO's biennial industrial development survey. Outlines the energy requirements of the manufacturing sector, the response of industry worldwide to changing conditions in energy markets and trends in structural changes resulting from them.

265 WORLD MINERAL STATISTICS 1972-76
Institute of Geological Science, London

English 1979 216pp £15.50

Includes data on coal and uranium. Production, imports and exports of coal, by main type. Production of uranium in nine key countries.

266 WORLD MINES REGISTER
Miller Freeman Publications Inc, San Francisco

English half-yearly 408pp

Industry guide to mining and mineral processing operations worldwide, noting type of operation, ores mined and processed, production capacities and equipment in use.

267 WORLD MINING
Miller Freeman Publications Inc, San Francisco

English monthly 90pp

Journal containing general news on the mining industry and economic and technical articles. Distributed free within the industry and at a subscription of $30-40.00 p.a. to other interested parties.

268 WORLD OIL
Gulf Publishing Co Ltd, Tulsa

English monthly c140pp

Journal with news on oil and gas industries and articles of an economic or technical nature. Distributed free within the industry and on subscription of $8-12.00 p.a. to interested parties outside the industry.

269 WORLD OIL STATISTICS
Institute of Petroleum, London

English 1978 8pp free

Statistics for crude oil reserves, petroleum production, oil refining, consumption, the world oil trade and tanker fleet.

270 WORLD PETROCHEMICALS
The Financial Times Ltd, London

English 1978 127pp £50.00

Collection of papers on the petrochemical industry since 1973. Subjects discussed include the role of producer states and the impact of rising feedstock costs.

271 WORLD TANKER FLEET REVIEW
John I Jacobs and Co Ltd, London

English half-yearly £25.00 p.a.

Statistical analysis of world tanker fleet with commentary on the previous six-months events relating to tankers, port and shipyard developments, technical aspects, LPG and LNG carriers, combined carriers, sale and purchase, market highlights and future outlook.

272 WORLD URANIUM POTENTIAL
Nuclear Energy Agency/International Atomic Energy Agency, Paris/Vienna

English 1978 176pp $16.00

Report of a study by experts of the NEA and the IAEA, extending the analysis of supply and demand in the regular NEA/IAEA surveys.

273 WORLD URANIUM POTENTIAL: AN INTERNATIONAL EVALUATION
OECD, Paris

English 1979 176pp $16.00

Describes the areas of the world containing significant uranium deposits and identifies other areas believed to be favourable for the discovery of uranium resources. Describes efforts currently being undertaken in uranium exploration.

274 WORLDWIDE OPERATING EXPERIENCE WITH COMMERCIAL PLANTS PRODUCING SNG FROM LIGHT PETROLEUM
International Gas Union, Paris

English 1979 20pp

Review of technologies and operating results of plants in Italy, France, the United Kingdom, Japan and the United States. Paper given at the IGU's 1979 international conference.

275 WORLDWIDE PETROCHEMICAL DIRECTORY
Penn Well Publishing Co, Tulsa

English 1982 273pp $45.00

Includes information on companies producing petrochemicals worldwide. Data on companies and individual plants.

276 WORLDWIDE PIPELINES AND CONTRACTORS DIRECTORY
Penn Well Publishing Co, Tulsa

English 1981 242pp $35.00

Guide to pipeline contractors, pipelines for crude oil, natural gas and finished products and pipeline companies. Also covers current pipeline construction activities and construction contracts.

277 WORLDWIDE REFINING AND GAS PROCESSING DIRECTORY
Penn Well Publishing Co, Tulsa

English annual 392pp $50.00

Details of companies active in crude oil and natural gas processing in many parts of the world. Also information on firms engaged in engineering and construction services.

278 WORLDWIDE SURVEY OF MOTOR GASOLINE QUALITY
Associated Octel Ltd, London

English annual c100pp

Contains data on average octane ratings and lead alkyl content of gasoline in countries throughout the world, with figures for total gasoline market in each country.

279 YEARBOOK OF WORLD ENERGY STATISTICS
United Nations, New York

English/French 1981 1209pp

Systematic statistics on UN member countries, covering production and consumption of primary energy forms during 1979, imports and exports, and consumption per capita.

Middle East

Bahrain

300 **BAHRAIN MARKET PROFILE**
Rediffusion International Ltd

English annual 14pp

Report on the Bahrain economy. Summary of exports and imports of oil, reports on the government owned Bahrain Petroleum Company (BAPCO), information on new refinery capacity, statistics on oil production, reserves and future prospects. Also contains information on natural gas production, electricity generated and liquid petroleum gas output.

301 **BAHRAIN MONETARY AGENCY—ANNUAL REPORT**
Bahrain Monetary Agency

English/Arabic annual 39pp

Contains a report on the economic situation of Bahrain and the role of oil exports. Includes statistics and reports on oil production, refining, exports and future developments. Also covers production and consumption of electric power and natural gas.

302 **BAHRAIN NATIONAL OIL COMPANY—ANNUAL REVIEW**
Bahrain National Oil Co

English annual

Review of the activities of the national oil company and of the oil industry in Bahrain.

303 **FOREIGN TRADE STATISTICS**
Directorate of Statistics

English/Arabic annual 400pp

Imports and exports classified by country and commodity and covering about six years. Includes data on coal, coke and briquettes, natural and manufactured gas, petroleum and petroleum products and exports of petroleum products, fuels and lubricants.

304 **STATISTICAL ABSTRACT**
Directorate of Statistics

English annual 223pp

Contains information on crude oil production and refining, Bahrain oil wells, petroleum export statistics for the last six years and monthly for the current year, reports on co-operation with Saudi Arabia and how much oil is produced and marketed locally. Also details natural gas production and distribution and the value of oil imports and exports. Statistics for electric power production and consumption for the last 25 years are included.

Iran

310 **CENTRAL BANK OF IRAN BULLETIN**
Central Bank of Iran

English/Arabic quarterly 320pp

Gives a summary of the economic situation with quarterly and yearly statistics where relevant. Includes: foreign trade, production, export and consumption of crude oil and refined products. Also covers the production, export and domestic consumption for each oil company in Iran and gives the relevant statistics for natural gas production.

311 ECONOMIC REVIEW OF THE YEAR
The Bank of Iran and the Middle East

English annual 90pp

Contains a report on Iran's energy situation. Covers: oil prices; National Iranian Oil Company; investment; production; new fields; pipelines; refineries; natural gas production. Also gives statistical information on exports and gives a report on Iran's gas and electricity industry together with information on Iran's Development Plan and nuclear energy projects.

312 ENERGY POLICY IN IRAN DOMESTIC CHOICES AND INTERNATIONAL IMPLICATIONS
Pergamon Press

English 1981 160pp $16.00

Evaluates the country's future energy production and consumption requirements in the light of the change in government and war with Iraq. Contents cover: oil; natural gas; hydro-electricity; solid and other fuels; delivered electricity; nuclear power and petrochemicals.

313 FOREIGN TRADE STATISTICS
Iran Customs Administration

English/Arabic annual

Official cumulative statistics of imports and exports, including details of crude oil, petroleum products and natural gas.

314 IRAN: A SURVEY OF BUSINESS OPPORTUNITIES
U.S. Department of Commerce

English 1977 296pp

Includes some basic facts about the oil industry, details of exports and imports, the revenues obtained from the oil industry, pipeline usage, refineries, wells drilled, investment, expansion and foreign involvement. Also covers coal production, natural gas production, the gas industry and the electricity supply.

315 IRAN: BUSINESS OPPORTUNITIES FOR THE 1980s
Metra Consulting Group

English 1979

Discusses the Fifth and Sixth Development Plans and gives the budgets for oil, gas and electricity. Outlines the organisation of the Iranian oil industry and discusses their activities, with statistics where appropriate. Gives export figures for oil and gas and outlines future plans both for conventional energy industries and for nuclear power and other alternative energy sources. Also gives tables and graphs on energy usage and consumption.

316 IRAN OIL JOURNAL
National Iranian Oil Co.

English quarterly

Magazine with information on the oil industry. Provides information in total and by company on production, consumption, new exploration, new plants, pipelines, refineries and natural gas production, consumption and flaring. Contains export figures for crude oil, refined oil and natural gas.

317 IRAN TRADE AND INDUSTRY JOURNAL
Echo

English monthly 50pp per issue

Journal of Iranian business and industry. Contains an oil section with information and statistics on oil production and exports and news and statistics about individual companies and projects. Also contains similar news of natural gas.

318 IRAN YEARBOOK
Kayhan Group of Newspapers

English annual 600pp

A compendium of statistical and other information on Iran. Contains information about Iran's development projects and mining industry. Includes a large section on the oil industry, crude oil production, major offshore oil fields, Iran's revenues from oil, petroleum products and pipelines.

Also describes Iran's petrochemical industry and gives statistics for electric power generation, distribution, consumption, rural electrification and investment.

Iraq

---

320 ANNUAL ABSTRACT OF STATISTICS
Central Statistical Organisation

English/Arabic annual 300pp

Includes statistical and other information about the petroleum industry, including number of employees, wages and value of output. Lists the quantity and value of electricity generated and gives trade figures for exports of petroleum products and import figures for coal, coke and briquettes and gas.

321 FOREIGN TRADE STATISTICS
Central Statistical Organisation

English/Arabic annual

Cumulative figures for imports and exports of crude oil and refined products, analysed by country of origin/destination.

322 IRAQ
International Communications Ltd

English 1982 $190.00

Contains information on the state of the oil industry in Iraq and proposals for future energy developments. Published in the series 'International Business Opportunities'.

323 IRAQ: COUNTRY STUDY
The American University

English 1979 317pp

General survey of Iraq dealing with the development of the oil industry before and after nationalisation and the position of Iraq in international oil relations.

324 IRAQ: INTERNATIONAL RELATIONS AND NATIONAL DEVELOPMENT
Westview Press Inc

English 1978

Information on Iraq's oil industry and its relationship to development of the economy.

325 LONG-TERM PROSPECTS OF INDUSTRIAL DEVELOPMENT IN IRAQ
United Nations

English 1980 195pp

Report from the Fifth Industrial Development Conference held in Algeria. It contains details of the Fifth Development Plan and information and statistics on output, consumption, demand and future developments in the oil, gas and electricity power industries. Gives a quarterly comparative summary of trade statistics.

326 STATISTICAL POCKET BOOK
Central Statistical Organisation

English/Arabic annual

Concise data on the Iraq economy, including basic energy information for petroleum and electricity.

Israel

330 BANK OF ISRAEL ANNUAL REPORT
Bank of Israel

English annual 371pp

Includes a report on Israel's industrial production and exports. Covers mining, chemicals and refined petroleum.

331 BUSINESS REVIEW AND ECONOMIC NEWS FROM ISRAEL
IBB Bankholding Corporation Ltd

English quarterly

Contains reports on energy production and requirements and details new projects such as hydro-electric schemes.

332 ENERGY IN ISRAEL 1970-79
Central Bureau of Statistics

English/Hebrew 1981 104pp IS20.00

Basic statistical series on energy covering: energy balances; pattern of energy consumption; production and imports of petroleum; sales of petroleum products; generation and sales of electricity; production and sales of natural gas; fuel consumed by power stations; use of automotive fuels; fuel prices. Published as Special Series No. 660.

333 ENERGY IN ISRAEL 1970-80
Central Bureau of Statistics

English/Hebrew 1981 30pp

Published as a supplement to the Monthly Bulletin of Statistics. Up-dates basic series in Energy in Israel 1970-79, with additional information on final consumption of energy.

334 FOREIGN TRADE STATISTICS ANNUAL
Central Bureau of Statistics

English/Hebrew annual 340pp

Detailed statistics for the most recent year with comparative data for preceding years on imports and exports of energy commodities by country of origin/destination. In two volumes, analysing trade by country and by commodity.

335 FOREIGN TRADE STATISTICS MONTHLY
Central Bureau of Statistics

English/Hebrew monthly 40pp

Cumulative figures on a monthly basis for imports and exports of main energy commodities.

336 ISRAEL ELECTRIC CORPORATION—ANNUAL REPORT
Israel Electric Corporation

English annual c42pp

Reviews operations of the state electricity undertaking, with notes on main power generation and transmission projects and statistics for generating capacity, number of consumers and electricity consumption by sector.

337 THE ISRAEL YEARBOOK
Israel Yearbook Publications

English annual 302pp

Contains statistical and other information on energy supply and demand in Israel and background information on major development projects.

338 MONTHLY BULLETIN OF STATISTICS
Central Bureau of Statistics

English/Hebrew monthly

Contains a summary of Israel's energy balances and statistics on electricity production and consumption and information on petroleum products supply.

339 PLAN FOR THE DEVELOPMENT OF INDUSTRY IN ISRAEL 1976-85
Ministry of Industry, Trade and Tourism

English 1977

Analyses activity and prospects of individual industry sectors with information on output, foreign trade, investment, research and development.

340 STATISTICAL ABSTRACT OF ISRAEL
Central Bureau of Statistics

English/Hebrew annual 830pp IS95.00

Summarised quantitative data on the population and economy of Israel. Includes details of foreign trade and electricity production and supply by the Electric Corporation. Gives a report on the current energy situation in Israel covering both primary and secondary energy and Israel's future energy requirements.

Jordan

345 EXTERNAL TRADE STATISTICS AND SHIPPING ACTIVITIES IN AQABA PORT
Department of Statistics

English/Arabic quarterly 200pp

Gives statistical details of all imports and exports. Contains details of imports of crude petroleum and petroleum gases.

346 FIVE YEAR PLAN
National Planning Council

Current economic development plan for the period 1981-85, including implications for energy supply and investment programmes by main state energy undertakings.

347 JORDAN: A COUNTRY STUDY
The American University

English 1980 308pp

Contains a report on energy and the goal for the 1980 Five Year Plan. Reviews the petroleum supply situation, with details of oil refining, electricity supply and the possibility of uranium reserves in the country.

348 THE JORDAN ECONOMY IN FIGURES
Department of Statistics

English/Arabic monthly 55pp

Contains a two year summary of statistics for all aspects of Jordan's economy. Includes figures for production of fuel oil, gas oil, kerosene and propane.

349 STATISTICAL YEARBOOK
Department of Statistics

English/Arabic annual 194pp

Contains figures for foreign trade in petroleum, statistics for supply of petroleum products and details of electricity generation and consumption.

Kuwait

355 ANNUAL STATISTICAL ABSTRACT
Central Statistical Office

English/Arabic annual 410pp

Section on petroleum contains information on production of crude oil (yearly and monthly statistics), exports according to company and country of destination. Compares the production of crude oil and refined products for each company. Also covers natural gas in a similar manner and compares Kuwait production and export figures with other countries.

356 ECONOMIC REPORT
National Bank of Kuwait

English/Arabic annual

Reports on the domestic economy, including the oil industry. Outlines the part oil plays in the Kuwait economy and gives domestic comparative statistics for the industry, with international comparisons. Statistics cover production and export of crude oil, oil refinery output, and activities of the gas and petrochemical industries.

357 ELECTRICAL ENERGY AND WATER STATISTICAL YEARBOOK
Central Statistical Office

English/Arabic annual 165pp

Summary of main electrical power sources, their development and capacity. Statistical and other information on development of the industry; individual power stations; electricity transmission and distribution; amount of electricity exported, electricity generated, per capita consumption of electricity and details of consumption by major customers. Most statistical information goes back about 25 years.

358 THE KUWAIT ECONOMY IN 10 YEARS: ECONOMIC REPORT FOR THE PERIOD 1969-1979
National Bank of Kuwait

English 1980 175pp

Has special section on the oil industry in which it gives statistical information covering ten years for: oil revenues (with international comparisons); oil markets; development of oil prices; crude oil production; refined oil production.

359 MONTHLY BULLETIN OF FOREIGN TRADE STATISTICS
Central Statistical Office

English/Arabic monthly 210pp

Lists imports, exports and re-exports for Kuwait. Includes petroleum, petroleum products, gas, coal, coke and briquettes.

360 QUARTERLY STATISTICAL BULLETIN
Central Statistical Office

English/Arabic quarterly 38pp

Includes information on electricity generated and consumed. Gives detailed oil statistics: production and export of crude oil; production of refined oil products; production and usage of natural gas; refined products of Kuwait National Petroleum Company; sales of oil products

361 QUARTERLY STATISTICAL BULLETIN
National Bank of Kuwait

English/Arabic quarterly 40pp

Contains comparative statistics for several years with monthly and quarterly figures for the last two years. Includes a statistical survey of the oil industry and information on: crude oil production and exports; production and utilisation of natural gas; exports of refined oil products.

362 SECOND NATIONAL DEVELOPMENT PLAN
Ministry of Planning

English 1977

Five-year development plan for the Kuwait economy, containing major programmes for economic and social development related to earnings from exports of oil and hydrocarbon based products.

Lebanon

365 L'ECONOMIE LIBANAISE
Chambre de Commerce et d'Industrie de Beyrouth

French/Arabic annual

Contains a section on energy dealing with refining and electricity supply. Gives comparative production figures for refineries in the Lebanon and for the production of electricity. Also includes import and export figures.

366 L'ECONOMIE LIBANAISE ET ARABE
Chambre de Commerce et d'Industrie de Beyrouth

French/Arabic monthly

Most recent statistics on a monthly basis for imports of crude oil and refined products.

367 ELECTRICITE DU LIBAN
Imprimerie Habid Eid

French/Arabic biennial

Report on the activities of the electricity supply industry. Contains the annual accounts and statistics on production for a number of years. Information on hydro-electric and thermal plant, production, consumption and distribution. Prepared by the Service Transport et Répartition of Electricité du Liban.

368 JOINT REPORT BY THE COMMERCIAL AND ECONOMIC COUNSELLORS OF THE EUROPEAN COMMUNITY COUNTRIES IN LEBANON
Commission of the European Communities

English 1980 42pp

Analyses the development of trade in the Lebanon. Contains a large section on energy giving statistics on the output from oil refineries. Also describes and gives statistics on the electricity supply system in Lebanon.

369 THE LEBANESE ECONOMY IN 1980
Middle East Consultants

English 1981 110pp

Gives an overall picture of the Lebanese economy and includes a section on power and energy requirements. Covers electric power, electricity production and consumption, upgrading and expansion of the electricity network, demand for petroleum production and figures for supply of crude oil and refined products.

370 STATISTIQUES DU COMMERCE EXTERIEUR
Chambre de Commerce et d'Industries de Beyrouth

French/Arabic annual 2 vols

Figures on an annual basis for imports and exports of crude oil and refined products and movements of solid fuel. Data collected by the Administration des Douanes and the Beirut Chamber of Commerce and Industry.

Oman

375 OMAN FACTS AND FIGURES
Directorate General of National Statistics

English/Arabic annual 32pp

A summary of statistics on the Oman economy, including information on oil production and electricity generation.

376 QUARTERLY BULLETIN ON MAIN ECONOMIC INDICATORS
Directorate General of National Statistics

English/Arabic quarterly 30pp

Contains economic statistics for the last four years, with the preceding year in quarters. Includes oil production and export figures, and imports of mineral fuels, petroleum and petroleum products.

377 STATISTICAL YEARBOOK
Directorate General of National Statistics

English annual 171pp

Contains statistics for a 15-year period for crude oil production and producing fields, price of oil, exports and sales of petroleum products. Also describes the electricity supply system, with details of production and distribution of electricity.

Qatar

380 STATISTICS ON COMMERCE AND INDUSTRY
Ministry of Economy and Commerce

English annual

Annual review of economic activity of key sectors of the Qatar economy.

381 YEARLY BULLETIN OF IMPORTS AND EXPORTS
Customs Department

English annual 99pp

Annual statistics of petroleum production and exports month by month and exports of oil by destination. Contains details of petroleum and petroleum products and manufactured and natural gas imports.

Saudi Arabia

385 ANNUAL IMPORTS AND EXPORTS
Central Department of Statistics

English/Arabic annual 2 vols

Figures on an annual basis for trade in crude oil and petroleum products. Two volumes analyse trade by commodity and by country of origin/destination.

386 ARAMCO—ANNUAL REPORT
Aramco

English annual c32pp

Reviews operations of the principal crude oil producing organisation in Saudi Arabia. Contains details of fields and facilities and statistics for production and refinery throughputs over a long period of years.

387 DEVELOPMENT OF THE INDUSTRIAL CITIES OF JUBAIL AND YANBU
Committee for Middle East Trade

English 1979

Outlines the programmes, including the gas-gathering system. Also covers the petrochemical industry, new terminals, gas and oil pipelines and expected exports.

388 FOREIGN TRADE STATISTICS
Central Department of Statistics

English/Arabic quarterly 340pp

Details of imports and exports of energy commodities, with information on disposals of crude oil and refined products.

389 A GUIDE TO INDUSTRIAL INVESTMENT IN SAUDIA ARABIA
The Industrial Studies and Development Centre

English 1977

Outlines the industrial structure of Saudi Arabia, the opportunities for investment, and the procedures and regulations. Includes a review of the economics of the oil industry and statistical information on the oil and electrical industries. Also contains information on the state organisation Petromin.

390 LONG-TERM PROSPECTS OF INDUSTRIAL DEVELOPMENT IN SAUDI ARABIA
United Nations Industrial Development Organisation

English 1979 173pp

Published following the Fifth Industrial Development Conference for Arab States. Gives an overview of the economy and Saudi Arabia's balance of payments. Discusses the progress of the 2nd Development Plan (1975/76-1979/80). Gives comparative statistical information on the oil companies and their output and exports, and details of the role and activities of government departments and state corporations in the energy supply industries.

391 PETROLEUM STATISTICAL BULLETIN
Ministry of Petroleum and Mineral Resources

English/Arabic annual 75pp

Complete oil statistics for a ten-year period. Includes crude oil production, production and consumption of refined products and natural gas, export figures for crude oil, reserves, tanker loadings, revenue from oil and payments to government.

392 SAUDI ARABIA
International Communications Ltd

English 1981 $190.00

General commentary with statistical and other information on developments in Saudi Arabia, covering energy projects and use of oil revenues. Chapters on: oil production and refining; future oil production; gas production and treatment; minerals exploration; electricity generation and distribution. Published in the series 'International Business Opportunities'.

393 SAUDI ARABIA
Middle East Economic Digest

English 1981 280pp

Includes information about energy projects in the third five-year plan and statistical information on total production, production of the main companies and details of refineries. Outlines the plans for a new gas gathering and processing programme, and new electricity projects. Export and import statistics are also given.

394 SAUDI ARABIA BUSINESS OPPORTUNITIES
Metra Consulting Group

English 1978 223pp

Contains a summary of oil exploration and production. Gives details of oil resources, concessions, disposal of crude oil, refined products, oil refining sites, plans for hydrocarbon-based industries, gas-gathering plans, NGL and crude oil pipelines and export figures.

395 SAUDI ARABIA: RUSH TO DEVELOPMENT
Holt-Saunders

English 1982 350pp £14.95

Analyses the problems and achievements of Saudi development and provides a detailed critique of the Third Development Plan. Includes an analysis of the role of oil in the country's development.

396 SAUDI ARABIAN MONETARY AGENCY—ANNUAL REPORT
Saudi Arabian Monetary Agency

English annual 174pp

Reviews the major developments in the country's economy during the year, including activities in oil, gas and petrochemicals. Gives statistics for Aramco and Petromin and compares them to other companies in Saudi Arabia. Information on the hydrocarbon and mineral-based industries and the new projects associated with them. Also includes information on electrical power, the Electricity Affairs Department, rural electrification and the General Electricity Organisation.

397 SAUDI ARABIAN YEARBOOK
Central Department of Statistics

Background information and statistics on the oil and gas industry, foreign trade in crude oil and petroleum products, energy consumption and electricity supply.

398 STATE, SOCIETY AND ECONOMY IN SAUDI ARABIA
Holt-Saunders

English 1982 320pp £12.95

Deals with general energy issues and policies in Saudi Arabia, including economic, financial and banking aspects.

399 STATISTICAL INDICATOR
Central Department of Statistics

English/Arabic annual 200pp

Comparative statistics for the last 10 years covering the oil industry's activities in production, consumption, refining and exports. Also gives posted prices.

400 STATISTICAL SUMMARY
Saudi Arabian Monetary Agency

English/Arabic annual 60pp

Reviews major economic developments during the year. Statistical tables include: crude oil production, production by Aramco, Getty Oil, Arabian Oil for 10 years with the preceding year by month. Also gives export figures and information on refined products, oil concession areas, reserves and consumption.

401 STATISTICAL YEARBOOK
Central Department of Statistics

English/Arabic annual 540pp

Contains official and non-official statistics about most of the activities for the year. Includes statistics for crude oil production, deliveries, domestic consumption, exports, refineries, reserves of oil, concession areas, refined products, total sales to the general public, prices. Also gives tables for electricity production.

402 THIRD DEVELOPMENT PLAN
Ministry of Planning

English 1980 503pp

Economic development plan for the period 1980-85, containing detailed background information on energy industries, energy supply, use of oil and gas revenues and key development projects and programmes.

403 THE THIRD SAUDI ARABIAN DEVELOPMENT PLAN
Committee for Middle East Trade

English 1981 47pp

Covers plans for the period to 1985 for: crude oil production expansion; replacement of existing equipment; exploration; construction of new pipelines; oil production, exports and reserves; natural gas production, utilisation and exports of LPG; electricity distribution plans.

Syria

410 LE PETROLE EN SYRIE
Office Arabe de Presse et de Documentation

French annual

Review of the oil and gas industry's operations in Syria, covering · production refining and export trade.

411 QUARTERLY SUMMARY OF FOREIGN TRADE
Central Bureau of Statistics

English/Arabic quarterly 75pp

Contains information on a yearly and monthly basis and by international block, main currencies and by countries and destinations for foreign trade. Covers mineral oils, other fuels and energy-related products.

412 STATISTICAL ABSTRACT
Central Bureau of Statistics

English/Arabic annual

Summary statistics for all sectors of the Syrian economy, including information on energy supply.

413 STATISTIQUE DU COMMERCE EXTERIEUR
Direction Générale des Douanes

French annual 2 vols

Includes trade statistics for crude oil, refined products and solid fuels analysed by commodity and by country of origin/destination.

414 SYRIA: A COUNTRY STUDY
The American University

English 1979 268pp

Contains a large section on the economy, industry and energy and energy resources. Includes details of the fourth five-year plan and its anticipated activity in mining and electricity production. Gives information on small crude oil reserves and fields, production and exports, pipelines capacity, natural gas production, the electric power grid and electricity production and distribution.

Turkey

415 ANNUAL FOREIGN TRADE STATISTICS
State Institute of Statistics

English/Turkish annual 550pp

Covers imports and exports by commodities and countries. Includes figures for crude oil, petroleum products, coal and coke, gas and electricity.

416 ECONOMIC SURVEY
OECD

English annual c70pp

Covers economic policy and developments and prospects in major sectors, including energy production, coal lignite and crude oil supplies, coke manufacture and electricity supply. Gives percentage changes from previous years, prices and statistical series over a number of years for trade in oil and solid fuels. Also contains background information on state energy enterprises.

417 MANUFACTURING INDUSTRY QUARTERLY
State Institute of Statistics

English/Turkish quarterly 96pp

Reports and statistics covering the operations of petroleum refineries, coke and briquette production and the electricity supply industry.

418 MONTHLY BULLETIN OF STATISTICS
State Institute of Statistics

English/Turkish or French/Turkish monthly 124pp

Comparative monthly statistics for: crude oil production; oil refining; imports and exports of crude oil and petroleum products; mineral energy sources; fuels used. Also covers electricity production by plants and by power sources.

419 MONTHLY ECONOMIC INDICATORS
State Institute of Statistics

English/Turkish monthly

Contains indicators for total output and activity levels in the oil, gas and electricity supply industries.

420 STATISTICAL ABSTRACT
Istanbul Chamber of Commerce

English annual 32pp

Abstracts of comparative statistics for last five years. Covers petroleum and crude oil production, oil refining, supply of fuels, foreign trade in petroleum products.

421 STATISTICAL YEARBOOK OF TURKEY
State Institute of Statistics

English/Turkish annual 430pp

Eight-year series of comparative statistics and graphs. Includes primary energy production and consumption. Gives details of power plant, production and consumption of electric power, consumption of fuels in electricity production. Lists consumption of electricity by economic activity and the anticipated life of the power plants. Includes production of gas, consumption of coal-gas and selling prices. Also covers imports of all energy-related products and exports of petroleum.

422 SUMMARY OF EXPORTS
State Institute of Statistics

English annual 70pp

Contains a summary of exports of petroleum products on a yearly, quarterly and monthly basis.

423 TURKEY: AN ECONOMIC SURVEY
Turkish Industrialists and Businessmen's Association

English annual

Annual survey of the situation in main industrial sectors. Contains statistical and other information on oil production and foreign trade in fuel products.

United Arab Emirates

425 ANNUAL INDUSTRIAL PRODUCTION STATISTICS
Central Statistics Department

English/Arabic annual 73pp

Includes detailed information on the petroleum industry: number of personnel; wages; output of crude petroleum and natural gas; value added. Covers each member country individually.

426 ANNUAL STATISTICAL ABSTRACT
Central Statistics Department

English/Arabic annual 463pp

Covers production of crude oil by producing company and by month; exports of crude oil by company and month and by country of destination; tankers loaded; consumption of petroleum products; production from refineries; overall crude oil production; production of natural gas.

427 **BULLETIN OF THE CURRENCY BOARD**
Currency Board of the United Arab Emirates

English annual 140pp

Contains comparative statistics for oil exports for each of the member countries of the United Arab Emirates. Also gives statistics for daily average oil production by month and by year and gives details of oil reserves, prices and crude oil exports.

428 **DUBAI ANNUAL TRADE**
External Development Services

English annual 124pp

Prepared for the government of Dubai, it contains trade figures plus reports and statistics for Dubai's main industrial enterprises. Includes comparative trade statistics for crude oil, petroleum products, other fuels, electricity production and consumption. Also carries a report on the Dubai Petroleum Company, giving news of individual fields, new exploration and natural gas production.

429 **DUBAI EXTERNAL TRADE STATISTICS**
Statistical Office, Dubai

English annual 44pp

Comparative studies of imports and exports for two years. Separate section for the Free Trade Zone. Includes statistics for petroleum and petroleum products, and for solid fuels.

430 **FOREIGN TRADE STATISTICS: HALF-YEARLY BULLETIN OF ABU DHABI**
Customs Department

English/Arabic half-yearly

Contains information on imports and exports classified by mode of shipping and by country of origin/destination. Covers minerals, fuels, lubricants and related materials: coal, coke and briquettes; petroleum and petroleum products; natural and manufactured gas.

431 **OIL STATISTICS REVIEW**
Ministry of Petroleum and Mineral Resources

English annual 143pp

Statistics for the UAE on oil and gas production, exploration and development, gas liquefaction, shipments of oil and gas, output of refined products.

432 **PETROLEUM AND THE ECONOMY OF THE UNITED ARAB EMIRATES**
Croom Helm Ltd

English 1977 288pp

Includes a general report on the oil industry in the United Arab Emirates: framework of the industry; oil companies operating in UAE; historical information; UAE's relations with other oil producing countries; the role of petroleum in the country's economic structure.

433 **THE PETROLEUM CONCESSION AGREEMENTS OF THE UNITED ARAB EMIRATES: ABU DHABI 1939-1981**
Croom Helm Ltd

English 1982 2 vols £75.00

Detailed information on concession terms operative in Abu Dhabi since 1939, and including the most recent agreements.

434 **STATISTICAL YEARBOOK**
Central Statistics Department

English/Arabic annual 292pp

Includes: details of exports of crude oil by company, and country of destination over 11 years; production by company; number of tankers loaded by company; local consumption of oil products; production of natural gas; refinery production. Also covers production and sales from the gas liquefaction plant and figures for electricity production and consumption.

Yemen Arab Republic

440 **CENTRAL BANK OF YEMEN—ANNUAL REPORT**
Central Bank of Yemen

English annual 145pp

Review of the Yemen economy and the Bank's activities, including information on imports of petroleum.

441 **FINANCIAL STATISTICAL BULLETIN**
Central Bank of Yemen

English/Arabic quarterly 56pp

Contains information on Yemen's imports and exports of petroleum and other fuels.

442 MARKET OPPORTUNITIES IN THE YEMEN ARAB REPUBLIC
Committee for Middle East Trade

English 1978 115pp

Includes information on electricity generating capacity and the state of the electricity network. Outlines Yemen General Electricity Group's (YGEG) future plans and finances.

443 STATISTICAL YEARBOOK
Central Planning Organisation

English/Arabic annual 115pp

Principal compilation of data on the Yemen economy and foreign trade of main commodities.

Africa

Algeria

450 ALGERIA: INTERNATIONAL BUSINESS OPPORTUNITIES
International Communications Ltd

English 1979 175pp $25.00

Contains information on: petroleum production, exploration, transportation and refining; the Hydrocarbons Development Plan, natural gas, manufactured gas and electricity. Published in the series 'International Business Opportunities'.

451 THE ALGERIAN FIVE YEAR PLAN
Committee for Middle East Trade

English 1981 £6.00

Summary of the main projects in the 1980-84 economic development plan as it affects the oil, gas, electricity supply and mining industries.

452 ANNUAIRE STATISTIQUE DE L'ALGERIE
Direction de Statistiques

French annual 460pp

Contains statistics on Algeria's electricity, gas, mining and hydrocarbons industries. These include figures for production, consumption and import/export trade.

453 BULLETIN TRIMESTRIEL DE STATISTIQUES
Direction de Statistiques

French quarterly c60pp

Contains quarterly statistics on Algeria's electricity, natural gas, mining and petroleum industries together wtih trade statistics.

454 L'ECONOMIE ALGERIENNE 1981
Ediafric-La Documentation Africaine

French 1981 249pp

Contains a summary of the country's energy situation and outlook, and statistics for production, consumption, reserves and trade of the petroleum, natural gas and electricity industries together with notes on other sources of energy such as hydro-electricity.

455 GENERAL REPORT ON THE 1980-84 FIVE YEAR PLAN
Ministère de la Planification et de l'Aménagement du Territoire National

English 1980

Includes information on the status of the investment programme for oil and gas development and the extension of electrification.

Angola

460 NATIONAL PLAN FOR 1981
Government of Angola

English 1981

Guidelines to individual ministries within the context of general policy for 1981-85, including development of hydro-electricity and improved organisation of the energy supply industries.

461 STATISTIQUES DU COMMERCE EXTERIEUR
Instituto Nacional de Estatistica

French/Portuguese annual 325pp

Cumulative statistics of imports and exports analysed by country and by commodity covering crude oil and petroleum products.

Cameroon

465 BULLETIN MENSUEL DE STATISTIQUE
Direction de la Statistique et de la Comptabilité Nationale

French monthly c40pp

Includes details of electricity production and consumption and imports of petroleum products.

466 COUTS ACTUALISES DES FACTEURS AU CAMEROUN
Ministère de l'Economie et du Plan

French 1980 60pp

Includes details of electricity installations, their output and new plants planned or under construction. Also reports on the hydrocarbons industry in Cameroon.

467 L'ECONOMIE CAMEROUNAISE 1981
Ediafric-La Documentation Africaine

French 1981 260pp FF380.00

Survey of the Cameroon economy and current developments. Includes statistical and other data on the petroleum and electricity supply industries.

468 GUIDE ECONOMIQUE
Editions Belibi & Co

French annual 304pp CFA F5,000

Contains information on companies and undertakings involved in the supply of petroleum products and electricity. Data for capacity, output and turnover.

469 LE 4ME PLAN QUINQUENNAL DE DEVELOPPEMENT ECONOMIQUE, SOCIAL ET CULTURAL
Ediafric—La Documentation Africaine

French 1977

The Plan covers the general and economic development of Cameroon, including the electrical industry. Published in 'Les Plans de Développement des Pays d'Afrique Noire'.

470 SO.NA.RA.
Société Nationale de Raffinage

English/French 20pp

Describes the background of the national oil refinery and gives details of plant and storage facilities, financing, company participation and role in the supply of oil in Cameroun.

Egypt

475 ANNUAL REPORT OF ELECTRICITY STATISTICS
Ministry of Electricity and Power

English annual c40pp

Contains information on Egypt's Five-Year Plan and its hydro-electric and thermal power stations, substations and transmission lines. Gives details and statistical tables on fuel consumed by power stations, energy output, electricity consumption and prices.

476 AREA HANDBOOK FOR EGYPT
The American University

English irregular 454pp

Prepared by the Foreign Area Studies group of the American University. It is a compilation of basic facts about Egypt and contains comparative statistics on all Egypt's energy-related industries.

477 EGYPT
International Communications Ltd

English 1982 $190.00

Review of the Egyptian economy, developments in the oil and gas industries and notes of new projects. Published in the series 'International Business Opportunities'.

478 EGYPTIAN COMPANIES DIRECTORY
The Middle East Observer

English annual 114pp

Published as an annual supplement to the Middle East Observer, containing basic information on ministries responsible for petroleum, minerals and electricity, state organisations and related private sector companies. Foreign companies operating in Egypt are listed.

479 THE FIVE YEAR PLAN 1978-82
Ministry of Planning

English 1979 16 vols

Detailed programmes for all sectors of the economy, including volumes dealing with energy projects and energy supply and demand.

480 MONTHLY BULLETIN OF FOREIGN TRADE
Central Agency for Public Mobilisation and Statistics

English monthly

Data on imports and exports of principal energy commodities.

481 STATISTICAL ABSTRACT
Central Agency for Public Mobilisation and Statistics

English annual

Includes basic statistics of supply and consumption of petroleum and electricity, with data on production and trade.

482 STATISTICAL YEARBOOK
Central Agency for Public Mobilisation and Statistics

English annual 280pp

Contains details of activities in the energy sector. Statistics for crude oil production, output of refined products in detail, production/consumption of electricity. In most cases series cover a number of years.

483 SUEZ CANAL REPORT
Suez Canal Authority Press

English monthly 40pp

Contains monthly details and news of all traffic through the canal, including oil and gas tankers. It gives information and statistics on their tonnage, origin and destination.

484 SUEZ CANAL YEARLY REPORT
Suez Canal Authority Press

English annual 160pp

Gives a statistical summary of the traffic through the Suez Canal throughout the year and its destination and tonnage, with comparative information for previous years.

485 YEAR BOOK
Federation of Egyptian Industries

English annual 726pp

The Yearbook is divided into four parts:
Part 1: the report of the Federation, which contains Egypt's Socio-Economic Plan and an overview of Egypt's fuel and other energy-related industries;
Part 2: the report of the Industrial Chambers which contains reports and statistics on petroleum and mining agreements, prospection and exploration, new oil discoveries, crude oil and natural gas production, refineries, consumption and trade together with the relevant details for electricity;
Part 3: comparative production and trade statistics;
Part 4: details of exploration and production agreements.

Ethiopia

490 ETHIOPIA STATISTICAL ABSTRACT
Central Statistical Office

English annual 270pp

Statistical data on the social and economic condition of Ethiopia. Contains current production and trade data together with comparative data for a 3-5 year period. Includes information on the mining industry, electricity generating capacity and electricity production by each station.

Gabon

495 L'ECONOMIE GABONAISE
Ediafric—La Documentation Africaine

French FF470.00

Survey of the Gabon economy and prospective developments. Includes statistical and other data on the energy supply industries.

496 REVUE MENSUELLE
Chambre de Commerce, d'Agriculture, d'Industrie et des Mines du Gabon

French monthly 95pp per issue

Bulletin of the Gabon Chamber of Commerce, Agriculture, Industry and Mining, containing information about uranium production.

Ghana

500 EXTERNAL TRADE STATISTICS
Central Bureau of Statistics

English annual 254pp

Contains comparative trade figures for imports of coal, coke and briquettes, petroleum products and gas. Export figures are given for petroleum products.

501 GHANA—AN OFFICIAL HANDBOOK
Ghana Information Services Department

English annual 540pp

Background information and statistics on the energy sector and in particular the roles and activities of the Volta River Authority, Ghana Oil Company and the public electricity supply system.

502 GHANA COMMERCIAL BANK—ANNUAL REPORT
Ghana Commercial Bank

English

Gives detailed information about commercial and economic activity in Ghana, including reports on energy supply and demand. Covers hydro-electric power, the oil industry and consumption of wood fuel and electricity. Special reports cover other energy sources, exploration and crude oil imports.

503 QUARTERLY ECONOMIC REVIEW
Ghana Commercial Bank

English quarterly c10pp

Gives quarterly reports of commercial and economic events in Ghana. Includes reports on the availability of oil products, electricity consumption, and figures for imports/exports of main fuel commodities.

Ivory Coast

505 LA COTE D'IVOIRE EN CHIFFRES
Ministère de l'Economie, des Finances et du Plan

French annual 324pp

Statistical information and analytical reports on every aspect of the Ivory Coast economy. Includes details of electricity production, consumption and distribution, petroleum exploration and production, and foreign trade in petroleum and petroleum related products.

506 L'ECONOMIE IVOIRIENNE 1980
Ediafric—La Documentation Africaine

French 1981 FF470.00

Survey of the latest year for which detailed economic information is available, covering energy supply industries and other sectors of the economy.

507 LES INDICES DE PRODUCTION INDUSTRIELLE
Direction de la Statistique

French annual 49pp

Includes comparative statistical information on the petroleum and electricity supply industries in the Ivory Coast.

508 LISTE DES ENTREPRISES INDUSTRIELLES
Chambre d'Industrie de Côte d'Ivoire

French annual

Second volume of the annual publication 'L'Industrie Ivoirienne', containing details of turnover, employment and other corporate information, notes on individual plant locations or products. Covers oil and gas production, oil refining and electricity supply. Published by the Chamber of Industry in conjunction with the Syndicat des Entreprises et Industriels.

Kenya

510 ANNUAL TRADE REPORT
Customs and Excise Department

English annual c800pp

Contains both summary and detailed tables for mineral fuels, lubricants and related materials, petroleum and related products, natural gas and manufactured gas.

511 ECONOMIC SURVEY
Central Bureau of Statistics

English annual 212pp Sh45.00

Annual review and analysis of the energy situation in Kenya. Detailed figures for supply and demand of petroleum products and electricity. Also prices of products and overall energy balances. Figures cover several years.

512 FOURTH DEVELOPMENT PLAN
Ministry of Finance and Planning

English 1979 2 vols

The Fourth Plan covers the period 1979-83. Contains background information and analysis of the existing economic situation, and details of government programmes for investment in energy and other infrastructure.

513 KENYA: FOREIGN INVESTMENT IN A DEVELOPING ECONOMY
Business International SA

English 1980 166pp

Survey of the Kenya economy and of the role of the energy sector within it. Contains details of energy supply, consumption of electricity, the role of energy developments and notes on uranium exploration.

514 KENYA STATISTICAL DIGEST
Central Bureau of Statistics

English quarterly 40pp

Regular data on a quarterly basis for movements of crude oil and petroleum products, production and consumption of oil products and electricity.

515 MONTHLY TRADE STATISTICS
Customs and Excise Department

English monthly c100pp

Contains information commodity by commodity and includes statistics on mineral fuels, lubricants and related materials, petroleum and related products.

516 QUARTERLY ECONOMIC REPORT
Central Bureau of Statistics

English quarterly

Contains quarterly comparative statistics of the oil and electricity supply industries, together with a report on the current economic and energy situation.

517 RURAL/URBAN HOUSEHOLD ENERGY CONSUMPTION SURVEY
Central Bureau of Statistics

English 1980 42pp Sh40.00

Results of a survey to establish the pattern of fuels consumed in the residential sector, analysed by region.

518 STATISTICS OF ENERGY AND POWER 1969-77
Central Bureau of Statistics

English 1978 61pp Sh30.00

Detailed statistics and commentary on energy supply and demand in Kenya. Tables for production, trade and consumption of primary energy, electricity production by source and consumption by class of consumer.

Liberia

520 EXTERNAL TRADE OF LIBERIA
Ministry of Planning and Economic Affairs

English annual 325pp

Includes statistical details country by country of imports of mineral fuels, petroleum and petroleum products, coal and related products, gas and electricity.

Libya

525 ANNUAL SURVEY OF THE PETROLEUM INDUSTRY
Libya Information

English annual

Summary information on the activities of the oil and gas sector in Libya.

526 EXTERNAL TRADE STATISTICS
Census and Statistics Department

English/Arabic annual c600pp

Detailed statistics for imports and exports of energy commodities analysed by commodity and by country, covering coal, coke, briquettes, fuel oil, gas and other petroleum products.

527 LIBYA: THE EXPERIENCE OF OIL
Croom Helm Ltd

English 1982 328pp £14.95

Analyses the impact of oil and gas exploitation on the Libyan economy. Reviews oil and gas developments and examines the use of revenues generated by energy exports. Assesses the constraints on utilisation of revenues.

528 OIL AND GAS STATISTICS
Ministry of Petroleum and Economic Affairs

English quarterly

Regular series of data on production of oil and gas in Libya.

529 STATISTICAL ABSTRACT OF LIBYA
Census and Statistics Department

English annual

Basic data on the Libyan economy, including production and supply of energy for the economy and for export.

530 SUMMARY OF THE SOCIO-ECONOMIC TRANSFORMATION PLAN 1981-85
Secretariat of Planning

English 1981 60pp

Contains: an evaluation of the socio-economic transformation plan 1976-80; summary of the sectorial strategies and objectives of the 1981-1985 plan, including the plans for the petrochemical, oil, gas and electricity supply industries; gross fixed investment; the development budget; external and internal trade in energy.

Malagasy Republic

535 BULLETIN MENSUEL DE STATISTIQUE
Ministère des Finances

French monthly

Figures on a monthly basis for electricity production and consumption.

536 SITUATION ECONOMIQUE
Ministère des Finances

French annual 77pp

Annual statistical review of the economy with details of imports of crude oil, coal and gas oil, output of principal oil products from refining, detailed figures for consumption of electricity by sector. Data for the last three years are given.

537 STATISTIQUE DU COMMERCE EXTERIEUR DE MADAGASCAR
Ministère des Finances

French

Regular cumulative figures for imports and exports of commodities, including crude oil and petroleum products.

Malawi

540 ANNUAL STATEMENT OF EXTERNAL TRADE
National Statistical Office

English yearly 280pp

Detailed statistical tables of foreign trade in energy products. Figures for most recent years cover coal, coke and petroleum products.

541 MONTHLY STATISTICAL BULLETIN
National Statistical Office

English monthly 20pp

Includes electricity production, consumption and sales and details of cross-border transfers of electricity.

542 STATISTICAL YEARBOOK
National Statistical Office

English annual 150pp

Includes details of electricity generating stations, the electrical energy generated and consumption and a brief report on the import statistics for oil products.

Mali

545 BULLETIN MENSUEL DE STATISTIQUE
Direction Nationale de la Statistique et d'Information

French monthly 48pp

Monthly bulletin providing recent series for production and consumption of electricity.

546 STATISTIQUES DOUANIERES DU COMMERCE EXTERIEUR
Direction des Douanes

French quarterly

Statistics of imports and exports, including trade in petroleum products.

Morocco

550 ANNUAIRE STATISTIQUE DU MAROC
Direction de la Statistique

French/Arabic annual 206pp

Statistical yearbook providing information on energy in the Moroccan economy. Figures for energy supply and imports/exports.

551 KOMPASS MAROC/MOROCCO 1979-80
Kompass Maroc-Veto

French 1979

Information on companies involved in oil refining and the supply of petroleum products and electricity distribution authorities.

552 LE MAROC EN CHIFFRES
Ministère du Plan et du Développement Régional

French annual 117pp D22.00

Annual figures for production of energy for electricity (by source), crude oil, natural gas, coal and individual petroleum products. Energy consumption and import/export of crude oil. Consumption of electricity by economic sector.

553 OFFICE NATIONAL DE L'ELECTRICITE—RAPPORT D'ACTIVITE
Office National de l'Electricité

French/Arabic annual c30pp

Financial and operational information on the supply and demand for electricity in Morocco, including: production by source; hydro-electric potential; fuels used in thermal power stations; generating capacity and electricity production; sales of electricity by type of consumer.

554 LA SITUATION ECONOMIQUE DU MAROC
Direction de la Statistique

French annual 127pp

Contains an overview and then detailed statistics of developments in energy supply industries. Covers coal, electricity, petroleum and petroleum products.

555 STATISTIQUES DU COMMERCE EXTERIEUR
Ministère des Finances

French/German annual 860pp

Includes statistical information on the export and import of coal, coke, petroleum, petroleum products, gas and other energy related products.

Niger

560 ANNUAIRE STATISTIQUE
Direction de la Statistique

French annual 210pp

Includes statistics of electricity, coal and petroleum supply. Figures for electricity production by fuel source, consumption of electricity and availability of coal and oil.

561 BULLETIN DE STATISTIQUE
Direction de la Statistique

French quarterly 60pp

Regular statistics on the economy, covering availability and consumption of electricity, petroleum and solid fuel.

562 COMMERCE EXTERIEUR
Direction de la Statistique

French annual

Statistics of foreign trade, including imports/exports of petroleum products and solid fuel.

563 MINISTERE DES MINES ET DE L'HYDRAULIQUE—RAPPORT ANNUEL
Ministère des Mines et de l'Hydraulique

French annual 70pp

Detailed information and statistics for the country's energy sector, covering location, output and consumption at individual mines and depots. Figures for uranium, petroleum and hydro-electric plant and general figures for foreign trade.

564 PLAN QUINQUENNAL DE DEVELOPPEMENT ECONOMIQUE ET SOCIAL 1979-83
Ministère du Plan

French 1979 666pp

Current five-year economic development plan, which identifies the requirement for energy and existing resources. Plan is in three volumes, dealing with: overall development; development at sector level; regional development.

Nigeria

570 CENTRAL BANK OF NIGERIA MONTHLY REPORT
Central Bank of Nigeria;

English monthly 45pp

Includes information on production, consumption and prices of petroleum products, electricity and fuel, including also firewood, charcoal and kerosene.

571 ECONOMIC AND FINANCIAL REVIEW
Central Bank of Nigeria

English annual c90pp

Review of the past year for the Nigerian economy. Commentary and key statistics for all aspects, including oil production, electricity supply, generating capacity and oil refineries. Statistics usually cover several years.

572 FOURTH NATIONAL DEVELOPMENT PLAN
Central Planning Office

English 1981

The Fourth National Plan covers the period 1981-85 and contains an analysis of the existing economic situation, the availability of resources, expected revenues to be generated from oil and gas exploitation and energy related projects to support development.

573 INTERNATIONAL BUSINESS OPPORTUNITIES IN NIGERIA
International Communications Ltd

English 1979 214pp £60.00

Includes a report and statistical tables on oil and natural gas in Nigeria. Also covers new exploration and analyses the prospects for the future.

574 NATIONAL ELECTRIC POWER AUTHORITY—ANNUAL REPORT
National Electric Power Authority

English annual 34pp

Contains a general review of the year and gives detailed reports and statistical tables for system operations, distribution, sales, customer service, customer relations, rural electrification and new facilities.

575 NIGERIA
Metra Consulting Group Ltd

English 1981 226pp

Review of the Nigerian economy. Sections deal with electricity generation, transmission and distribution. Statistical and other information on projects and other commercial aspects.

576 NIGERIA
U.S. Department of Commerce

English 1976 190pp

Sub-titled 'A Survey of Business Opportunities'. Includes a report on Nigeria's electric power and the development of the national grid, details of hydro-electric schemes, rural electrification, petroleum exploration, production and refining, natural gas production and a report on the mining situation and new mines opened.

577 NIGERIA: ITS PETROLEUM GEOLOGY, RESOURCES AND POTENTIAL
Graham & Trotman Ltd

English 1982 2 vols £50.00

A complete description of resource potential for oil and gas in Nigeria, with historical review of exploration and development and survey of prospects.

578 STRUCTURE OF THE NIGERIAN ECONOMY
Macmillan Press Ltd

English 1979 270pp £8.95

Study of the Nigerian economy, with sections on mining and resources. Contains data on the location of energy resources and background information on companies involved in energy production and supply.

Senegal

580 BULLETIN STATISTIQUE ET ECONOMIQUE MENSUEL
Direction de la Statistique

French monthly 40pp

Includes comparative statistics on the mining and electricity supply industries and import/export figures for main energy products.

581 L'ECONOMIE SENEGALAISE
Ediafric—La Documentation Africaine, Paris

French annual

Review of the economic situation in Senegal, policy measures and projected development. Data on basic industries, including the supply of electricity and other energy commodities.

582 LA SITUATION ECONOMIQUE DU SENEGAL
Direction de la Statistique

French annual

Annual review and statistics of the Senegal economy, covering activities of the energy supply industries. Data for electricity production and generating capacity, fuels used in generation, imports/exports and availability of petroleum products.

Sierra Leone

585 ANNUAL STATISTICAL DIGEST
Central Statistical Office

English annual 70pp

Statistics for all main aspects of the economy. Includes details of Sierra Leone's imports of petroleum.

586 BANK OF SIERRA LEONE—ANNUAL REPORT
Bank of Sierra Leone

English annual 100pp

Review of developments in the Sierra Leone economy in the past year. Contains some statistical information on energy supply and imports of petroleum.

South Africa

590 AMCOAL—ANNUAL REPORT
Anglo American Coal Corporation Ltd

English annual

Report on the past year's operations by one of South Africa's principal coal mining groups and the general domestic and international context within which it has been operating.

591 ANGLO AMERICAN CORPORATION—ANNUAL REPORT
Anglo American Corporation of South Africa Ltd

English or Afrikaans annual 80pp

Detailed information on individual coal and uranium mining companies of the AAC group or those in which it has interests. Statistics include output, tonnage processed, yield of uranium, and revenues. Consolidates information relating to Anglo American Coal Coporation and gives a review of developments in the energy sector.

592 ATOMIC ENERGY BOARD—ANNUAL REPORT
Atomic Energy Board

English or Afrikaans annual 56pp

Review of activities in the nuclear sector and the Board's participation in international activities. Contains details and statistics of the uranium resource and production situation, and information on nuclear power development and related issues.

593 BULLETIN OF STATISTICS
Department of Statistics

English or Afrikaans quarterly c300pp

Includes statistics on South Africa's mining and electricity supply industries together with details of internal and external trade in coal.

594 CHAMBER OF MINES NEWSLETTER
Chamber of Mines of South Africa

English monthly c8pp

News of developments in the mining sector in South Africa—statistical, technological, market and policy information. Published by the mining industry's organisation.

595 CHAMBER OF MINES OF SOUTH AFRICA—ANNUAL REPORT
Chamber of Mines of South Africa

English annual 120pp

Contains an overview of the year's mining activities, including details of coal and uranium production. Details of research activities, safety performance, future outlook, mining acts, conferences and exhibitions. The statistical tables include coal and uranium production, distribution, revenue and other operational data.

596 CHAMBER OF MINES YEAR-END REVIEW
Chamber of Mines of South Africa

English annual 20pp

Review by the principal mining industry organisation of developments in the sector, market trends and new projects. Contains statistics on output etc for the coal industry and uranium producing mines.

597 COAL, GOLD AND BASE METALS OF SOUTH AFRICA
The Pithead Press (Pty) Ltd

English monthly 225pp

Contains details of the mining industry in South Africa with news of new exploration activity and companies. Includes statistics of coal production and sales.

598 COAL MINES IN THE REPUBLIC OF SOUTH AFRICA
Minerals Bureau

English 1980 62pp

A directory of information about coal mines in South Africa, including ownership, output, notes on disposal and other technical data. Also data on coal grades and qualities and industry groupings.

599 ECONOMIC DEVELOPMENT PROGRAMME FOR THE REPUBLIC OF SOUTH AFRICA 1978-1987
Office of the Economic Adviser to the Prime Minister

English 1979 125pp

Sub-titled 'A Strategy for Growth'. Contains an analysis of the economy and the role of energy within it. Information on imports and exports of electricity and coal and projections for these sectors.

600 ELECTRICITY SUPPLY COMMISSION-ANNUAL REPORT
Electricity Supply Commission

English annual 68pp

Contains statistical highlights for the year and details of electricity supply in South Africa, sales information, financial information, system operation, power stations under construction, transmission, manpower, environmental factors, technical advances and maps showing licensed areas of supply.

601 FOREIGN TRADE STATISTICS
Department of Statistics

English/Afrikaans monthly

Contains the foreign trade statistics for South Africa commodity by commodity and country by country. Gives details of trade in coal and coal products.

602 GENCOR—ANNUAL REPORT
General Mining Union Corporation

English/Afrikaans annual c80pp

Detailed review of the operations of the main divisions of the major South African mining group. Information on individual collieries with data for turnover, production and profitability.

603 GENCOR PROFILE
General Mining Union Corporation

English 1981 14pp free

Information publication relating to the operations of the Gencor group, formed in 1980 on the merger of General Mining and Finance Corporation and Union Corporation.

604 JOHANNESBURG CONSOLIDATED INVESTMENT COMPANY—ANNUAL REPORT
Johannesburg Consolidated Investment Co Ltd

English annual 72pp

Contains a review of the markets for coal and uranium and information on the operations and plans of coal and uranium producing subsidiaries and associates.

605 JOURNAL OF THE SOUTH AFRICAN INSTITUTE OF MINING AND METALLURGY
Institute of Mining and Metallurgy

English/Afrikaans monthly

Official journal of the Institute. Contains articles about South Africa's mining activities and research and includes information on coal and uranium.

606 MINERAL SUPPLIES FROM SOUTH AFRICA
Economist Intelligence Unit

English 1978 64pp

Special report on the availability of mineral resources in South Africa and the country's role in a global context. Reviews developments and issues. Contains statistics of output and sales and background information for each mineral commodity.

607 MINERALS
Department of Mineral and Energy Affairs

English/Afrikaans quarterly 48pp R1.30 per issue

Contains cumulative quarterly statistics for production of bituminous coal and anthracite by state in volume and value for local sales and for export.

608 MINING AND MINERALS
Van Rensburg Publications (Pty) Ltd

English annual c28pp

Reprint of the mining section of the South African yearbook containing detailed background information and statistics on production of coal and uranium and information about related organisations and legislation.

609 MINING STATISTICS
Department of Mineral and Energy Affairs

English/Afrikaans annual 51pp

Includes detailed figures of output of coal and coke and total revenues by commodity.

610 MINING SURVEY
Chamber of Mines of South Africa

English or Afrikaans quarterly 56pp

Detailed articles, statistics and news of mining sector activities and prospects. Information about individual companies and mines. Special surveys of commodities from time to time. Issued free to shareholders of member companies.

611 PRINCIPLES OF ENERGY CONSERVATION
Department of Planning and the Environment

English 1978 88pp

Examines aspects of energy conservation including a comparative study with other countries. Deals with the availability of energy sources, demand for and supply of crude oil and coal, world energy consumption with reference to South Africa and energy consumption sector by sector. Covers less conventional forms of energy such as solar energy and wind energy, and gas from natural wastes.

612 PROSPECTS FOR 19—
Bureau for Economic Research

English annual c50pp

Survey of the economic situation and assessment of prospects for the coming year. Covers both mining and electricity supply industries.

613 REPORT OF THE DEPARTMENT OF MINERAL AND ENERGY AFFAIRS
Department of Mineral and Energy Affairs

English annual c24pp

Includes the official reports of the Government Mining Engineer, the Geological Survey, the Minerals Bureau and the Energy Branch. Provides a survey of all activities in which the department is involved and information on output, consumption, exports, prices and duties.

614 SASOL LIMITED—ANNUAL REPORT
Sasol Ltd

English/Afrikaans annual 64pp

General review of Sasol's operations and performance, with background information on all energy sectors with which the company is involved. Details of Sasol's oil-from-coal plants, related coal production facilities and offtakes of gas and feedstocks.

615 SOLAR ENERGY IN SOUTH AFRICA
Department of Planning and the Environment

English 1978 42pp

Report attempting to make an objective evaluation of solar energy in South Africa and to provide the general public with an information document about solar energy. Covers availability of sunlight, the state of solar energy in South Africa and energy saving by means of present solar technology.

616 SOUTH AFRICA
Van Rensburg Publications (Pty) Ltd

English annual c1,040pp

The official yearbook for South Africa, containing substantial background information on past and current development of the energy supply industries and statistics of production/consumption.

617 SOUTH AFRICAN MINING AND ENGINEERING JOURNAL
Thomson Publications South Africa (Pty) Ltd

English monthly 130pp

Contains details of mining and mining processing in South Africa, and articles about current issues. Details of specific mines are carried from time to time.

618 SOUTH AFRICAN MINING AND ENGINEERING YEARBOOK
Thomson Publications South Africa (Pty) Ltd

English annual 334pp

News, articles and statistical information relating to South African mining companies and their operations. Details of South African production activities by sector and the international situation. Includes information on mining in neighbouring countries and profiles of South African mining companies.

619 A SURVEY OF THE ACCOUNTS OF MINING COMPANIES
Department of Statistics

English/Afrikaans annual 16pp

Summary statistics and analysis of the financial performance and situation of South African mining companies.

620 TOP COMPANIES
South African Financial Mail

English annual 160pp

Annual supplement to the Financial Mail including sections on companies involved in coal and uranium production and its impact on their performance.

621 THE OUTLOOK FOR ENERGY IN SOUTH AFRICA
Department of Planning and the Environment

English 1977 146pp

Report to the Planning Advisory Council on the energy resource of South Africa and potential demand. Analysis of consumption by fuel type by sector, trade in energy and primary energy supply. Forecasts for the period up to year 2000.

Sudan

630 THE ECONOMIC AND POLITICAL DEVELOPMENT OF THE SUDAN
MacMillan Press Ltd

English 1977 172pp

Survey of the Sudan economy, including the supply and demand situation of the energy sector.

631 FOREIGN TRADE STATISTICAL DIGEST
Bank of Sudan

English annual c70pp

Covers all exports and imports channelled through Port Sudan, Juba and other points of entry. Figures given for crude oil, petroleum products, coal and coal products.

632 FOREIGN TRADE STATISTICS
Department of Statistics

English annual c400pp

Includes trade statistics month by month, commodity by commodity and country by country for crude oil, lubricants and related materials, coal, coke and briquettes, and petroleum products.

633 SIX YEAR PLAN OF ECONOMIC AND SOCIAL DEVELOPMENT
Ministry of National Planning

English 1972 2 vols

Framework for resource allocation and economic planning: Vol 1 contains an analysis of the existing economic situation, an outline of the plan and programmes for regional development. Vol 2 gives proposals by sector.

634 SUDAN: THE COUNTRY AND ITS MARKET
Jeune Afrique

English 1979 148pp

Contains details of the current Six Year Plan and additional information on plans for hydro-electric power generation, new thermal power stations and rural electrification programmes.

Tanzania

635 AREA HANDBOOK FOR TANZANIA
The American University

English 1978

Prepared by the Foreign Area Studies group of the University. A compilation of basic information about Tanzania including statistical and other data on the energy sector.

636 QUARTERLY STATISTICAL BULLETIN
Bureau of Statistics

English 45pp

Includes details of output, consumption, sales, and, where appropriate, exports and imports of energy commodities.

637 THIRD FIVE YEAR DEVELOPMENT PLAN
Ministry of Finance

English 1979 2 vols

Survey of the existing economic situation in Tanzania and programmes for development of infrastructure, energy supply and industrial development.

Togo

640 ANNUAIRE DES STATISTIQUES DU COMMERCE EXTERIEUR
Direction de la Statistique

French annual 370pp

Figures for imports and exports by country and commodity, including coverage of crude oil and petroleum products.

641 BULLETIN MENSUEL DE STATISTIQUE
Direction de la Statistique

French monthly 73pp

Contains figures on a monthly basis for many aspects of the economy and includes series for production/consumption of electricity.

Tunisia

645 ANNUAIRE STATISTIQUE DE LA TUNISIE
Institut National de la Statistique

French annual

Collected statistics for the most recent year for many aspects of the Tunisian economy, with basic data on energy supply industries.

646 AREA HANDBOOK FOR TUNISIA
The American University

English 1979

Prepared by the Foreign Area Studies group of the University. A compilation of basic information about Tunisia, including statistical and other data on the energy sector.

647 BANQUE CENTRALE DE TUNISIE—RAPPORT ANNUEL
Banque Centrale de Tunisie

French annual 238pp

Includes comparative statistics on: mineral production, electricity production, consumption and trade with Algeria; crude oil and natural gas production; processing and imports of fuel; production of manufactured gas.

648 BULLETIN MENSUEL DE STATISTIQUE
Institut National de la Statistique

French monthly 35pp

Includes comparative statistics for the quarter and previous quarters for electricity, natural gas, towns gas, crude oil and petroleum products. Also gives details of output of refined products and exports and imports of energy commodities.

649 STATISTIQUE FINANCIERE
Banque Centrale de Tunisie

French quarterly 77pp

Includes statistics on imports of energy products and information on mining, electricity supply and oil industries.

650 STATISTIQUE RESUMEE DU COMMERCE EXTERIEUR
Institut National de la Statistique

French monthly

Monthly series of statistics of imports and exports, distinguishing trade in crude oil and petroleum products, solid fuels and cross-border transfers of electricity.

651 STATISTIQUES DU COMMERCE EXTERIEUR
Institut National de la Statistique

French annual 320pp

Statistics of foreign trade, including movements of crude oil, petroleum products and solid fuels, analysed by country of origin/destination.

652 TUNISIE
Commission of the European Communities

French 1979

Report by the economic missions of EEC member states in Tunis. Contains sections on: the mining industry and its history; energy and the political situation associated with the supply of energy; petroleum production in Tunisia; petroleum consumption; natural and manufactured gas, electricity; alternative sources of energy such as solar energy.

Upper Volta

655 BULLETIN MENSUEL D'INFORMATION STATISTIQUE ET ECONOMIQUE
Institut National de la Statistique et de la Démographie

French monthly 64pp

Detailed figures for production of electricity and imports of individual petroleum products. Figures on a monthly or quarterly basis and for last few years.

Zaire

660 BULLETIN TRIMESTRIEL
Banque de Zaire

French quarterly 80pp

Includes statistics on mining, electricity and petroleum production with trade figures where appropriate.

661 CONJONCTURE ECONOMIQUE
Département de l'Economie Nationale, de l'Industrie et du Commerce

French annual 500pp

Yearbook containing comparative statistics and information on electrical energy, petroleum production and consumption zone by zone and mineral production, including coal and crude oil. Also includes, where appropriate, the trade figures for these products.

662 PROGRAMME DE RELANCE ECONOMIQUE
Secrétariat Général au Plan

French 1979 3 vols

Programme of development for the Zaire economy. Vol 1 contains analysis of the existing situation. Vol 3 covers specific requirements in energy supply.

Zambia

665 AREA HANDBOOK FOR ZAMBIA
The American University

English 1979

Prepared by the Foreign Area Studies group of the University. A compilation of basic information about Zambia, including statistical and other data on the energy sector.

666 BANK OF ZAMBIA—REPORT AND STATEMENT OF ACCOUNTS
Bank of Zambia

English annual c114pp

Detailed review of the Zambian economy, with sections on energy supply and demand, coal mining, supply of oil products, electricity production, and manufacture of coke. Statistics cover production, sales and prices.

667 ECONOMIC REPORT
Office of the President

English annual 164pp

Transcript of the government's annual report to the National Assembly. It contains information on coal and coke imports, production and consumption and details of exploration activity. Also gives a report and statistical information on electrical energy production, consumption, exports, prices and new work being carried out by the state electricity supply undertaking. Petroleum imports, products from the refinery, marketing of oil products, crude oil prices, new plants, pipelines and storage are detailed. Other energy sources such as solar, wind and biomass, are noted.

668 MONTHLY DIGEST OF STATISTICS
Central Statistical Office

English monthly 60pp

Includes details of electricity supply and consumption and information on coal production. Gives figures for movement of coal and electricity between regions and imports of coal, coke, briquettes, petroleum and petroleum products.

669 THIRD NATIONAL DEVELOPMENT PLAN
National Commission for Development Planning

English 1979 454pp

Economic plan for the period 1978-83 setting out objectives and programmes for public and private sector investment and output in energy supply, and infrastructure.

670 ZESCO—ANNUAL REPORT
Zambia Electricity Supply Corporation Ltd

English annual c28pp

Annual review of the electricity supply situation in Zambia. Statistical sections give figures by month and for individual regions, covering generation by location, bulk deliveries, retail sales and import/exports of electricity.

671 ZESCO—TEN YEARS OF DEVELOPMENT
Zambia Electricity Supply Corporation Ltd

English 1980 24pp

Background information on the development of the national electricity supply authority, containing details of production, transmission distribution and retailing activities.

Zimbabwe

675 ECONOMIC SURVEY OF ZIMBABWE
Ministry of Finance

English annual 24pp

Report with statistical details on the economy. Includes coverage of mining, electricity supply and oil industries. Figures for mine output, electricity generation and consumption, production of oil products.

676 ELECTRICITY SUPPLY COMMISSION—ANNUAL REPORT
Electricity Supply Commission

English annual c50pp

Review of the financial and operational performance of the national electricity supply undertaking, with details of capacity and electricity production, the transmission system and sales to main economic sectors.

677 MINING IN ZIMBABWE
Thomson's Publications

English annual 92pp

Annual review of the mining industry sector by sector, giving details of output. Also lists mining companies and contains a schedule of information on mines in operation.

678 MONTHLY DIGEST OF STATISTICS
Central Statistical Office

English monthly 83pp

Detailed monthly statistics covering mining production, electricity production and consumption detailed by industrial and domestic sectors. Also gives trade figures for coal and coke exports and imports of energy commodities.

679 STATEMENT OF EXTERNAL TRADE BY COMMODITIES
Central Statistical Office

English annual

Includes an historical summary of the trade figures and then statistical details of exports, imports and re-exports of energy commodities.

680 WANKIE COLLIERY COMPANY—ANNUAL REPORT
Wankie Colliery Co Ltd

English annual 20pp

Operational and other information on the key coal producing company of Zimbabwe. Statistics cover production at each mine, saleable output, production of coke and by-products and prices. Also gives information on plans for productive investment.

Asia/Pacific

Australia

700 ATOMIC ENERGY IN AUSTRALIA
Australian Atomic Energy Commission

English quarterly 32pp A$6.00 p.a.

Published by the Commission with the purpose of providing information on the uranium and nuclear power industry. Includes articles of a general, technical and technico-economic nature.

701 AUSTRALIAN ATOMIC ENERGY COMMISSION—ANNUAL REPORT
Australian Government Publishing Service

English annual

Contains details of activity and developments in all stages of the nuclear fuel cycle, reports on uranium exploration and production and information on uranium projects. Published annually as a Parliamentary Paper.

702 AUSTRALIAN COAL MINER
Strand Publishing Pty Ltd

English monthly A$12.00 p.a.

Regular journal for the Australian coal mining industry, with news of technical and economic developments.

703 AUSTRALIAN ENERGY POLICY—A REVIEW
Department of National Development and Energy

English 1979 46pp

Prepared for the National Energy Conservation Conference of 1979. Sections deal with: international context; Australian energy resources; supply and demand; oil exploration and development; new energy sources; R&D; energy prices; taxation; conservation; policy co-ordination.

704 AUSTRALIAN GAS JOURNAL
Objective Publications Pty Ltd

English quarterly 60pp

Published as the official journal of the Australian Gas Association, containing news of developments in the gas industry, activities of government and other organisations, technical news and other articles and data.

705 AUSTRALIAN MINERAL INDUSTRY ANNUAL REVIEW
Bureau of Mineral Resources, Geology and Geophysics

English annual 325pp A$18.00

Record of the development of the Australian mining industry. Separate preprints of part of the review are available on such topics as black coal, petroleum and uranium.

706 AUSTRALIAN MINERAL INDUSTRY REVIEW
Australian Bureau of Statistics

English quarterly 38pp A$2.00 per issue

Part 1: Commodity review. Part 2: Quarterly Statistics production of black coal, brown coal, metallurgical coke, briquettes, crude oil, natural gas, refinery gas, individual petroleum products, uranium. Imports and exports in detail.

707 AUSTRALIAN MINING AND PETROLEUM LAWS
Butterworth & Co Ltd

English 1979 288pp

Covers all Australian laws in detail, from the introduction of the earliest mining legislation to recent enactments, affecting the mining of Australia's mineral wealth on land and under the sea. Also covers environmental controls and gives an historical introduction.

708 AUSTRALIAN NATIONAL ACCOUNTS
Australian Bureau of Statistics

English annual A$2.70

Includes annual series for mining, electricity and gas/water industries of output, costs, investment. Consumers expenditure on fuels by year and by state.

709 AUSTRALIAN PETROLEUM STATISTICS
Department of National Development and Energy

English half-yearly free

Summary statistics of oil and gas exploration and production, and petroleum products supply.

710 BHP—ANNUAL REPORT
BHP Ltd

English annual

Contains details of exploration and production of the leading Australian based oil and gas producing company.

711 BHP JOURNAL
BHP Ltd

English half-yearly free

Published mainly as an internal information bulletin by one of Australia's leading oil and gas exploration and production companies, containing news of BHP's activities and related energy developments.

712 BMR YEARBOOK
Bureau of Mineral Resources, Geology and Geophysics

English annual

Contains an introductory summary of the Bureau's work during the year followed by articles describing certain projects and lists of all publications, maps and unpublished data made available during that year.

713 CENSUS OF MINING ESTABLISHMENTS
Australian Bureau of Statistics

English annual

Data on employment, turnover, energy consumption capital expenditure, stocks and royalty payments in the black coal, brown coal, oil and gas production industries.

714 CENSUS OF MINING ESTABLISHMENTS: SELECTED ITEMS OF DATA
Australian Bureau of Statistics

English irregular

Items of data classified by employment size and industry class: number of establishments, and numbers employed in coal, oil and gas production. Turnover and value added data by employment size and industry class.

715 CRA—ANNUAL REPORT
CRA Ltd

English annual

Includes background information to energy markets and specific details of developments of Mary Kathleen Uranium Ltd and major coal projects in which the group is involved.

716 CSIRO—ANNUAL REPORT
Australian Government Publishing Service

English annual A$2.00

Report by the principal government research organisation concerned with energy R & D on activity under its mineral, energy and water resources programme. Published annually as a Parliamentary Paper.

717 CSR—ANNUAL REPORT
CSR Ltd

English annual

Includes statistical information on output of coal at numerous collieries in which CSR has interests and news of new project developments.

718 DAMPIER TO PERTH NATURAL GAS PIPELINE
State Energy Commission of Western Australia

English 1980 16pp

Background information on the North West Shelf gas project, the use of natural gas in Western Australia and the programme for utilising part of the offshore resource for development in the state.

719 DEPARTMENT OF NATIONAL DEVELOPMENT—ANNUAL REPORT
Australian Government Publishing Service

English annual

Detailed report on energy policy, energy R & D, the petroleum sector, uranium exploration and development. Statistical data on oil prices and levies. Contains commentary on the economic situation of the energy sector and developments in exploration, production and technology. Published as a Parliamentary Paper.

720 DEPARTMENT OF TRADE AND RESOURCES—ANNUAL REPORT
Australian Government Publishing Service

English annual

Contains reports on the activities of the Uranium Export Office, the Uranium Advisory Council and other official advisory bodies. Brief survey of trade in coal and uranium. Published as a Parliamentary Paper.

721 DIGEST OF CURRENT ECONOMIC STATISTICS
Australian Bureau of Statistics

English monthly

Includes series of monthly data for production of black coal, electricity and gas.

722 DIRECTORY OF AUSTRALIAN SOLAR ENERGY RESEARCH AND DEVELOPMENT
CSIRO

English 1978 84pp

Compilation by the principal government body concerned with technological development of activities by departments, agencies, institutions and other bodies on developing solar energy as a useful energy form.

723 ELECTRICITY AND GAS ESTABLISHMENTS
Australian Bureau of Statistics

English irregular

Survey of the electricity and gas industries with details of operations, by states and territories. Numbers of establishments, employment turnover, capital expenditure and sales of electricity and gas.

724 THE ELECTRICITY SUPPLY INDUSTRY IN AUSTRALIA
Electricity Supply Association of Australia

English annual c74pp

Collected information on the status and activities of electricity undertakings in each state. Maps of systems and statistics on power stations, output, consumption and financial performance.

725 ENERGY
Objective Publications Pty Ltd

English monthly 16pp A$75.00 p.a.

Regular newsletter covering all energy sectors, containing information on projects, policy developments, production and trade.

726 ENERGY IN AUSTRALIA 1977-2000
Committee for Economic Development of Australia

English 1977 52pp

Subtitled 'A Statement of Policy'. Produced in the Committee's series of Information and Position Papers.

727 AN ENERGY POLICY FOR WESTERN AUSTRALIA
Government of Western Australia

English 1980 30pp

Statement of the broad lines of the state government's energy policy, review with statistics of the energy situation of Australia and the state, summary of state energy resources and use and forecast of the Western Australian market to 1997.

728 ENERGY: PRODUCTION, CONSUMPTION AND RESERVES—QUEENSLAND
Australian Bureau of Statistics

English annual

Detailed information on black coal, crude oil, natural gas, uranium, hydro-electric power and potential and bagasse in mass, volume, power or energy units. Coal production distinguished by type of operation. Installed capacity and fuel consumed in the electricity sector. Consumption of main petroleum products.

729 ENGINEERS AUSTRALIA
Miadna Pty Ltd

English bi-monthly c72pp

Reports latest engineering and energy news of interest to the Australian engineer. Incorporates the Journal of the Institution of Engineers.

730 ENTERPRISE STATISTICS
Australian Bureau of Statistics

English annual c150pp A$5.50

Details and statistics of operations by industry class for enterprises engaged in mining, oil refining, electricity and gas production and distribution.

731 EXPORTS AUSTRALIA
Australian Bureau of Statistics

English quarterly

Exports of commodities at SITC group level. Monthly series for coal, coke, briquettes, crude oil and main petroleum products. Role of energy within total exports.

732 EXPORTS BY COMMODITY DIVISIONS
Australian Bureau of Statistics

English monthly

Value of major commodities—monthly series for coal/coke/briquettes, petroleum and petroleum products. Export volumes of main energy commodity groups.

733 FACTS ABOUT CRA
CRA Ltd

English 1982 72pp

Contains details of CRA's operations in Australia and worldwide, participations, projects, plant and production in metal and mineral extraction and processing.

734 IMPORTS AUSTRALIA
Australian Bureau of Statistics

English monthly

Includes monthly series for imports of coal, coke, briquettes, crude oil and petroleum products. Role of energy within total imports.

735 IMPORTS BY COMMODITY DIVISIONS
Australian Bureau of Statistics

English monthly

Value of imports of main fuel groups by month. Import volumes of coal/coke/briquettes, petroleum and petroleum products.

736 LIQUEFIED PETROLEUM GAS
Australian Liquefied Petroleum Gas Association

English bi-monthly

Regular bulletin of the Australian LPG supply industry, with information on general technical and commercial aspects.

737 MINERAL EXPLORATION, AUSTRALIA
Australia Bureau of Statistics

English annual

Details of exploration expenditure by type of activity, by mineral sought and by type of company. Drilling activity, private and government and by state. Number of wells.

738 MINERAL INDUSTRY QUARTERLY
Department of Minerals and Energy, South Australia

English quarterly free

Regular summary information on activities and operations of energy industries in South Australia.

739 MINERAL PRODUCTION, AUSTRALIA
Australian Bureau of Statistics

English annual c20pp

Details of quantity and value of minerals produced in the current year and some comparative statistics relating to preceding years. Covers metallic materials, coal, oil and gas. Data for individual states. Estimates of world production.

740 MINERAL RESOURCES REVIEW
Department of Minerals and Energy, South Australia

English half yearly free

Cumulative data on energy supply industries in South Australia.

741 MINERALS AND MINERAL PRODUCTS
Australian Bureau of Statistics

English monthly

Statistics of mine output of principal minerals, including black coal, brown coal and petroleum. Also figures for manufacture of brown coal briquettes.

742 MINING REVIEW
Australian Mining Industry Council

English monthly

Regular review of activities of mining industries in Australia, including statistical and other information on coal production and shipments.

743 MONTHLY SUMMARY OF STATISTICS
Australian Bureau of Statistics

English monthly 46pp

Contains monthly statistics on the energy industries production of black coal, brown coal, gas, crude oil, electricity, petroleum products.

744 THE MOONIE OIL COMPANY LIMITED—ANNUAL REPORT
Moonie Oil Co Ltd

English annual

Contains details of production of the Moonie (Queensland) oil field and of exploration and development at other fields in the area.

745 NEW SOUTH WALES YEARBOOK
Australian Bureau of Statistics, New South Wales

English annual c750pp

Contains sections dealing with the background and current developments of energy production and supply industries in New South Wales, covering oil, natural gas, coal and electricity.

746 OIL AND AUSTRALIA: THE FIGURES BEHIND THE FACTS
Australian Institute of Petroleum Ltd

English annual 24pp

Compendium of statistics about petroleum exploration, reserves, production, refining, transportation, trade and consumption.

747 PETROLEUM GAZETTE
Australian Institute of Petroleum Ltd

English monthly 24pp

Regular news journal of the principal oil industry association. Contains articles and analysis of Australian energy issues and their international context, news of project developments and oil and gas discoveries.

748 PETROLEUM NEWSLETTER
Bureau of Mineral Resources, Geology and Geophysics

English quarterly free

Contains summary information and statistics on petroleum activity throughout Australia.

749 PETROLEUM STATISTICS
Department of National Development and Energy

English 1978 28pp

Compilation of statistics on the petroleum sector covering: production of crude oil and natural gas, oil refining, output of petroleum products, consumption, imports and exports, reserves, refining capacity, investment and prices.

750 PEX
Lipscombe & Associates Ltd

English monthly A$84.00 p.a.

Newsletter carrying information about activities and developments in oil and gas exploration in Australia and other related information.

751 POCKET YEAR BOOK AUSTRALIA
Australian Bureau of Statistics

English annual A$1.90

Summarised data on: mining establishments; production of coal, oil and natural gas over several years; electricity generation and gas supply; imports and exports of major fuel commodities.

752 PORT STATISTICS
Maritime Services Board of New South Wales

English annual 36pp

Detailed statistics of movements of bulk commodities through New South Wales ports and terminals. Figures for coal and oil for inter-state or overseas shipment, with analysis by destination.

753 PRODUCTION STATISTICS
Australian Bureau of Statistics

English monthly

Includes monthly series of output by the gas and electricity industries.

754 QUARTERLY SUMMARY OF AUSTRALIAN STATISTICS
Australian Bureau of Statistics

English quarterly 159pp

Contains information and statistics of Australian industries, including coal, petroleum, electricity and gas.

755 QUEENSLAND GOVERNMENT MINING JOURNAL
Department of Mines, Queensland

English monthly A$12.00 p.a.

Review of developments in the state's mining industries, including output, prices and other statistical information.

756 QUEENSLAND YEARBOOK
Australian Bureau of Statistics, Queensland

English annual c625pp

Chapters on primary industries and manufacturing include background information, statistics and surveys of current developments in oil, coal, gas and electricity exploration, production and supply industries.

757 THE RANGER URANIUM PROJECT
Australian Government Publishing Service

English 1979

Full details of the agreement between the Commonwealth government and participants in the Ranger uranium project concerning shareholdings, financing, development programme etc. Published as Parliamentary Paper No 7/1979.

758 REPORT OF THE SOUTH AUSTRALIAN STATE ENERGY COMMITTEE
State Energy Committee, South Australia

English 1976 178pp

Details the State's energy requirements of fossil fuels for the next 25 years; evaluates the known energy reserves of coal, natural gas and crude oil and the potential of nuclear fuel resources.

759 A RESEARCH AND DEVELOPMENT PROGRAMME FOR ENERGY
National Energy Advisory Committee

English 1978 20pp

Statement by the government's principal advisory group on energy of the main themes and programmes which should be addressed. Report No 3 of the Advisory Committee.

760 SOLAR FOCUS
Australian Syndicators Pty Ltd

English bi-monthly 24pp

Periodical dealing with solar energy developments both technological and economic in Australia, New Zealand and internationally. Also includes information on other renewable energy sources.

761 SOUTH AUSTRALIA YEARBOOK
Australian Bureau of Statistics, South Australia

English annual c660pp

Includes chapters on exploration and development of energy resources and detailed background and statistics of the energy supply industries.

762 STATE ENERGY COMMISSION OF WESTERN AUSTRALIA—ANNUAL REPORT
State Energy Commission of Western Australia

English annual c50pp

Includes a survey of energy developments in western Australia and details of the Commission's activities in the supply of gas and electricity.

763 SURVEY OF MOTOR VEHICLE USAGE
Australian Bureau of Statistics

English irregular

Results of the survey include: annual fuel consumption by type of vehicle and type of fuel, average rate of fuel consumption.

764 SURVEY OF MOTOR VEHICLE USAGE—COMMERCIAL VEHICLE USAGE
Australian Bureau of Statistics

English triennial

Includes detailed statistics of tonnage and tonne-kilometres by industry and by commodity and total annual fuel consumption by commercial vehicles, analysed by type of vehicle and type of fuel.

765 TASMANIAN YEARBOOK
Australian Bureau of Statistics

English annual 580pp

Contains detailed chapters dealing with mining and energy sectors, with statistical and other information on developments and plans.

766 VICTORIAN YEARBOOK
Australian Bureau of Statistics, Victoria

English annual 800pp

Presents a detailed and statistical account of Victoria. Includes a section on energy and materials which covers the coal, electricity, petroleum and gas industries.

767 WESTERN AUSTRALIAN YEAR BOOK
Australian Bureau of Statistics, Western Australia

English annual c580pp

Contains substantial sections dealing with the development of oil, gas and coal resources, and statistical and other background information on electricity and gas supply industries.

768 YEAR BOOK AUSTRALIA
Australian Bureau of Statistics

English annual 796pp A$21.00

Sections on the mineral industry and energy. Detailed coverage of developments and statistics on exploration, production, processing and oil refining. Series for black coal, brown coal, oil, natural gas, oil shale, uranium, electricity and gas. Other information on hydro-electric schemes, electricity and gas establishments, renewable resources and research and development programmes.

Bangladesh

775 BANGLADESH ECONOMIC SURVEY
Ministry of Finance

English annual 300pp

Detailed review of developments in all sectors of the economy. Background to energy reserves, exploration and production, state and other organisations involved in energy supply. Information on consumption of energy and projects. Statistics of production and consumption cover a number of years.

776 ELECTRIC POWER STATISTICS
Bangladesh Power Development Board

English annual

Annual summary of electricity statistics for Bangladesh by the State generating and supply authority, covering generating capacity, electricity production and sales and other data on the transmission/distribution system.

777 INDUSTRIAL INVESTMENT SCHEDULE
Ministry of Industries

English 1980

Detailed breakdown of the investment programmes

reflecting the objectives of the Second Five Year Plan. Includes projects and investment totals for the mining and petroleum industries.

778 **MONTHLY BULLETIN OF STATISTICS**
Bangladesh Bureau of Statistics

English monthly 100pp

Figures on a monthly basis for industrial output, including electricity production, supply of natural gas and availability of petroleum products.

779 **PERFORMANCE REPORT**
Jamuna Oil Co Ltd

English annual 117pp

The annual report of the Jamuna Oil Company contains detailed information on all sectors of energy supply in Bangladesh, including the activities of Eastern Refining Ltd, Bangladesh Petroleum Corporation, Meghna Petroleum, Burmah Eastern, and the Bangladesh Power Development Board.

780 **SALES REPORT**
Bangladesh Petroleum Corporation

English annual

Report by the state oil company on its sales of petroleum products for the past year.

781 **SECOND FIVE YEAR PLAN**
Planning Commission

English 1980 1,000pp

Detailed statement of the economic position of Bangladesh and programmes for investment and development of natural gas, hydro-electric power, projections of energy demand and expectations of oil and gas exploration activity.

782 **STATISTICAL YEARBOOK OF BANGLADESH**
Bangladesh Bureau of Statistics

English annual 700pp

Sections on energy and industry provide statistics covering several years for production of natural gas, electricity generating capacity and output, energy supply and demand.

Brunei

785 **BRUNEI STATISTICAL YEARBOOK**
Economic Planning Unit

English annual 124pp

Contains figures covering several years for production of crude oil, natural gasoline, petroleum products and electricity, consumption of electricity and production and use of natural gas. Also data for availability and consumption of LPG.

786 **STATISTICS OF EXTERNAL TRADE**
Economic Planning Unit

English annual

Detailed figures of Brunei's imports and exports of crude oil, petroleum products and natural gas. Data for volume and value, distinguishing principal destinations/sources.

Burma

790 **BULLETIN OF EXPORT TRADE**
Central Statistical Organisation

English annual

Annual statistics of exports of coal and oil, analysis by country of destination. Figures for volume and value of exports.

791 **BULLETIN OF IMPORT TRADE**
Central Statistical Organisation

English annual

Figures on an annual basis for imports of energy, specifying coal, crude oil and petroleum products, analysed by country of origin. Figures given for both volume and value.

792 **REPORT TO THE PYITU HLUTTAW**
Ministry of Planning and Finance

English annual c300pp

Annual report by the Ministry on financial, economic and social conditions in Burma. Detailed sections on oil exploration activity, production of oil and natural gas, electricity generating plant, electricity output and electricity transmission system.

793 **SELECTED MONTHLY INDICATORS**
Central Statistical Organisation

English monthly c46pp K3.50 per issue

Statistical information for recent months and several past years, covering: generation of electricity, consumption of electricity by main consumer groups; production of crude oil; output of refined products.

China (People's Republic of)

795 CHINA
Metra Consulting Group Ltd

English 1980 310pp

Published in the series 'Business Opportunities in the 1980s'. Reports on developments in the petroleum, coal mining and electricity supply industries, proposals for the energy sector under the state plans, and other background statistics, information and analysis.

796 THE CHINA BUSINESS REVIEW
National Council for US-China Trade

English bi-monthly 64pp $60.00 p.a.

Contains articles on various aspects of China's economic life. Recent issue contained reviews of China's electric power capacity.

797 CHINA: ENERGY BALANCE PROJECTIONS
Central Intelligence Agency

English 1975 33pp

Analyses energy supplies in China and the scope for exports. Details of internal energy production and consumption, with consumption analysed by sector.

798 CHINA FACTS AND FIGURES
Academic International Press

English annual 422pp

Contains information and statistics on all aspects of China's energy situation. Data on oil, coal and gas industries, hydro-electric and nuclear power, conservation activities and renewable energy sources.

799 CHINA NEWSLETTER
Japan External Trade Organisation

English bi-monthly 32pp $60.00 p.a.

Covers China's economy, trade, government policies and working level information. Frequently contains articles on energy.

800 CHINA, OIL AND ASIA: CONFLICT AHEAD
Columbia University Press

English 1977 317pp

Chapters deal with: China's oil potential, offshore oil and gas exploration plans, relationships with Taiwan, Korea and Japan, and the impact of major oil discoveries in the South China Sea.

801 CHINA OIL PRODUCTION PROSPECTS
Central Intelligence Agency

English 1978 28pp

Section 1 presents information on the size, distribution, and characteristics of China's oil reserves. Section 2 provides estimates for national production of crude oil since 1949 and Section 3 discusses each of the major fields. Section 4 assesses China's potential to expand crude oil production.

802 CHINA'S COAL
Petroplan International

English 1981 88pp $100.00

Subtitled 'A Force in the World Market in the 1980s'. Studies the state of China's coal industry, current development investment and potential for future coal export. Statistics and analysis of reserves, production, domestic consumption and export capability. Information on plans and projects to modernise the industry and transport infrastructure.

803 CHINA'S ENERGY ACHIEVEMENTS, PROBLEMS, PROSPECTS
Praeger Publishers Inc

English 1976 246pp

Contains a study of China's fossil fuels, hydro-electric energy, traditional and non-conventional sources, trade, and energy in Chinese society. It also makes international comparisons and forecasts for the future.

804 CHINA'S OIL FUTURE
Westview Press Inc

English 1978 148pp

Subtitled 'A Case of Modest Expectations'. Examines the political and technical obstacles to China becoming a major oil producing country. Contains details of oil fields.

805 CHINA'S PETROLEUM INDUSTRY
Praeger Publishers Inc

English 1976 245pp

Examines the output, growth and export potential of China's petroleum industry. Contains details of the First Five Year Plan, growth of crude oil output, main oil fields, reserves, refineries and transportation, the supply of petroleum equipment, the contribution to the national economy, potential trade, together with other relevant statistics.

806 ECONOMIC GROWTH IN CHINA AND INDIA 1950-80
Economist Intelligence Unit

English 1976 95pp

Contains details of the growth of output of electric power, coal and crude oil since 1950 and details of China's trade in fuels and petroleum products.

807 ENCYCLOPEDIA OF CHINA TODAY
Macmillan Press Ltd

English 1979 336pp

Contains sections on the background to development of the coal and petroleum industries, oil refining and electricity supply. Outlines of the power development and electrification programmes.

808 THE INTERNATIONAL ENERGY RELATIONS OF CHINA
Stanford University Press

English 1980 717pp

Part 1 outlines the energy policy of China, internally and with regard to other Far East countries. Part 2 provides a statistical profile of primary energy reserves, production, consumption and foreign trade by commodity. Foreign trade analysed by country. Also covers trade in energy plant and equipment.

809 OIL IN THE PEOPLE'S REPUBLIC OF CHINA
C Hurst & Co (Publishers) Ltd

English 1977 125pp

Prepared by the Institute of Asian Affairs, Hamburg. Contains information and statistics on China's oil industry in general, oil production by regions, the refineries and the main government officials connected with the oil industry.

810 THE PETROLEUM INDUSTRY OF THE PEOPLE'S REPUBLIC OF CHINA
Hoover Institution Press

English 1975 264pp

Contains details of the major sources of energy; crude oil production, principal oil fields and refineries; production of coal and natural gas; imports and consumption of oil. Also provides information on hydro-electric capacity.

811 REPORT OF THE STATE PLANNING COMMISSION
BBC Monitoring Services

English 1980

Text of the report by the Chairman of the State Planning Commission to the Communist Party Congress in 1980, detailing plans for the energy industries, energy conservation and exploration activity.

812 SMALL HYDROELECTRIC PROJECTS FOR RURAL DEVELOPMENT
Pergamon Press

English 1981 175pp $17.50

Published in cooperation with the East-West Center in Hawaii, it contains a case study of a small multiple-use hydro-electric energy source in China which also supplies water, flood control and crop irrigation.

Hong Kong

820 CHINA LIGHT AND POWER COMPANY—ANNUAL REPORT
China Light & Power Co

English annual

Provides detailed information on the electricity supply situation on the Kowloon Peninsula and New Territories. Data on generating plant in use, electricity production and customers.

821 HALF-YEARLY ECONOMIC REPORT
Government Printer

English half-yearly 81pp

Review of recent economic developments and commentary with statistics on operations of individual industries including electricity production and gas supply.

822 HONG KONG 19—
Government Printer

English annual 307pp

Information and statistics on all aspects of the Hong Kong economy. Background information on public works and utilities, with review of recent developments and investment plans. Data on capacity and production.

823 HONG KONG ENERGY STATISTICS 1970-80
Census and Statistics Department

English 1981 HK$7.00

Brings together basic information on all principal aspects of energy supply, consumption and stocks.

824 HONG KONG EXTERNAL TRADE
Census and Statistics Department

English monthly 25pp

Includes statistics of imports/exports by month for gas oil, diesel oil, distillate fuel and heavy fuel oils.

825 MONTHLY DIGEST OF STATISTICS
Census and Statistics Department

English monthly

Detailed statistics on a monthly basis of production by electricity and gas undertakings and figures for trade in petroleum.

India

830 BIOMASS ENERGY PROJECTS
Pergamon Press

English 1980 300pp £12.50

Examines the detailed results of biomass energy production plants in India on the basis of case study analysis.

831 CCAI NEWS LETTER
Coal Consumers Association of India

English monthly

Current information concerning coal availability, markets and technologies of interest to coal users.

832 COAL BULLETIN
The Director-General of Mines Safety,

English monthly 27pp

Contains details for the coal mining industry of: employment; wages; hours of work and productivity; production; stocks; machinery; labour disputes and accidents.

833 ECONOMIC GROWTH IN CHINA AND INDIA 1950-1980
Economist Intelligence Unit

English 1976 95pp

Contains information on India's energy resources and consumption and compares the figures for the period since 1950. Also includes details of India's trade in petroleum products and mineral fuels.

834 ELECTRICITY CONSERVATION QUARTERLY
Devki R & D Engineers

English quarterly 16pp Rs50 p.a.

Articles on various aspects of energy conservation, case studies, management techniques and available equipment. News of relevant developments in policy.

835 HANDBOOK OF STATISTICS
Association of Indian Engineering Industry

English annual c380pp

Includes a review of developments in Indian industry and the impact on electric power supply and demand.

836 HARYANA ELECTRICITY
Haryana State Electricity Board

English quarterly

Regular report by the electricity supply and distribution authority for Haryana of its activities and plans.

837 JOURNAL OF MINES, METALS AND FUELS
Books and Journals (Pte) Ltd

English monthly Rs40.00 p.a.

Incorporating the Indian Mining Journal and the official Organ of Indian Mine Managers Association, it contains information on India's mining and energy operations.

838 KOTHARI'S ECONOMIC AND INDUSTRIAL GUIDE OF INDIA
Kothari and Sons

English annual

Contains an economic review and general information, statistics and analysis of annual reports of approximately 2,000 businesses in India, including leading companies in mining, electric power production and petrochemicals.

839 MINISTRY OF STEEL AND MINES REPORT
Ministry of Steel, Mines and Coal

English annual 70pp

Contains an overview of coal production and productivity for the year and reports and statistical tables on topics which include: coal planning, conservation and safety; production; distribution; industrial relations; research and development; coal mines and their staff.

840 PERFORMANCE REPORT OF PUBLIC ENTERPRISES
Bureau of Public Enterprises

English annual 3vols

Detailed review of each state-owned corporation. Commentary on financial performance, operating data and note of plans for investment. Vol 1: Financial summary data Vol 2: Operational performance Vol 3: Detailed financial reports.

841 PUBLIC ELECTRICITY SUPPLY: ALL INDIA STATISTICS
Ministry of Energy

English annual c230pp

Contains data concerning the generation, distribution and utilisation of electric power throughout India.

842 QUICK RELEASE TO MINERAL STATISTICS OF INDIA
Indian Bureau of Mines

English monthly Rs42.00 p.a.

Up-to-date statistics on a monthly basis for main minerals, including production of coal.

843 RURAL ELECTRIFICATION CORPORATION—ANNUAL REPORT
Rural Electrification Corporation Ltd

English annual c68pp

Review of the state of electrification in India, with details of projects and investments by region.

844 SIXTH FIVE YEAR PLAN
Planning Commission

English 1980 £4.67 463pp

The plan document is composed of two parts. Part 1 is on India's objectives, resources and policies. Part 2 deals with various programmes of development, including energy. An appendix summarises its main objectives and the financial resources available to implement it.

845 SUN: MANKIND'S FUTURE SOURCE OF ENERGY
Pergamon Press

English 1978 2vols £140.00

Proceedings of the International Solar Energy Congress, New Delhi, India in 1978. Contains India's national plans and programmes for solar energy.

846 TIMES OF INDIA INDUSTRIAL REVIEW
Times of India

English annual

Special supplement on Indian industry reviewing in detail the state and prospects for individual sectors. Covers petroleum, coal mining, petrochemicals and electricity supply.

847 TIMES OF INDIA YEARBOOK
Times of India

English annual c1200pp

Compendium of information on state corporations and private companies. Includes most of the key undertakings involved in energy production and supply—coal mining, oil and gas production, oil refining, electricity generation, petrochemicals.

848 TRADE STATISTICS OF INDIA
Department of Commercial Intelligence and Statistics

English monthly

Figures for the most recent month and cumulatively for year to date commencing July. Volume and values for individual fuel products—coal, coke, briquettes, crude oil, main petroleum products. Two series of volumes for imports and exports.

Indonesia

850 EKSPOR
Biro Pusat Statistik

English/Indonesian annual

Detailed statistics of Indonesia's export trade. Value and volume of exports of principal energy products for main destinations.

851 ELECTRICITY GAS AND WATER
Biro Pusat Statistik

English/Indonesian 1979 44pp

Detailed figures from an analysis of the activities of establishments engaged in electricity and gas production and supply. Includes figures for plant, output and sales.

852 IMPOR
Biro Pusat Statistik

English/Indonesian annual

Statistics on an annual basis for imports, by volume and value. Details of imports for principal fuel products, analysed by source.

853 INDIKATOR EKONOMI
Biro Pusat Statistik

English/Indonesian monthly

Regular bulletin of current economic statistics, including production by the oil and gas industries, generation of electricity and availability of petroleum products.

854 INDONESIA
Metra Consulting Group Ltd

English 1977

Published in the series 'Business Opportunities in the 1980s'. Contains a review of developments and plans in the energy sectors, with statistics on production/consumption.

855 INDONESIA'S OIL
Centre for Strategic and International Studies

English 1976 89pp

Provides background to the oil and gas industry in Indonesia; analyses the country's energy position and the role of oil in the economy, and examines Indonesia's relationship to other oil producers and fellow ASEAN member countries.

856 KOMPASS INDONESIA
Kompass Publishers Ltd

English annual

Directory of Indonesian firms and their principal fields of activity. Includes firms engaged in the supply of energy and related goods and services.

857 MONTHLY BULLETIN OF THE PETROLEUM AND NATURAL GAS INDUSTRY
Directorate-General of Oil and Gas

English monthly

Detailed statistics of oil and gas exploration and production for a series of months and most recent years. Production by company and by field, transfers to refineries and refinery output. Utilisation of natural gas output.

858 STATISTIK INDONESIA
Biro Pusat Statistik

English/Indonesian annual

Annual compendium of statistics for Indonesia, with sections dealing with oil and gas production, coal mining and the gas and electricity supply industries. Figures for national consumption of energy and activities of oil refineries.

859 THIRD FIVE YEAR DEVELOPMENT PLAN
National Development Planning Agency

English 1979 61pp

Summary of the economic development plan for 1979-84, detailed programmes for development of natural resources and investment in mining, energy production and electricity generation.

Japan

865 ANNUAL REPORT ON ATOMIC ENERGY
Atomic Energy Commission

English annual

Report on the state of nuclear development in Japan and the activities of the Atomic Energy Commission and related organisations.

866 ATOMS IN JAPAN
Japan Atomic Industry Forum

English monthly

Published by the Atomic Industry Forum as part of its programme of information dissemination on nuclear power developments.

867 CURRENT ENERGY SITUATION
Foreign Press Center

English 1980 16pp

Presentation by the President of the Institute of Energy Economics on the general energy situation in relation to Japan, with projections for 1990.

868 CURRENT INFORMATION
Kansai Electric Power Co

English irregular 20pp

Contains background information and statistical material on electricity supply and demand in Japan, including results of Kansai Electric's operations.

869 ENERGY CONSERVATION IN JAPAN
Ministry of International Trade and Industry

English 1979 50pp

Background information on the existing position regarding energy conservation, scope for improvements and government measures to promote conservation. Chapter 1: The Present Status of Energy Conservation in Japan. Chapter 2: Energy Conservation Measures. Annexes contain details of the Law on Rationalisation of the Use of Energy, measures to cut back oil consumption, and a survey of energy saving measures implemented.

870 ENERGY DEVELOPMENTS IN JAPAN
Rumford Publishing Co

English quarterly $98.50

Regular review of activities and projects in energy production supply and technology in Japan.

871 ENERGY IN JAPAN
Institute of Energy Economics

English quarterly 20pp $50.00 per annum

Commentary and economic analysis of energy issues with respect to the Japanese energy situation. Contains economic and statistical data, covering all forms of energy. Articles and reports on: Japan's energy balance; solar energy; coal use in power stations; economics of nuclear power, living standards and energy demand etc.

872 ENERGY IN JAPAN—FACTS AND FIGURES
Agency of Natural Resources and Energy

English 1981 30pp

Special survey of the current situation and outlook. Part 1: Recent Energy Situation and Japan. Part 2: Japanese Energy Policy. Annexes on Provisional Long Term Energy Supply and Demand Outlook, Demand and Supply Plan for 1981-85, Electric Power Supply.

873 ENERGY POLICY FOR THE 1980s
Industrial Structure Council

English 1980

Analysis, comment and policy recommendations of the Industrial Structure Council, which advises the Ministry of International Trade and Industry. Excerpt from the Council's report 'Industrial Policy in the 1980s'.

874 EXPORTING TO JAPAN
Japan External Trade Organisation

English 1980

Contains details of Japan's actual and anticipated trade in steam coal, coking coal and metallurgical coke.

875 INDUSTRIAL GROUPINGS IN JAPAN
Dodwell Marketing Consultants

English 1980

Details of Japan's industrial groupings and analyses of the achievements of each industry. Contains a detailed analysis of Japan's petroleum and coal products industries.

876 INDUSTRIAL REVIEW OF JAPAN
The Japan Economy Journal

English annual c150pp $19.50

Analysis of the state of the Japanese economy, with commentary on key issues affecting energy supply and demand. Statistics on oil, coal, electricity and petrochemicals industries.

877 INDUSTRIAL STATISTICS MONTHLY
Ministry of International Trade and Industry

English/Japanese monthly 200pp

Statistics by month and for the previous two years on production of electricity (hydro and thermal), gas production, deliveries of main oil products, energy consumption and stocks by product.

878 JAPAN ECONOMIC YEARBOOK
The Oriental Economist

English annual 320pp

Sections deal with coal, electric power, petroleum and nuclear power. Summarises the current situation, developments and projects, with key statistics.

879 JAPAN PETROLEUM WEEKLY
Japan Petroleum Consultants Ltd

English weekly c20pp yen100,000 p.a.

Newsletter containing up-to-date statistical information on all aspects of oil and gas exploration, production, refining distribution and marketing in Japan. Detailed analysis of oil imports. Information on energy policy developments and other news.

880 JAPAN STATISTICAL YEARBOOK
Japan Statistical Association

English/Japanese annual

Contains detailed statistics for the economic sector, including indigenous production of primary energy forms and operations of the gas and electricity supply industries.

881 JAPAN'S ENERGY SITUATION
Foreign Press Center

English 1981 32pp

Produced in the Foreign Press Center's series 'About Japan', containing background information and statistics on the energy situation of Japan, government plans and individual supply industries.

882 JAPAN'S ENERGY STRATEGY TOWARD THE TWENTY-FIRST CENTURY
Ministry of International Trade and Industry

English 1979 37pp

Report of the Advisory Committee for Energy's Conference on Fundamental Issues. Outlines the country's basic energy problems and long term supply and demand outlook, defines the tasks of energy policy, and sets out proposed measures relating to conservation, petroleum supply, alternative energy sources, new technologies and the siting of power stations.

883 JAPAN'S SEARCH FOR OIL
Hoover Institution Press

English 1977 116pp

Study of Japan's efforts to establish secure energy supplies, the country's response to oil supply difficulties and rising prices and the implications for Japan's relationships with neighbouring countries, especially China.

884 JNOC
Japan National Oil Corporation

English 1981 20pp

Describes the history, functions and activities of the Corporation with details of the organisation, projects being given financial support and availability of overseas crude oil from JNOC-assisted companies.

885 NEW ECONOMIC AND SOCIAL SEVEN YEAR PLAN
Economic Planning Agency

English 1979 201pp

Outlines the programmes and policies for economic development and energy supply, with projections for oil, coal, nuclear power, hydro-electric generation, geothermal power, total electricity supply and demand. Details policies towards energy conservation and waste management.

886 NIPPON: A CHARTED SURVEY OF JAPAN
Kokusei-Sha Co Ltd

English annual 347pp

Provides statistical surveys of Japan. There is a large section which covers petrochemicals, coal, petroleum, natural gas, liquefied petroleum gas, city gas, electric power and atomic energy.

887 NIPPON OIL GROUP
Nippon Oil Co Ltd

English annual 28pp

Issued in conjunction with the company's annual report and accounts. Summarises Nippon Oil's operational activities and financial performance with detailed statistics of oil refining and product consumption in Japan.

888 AN OUTLINE OF THE 1979 REPORT ON ENERGY RESEARCH
Foreign Press Center

English 1980 8pp

Translation of a summary issued by the Prime Minister's Office detailing expenditure on energy research by main subject.

889 PETROLEUM INDUSTRY IN JAPAN
Japanese National Committee of the World Petroleum Congresses

English 1981 60pp

Reviews recent and historical developments in each sector of the oil industry, with comment on future plans and projects. Separate sections on exploration and production, supply and demand, refining, marketing, transportation and storage, safety and environmental protection, petrochemicals. Includes detailed statistics of product sales, refining plant etc.

890 PROMOTION OF CONSISTENT AND EFFECTIVE COMPREHENSIVE ENERGY POLICIES
Ministry of International Trade and Industry

English 1977 14pp

Sub-titled 'Energy Program for Future Japan'. Contains an analysis of the Japanese energy situation and sets out the guidelines and policies towards conservation, the development of alternative energy sources and energy technology.

891 PROMOTION OF THE MOONLIGHT PROJECT
Ministry of International Trade and Industry

English 1979 37pp

Details of the government's policies and measures relating to research, development and demonstration of energy conservation technologies.

892 STATISTICAL SURVEY OF JAPAN'S ECONOMY
Economic and Foreign Affairs Research Association

English annual 84pp

Includes sections dealing with natural resources: production of principal fuels; import data; energy consumption by fuel.

893 SUNSHINE PROJECT
Ministry of International Trade and Industry

English 1978 40pp

Sub-titled 'New Energy Research and Development in Japan'. Deals with the current state of technological development, the need for research and development and proposed measures in the fields of solar energy, geothermal energy, coal gasification and liquefaction and hydrogen energy.

894 THE WORLD ENERGY OUTLOOK IN THE 1980s AND JAPAN
Japan Institute of International Affairs

English 1980 31pp

Sections deal with: Foreign Policy Guidelines; World Energy Outlook to 2000; Energy Outlook for Japan to 2000; and International Oil Relations.

Korea (Republic of)

900 FIVE YEAR ECONOMIC AND SOCIAL DEVELOPMENT PLAN
Economic Planning Board

English 1981

The Development Plan examines the resources, potential and issues of each main sector of the economy, with plans for individual energy industries, projections of energy demand and details of major projects.

901 FOREIGN TRADE STATISTICS
National Bureau of Statistics

English/Korean monthly

Monthly, cumulative and annual statistics of imports and exports by volume and value for coal, coke, briquettes, crude oil and petroleum products. Imports/exports distinguished by principal countries of origin/destination.

902 IMPLEMENTATION PLAN OF ENERGY AND RESOURCES SECTOR
Ministry of Energy and Resources

English 1981 57pp

Details of projects and plans in the energy sector under the Fifth Five Year Plan for the period 1982-86.

903 INDUSTRY IN KOREA
Korea Development Bank

English annual 316pp

Survey of recent developments and the current situation of Korea's industries. Covers electricity supply, coal mining, oil refining and petrochemicals industries.

904 KOREA ANNUAL
Yonhap News Agency

English annual 750pp

Yearbook giving political economic and social background information and review of developments and plans. Includes sections dealing with energy situation and the petroleum, electricity and coal industries.

905 KOREA MINING PROMOTION CORPORATION
Korea Mining Promotion Corporation

English 26pp

Describes the activities of the State mining organisation in improving the performance capability of the mining industry. Aspects covered include: exploration programme; mining funds; safety; equipment supply; overseas development programme.

906 KOREAN STATISTICAL HANDBOOK
National Bureau of Statistics

English annual 150pp

Statistical review of the Korean economy. Summary data on production and consumption of coal, oil, gas and electricity.

907 KOREAN STATISTICAL YEARBOOK
National Bureau of Statistics

English annual

Statistics on all aspects of the Korean economy, including: production of coal; supply of oil products; generation of electricity; electricity consumption by sector; gas production and consumption; electricity generating plant by type. Figures cover a series of years.

908 LONG-TERM PROSPECTS FOR ECONOMIC AND SOCIAL DEVELOPMENT 1977-91
Korea Development Institute

English 1978 305pp

Detailed survey and analysis of the Korean economy with projections to 1991. Examines resources and investment needs and issues relating to the supply of energy for economic growth.

909 MONTHLY STATISTICS OF KOREA
National Bureau of Statistics

English monthly 166pp

Monthly data on electricity production, electricity consumption by industry, production series for coal and petroleum products.

910 STATISTICS OF ELECTRIC POWER IN KOREA
Korea Electric Company

English/Korean annual 75pp

Covers a variety of monthly statistical data concerning electric utility operations in Korea. Divided into four sections: electric power facilities; electric power generation; power sold; management.

Malaysia

915 BERITA SHELL
Shell Eastern Petroleum Ltd

English/Malay irregular free

Publication of the Shell group, with news of the group's activities and operations in Malaysia and Singapore.

916 ESSO IN MALAYSIA
Esso Malaysia Bhd

English or Malay irregular free

Produced as an information bulletin on activities of Exxon group companies operating in Malaysia.

917 FOURTH MALAYSIAN PLAN
Prime Ministers Department

English 1981

Fourth five-year plan, adopted in 1981, for the period 1981-85, includes plans for development of indigenous energy resources and the shift away from imported oil.

918 GAS OPTION FOR MALAYSIA
Petronas

English

Report on the expected resources of natural gas available from recent exploration work and the various uses to which it might be put.

919 INFORMATION MALAYSIA
Berita Publishing Sdn Bhd

English annual 770pp

Incorporates Malaysia Yearbook. Contains statistical and other information on all aspects of the Malaysian economy, including the mining and extractive industries. Background to oil and gas development, and energy requirements of the economy.

920 NATIONAL OIL POLICY
Petronas

English

Publication of the state oil and gas corporation dealing with policy issues for oil.

921 POLICY OPTIONS FOR OIL SUBSTITUTION IN MALAYSIA
Petronas

English

Examines the scope for substituting natural gas and solid fuel for some of the country's energy requirements currently met from oil resources.

922 PROFILE OF PETROLEUM RESOURCES IN MALAYSIA
Petronas

English

Survey by the state oil corporation of petroleum resources evaluated to date in Malaysia.

923 SARAWAK ELECTRICITY SUPPLY CORPORATION—ANNUAL REPORT
Sarawak Electricity Supply Corporation

English annual

Reports on activities of the Corporation, including electricity generated, details of plant, number of customers and sales by sector. Data cover several years.

924 SHELL IN MALAYSIA
Shell Malaysia Ltd

English 38pp free

Information publication on all Shell companies operating in Malaysia, Sarawak and Sabah, with reference to their activities in the Malaysian context and companies associated with Shell in the energy sector.

New Zealand

930 ANNUAL RETURNS OF PRODUCTION FROM QUARRIES AND MINERAL PRODUCTION STATISTICS
Mines Division, Ministry of Energy

English annual 32pp

Statistics for the production of individual commodities, including production of coal at private and state mines, natural gas, LPG, natural gasoline and condensate.

931 BIOMASS ENERGY PROJECTS
Pergamon Press

English 1980 300pp £12.50

Contains case studies of biomass energy projects in New Zealand.

932 ELECTRIC POWER DEVELOPMENT AND OPERATION
Ministry of Energy

English annual c35pp

Contains statistics of operations by New Zealand Electricity, the Electricity Division of the Ministry of Energy, and the Electrical Supply Authorities which purchase electricity in bulk for retail distribution.

933 **1981 ENERGY PLAN**
Government Printer

English 1981 86pp NZ$2.90

Detailed assessment and policy proposals of the Ministry of Energy under the country's first comprehensive energy plan. Sections deal with: Background; Energy demand; Proposed plans for energy supply; Energy research and development: Conclusions—Long term trends and summary.

934 **ENERGY PLANNING '81**
Ministry of Energy

English 1981 36pp

Subtitled 'An Introduction to the 1981 Energy Plan', containing much detailed information on energy resources in New Zealand, the pattern of energy consumption and the background to key energy production and consuming projects. Also summarises energy R & D programmes and related agencies and institutions.

935 **EXTERNAL TRADE**
Department of Statistics

English annual c400pp

Shows statistical tables of external trade by commodity by country and country by commodity. The section on mineral fuels, lubricants and related materials includes coal, coke, briquettes, petroleum and related products. Separate series for imports and exports.

936 **IMPLICATIONS OF NEW ENERGY DEVELOPMENTS**
New Zealand Planning Council

English 1979

An examination of the long-term implications of changes in the New Zealand energy supply situation, providing a basis for development of the government's future energy policy.

937 **LIQUID FUELS TRUST BOARD—REPORT**
Government Printer

English annual NZ$0.65

The Board's annual report to Parliament containing information on the development of alternative transport fuels and details of projects and programmes being supported by the Board.

938 **MINISTRY OF ENERGY—ANNUAL REPORT**
Government Printer

English annual NZ$3.65

Annual report to Parliament containing detailed background information on New Zealand's energy situation, resources and projects. Financial and operating statistics of NZ Electricity, state coal mines and gas undertakings.

939 **MONTHLY ABSTRACT OF STATISTICS**
Department of Statistics

English monthly c130pp

Includes figures for production of coal and gas, deliveries and stocks, electricity generated and sold.

940 **NEW ZEALAND COAL**
Coal Mining Industries Welfare Council

English quarterly NZ$4.00

Journal dealing with general developments in the New Zealand coal sector.

941 **NEW ZEALAND ENERGY JOURNAL**
Technical Publications Ltd

English monthly NZ$8.00 p.a.

Regular journal dealing with developments in New Zealand and related issues.

942 **NEW ZEALAND OFFICIAL YEARBOOK**
Department of Statistics

English annual c1,050pp NZ$3.25

Provides a statistical survey of the economy of New Zealand with a back-ground of text. Includes a large section on production, which contains details of energy demand and supply, electric power, oil, gas and mining industries.

943 **NEW ZEALAND REFINING COMPANY—ANNUAL REPORT**
New Zealand Refining Co Ltd

English annual

Report containing details of the refinery's operations, including crude oil imports and disposal of finished products to the domestic market, and a review of plant capabilities.

944 **NUCLEAR POWER GENERATION IN NEW ZEALAND**
Ministry of Energy

English 1978 NZ$4.00

An examination of the possible need for development of nuclear power in New Zealand, preparatory to formulation of the government's overall energy plan. Published as a Parliamentary paper.

945 **PETROCORP REVIEW**
Petroleum Corporation of New Zealand Ltd

English irregular 20pp

Contains news of Petrocorp activities and general energy developments in New Zealand, including oil and gas exploration, gas utilisation, petrochemicals and policy developments.

946 POCKET DIGEST OF STATISTICS
Department of Statistics

English annual 300pp NZ$2.50

Summary of statistical data on energy supply and consumption, including production of coal, petroleum, natural gas, manufactured gas and electricity.

947 PROJECTIONS OF ELECTRIC POWER GENERATION, CAPACITY AND CONSUMPTION
Ministry of Energy

English annual

Comprises the annual reports of the Electricity Sector Forecasting Committee and the Electricity Sector Planning Committee.

948 REPORT OF THE COMMITTEE TO REVIEW POWER REQUIREMENTS
Ministry of Energy

English annual 30pp

Annual reports on energy consumption, with particular reference to electricity, dealing with conservation, sources and technologies of generation, and prospective energy requirements for up to 15 years. Published as a Parliamentary Paper.

949 REPORT OF THE NEW ZEALAND GAS COUNCIL
Ministry of Energy

English annual

Annual report to Parliament on the operations of, and situation in, the New Zealand gas industry. Published as a Parliamentary Paper.

950 SMALL HYDROELECTRIC PROJECTS FOR RURAL DEVELOPMENT
Pergamon Press

English 1981 175pp $17.50

Published in cooperation with the East-West Center in Hawaii, it contains a case study of a small multiple-use hydroelectric energy schemes in New Zealand.

951 WAIKATO COAL
Mines Division, Ministry of Energy

English 14pp

Specially prepared pamphlet detailing the history and current developments in New Zealand's principal coalfields.

Pakistan

955 ENERGY YEAR BOOK
Directorate-General of Energy Resources

English annual c56pp

Energy balances and detailed statistics for production, consumption and foreign trade in oil, gas, coal and electricity. Information on individual fields, regions and sectors of consumption. Also includes prices of fuels.

956 FIFTH FIVE YEAR PLAN
Planning Commission

English 1978 266pp

Economic development plan for the period 1978-83, setting out the economic programme within which energy supply industries will operate. Specific programmes detailed for each industrial sector. Statistical section includes data on energy supply and consumption.

957 INDUSTRIAL INVESTMENT SCHEDULE
Planning Commission

English 1978 75pp

Programme of expected investment under the Fifth Five Year Plan, covering the period 1978-83. Establishes the requirements for the petroleum, petrochemicals, electricity supply and related industries.

958 KARACHI ELECTRIC SUPPLY CORPORATION LIMITED—ANNUAL REPORT
Karachi Electric Supply Corp Ltd

English annual c40pp

Report of the electricity supply authority for the Karachi region. Contains data on operations and plant capacity, new projects and historical series of electricity sales.

959 PAKISTAN ECONOMIC SURVEY
Ministry of Planning and Development

English annual c300pp

Review of each sector of the economy with detailed information on energy supply industries. Statistics of production, foreign trade and consumption. Information on planned investment projects.

960 PAKISTAN PETROLEUM LTD—ANNUAL REPORT
Pakistan Petroleum Ltd

English annual

Details on the operation of Pakistan's principal gas field and related facilities. Data on gas reserves, production of gas and condensate for a ten-year period.

961 PAKISTAN YEAR BOOK
Ministry of Planning and Development

English annual c900pp

Review in detail of the state of the economy and developments and projects in energy sectors. Information includes: statistics of production and consumption of coal, oil, natural gas and electricity; capacity of oil refineries, generating plant and mines; exploration activity by state and private companies; projects for power stations and electricity transmission systems.

962 STATISTICAL BULLETIN
Statistics Division

English monthly 150pp Rs10.00 p.a.

Regular statistics on a monthly basis for production of coal, oil and gas, generation of electricity and availability of petroleum products.

Papua New Guinea / Pacific Islands

965 ABSTRACT OF STATISTICS
Papua New Guinea National Statistical Office

English quarterly 20pp

Includes statistics on a monthly basis for electricity generation.

966 ANNUAL ABSTRACT OF STATISTICS
Statistical Office (Solomon Islands)

English annual 170pp

Collected statistics on the Solomon Islands economy, including information on extractive industries and details of the electricity supply industry.

967 EIGHTH DEVELOPMENT PLAN
Central Planning Office (Fiji)

English 1983 370pp

Economic development plan, covering the period 1981-85. Sets out policies and programmes for each key section of the economy and the framework for energy demand in the plan period.

968 ELECTRIC POWER CORPORATION—ANNUAL REPORT
Electric Power Corporation (Western Samoa)

English/Samoan annual 46pp

Contains a report on activities and operations in the public supply of electricity with data on generating plant, transmission equipment, production and sales and number of consumers by category.

969 ENERGY IN PAPUA NEW GUINEA 1981
Department of Minerals and Energy (Papua New Guinea)

English 1981 44pp

Summary of the policy and state of implementation of the Papua New Guinea Energy Programme. Sections deal with: Energy Policy; Energy Projects and Activities; Oil Import Data; Energy Pricing Data; Electricity Statistics.

970 FIJI ELECTRIC POWER STUDY
Enex (New Zealand)

English 1975 6vols

Reports on investigations into electricity supply questions in Fiji. Volumes deal with: engineering investigations and outline project to exploit hydroelectric potential; electricity demand; financial and economic evaluation; organisation and management of the electricity supply industry.

971 FOURTH FIVE YEAR DEVELOPMENT PLAN
Economic Development Department (Western Samoa)

English 1980 500pp

Economic development plan for the period 1980-84, providing an analysis of the economic situation, objectives of the development plan and a description of projects in the energy sector in order to meet the energy requirements of economic growth.

972 NATIONAL DEVELOPMENT PLAN
Central Planning Office (Solomon Islands)

English 1980/1981

Economic development plan for 1980-84 published in 1980 and revised in 1981 to cover the period 1981-85. Provides historical background, an analysis of prospects and sets out policies for key sections of the economy and the framework for energy demand.

973 NATIONAL INVESTMENT PRIORITIES SCHEDULE
National Investment and Development Authority (Papua New Guinea)

English 50pp

Provides up-to-date information on areas for investment in Papua New Guinea and the policy towards foreign investment. Identifies areas being actively promoted including mineral and petroleum exploration and development.

974 THE NATIONAL PUBLIC EXPENDITURE PLAN 1980-83
National Planning Office (Papua New Guinea)

English 1979 264pp

Includes details of public expenditure programmes by sector, with details of proposals in the energy sector, including hydro-electric schemes and possible biomass projects. Also covers in detail the activities of the Electricity Commission.

975 PAPUA NEW GUINEA ELECTRICITY COMMISSION—ANNUAL REPORT
Papua New Guinea Electricity Commission

English annual 42pp

Details of generation, distribution and consumption of electricity and notes on the status of new projects.

Philippines

980 ENERGY FORUM
Philippine National Oil Co

English quarterly 32pp free

Periodical publication of the national oil company, providing information and discussion of topical energy issues of particular relevance to the Philippines, including news and articles on geothermal energy and renewable energy forms.

981 FOREIGN TRADE STATISTICS OF THE PHILIPPINES
National Economic and Development Authority

English annual

Figures on an annual basis for imports and exports of individual fuel commodities on a volume and value basis, analysed by country of origin/destination. Covers coal, coke, crude oil and individual petroleum products.

982 INTERLOCK
National Power Corporation

English bi-monthly c36pp free

Journal of the national electricity generating authority, containing news of developments relating to electricity supply and demand, and reports of the Corporation's operations.

983 MONTHLY BULLETIN OF STATISTICS
National Economic and Development Authority

English 32pp

Up-to-date statistics on a monthly basis for production of coal by region, electricity generation, electricity consumption by sector, gas production and consumption by sector.

984 NATIONAL ELECTRIFICATION ADMINISTRATION—ANNUAL REPORT
National Electrification Administration

English annual 24pp

Reviews the state of electrification in the Philippines, with details of projects underway. Also provides information on small-scale hydro-electric development and dendrothermal power projects.

985 THE NATIONAL ENERGY PROGRAM 1981-86
Ministry of Energy

English 1981 68pp

Detailed background information and statistics for all energy forms in the Philippines. Sections deal with: Performance and Accomplishments 1973-80; Sectoral Targets 1981-86; Factor Requirements.

986 PHILIPPINE NATIONAL OIL COMPANY—ANNUAL REPORT
Philippine National Oil Co

English annual 36pp

Details of activity of PNOC and its subsidiary companies in oil exploration, geothermal energy, coal exploration and production, oil supply, refining and marketing, and development of renewable energy sources.

987 PHILIPPINE STATISTICAL YEARBOOK
National Economic and Development Authority

English annual c700pp

Detailed statistics for the mining and electric power sectors. Information on electricity generation and the supply of main petroleum products. Data cover a number of years.

988 SMALL HYDRO-ELECTRIC PROJECTS FOR RURAL DEVELOPMENT
Pergamon Press

English 1981 175pp $17.50

Published in cooperation with the East-West Center in Hawaii, contains a case study of small scale multiple-use hydro-electric energy schemes in the Philippines.

Singapore

990 BERITA SHELL
Shell Eastern Petroleum Ltd

English irregular free

Published by the local affiliate of the Royal Dutch/ Shell group, containing news of the group's activities in the Singapore and Malaysia area.

991 ECONOMIC SURVEY OF SINGAPORE
Ministry of Trade and Industry

English annual 124pp

Contains sections relating to energy developments, with statistics on electricity consumption by sector, and prices for automotive fuels.

992 KOMPASS SINGAPORE
Kompass Publishers Ltd

English annual

Directory of Singapore firms and their principal fields of activity. Includes firms engaged in the supply of energy and related goods and services.

993 MONTHLY DIGEST OF STATISTICS
Department of Statistics

English monthly 134pp

Figures for recent months on: production and sales of gas and electricity; consumption of gas and electricity by sector; numbers of consumers of gas and electricity; imports of crude oil and petroleum products; exports of petroleum products.

994 PUBLIC UTILITIES BOARD—ANNUAL REPORT
Public Utilities Board

English annual 54pp

Contains data on the supply of electricity and gas in Singapore, with details of production capacity, output, sales and new projects.

995 SINGAPORE TRADE STATISTICS
Department of Statistics

English monthly

Figures on a monthly and cumulative basis for the year of imports and exports of crude oil and oil products. Figures for volume and value analysed by main sources/destinations.

996 YEARBOOK OF STATISTICS
Department of Statistics

English annual 266pp

Energy sections contain statistics for the electricity and gas industries over several years. Figures for production and sales of electricity and gas, with sales analysed by main consumer groupings.

Sri Lanka

1000 CENTRAL BANK OF CEYLON—ANNUAL REPORT
Central Bank of Ceylon

English annual Rs10.00

Contains sections on fuel and power. Commentary and key statistics for the energy sector. Consumption, prices of petroleum products, electricity generating capacity, sales of electricity by sector.

1001 EXTERNAL TRADE STATISTICS
Sri Lanka Customs

English half yearly 650pp

Figures for imports and exports on a cumulative basis for coal/coke and petroleum products.

1002 MONTHLY BULLETIN
Central Bank of Ceylon

English monthly 108pp

Analytical articles, notes and statistical appendices covering: electricity generation in the most recent month and change over corresponding period; cumulative data on electricity usage by sector of industry; sales of industrial fuel oil.

1003 REVIEW OF THE ECONOMY
Central Bank of Ceylon

English annual

Review with substantial statistical sections covering activities of the Ceylon Petroleum Corporation and Ceylon Electricity Board, investment, imports by product, sales, domestic consumption of oil products, electricity generating capacity and output, energy consumption by sector. Figures given for several years.

1004 STATISTICAL ABSTRACT OF THE DEMOCRATIC SOCIALIST REPUBLIC OF SRI LANKA
Department of Census and Statistics

Multi-language annual c540pp

Contains many statistics relating to the production and supply of electricity and gas in Sri Lanka.

1005 STATISTICAL POCKET BOOK
Department of Census and Statistics

English annual 186pp

Statistical series covering several years for: electricity generating capacity, production, consumption by sector, number of consumers; output of state corporations; production of petroleum products; gas production of Colombo Gas and Water Co.

Taiwan

1010 CHINA YEARBOOK
China Publishing Co

English annual 740pp

Includes section with background information and statistics on major industrial sectors—petroleum production, electricity generation, hydro-electric schemes.

1011 THE ENERGY SITUATION IN TAIWAN
Energy Committee

English 1981 30pp

Background information and statistics on development of the current energy situation, energy supply industries and energy supply and demand with forecast for the period to 1989.

1012 INDUSTRY OF FREE CHINA
Council for Economic Planning and Development

English/Chinese monthly 212pp NT$200.00 p.a.

Commentary and analysis with statistics on industry activity. Details of supply and demand for commercial energy, electricity consumption by industry sector, production of coal, crude oil, natural gas and hydro-electric plant.

1013 MONTHLY BULLETIN OF STATISTICS
Directorate-General of Budget, Accounting and Statistics

English monthly 26pp

Statistical information is included on production of coal, lignite and natural gas. Series for average prices of coal and crude oil. Indicators of energy sector output.

1014 MONTHLY STATISTICS OF THE REPUBLIC OF CHINA
Directorate-General of Budget, Accounting and Statistics

English monthly 270pp

Detailed national statistics. Energy sector coverage includes: production of coal, some oil products, natural gas, manufactured gas and electricity; consumption of electricity by sector of industry.

1015 MONTHLY STATISTICS OF TRADE
Inspectorate-General of Customs

English/Chinese monthly

Figures by volume and value on a monthly and cumulative basis for main energy commodities, analysed by country of origin/destination.

1016 NATIONAL CONDITIONS
Directorate-General of Budget, Accounting and Statistics

English quarterly 37pp

Includes statistical data for production of coal, gas, electricity and oil.

1017 STATISTICAL YEARBOOK OF THE REPUBLIC OF CHINA
Directorate-General of Budget, Accounting and Statistics

English annual c740pp

Includes sections covering energy production and supply. Indices of fuel production. Data for electricity and gas production and generating capacity.

1018 TAIWAN INDUSTRIAL PRODUCTION STATISTICS
Department of Statistics

English monthly c140pp

Contains monthly statistics on sales of electricity, gas and LPG, and for production of crude oil, natural gas and coal.

1019 TAIWAN STATISTICAL DATA BOOK
Council for Economic Planning and Development

English annual 318pp

Compendium of statistics, including coverage of energy supply and demand, energy consumption by sector, electricity generating capacity and output, mineral and energy reserves and industry output.

1020 TEN-YEAR ECONOMIC DEVELOPMENT PLAN FOR TAIWAN
Council for Economic Planning and Development

English 1980 89pp

Development plan for the 1980s providing projections and programmes for the economy. Assesses the availability and demand for energy by sector and the investment in electricity generating plant and other energy supply industries required.

1021 THE TRADE OF CHINA
Inspectorate-General of Customs

English/Chinese annual 115pp

Statistics on an annual basis by volume and value for imports and exports of commodities, analysed by country of origin/destination and covering coal, coke, crude oil and petroleum products.

Thailand

1025 EGAT—ANNUAL REPORT
Electricity Generating Authority of Thailand

English annual 80pp

The report of the state generating and distribution authority contains detailed information on electricity supply and consumption and news of new power station and hydro-electric projects.

1026 ELECTRIC POWER IN THAILAND
National Energy Administration

English annual

Regular annual report on the electricity supply industry, with data on: installed generating capacity; transmission lines; electricity production; financial performance; electricity prices and revenue; fuel consumption; hydro-electric development.

1027 FOREIGN TRADE STATISTICS OF THAILAND
Department of Customs

English monthly c450pp

Covers imports and exports of the whole country; December issue is the cumulative total. Includes coal, electrical equipment and machinery, electric and gas generators, oil and uranium.

1028 FOURTH NATIONAL ECONOMIC AND SOCIAL DEVELOPMENT PLAN
National Economic and Social Development Board

English 1977 446pp

Establishes national economic development strategy and programmes identifying implications for individual industrial sectors and the basis of future energy demand in the economy.

1029 GEOLOGY OF GEOTHERMAL RESOURCES OF NORTHERN THAILAND
Department of Mineral Resources

English 1980

Report by the Geological Survey Division on the potential geothermal energy availability in the Sakamphaeng, Fang and Mae Chan areas of northern Thailand.

1030 MEA—ANNUAL REPORT
Metropolitan Electricity Authority

English annual 50pp

Annual report on electricity supply in the Bangkok area. Details of availability of electricity, sales by economic sector and distribution programme.

1031 OIL AND THAILAND
National Energy Administration

English annual

Annual report on the oil industry in Thailand. Sections deal with: consumption; refining; imports and exports; offshore exploration; taxes on petroleum products; wholesale and retail prices; general and international data.

1032 PETROLEUM ACTIVITIES IN THAILAND
Department of Mineral Resources

English annual

Mimeographed report by the Mineral Fuels Division on exploration activities in Thailand, including information on concessions.

1033 PETROLEUM AUTHORITY—ACTIVITY REPORT
Petroleum Authority of Thailand

English annual

Report by the state oil and gas undertaking on its activities and operations.

1034 PROVINCIAL ELECTRICITY AUTHORITY—ANNUAL REPORT
Provincial Electricity Authority

English annual

Report of the electricity generating and supply authority for the area outside Bangkok. Reviews the state of electrification, existing and planned plant and transmission system and details of projects. Data on sales and number of customers, with some data covering a ten-year period.

1035 STATISTICAL YEARBOOK
National Statistical Office

English/Thai annual 560pp

Includes data for a number of years covering output from lignite mining, electricity generating capacity, production and sales, electricity consumption by sector of industry, generating equipment in the private sector.

1036 THAILAND ENERGY SITUATION
National Energy Administration

English annual 44pp

Energy balance sheets analysed by fuel type. Detailed statistics of production from indigenous and imported oil, electricity generated and plant capacities, energy consumption and reserves. Data cover a number of years, with figures for value and volume.

1037 THAILAND PAPERS ON ENERGY
National Energy Administration

English 1981 71pp

Collected information on energy availability and use in Thailand, prepared for the Committee on Natural Resources of the United Nations Economic and Social Commission for Asia and the Pacific.

Indexes to Publications

Index 1: Publications A-Z

A

B

C

D

E

F

G

H

I

J

K

O

P

Q

R

S

T

U

V

W

Y

Z

Index 2: Publications by Subject and Country

2.1 TRADE DIRECTORIES AND BUSINESS HANDBOOKS

OMAN

PAKISTAN

PAPUA NEW GUINEA/PACIFIC ISLANDS

QATAR

SAUDI ARABIA

SENEGAL

SIERRA LEONE

SINGAPORE

SOUTH AFRICA

SRI LANKA

SUDAN

SYRIA

TAIWAN

TANZANIA

THAILAND

TUNISIA

TURKEY

UNITED ARAB EMIRATES

YEMEN ARAB REPUBLIC

ZAIRE

ZAMBIA

2.2 GENERAL ENERGY STATISTICS

IRAN

IRAQ

ISRAEL

IVORY COAST

JAPAN

JORDAN

KENYA

KOREA (REPUBLIC OF)

KUWAIT

LEBANON

LIBYA

MALAGASY REPUBLIC

MALAWI

MALAYSIA

MALI

MOROCCO

NEW ZEALAND

NIGER

NIGERIA

OMAN

PAKISTAN

PAPUA NEW GUINEA/PACIFIC ISLANDS

PHILIPPINES

2.3 ENERGY TRANSPORTATION AND TRADE

(for individual fuels see Sections 2.6.1–2.6.5)

2.4 ENERGY MARKET ANALYSES

2.5 ENERGY POLICIES

BANGLADESH

CHINA (PEOPLE'S REPUBLIC OF)

EGYPT

INDIA

INDONESIA

IRAN

IRAQ

ISRAEL

JAPAN

JORDAN

KOREA (REPUBLIC OF)

LIBYA

MALAYSIA

NEW ZEALAND

NIGERIA

PAKISTAN

PAPUA NEW GUINEA/PACIFIC ISLANDS

PHILIPPINES

SAUDI ARABIA

SOUTH AFRICA

SRI LANKA

TAIWAN

THAILAND

ZAIRE

ZAMBIA

2.6 ENERGY INDUSTRIES

2.6.1 OIL

INTERNATIONAL

ALGERIA

AUSTRALIA

2.6.2. COAL AND THE MINING INDUSTRY

IRAQ

ISRAEL

IVORY COAST

JAPAN

JORDAN

KOREA (REPUBLIC OF)

KUWAIT

LIBYA

MALAYSIA

MOROCCO

NEW ZEALAND

NIGERIA

PAKISTAN

PHILIPPINES

QATAR

SAUDI ARABIA

SINGAPORE

SRI LANKA

SYRIA

TAIWAN

THAILAND

TUNISIA

2.6.4. ELECTRICITY SUPPLY

2.6.5. NUCLEAR POWER

2.6.6. ENERGY RELATED INDUSTRIES

IRAN

ISRAEL

JAPAN

KOREA (REPUBLIC OF)

KUWAIT

LIBYA

NIGERIA

PAKISTAN

PHILIPPINES

SAUDI ARABIA

2.7 EXPLORATION AND DEVELOPMENT

INTERNATIONAL

AUSTRALIA

BANGLADESH

BURMA

CHINA (PEOPLE'S REPUBLIC OF)

EGYPT

INDIA

INDONESIA

IRAN

IVORY COAST

JAPAN

KOREA (REPUBLIC OF)

LIBYA

MALAYSIA

NEW ZEALAND

NIGER

NIGERIA

PAKISTAN

PAPUA NEW GUINEA/PACIFIC ISLANDS

PHILIPPINES

SAUDI ARABIA

SOUTH AFRICA

SYRIA

THAILAND

UNITED ARAB EMIRATES

ZAMBIA

ZIMBABWE

2.8 CONSERVATION/UTILISATION

INTERNATIONAL

2.9 NEW ENERGY TECHNOLOGIES

Index 3: Publishing Bodies

A

ASIAN DEVELOPMENT BANK
Manila
Philippines

Entries: 016 126

ASSOCIATED OCTEL LTD
20 Berkeley Square
London W1
England
Tel: (01) 499 6030

Entry: 278

ASSOCIATION OF INDIAN ENGINEERING INDUSTRY
172 Jorbagh
New Delhi 110003
India

Entry: 835

ATOMIC ENERGY BOARD
Private Bag X256
Pretoria 0001
Transvaal
South Africa

Entry: 592

ATOMIC ENERGY COMMISSION
2-2-1 Kasumigaseki
Chiyoda-ku
Tokyo 100
Japan
Tel: (03) 581 5271

Entry: 865

AUSTRALIAN ATOMIC ENERGY COMMISSION
Cliffbrook
45 Beach Street
Coogee
New South Wales 2034
Australia
Tel: (02) 665 1221

Entry: 700

AUSTRALIAN BUREAU OF STATISTICS
PO Box 10
Belconnen
Canberra ACT 2616
Australia
Tel: 52 79 11

Entries: 706 708 713 714 721 723 728 730 731 732 734 735 737 739 741 743 751 753 754 763 764 765 768

AUSTRALIAN BUREAU OF STATISTICS, NEW SOUTH WALES
St Andrew's House
Sydney Square
Sydney
New South Wales 2000
Australia
Tel: (02) 236 6611

Entry: 745

AUSTRALIAN BUREAU OF STATISTICS, QUEENSLAND
Statistics House
345 Ann Street
Brisbane
Queensland 4000
Australia
Tel: (07) 33 54 84

Entry: 756

AUSTRALIAN BUREAU OF STATISTICS, SOUTH AUSTRALIA
195 North Terrace
Adelaide
South Australia 5000
Australia
Tel: (08) 228 9439

Entry: 761

AUSTRALIAN BUREAU OF STATISTICS, VICTORIA
Commonwealth Bank Building
Corner Elizabeth and Flinders Streets
Melbourne
Victoria 3000
Australia
Tel: (03) 652 6177

Entry: 766

AUSTRALIAN BUREAU OF STATISTICS, WESTERN AUSTRALIA
1-3 St George's Terrace
Perth
Western Australia
Australia 6000
Tel: (09) 323 5140

Entry: 767

AUSTRALIAN GOVERNMENT PUBLISHING SERVICE
PO Box 84
Canberra ACT 2600
Australia

Entries: 701 719 720 757

AUSTRALIAN INSTITUTE OF PETROLEUM LTD
227 Collins Street
Melbourne
Victoria 3000
Australia
Tel: (03) 63 27 56

Entries: 746 747

AUSTRALIAN LIQUEFIED PETROLEUM GAS ASSOCIATION
12 O'Connell Street
Sydney
New South Wales 2000
Australia

Entry: 736

AUSTRALIAN MINING INDUSTRY COUNCIL
PO Box 363
Dickson ACT 2602
Australia

Entry: 742

B

C

CENTER FOR ECONOMIC, FINANCIAL AND SOCIAL RESEARCH AND DOCUMENTATION SAL
Gefinor Center
PO Box 11-6068
Beirut
Lebanon
Tel: 36 40 10

Entry: 010

CENTRAL AGENCY FOR PUBLIC MOBILISATION AND STATISTICS
Nasr City
Cairo
Egypt
Tel: 83 31 99

Entries: 480 481 482

CENTRAL BANK OF CEYLON
Janadhipathi Mawatha
Colombo 1
Sri Lanka

Entries: 1000 1002 1003

CENTRAL BANK OF IRAN
Tehran
Iran

Entry: 310

CENTRAL BANK OF NIGERIA
PMB 12194
Tinubu Square
Lagos
Nigeria

Entries: 570 571

CENTRAL BANK OF YEMEN
Sana'a
Yemen

Entries: 440 441

CENTRAL BUREAU OF STATISTICS (GHANA)
PO Box 1098
Accra
Ghana
Tel: 66512

Entry: 500

CENTRAL BUREAU OF STATISTICS (ISRAEL)
Hakirya-Romema
PO Box 13015
Jerusalem 91130
Israel
Tel: (02) 52 61 61

Entries: 332 333 334 335 338 340

CENTRAL BUREAU OF STATISTICS (KENYA)
PO Box 30266
Nairobi
Kenya
Tel: 33 39 70

Entries: 511 514 516 517 518

CENTRAL BUREAU OF STATISTICS (SYRIA)
Abdel-Malek Bin Marwan Street
Malki Quarter
Damascus
Syria
Tel: 33 58 30

Entries: 411 412

CENTRAL DEPARTMENT OF STATISTICS
Ministry of Finance and National Economy
PO Box 3735
Riyadh
Saudi Arabia
Tel: 29120

Entries: 385 388 397 399 401

CENTRAL INTELLIGENCE AGENCY
Washington DC 20505
USA

Entries: 119 797 801

CENTRAL PLANNING OFFICE (FIJI)
Suva
Fiji

Entry: 967

CENTRAL PLANNING OFFICE (NIGERIA)
Federal Ministry of National Planning
Block 1
New Secretariat
Ikoyi
Nigeria

Entry: 572

CENTRAL PLANNING OFFICE (SOLOMON ISLANDS)
Honiara
Solomon Islands

Entry: 972

CENTRAL PLANNING ORGANISATION
Statistics Department
PO Box 175
Sana'a
Yemen
Tel: 3506

Entry: 443

CENTRAL STATISTICAL OFFICE (ETHIOPIA)
PO Box 1143
Addis Ababa
Ethiopia
Tel: 11 30 10

Entry: 490

CENTRAL STATISTICAL OFFICE (KUWAIT)
Ministry of Planning
PO Box 15
Kuwait
Tel: 64 07 91

Entries: 355 357 359 360

D

E

F

HOLT-SAUNDERS LTD
1 St Anne's Road
Eastbourne BN21 3UN
England
Tel: (0323) 638221

Entries: 101 110 180 181 395 398

HOOVER INSTITUTION PRESS
Stanford University
Stanford
California CA 94305
USA
Tel: (415) 497 3373

Entries: 810 883

C HURST & CO (PUBLISHERS) LTD
38 King Street
London WC2E 8JC, England
Tel: (01) 240 2666

Entry: 809

I

IC MAGAZINES
PO Box 261
Carlton House
69 Great Queen Street
London WC2B 5BZ
England

Entries: 004 136 145

IMPRIMERIE HABID EID
Furn-el-Chabbak
Lebanon

Entry: 367

INDIAN BUREAU OF MINES
New Secretariat Building
Nagpur 400001
India

Entry: 842

INDUSTRIAL STRUCTURE COUNCIL
Ministry of International Trade and Industry
3-1 Kasumigaseki 1-chome
Chiyoda-ku
Tokyo 100
Japan

Entry: 873

INDUSTRIAL STUDIES & DEVELOPMENT CENTRE
PO Box 1267
Riyadh
Saudi Arabia

Entry 389

INSPECTORATE-GENERAL OF CUSTOMS
Ministry of Finance
85 Hsin-Hseng South Road, Section 1
Taipei
Taiwan
Tel: 741 3181

Entries: 1015 1021

INSTITUT NATIONAL DE LA STATISTIQUE
27 Rue du Liban
BP 65
Tunis
Tunisia
Tel: 28 25 00

Entries: 645 648 650 651

INSTITUT NATIONAL DE LA STATISTIQUE ET DE LA DEMOGRAPHIE
Direction des Comptes Economiques et de la Conjoncture
Ministère du Plan et de la Co-opération
Ouagadougou
Upper Volta

Entry: 655

INSTITUTE OF ENERGY
18 Devonshire Street
London W1N 2AU
England
Tel: (01) 580 7124

Entry: 083

INSTITUTE OF ENERGY ECONOMICS
10 Mori Building
1-18-1 Toranomon
Minato-ku
Tokyo
Japan

Entry: 871

INSTITUTE OF GEOLOGICAL SCIENCE
Available from:
HMSO
PO Box 509
London SE1 9NH
Tel: (01) 928 1321

Entry: 265

INSTITUTE OF MINING AND METALLURGY
PO Box 61019
Marshalltown 2107
South Africa

Entry: 605

INSTITUTE OF PETROLEUM
61 New Cavendish Street
London W1M 8AR
England
Tel: (01) 636 1004

Entry: 269

L

M

McGRAW HILL PUBLICATIONS
457 National Press Building
Washington DC 20045
USA
Tel: (202) 624 7561

Entries: 032 043 068 123 150 193 194

MACMILLAN PRESS LTD
4 Little Essex Street
London EC2R 3LF
England
Tel: (01) 836 6633

Entries: 250 578 630 807

MARITIME SERVICES BOARD OF NEW SOUTH WALES
GPO Box 32
Sydney
New South Wales 2001
Australia
Tel: 20545

Entry: 752

METRA CONSULTING GROUP LTD
23 Lower Belgrave Street
London SW1W 0NS
England
Tel: (01) 730 0855

Entries: 108 147 315 394 575 795 854

METROPOLITAN ELECTRICITY AUTHORITY
121 Chakraphet Road
Bangkok 2
Thailand
Tel: 21 21 11

Entry: 1030

MIADNA PTY LTD
11 National Circuit
Barton ACT 2600
Australia

Entry: 729

MICROINFO LTD
PO Box 3
Alton GU34 2PG
England
Tel: (0420) 84300

Entry: 079

MIDDLE EAST CONSULTANTS
PO Box 11-7323
Beirut
Lebanon

Entry: 369

MIDDLE EAST ECONOMIC DIGEST
MEED House
21 John Street
London WC1N 2BP
England
Tel: (01) 404 5513

Entries: 001 133 393

MIDDLE EAST OBSERVER
8 Chawarby Street
Cairo
Egypt

Entry: 478

MIDDLE EAST PETROLEUM & ECONOMIC PUBLICATIONS
PO Box 4940
Nicosia
Cyprus

Entry: 134

MILLER FREEMAN PUBLICATIONS INC
500 Howard Street
San Francisco CA 94105
USA
Tel: (415) 397 1881

Entries: 072 210 254 255 266 267

MINERALS BUREAU
Private Bag X4
Braamfontein 2017
Transvaal
South Africa
Tel: 725 3360

Entry: 598

MINES DIVISION, MINISTRY OF ENERGY
Anvil House
Wakefield Street
PO Box 6342
Wellington
New Zealand
Tel: 735 755

Entries: 930 951

MINING JOURNAL LTD
PO Box 10
Edenbridge TN8 5NE
England
Tel: (0732) 864333

Entries: 138 140 141

MINING JOURNAL BOOKS LTD
PO Box 10
Edenbridge TN8 5NE
England
Tel: (0732) 864333

Entry: 245

OFFICE NATIONAL DE L'ELECTRICITE
65 Rue Aspirant Lafuente
Casablanca
Morocco
Tel: 22 41 65

Entry: 553

OFFICE OF ECONOMIC ADVISER TO THE PRIME MINISTER
Economic Planning Branch
Pretoria
Transvaal
South Africa

Entry: 599

OFFICE OF THE PRESIDENT
National Commission for Development Planning
Lusaka
Zambia

Entry: 667

OIL AND GAS DIRECTORY
PO Box 13508
Houston
Texas TX 77019
USA

Entry: 163

ORGANISATION OF ARAB PETROLEUM EXPORTING COUNTRIES
PO Box 20501
Kuwait

Entries: 056 156 184

ORGANISATION OF PETROLEUM EXPORTING COUNTRIES
Obere Donaustrasse 93
A-1020 Vienna
Austria
Tel: (0222) 26 55 11

Entries: 176 177

THE ORIENTAL ECONOMIST
1-4 Hangokucho
Nihonbashi
Chuoku
Tokyo 103
Japan
Tel: (03) 270 4111

Entry: 878

OXFORD UNIVERSITY PRESS
Walton Street
Oxford OX2 6DP
England
Tel: (0865) 56767

Entry: 256

P

PAKISTAN PETROLEUM LTD
PO Box 3942
PIDC House
Dr Ziauddin Ahmed Road
Karachi
Pakistan
Tel: 51 13 30

Entry: 960

PAPUA NEW GUINEA ELECTRICITY COMMISSION
PO Box 1105
Boroko
Papua New Guinea
Tel: 24 32 00

Entry: 975

PAPUA NEW GUINEA NATIONAL STATISTICAL OFFICE
PO Ward Srips
Waigani
Papua New Guinea
Tel: 27 17 05

Entry: 965

PENN WELL PUBLISHING CO
1421 South Sheridan Road
Tulsa
Oklahoma
USA

Entries: 003 014 122 159 164 275 276 277

PERGAMON PRESS
Headington Hill Hall
Oxford OX3 0BW
England
Tel: (0865) 64881

Entries: 021 031 044 045 050 059 062 067 098 099 105 109 111 112 130 178 206 220 221 222 223 235 236 312 812 830 845 931 950 988

PETROLEUM AUTHORITY OF THAILAND
1 Soi Yasoob 1
Viphavadi Rangsit Road
Bangkok 9
Thailand
Tel: 279 3742

Entry: 1033

PETROLEUM CORPORATION OF NEW ZEALAND LTD
Petrocorp House
86 Lambton Quay
PO Box 5082
Wellington
New Zealand
Tel: 739 812

Entry: 945

R

U

V

Appendix: Energy Units and Terms

ABBREVIATIONS

KW	kilowatt	KWh	kilowatt hour
MW	megawatt	MWh	megawatt hour
GW	gigawatt	GWh	gigawatt hour
TW	terawatt	TWh	terawatt hour
Kcal	kilocalorie	Kj	kilojoule
Mcal	megacalorie	Mj	megajoule
Gcal	gigacalorie	Gj	gigajoule
Tcal	teracalorie	Tj	terajoule
cf	cubic foot/feet	cfd	cubic feet per day
M^3	cubic metre	bbl	barrel
b/d	barrels per day	Btu	British thermal unit

MULTIPLE UNITS

kilo- = x 1,000 (10^3)
mega- = x 1,000,000 (10^6)
giga- = x 1,000,000,000 (10^9)
tera- = x 1,000,000,000,000 (10^{12})

CONVERSION FACTORS

Heat and Power

1 calorie = 4.187 joules
1 Btu = 252 calories or 1,055.06 joules
1 therm = 100,000 Btus or 25,200 Kcals
1 KWh = 859.6 Kcals or 3,411 Btus

Gas

1 M^3 = 35.32 cubic feet
1 bbl = 5.61 cubic feet
$10^9 M^3$p.a. = 96.7 x 10^6 cfd

Oil

1 bbl	=	35 Imperial gallons or 42 US gallons
1 tonne of crude oil	=	7.3 bbls (average)
1 M^3 motor gasoline	=	0.75 tonne
1 M^3 gas oil	=	0.84 tonne
1 M^3 fuel oil	=	0.94 tonne
1 million b/d	=	50 million tonnes per annum (average)

APPROXIMATE CALORIFIC VALUES

	Mcals per tonne
crude oil	10,000-11,000
hard coal	5,000-7,500
lignite	1,700-4,500
coke	6,300-7,600
firewood	3,300
bagasse	1,600
vegetable wastes	2,700
	Kcals per M^3
natural gas	
—Algeria	10,000
—Australia	9,300
—Bangladesh	8,400
—Brunei	9,900
—Indonesia	9,300
—Iran	9,400
—Malaysia	9,300
—Pakistan	8,700
manufactured gas	3,800-4,500

DEFINITIONS

'billion' means thousand million (10^9)
'tonne' means 1,000 kilogrammes

SIGNS

– means nil
.. means negligible